SURVEYOR
Program Results

NASA SP-184

SURVEYOR

Program Results

Compiled by
Surveyor Program
Lunar and Planetary Programs Division
Office of Space Science and Applications

Scientific and Technical Information Division
OFFICE OF TECHNOLOGY UTILIZATION   1969
NATIONAL AERONAUTICS AND SPACE ADMINISTRATION
Washington, D.C.

Preface

This, the final report under Surveyor Program auspices by the Surveyor Scientific Team, presents the key findings resulting from five Surveyor landings on the Moon. The vast amount of new lunar data generated by these missions will doubtless be studied for a long time to come, and additional individual papers can be expected in the future.

Arrangements have been made to provide interested scientists, not connected with the Surveyor Program, with Surveyor data on request; pictures and other information can be obtained from the National Space Science Data Center, Goddard Space Flight Center, Greenbelt, Md. 20771.

The Surveyor program was planned to achieve soft landings on the Moon by automated spacecraft capable of transmitting scientific and engineering measurements from the lunar surface. The program had three major objectives: (1) to develop and validate the technology for landing softly on the Moon, (2) to provide data on the compatibility of the Apollo manned lunar-landing spacecraft design with conditions to be encountered on the lunar surface, and (3) to add to our scientific knowledge of the Moon. All of these objectives have been achieved to a degree far beyond original expectations.

Surveyor I, the first U.S. spacecraft to land softly on the Moon, returned a large quantity of scientific data during its first 2 lunar days of operation on the lunar surface. Following its landing on June 2, 1966, in the southwest portion of Oceanus Procellarum, the spacecraft transmitted 11 240 high-resolution television pictures. Surveyor I completed its primary mission successfully on July 14, 1966, after transmitting, in addition to the television pictures, data on the bearing strength, temperatures, and radar reflectivity of the Moon. Subsequent engineering interrogations of the spacecraft were conducted through January 1967.

Surveyor II, launched on September 20, 1966, was intended to land in Sinus Medii, a different area of the Apollo zone. When the midcourse maneuver was attempted, one vernier engine failed to ignite, and the unbalanced thrust caused the spacecraft to tumble. Although repeated efforts were made to salvage the mission, none was successful.

Surveyor III successfully landed on the Moon on April 20, 1967, touching down in the eastern part of Oceanus Procellarum. This spacecraft, like its predecessors, carried a survey television camera and other instrumentation for determining various properties of the lunar-surface material. In addition, it carried a surface-sampler instrument for digging trenches, making bearing tests, and otherwise manipulating the lunar material in the view of the television system.

In its operations, which ended on May 4, 1967, Surveyor III acquired a large volume of new data and took 6326 pictures. In addition, the surface sampler accumulated 18 hours of operation, which yielded significant new information on the strength, texture, and structure of the lunar material to a depth of 17.5 cm.

Surveyor IV, carrying the same payload as Surveyor III, was launched on July 14, 1967. After a flawless flight to the Moon, radio signals from the spacecraft abruptly ceased during the terminal-descent phase, approximately 2½ minutes before touchdown. Contact with the spacecraft was never reestablished.

Surveyor V landed in Mare Tranquillitatis on September 11, 1967. This spacecraft was basically similar to its predecessor, except that the surface sampler was replaced by an alpha-backscatter instrument, a device for determining the relative abundance of the chemical elements in the lunar material. In addition, a small bar magnet was attached to one of the footpads to indicate the presence of magnetic material in the lunar soil.

Because of a critical helium-regulator leak in flight, a radically new descent profile was engineered in real time, and Surveyor V performed a flawless descent and soft landing within the rimless edge of a small crater, Mare Tranquillitatis, on a slope of about 20°.

During its first lunar day, which ended at sunset on September 24, 1967, Surveyor V took 18 006 television pictures. The alpha-backscatter instrument provided data from which the first *in situ* chemical analysis of an extraterrestrial body has been derived. This analysis showed that the lunar sample was similar to terrestrial basalt. Results of the bar-magnet test were compatible with this finding. Surveyor V also performed a rocket-erosion experiment on the Moon, in which its engines were fired for 0.55 second to determine the effects of high-velocity exhaust gases impinging on the lunar surface.

On October 15, 1967, after having been exposed to the 2-week deep freeze of the lunar night, Surveyor V responded immediately to the first turn-on command and operated until sunset of the second lunar day, October 24, 1967, during which time it transmitted over 1000 additional pictures.

Surveyor VI landed on the Moon on November 10, 1967. The landing site was in Sinus Medii, essentially in the center of the Moon's visible hemisphere, the last of four potential Apollo landing areas designated for investigation by the Surveyor program.

The performance of Surveyor VI on the lunar surface was virtually flawless. From touchdown until a few hours after sunset on November 24, 1967, the spacecraft transmitted 29 952 television pictures and the alpha-scattering instrument acquired 30 hours of data on the chemical composition of the lunar material. On November 17, 1967, the vernier rocket engines of Surveyor VI were fired for 2.5 seconds and the spacecraft lifted off the lunar surface and translated laterally about 8 feet to a new location, the first such known excursion on the Moon. This "lunar hop" provided excellent views of the surface disturbances produced by the initial landing and furnished significant new information on the effects of firing rocket engines close to the lunar surface. The displacement provided a baseline for stereoscopic viewing and photogrammetric mapping of the surrounding terrain and surface features.

Other data provided by Surveyor VI include pictures of a bar magnet installed on a footpad to determine the concentration of magnetic material in the lunar surface; views of the stars, Earth, and the solar corona; lunar-surface temperatures up to 41 hours after sunset; radar reflectivity data during landing; touchdown-dynamics data during the initial landing and the lunar hop which provided additional information on the mechanical properties of the lunar surface material; and on-surface doppler tracking data for refining existing information on the motions of the Moon.

On November 26, 1967, Surveyor VI was placed in hibernation for the 2-week lunar night. Contact with the spacecraft was resumed for a short period on December 14, 1967.

The successful accomplishment of the Surveyor VI mission not only satisfied all Surveyor obligations to Apollo, but completed the scientific investigation of four widely separated mare regions in the Moon's equatorial belt, spaced roughly uniformly across a longitude range between 43° W and 23° E, from which important generalizations regarding the lunar maria have been derived.

The investigations in the Apollo zone having been satisfactorily accomplished, Surveyor VII could be sent to an area of primary scientific interest, the rugged, rock-strewn ejecta blanket near the prominent, comparatively young, ray crater Tycho.

The area selected for investigation, a site about 18 miles north of Tycho, differs considerably from those examined by previous Surveyors. This region was chosen because it is in the highlands, well removed from the maria, and was expected to be covered with debris excavated from beneath the surface of the highlands when Tycho was formed. It thus provides a significantly different type of lunar sample for comparison with those of previous missions.

In view of the very rugged nature of the terrain in the landing area north of Tycho and the consequent hazards to landing, in preflight mission planning it was necessary to reduce the target area from the 30-km-radius circle used for previous Surveyor missions to one with a 10-km radius; that is, one-ninth the usual area. The mission was planned to employ two midcourse maneuvers in order to maximize the probability of landing within the small target circle.

Because of the excellent performance of the launch vehicle and the spacecraft, however, only one midcourse maneuver was found to be necessary. Surveyor VII landed less than 1½ miles from the center of the target circle, about 18 miles north of the rim of Tycho. Touchdown occurred at 01:05:36.3 GMT on January 10, 1968.

During the first lunar day, 20 993 television pictures were obtained. An additional 45 picture were obtained during the second lunar day.

The alpha-scattering instrument, after completing its background count in the intermediate position, failed to deploy the remainder of the distance to the lunar surface. The surface sampler was then brought into action and, by means of a series of intricate maneuvers, was able to force the alpha-scattering instrument to the surface. The surface sampler was later used to pick up the alpha-scattering instrument after the first chemical analysis had been completed and to move it to two other locations for additional analyses.

These delicate operations demonstrated the versatility of the surface sampler as a remote manipulation device and the precision with which its operations can be controlled from the Earth.

Approximately 66 hours of alpha-scattering data were obtained during the first lunar day on three samples: the undisturbed lunar surface, a lunar rock, and an area dug up by the surface sampler. An additional 34 hours of data were obtained on the third sample during the second lunar day.

The surface sampler dug a number of trenches, conducted static and dynamic bearing-strength tests, picked up rocks, fractured a rock, weighed a rock and performed various other manipulations of the lunar material. The performance of the instrument and its controllers was outstanding.

In addition to acquiring a wide variety of lunar-surface data, Surveyor VII also obtained pictures of the Earth and performed star surveys. Laser beams from the Earth were successfully detected by the spacecraft's television camera in a special test of laser-pointing techniques.

Postsunset operations were conducted for 15 hours after local sunset at the end of the first lunar day at 06:06 GMT on January 25, 1968. During these operations, additional Earth and star pictures were obtained, as were observations of the solar corona out to 50 solar radii.

Operation of the spacecraft was terminated at 14:12 GMT on January 26, 1968, 80 hours after sunset. Second-lunar-day operations began at 19:01 GMT on February 12, 1968, and continued until 00:24 GMT on February 21, 1968.

In summary, five Surveyors have landed and operated successfully on the lunar surface. Four of these examined widely separated mare sites in the Moon's equatorial belt; the fifth investigated a region deep within the southern highlands. Four spacecraft survived the extreme cold of the lunar night and operated for more than one lunar day/night cycle. In total, the five Surveyors operated over a combined elapsed time of about 17 months on the Moon, transmitted more than 87 000 pictures, performed 6 separate chemical analyses of surface and near-subsurface samples, dug into and otherwise manipulated and tested the lunar material, measured its mechanical properties, and obtained a wide variety of other data which have greatly increased our knowledge of the lunar surface and the processes that have been acting on it.

The scientific and technological contributions of Surveyor to the future exploration of the Moon and planets have provided a major step forward in man's drive to explore the universe. This final report stands as a tribute to the hundreds of engineers, scientists, technicians, and managers—in industry, Government, and the scientific community—whose devotion and professional excellence made Surveyor's remarkable achievements possible.

BENJAMIN MILWITZKY
Assistant Director for Automated Systems
Apollo Lunar Exploration Office

Contents

		PAGE
1	INTRODUCTION	1
	L. D. Jaffe and R. H. Steinbacher.	
2	PRINCIPAL SCIENTIFIC RESULTS FROM THE SURVEYOR PROGRAM	13
	L. D. Jaffe (Chairman), C. O. Alley, S. A. Batterson, E. M. Christensen, S. E. Dwornik, D. E. Gault, J. W. Lucas, D. O. Muhleman, R. H. Norton, R. F. Scott, E. M. Shoemaker, R. H. Steinbacher, G. H. Sutton, and A. L. Turkevich.	
3	TELEVISION OBSERVATIONS FROM SURVEYOR	19
	E. M. Shoemaker (Principal Investigator), E. C. Morris, R. M. Batson, H. E. Holt, K. B. Larson, D. R. Montgomery, J. J. Rennilson, and E. A. Whitaker.	
4	LUNAR SURFACE MECHANICAL PROPERTIES	129
	R. Choate, S. A. Batterson, E. M. Christensen (Chairman), R. E. Hutton, L. D. Jaffe, R. H. Jones, H. Y. Ko, R. F. Scott, R. L. Spencer, F. B. Sperling, and G. H. Sutton.	
5	SOIL MECHANICS SURFACE SAMPLER	171
	R. F. Scott (Principal Investigator) and F. I. Roberson.	
6	LUNAR SURFACE TEMPERATURES AND THERMAL CHARACTERISTICS	181
	J. W. Lucas (Chairman), W. A. Hagemeyer, J. M. Saari, L. D. Stimpson, and J. M. F. Vickers.	
7	LUNAR SURFACE ELECTROMAGNETIC PROPERTIES	203
	D. O. Muhleman, W. E. Brown, Jr., L. Davids, J. Negus de Wys, and W. H. Peake.	
8	THE ALPHA-SCATTERING CHEMICAL ANALYSIS EXPERIMENT ON THE SURVEYOR LUNAR MISSIONS	271
	A. L. Turkevich (Principal Investigator), W. A. Anderson, T. E. Economou, E. J. Franzgrote, H. E. Griffin, S. L. Grotch, J. H. Patterson, and K. P. Sowinski.	
9	LUNAR THEORY AND PROCESSES	351
	D. E. Gault (Chairman), J. B. Adams, R. J. Collins, T. Gold, G. P. Kuiper, H. Masursky, J. A. O'Keefe, R. A. Phinney, and E. M. Shoemaker.	
10	SURVEYOR POSTTOUCHDOWN ANALYSES OF TRACKING DATA	369
	F. B. Winn.	
11	LASER BEAM POINTING TESTS	397
	C. O. Alley (Chairman) and D. G. Currie.	
12	ASTRONOMY: SOLAR CORONA OBSERVATIONS	405
	R. H. Norton.	
	APPENDIX A. EFFECTS OF LUNAR PARTICLES ON SPACECRAFT MIRROR SURFACES	407
	L. D. Jaffe and J. J. Rennilson.	
	APPENDIX B. THE LUNAR SUNSET PHENOMENON	413
	L. H. Allen.	
	APPENDIX C. SURVEYOR SCIENCE TEAMS AND COGNIZANT PERSONNEL	421
	APPENDIX D. SURVEYOR MANAGEMENT ORGANIZATION	425

1. Introduction

L. D. Jaffe and R. H. Steinbacher

Preliminary science results from each successful Surveyor have been presented in separate mission reports, issued just subsequent to the end of each mission (refs. 1-1 to 1-5). In this Surveyor report, these preliminary findings are reviewed and summarized, the results from each mission are compared and, in some cases, the results are given of analyses made after the publication of the mission reports.

This section presents briefly those spacecraft characteristics most necessary to an understanding of the scientific data obtained and the corresponding characteristics of spacecraft operations. An account by the Surveyor Scientific Evaluation Advisory Team of the major findings from Surveyor is presented in chapter 2. Subsequent chapters, prepared by Surveyor Investigator Teams and Working Groups, provide further information in individual technical areas.

Selected lunar pictures, along with appropriate explanatory material, are presented in part 3 of the mission reports (refs. 1-6 to 1-10).

Spacecraft

The Surveyor I spacecraft configuration is shown in figure 1-1. The spaceframe structure was of tubular aluminum; hinged to the spaceframe were three landing legs, each with a shock absorber and a hinged footpad. The footpads and blocks, attached under the spaceframe near each leg hinge, were constructed of energy-absorbing aluminum honeycomb to reduce landing shock. Two thermally controlled compartments housed the electronic equipment. A vertical mast carried the movable solar panel and planar-array antenna (high gain). Two deployable omnidirectional antennas (low gain) were also available for communication.

The main retro engine of the spacecraft utilized solid propellant. Each of the three liquid-fueled vernier engines was throttleable from about 460- to 120-N[1] thrust. Nitrogen gas jets provided attitude control when the engines were off. For attitude reference, the spacecraft carried Sun and Canopus sensors, and gyroscopes. A radar altimeter furnished an altitude mark to initiate main retrofiring during descent to the lunar surface. Another radar, providing measurements of velocity and altitude, was used with the vernier engines in a closed loop under control of an onboard computer for the final phases of the descent.

Surveyor I, designed to attain the engineering objectives of the Surveyor program, carried a television system for operation on the lunar surface and over 100 engineering sensors, such as resistance thermometers, voltage sensors, strain gages, accelerometers, and position indicators for movable spacecraft parts. No instrumentation was carried specifically for scientific experiments.

The television camera flown on Surveyor I is shown in figure 1-2. The vidicon tube, lenses, shutter, filter, and iris were mounted along an optical axis inclined approximately 16° to the central axis of the spacecraft; they were topped by a mirror that could be turned in azimuth and elevation. The azimuth, elevation, focal length, focus, exposure, iris, and filter were adjusted as needed by commands from Earth. Focal length adjustment provided either narrow-angle (6.4°) or wide-angle (25°) fields of view. The vidicon could be scanned to provide either a 200- or a 600-line picture. The 200-line pictures could be transmitted over an

---

[1] N (newton) is a standard international unit. 1 N is equal to 10^5 dynes.

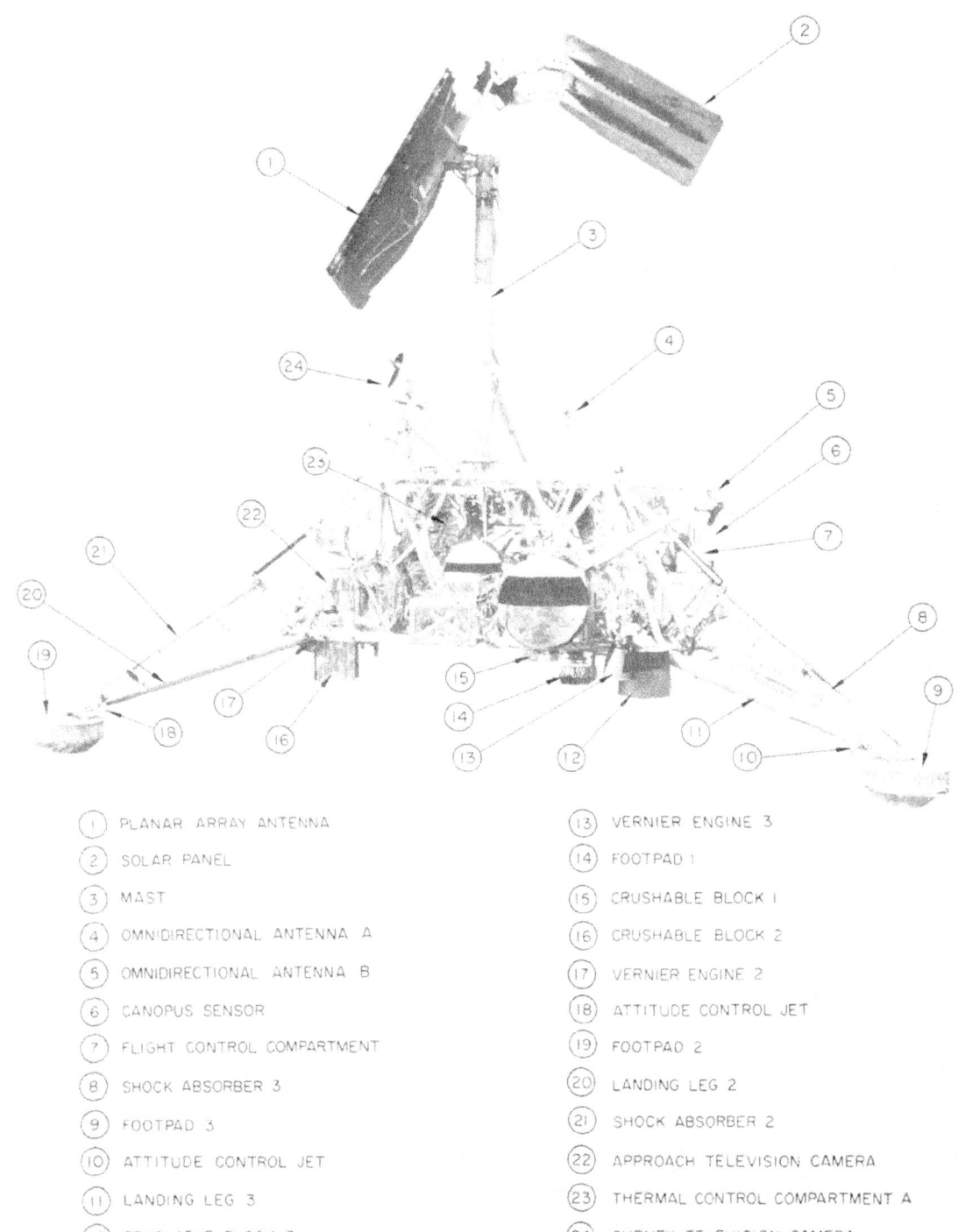

FIGURE 1-1.—Surveyor I spacecraft configuration.

INTRODUCTION

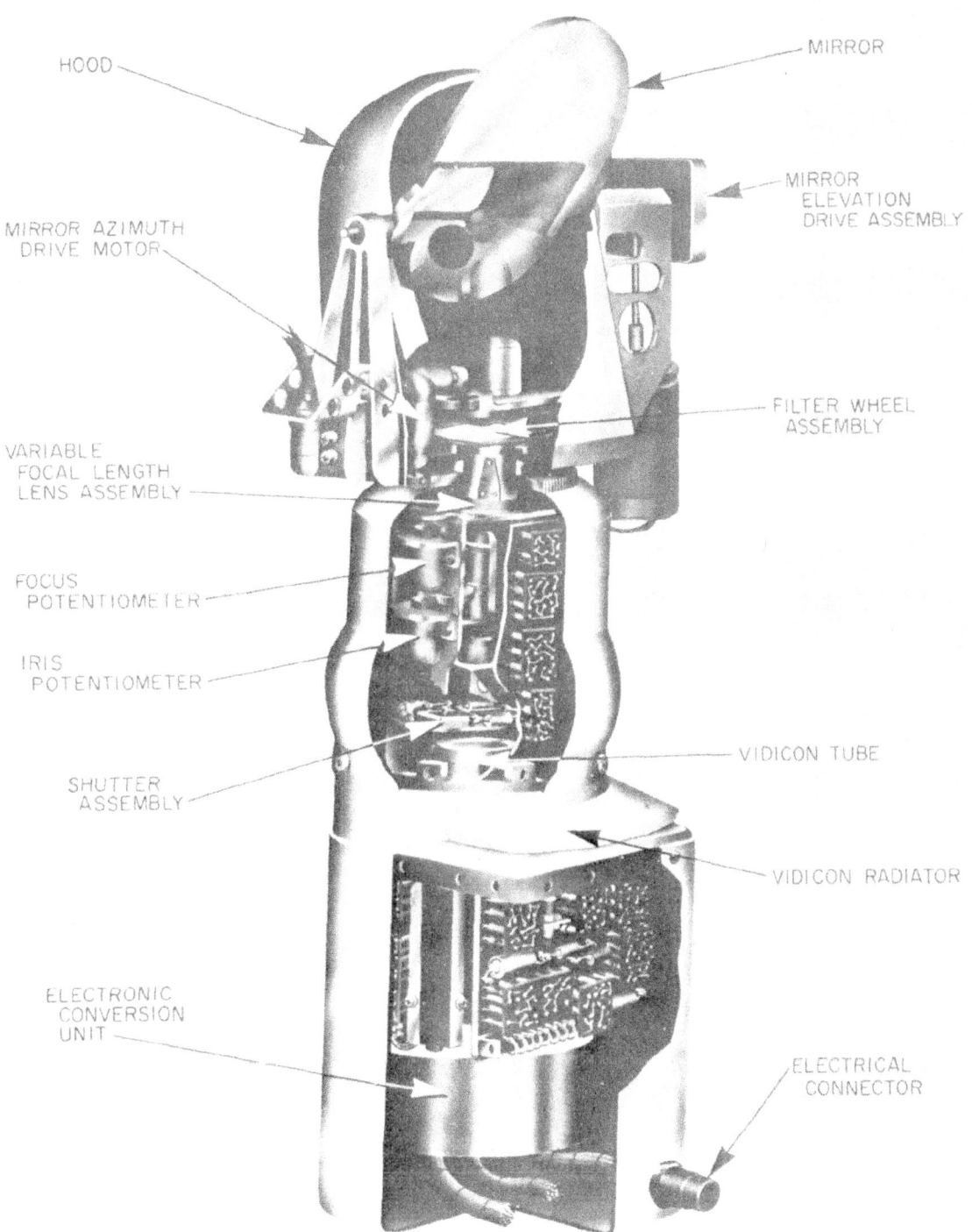

FIGURE 1-2.—Surveyor I television camera.

omnidirectional antenna or the planar-array antenna; the planar-array antenna was used for all 600-line pictures. The observed resolution for 600-line pictures was 0.5 mm at 1.6 meters from the camera.

At the time of the Surveyor I landing, additional spacecraft were in various stages of fabrication as part of the engineering effort. Changes were made to the design of these spacecraft to accommodate a scientific payload commensurate with spacecraft capability and schedule (fig. 1-3). Table 1-1 lists the scientifically significant spacecraft variations used. Many of the engineering changes made to increase Surveyor's reliability and performance, though not listed here, resulted in a better quality and greater amount of data (ref. 1-11).

On Surveyor III, a simplified surface-sampler instrument replaced the approach (downward-looking) television camera which had been carried, but not used, on Surveyor I. The surface sampler consisted primarily of a scoop, approximately 12 cm long and 5 cm wide, with a motor-operated door. The scoop was mounted on a pantograph arm that could be extended about 1.5 meters or retracted close to the spacecraft by a motor drive. The arm could also be moved in azimuth or elevation by motor drives, or dropped onto the lunar surface under force provided by gravity and a spring. The surface sampler could manipulate the lunar surface material in a number of ways, and the results could be observed by the television camera.

The hood on the Surveyor III television camera was fitted with a bonnet extension to provide additional shading for the lens and filter, thus extending the area of glare-free operation. Two auxiliary mirrors were attached to the spaceframe so that the camera could provide a better view of surface alterations produced by the vernier engines and by a crushable block.

Surveyors V and VI carried a television camera and an alpha-scattering instrument to

TABLE 1-1. *Payload differences among Surveyor spacecraft*

Instrumentation	Surveyor I	Surveyor III	Surveyor V	Surveyor VI	Surveyor VII
Television camera	Yes	Yes	Yes	Yes	Yes
Filters	Color	Color	Color	Polarizing	Polarizing
Glare hood	Standard	Extended	Extended	Box	Box
Elevation limit	35°	35°	35°	70°	70°
Photometric targets	2	2	2	2	3
Surface sampler	No	Yes	No	No	Yes
Azimuth range		+40° to −72° [a]			+71° to −41° [a]
Alpha-scattering instrument	No	No	Yes	Yes	Yes
Auxiliary mirrors	None	2 (flat)	2 (convex)	3 (convex)	3 (convex)
View under alpha-scattering instrument			Yes	Yes	Yes
View under vernier engine		Engines 2 and 3.	Engine 3	Engines 2 and 3.	Engines 2 and 3.
View under crushable block		Block 3	Block 3	Blocks 2 and 3.	Blocks 2 and 3.
Magnets:					
Footpads			Block 2	Block 2	Blocks 2 and 3.
Surface sampler					Yes
Stereoscopic and dust-detection mirrors.	No	No	No	No	Yes

[a] Arm movement CW(+) and CCW(−) rotation from the perpendicular to the spaceframe.

INTRODUCTION

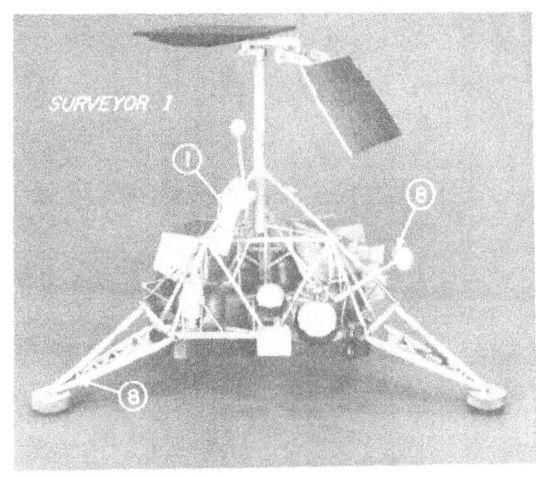
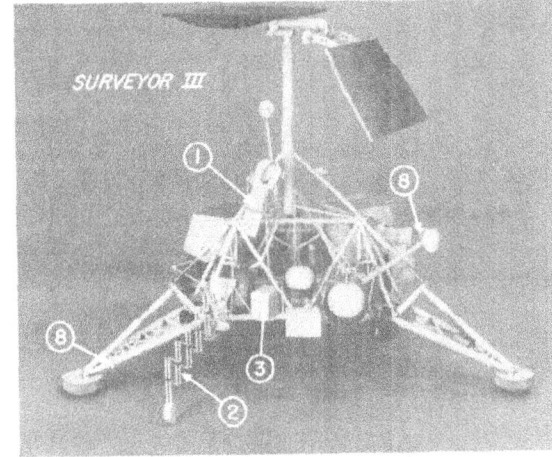
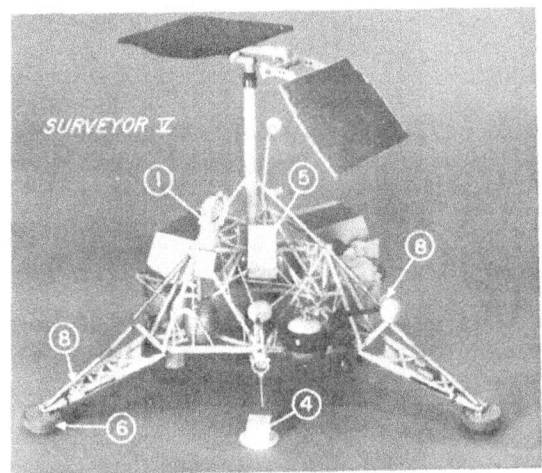

1. TELEVISION CAMERA
2. SURFACE-SAMPLER INSTRUMENT
3. SURFACE-SAMPLER AUXILIARY
4. ALPHA-SCATTERING-INSTRUMENT SENSOR HEAD
5. ALPHA-SCATTERING-INSTRUMENT ELECTRONIC COMPARTMENT
6. FOOTPAD MAGNET BRACKET
7. STEREO MIRROR
8. PHOTOMETRIC TARGET

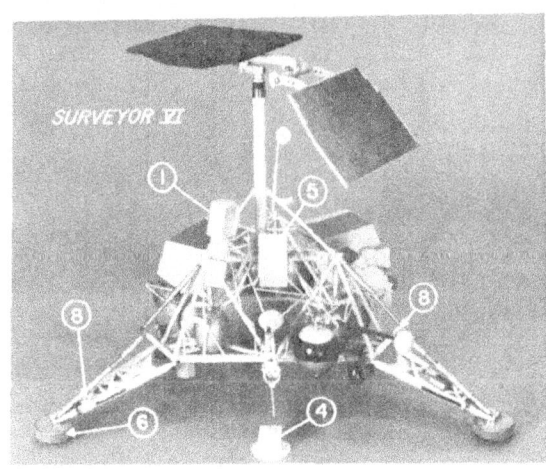
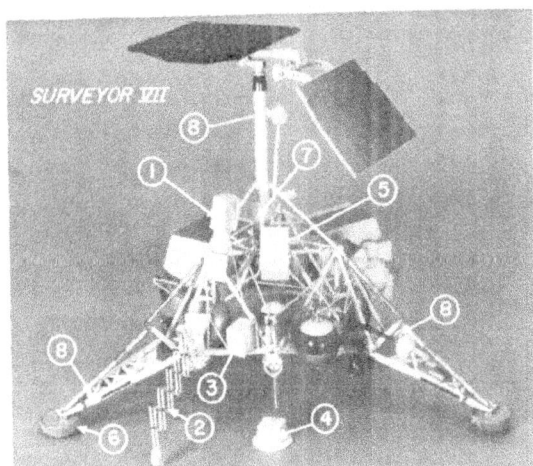

FIGURE 1-3.—Configuration of various Surveyor spacecraft showing changes in scientific payloads (models).

analyze chemically the lunar soil. This instrument was designed to irradiate the lunar surface with alpha particles from curium-242 sources and to measure the spectra of alpha particles scattered back. It also provided spectral data on protons produced by (α, p) reactions with the atoms of the lunar surface. These spectra could be interpreted in terms of the kind and quantity of elements present in the surface. Physically, the instrument consisted of a sensor head, a cube about 15 cm on a side, which, on command, could be lowered to the lunar surface by a nylon cord; and an alpha-scattering electronics compartment located on the spaceframe (refs. 1-12 and 1-13).

Surveyor VII carried, in addition to the camera and the alpha-scattering instrument, a soil mechanics surface sampler similar to that on Surveyor III. The Surveyor VII surface sampler, however, was mounted with a slightly different orientation so that it could reach the alpha-scattering instrument after deployment; because of the orientation change, it was not able to reach a spacecraft footpad.

The Surveyor V television camera was the same as that on Surveyor III. However, on Surveyors VI and VII, a new hood was added to provide better protection from glare and a better range in azimuth and elevation for the mirror. Polarizing filters were used on the Surveyor VI and VII cameras rather than the color filters used on the early missions. For camera calibration, Surveyors I, III, V, and VI carried photometric targets on a footpad and on an omnidirectional antenna boom; Surveyor VII carried an additional target on the other omnidirectional antenna boom.

To give greater area coverage, the auxiliary viewing mirrors were changed from flat to convex for Surveyors V, VI, and VII. For Surveyor V, one mirror was oriented to provide a view of the alpha-scattering-instrument sensor head on the surface rather than of the area under vernier engine 2. An additional viewing mirror was used on Surveyors VI and VII so that the area under vernier engine 2 and crushable block 2 could be observed. Surveyor VII also carried a mirror on the mast to provide a stereoscopic view of an area intersecting the arc of surface-sampler reach, and seven small mirrors to detect adhering dust.

Surveyors V and VI carried a bar magnet and a nonmagnetic control bar on one footpad to indicate the presence of lunar surface material with high magnetic susceptibility. Surveyor VII carried a similar magnet and control bar on two footpads and two small, horseshoe-shaped magnets placed in the pressure pad of the door of the surface-sampler scoop. All magnets could be observed by the television camera.

Surveyors V, VI, and VII incorporated some paint-pattern changes to reduce the brightness contrast between the dark lunar surface and some spacecraft parts painted white on earlier missions. The footpad tops were painted in stripes to reveal clearly any lunar material that might be deposited on them.

Landing Sites

Four Surveyor spacecraft landed in the lunar maria, near the equator. These sites were selected primarily because they were being considered for Apollo manned lunar landings. Surveyor VII, the last in the series, landed in the highland region close to the crater Tycho, a site chosen primarily for its scientific interest; it was thought to be a sample of very young highland material, which could have originated at considerable depth. The availability of Lunar Orbiter photographs of the area was considered in selecting landing sites for all Surveyors except Surveyor I, which preceded the Lunar Orbiter flights. The suitability of each site for making a safe landing was evaluated as part of the site-selection process (refs. 1-14 and 1-15).

Table 1-2 lists the varieties of terrain on which the Surveyors landed and their selenographic locations. Four sets of coordinates are given: The first set is based on radio tracking of the spacecraft during its flight to the Moon; the second is based on radio doppler tracking of the landed spacecraft from Earth. Both of these methods locate the site in inertial coordinates relative to the center of gravity of the Moon. The third set is a listing of the selenographic coordinates, in the system used in the Orthographic Atlas and in the Lunar Charts of the Aeronautical Chart and Information Center

TABLE 1-2. *Surveyor landing sites*

Spacecraft	Lunar region	Site characteristics	Spacecraft locations							
			Inertial coordinates from in-flight tracking [a]		Inertial coordinates from on-surface tracking [b]		Selenographic coordinates, Atlas ACIC system, from surface features [c]		Selenographic coordinates, Mills Arthur system, from surface features [c]	
			Longitude	Latitude	Longitude	Latitude	Longitude	Latitude	Longitude	Latitude
I	Southwest part of Oceanus Procellarum.	Level mare floor of Flamsteed ring.	43.34° W	2.44° S	43.32° W	2.46 –2.50 S	43.23° W	2.46° S	43.22° W	2.45° S
III	Southeast part of Oceanus Procellarum.	Wall of 200-meter crater.	23.41° W	3.0° S	23.32° W	3.06° S	23.34° W	2.99° S	23.34° W	2.97° S
V	Southwest of Mare Tranquillitatis.	Near top of 9- by 12-meter crater.	23.20° E	1.5° N	23.20° E	1.42° N	Not located		Not located	
VI	Sinus Medii	Level mare area, near mare ridge.	1.38° W	0.42° N	1.37° W	0.46° N	1.39° W	0.51° N	1.40° W	0.53° N
VII	Ejecta blanket, north of Tycho rim.	Hilly highland	11.41° W	41.01° S	11.44° W	40.97° S	11.45° W	40.88° S	11.47° W	40.86° S

[a] See ref. 1–11.   [b] See ch. 10 of this report.   [c] See ch. 3 of this report.

FIGURE 1-4.—Lunar Orbiter III photograph of Surveyor I on the lunar surface (enlarged from high-resolution frame H-194, framelet 248). The superimposed image is a model of the Surveyor spacecraft in a similar configuration and lighting to the Lunar Orbiter photograph. The image of the model and shadow was photoreduced on negative film to about the same size as it is on the Lunar Orbiter film. The resulting print was rephotographed through a television camera to produce a scan pattern.

(refs. 1-16 and 1-17).[2] These coordinates were obtained by determining the position of the landed Surveyors on Lunar Orbiter photographs by matching features shown in Surveyor pictures with corresponding features in the Lunar Orbiter photographs (refs. 1-2, 1-4, 1-5, and 1-19 (see ch. 3)). The Lunar Orbiter photographs were, in turn, related to the Orthographic Atlas and Mills/Arthur coordinates by matching large features, visible from Earth. Surveyor I was photographed on the lunar surface by Lunar Orbiters I and III; figure 1-4 is an enlargement of a Lunar Orbiter photograph of Surveyor I, together with a laboratory photograph of a Surveyor model with similar configuration and lighting. The Lunar Orbiter photographs of the other Surveyor landing sites were made before the Surveyors landed. Surveyor V has not yet been identified in Lunar Orbiter photographs; it may be outside the

---

[2] The fourth set is a listing in a more recent selenographic coordinate system, based on the catalogs of Mills (ref. 1-18) and of Arthur (ref. 1-19).

area of Lunar Orbiter high-resolution coverage.

The differences among the four sets of coordinates result, in part, from random errors and, in part, from systematic errors in the models used. Differences between the inertial and the selenographic coordinates arise from uncertainties in the selenographic grid and from differences between the center of figure and the center of gravity of the Moon.

Spacecraft Operations

Surveyor spacecraft were launched from Cape Kennedy, Fla., by Atlas/Centaur launch vehicles. After injection on a trajectory intersecting the Moon, the spacecraft were separated from the launch vehicles. Midcourse maneuvers, utilizing the vernier engines, were performed to bring the spacecraft within the desired landing areas. For the terminal descent, the main retro engine was ignited to provide most of the braking. After the main engine burned, nominally at about 10-km altitude, it was jettisoned, and the vernier engines continued to slow the spacecraft. To reduce disturbance of the lunar surface by engine exhaust, the vernier engines were turned off (except for Surveyor III) when the spacecraft altitude was about 4 meters and approach velocity was about 1.5 m/sec. The spacecraft then fell freely to the surface. The velocity components at touchdown were in the range of 3 to 4 m/sec vertically and less than 0.5 m/sec horizontally. The spacecraft masses at injection were 995 to 1040 kg; at touchdown, 294 to 306 kg.

The vernier engines on Surveyor III did not shut down before initial touchdown, but continued to burn, lifting the spacecraft from the surface. It landed again about 20 meters from the initial position, with engines still on, and lifted off a second time. The engines were then turned off, and the spacecraft touched down again 11 meters from the position of the second touchdown. The vertical velocity component for the three touchdowns was 1 to 2 m/sec; the horizontal component, 0.3 to 0.9 m/sec.

Table 1-3 shows the times at which the Surveyors landed. After touchdown, an engineering interrogation was made, then an

TABLE 1-3. *Times for Surveyor touchdowns and last data returns*

Spacecraft	Touchdown	Last data return
I	June 2, 1966; 06:17:36 GMT.	Jan. 7, 1967; 07:30 GMT.
III	Apr. 20, 1967; 00:04:17 GMT.	May 4, 1967; 00:04 GMT.
V	Sept. 11, 1967; 00:46:42 GMT.	Dec. 17, 1967; 04:30 GMT.
VI	Nov. 10, 1967; 01:01:04 GMT.	Dec. 14, 1967; 19:14 GMT.
VII	Jan. 10, 1968; 01:05:36 GMT.	Feb. 21, 1968; 00:24 GMT.

initial series of 200-line television pictures was transmitted through an omnidirectional antenna. The planar-array antenna was then pointed toward the Earth, and the transmission of 600-line television pictures was initiated. Engineering interrogations were interspersed with the various science operations.

Operations continued and some data were received for periods of 2 weeks to 8 months after landing, as shown in table 1-3. Operations were not always continuous; e.g., the spacecraft were shut down a few hours or days after each local sunset.

Lunar operations were commanded from, and spacecraft data were received through, the Deep Space Station 25-meter communications antennas at Goldstone, Calif.; Tidbinbilla (near Canberra), Australia; Robledo (near Madrid), Spain; and Johannesburg, South Africa. Occasionally, the 65-meter antenna at Goldstone and the 25-meter antenna at Honeysuckle Creek (near Canberra), Australia, were used.

A variety of data, not all from scientific instruments, was received from each spacecraft; sensors carried aboard the spacecraft for engineering information also provided scientific data about the Moon. The types of measurements and the scientific instrument or sensor that provided them are listed in table 1-4.

There were 87 674 television pictures obtained from the lunar surface (table 1-5). Photogrammetric, photometric, polarimetric, and

TABLE 1-4. *Measurement sources and types of scientific information derived from Surveyor spacecraft*

Instrument or sensor	Scientific information obtained
Television camera	Lunar topography, surface structure and geology; lunar photometry, polarization, and color; cohesion of lunar surface material; terrestrial photometry, polarization, and color; terrestrial atmospheric transmission, color, and cloud patterns; solar corona extent, photometry, and polarization; ability to point lasers from Earth to lunar locations.
Alpha-scattering instrument	Chemical composition of lunar surface and near surface; radiation level on lunar surface.
Soil mechanics surface sampler	Bearing and shear strengths of lunar surface and near-surface material; cohesion, internal friction, and density of lunar surface material; surface-rock strength and density.
Spacecraft leg strain gages, landing radar system, flight control gyros, and accelerometers	Bearing strength, cohesion, internal friction of lunar surface material; radar reflectivity and dielectric constant of lunar surface material; density and elastic velocity of lunar surface material.
Spacecraft resistance thermometers, solar panels, Sun sensors, and directional antenna	Lunar surface temperatures, thermal inertia, and directional infrared emission.
Magnets	Content of magnetic particles in lunar surface material.
Spacecraft vernier-propulsion system and attitude-control jet system	Lunar surface permeability to gases, cohesion, adhesion, response to gas erosion.

colorimetric data were included. Among the objects observed were

Lunar surface	5 sites in maria and highlands; objects 1½ meters to 30 km from camera; undisturbed surface and that disturbed by spacecraft; at Sun angles of 0° to 90°, and in earthlight.
Earth	In sunlight, at various phases and time intervals; eclipsing Sun.
Lasers on Earth	At 1-watt output.
Solar corona	Inner and outer, to 50 solar radii.
Planets	Mercury, Venus, and Jupiter.
Stars	To 6th magnitude.

By means of the alpha-scattering instrument, chemical analyses, covering major constituents with atomic numbers from carbon to iron, were made on six surface and slightly subsurface samples at two mare sites and at one highland site. Further analysis of the data obtained may permit some extension of this range of elements and determination of the flux on the lunar surface of protons with energies on the order of 100 MeV.

TABLE 1-5. *Number of Surveyor television pictures*

Spacecraft	Pictures taken per lunar day [a]			Total pictures taken
	1st day	2d day	4th day	
I	10 341	899		11 240
III	6 326			6 326
V	18 006	1048	64	19 118
VI	29 952			29 952
VII	20 993	45		21 038
Total				87 674

[a] These numbers reflect later information than those given in the mission reports (refs. 1-1 through 1-5).

With the surface sampler, measurements were made of the bearing load versus penetration curve for the granular lunar surface material and of its bearing capacity and shear resistance as a function of depth in the depth range of 1 to 20 cm. The nature of surface deformation was observed from trenching, static bearing, and impact; data were obtained on cohesion, internal friction, and porosity of the near-surface granular material at one mare and one highland site. Strength and density of individual rocks were also measured.

Strain gages on the leg shock absorbers of the spacecraft provided records of the leg loads during touchdown. In conjunction with flight control data on the spacecraft position and velocity, and television observations of the surface disturbances associated with landings, these records furnished information on bearing strength, cohesion, internal friction, elastic velocity, and porosity of the material at, and just below, the surface at all Surveyor landing sites.

Thermal sensors, measuring temperatures of spacecraft components radiatively coupled to the surface, permitted determination of surface temperatures, thermal inertia, and directional infrared emission at all sites.

The magnets carried by Surveyors V, VI, and VII provided information on the content of ferromagnetic and ferrimagnetic particles at two mare sites and one highland site.

At the mare landing sites, firings against the lunar surface of the vernier rocket engines on Surveyors III, V, and VI and of the attitude-control jets on Surveyors I and VI provided information on the permeability of the surface to gases, the lunar surface cohesion, response to gas erosion, and adhesion to terrestrial materials. On the Surveyor VI mission, 8 days after landing, the engines were fired in such a way as to lift the spacecraft 3.5 meters from the surface; it landed 2.4 meters from its original position.

Television pictures and other scientific data from Surveyor may be obtained from the National Space Science Data Center, Greenbelt, Md. 20771. Individual pictures can best be identified by the spacecraft mission and GMT day of year and time at which they were taken. Mosaics can best be identified by catalog number.

References

1-1. *Surveyor I Mission Report. Part II: Scientific Data and Results*, Tech. Rept. 32-1023, Jet Propulsion Laboratory, Pasadena, Calif., June 1, 1966. (See also NASA SP-126.)

1-2. *Surveyor III Mission Report. Part II: Scientific Results*, Tech. Rept. 32-1177, Jet Propulsion Laboratory, Pasadena, Calif., June 1, 1967. (See also NASA SP-146.)

1-3. *Surveyor V Mission Report. Part II: Science Results*, Tech. Rept. 32-1246, Jet Propulsion Laboratory, Pasadena, Calif., Nov. 1, 1967. (See also NASA SP-163.)

1-4. *Surveyor VI Mission Report. Part II: Science Results*, Tech. Rept. 32-1262, Jet Propulsion Laboratory, Pasadena, Calif., Jan. 10, 1968. (See also NASA SP-166.)

1-5. *Surveyor VII Mission Report. Part II: Science Results*, Tech. Rept. 32-1264, Jet Propulsion Laboratory, Pasadena, Calif., Mar. 15, 1968. (See also NASA SP-173.)

1-6. *Surveyor I Mission Report. Part III: Television Data*, Tech. Rept. 32-1023, Jet Propulsion Laboratory, Pasadena, Calif., Nov. 1, 1966.

1-7. *Surveyor III Mission Report. Part III: Television Data*, Tech. Rept. 32-1177, Jet Propulsion Laboratory, Pasadena, Calif., Nov. 10, 1967.

1-8. *Surveyor V Mission Report. Part III: Television Data*, Tech. Rept. 32-1246, Jet Propulsion Laboratory, Pasadena, Calif., July 15, 1968.

1-9. *Surveyor VI Mission Report. Part III: Television Data*, Tech. Rept. 32-1262, Jet Propulsion Laboratory, Pasadena, Calif., Aug. 1, 1968.

1-10. *Surveyor VII Mission Report. Part III: Television Data*, Tech. Rept. 32-1264, Jet Propulsion Laboratory, Pasadena, Calif. (to be published).

1-11. *Surveyor Project Report. Part I: Mission Description and Performance*, Tech. Rept. 32-1265, Jet Propulsion Laboratory, Pasadena, Calif. (to be published).

1-12. TURKEVICH, A. L.; KNOLLE, K.; EMMERT, R. A.; ANDERSON, W. A.; PATTERSON, J. H.; AND FRANZGROTE, E. J.: Instrument for Lunar Surface Chemical Analysis. *Rev. Sci. Inst.*, vol. 37, no. 12, 1966, pp. 1681-1686.

1-13. TURKEVICH, A. L.; KNOLLE, K.; FRANZGROTE, E. J.; AND PATTERSON, J. H.: Chemical Analysis Experiment for the Surveyor Lunar Mission. *J. Geophys. Res.*, vol. 72, no. 2, 1967, pp. 831-839.

1-14. WILLINGHAM, D. E.: Lunar Surface Generation and Surveyor Landing Analysis. Project Document 602-4, Jet Propulsion Laboratory, Pasadena, Calif., 1967.

1-15. FILICE, A. L.; THORNTON, T. H.; AND WILLINGHAM, D. E.: Surveyor Landing Site Selection. Project Document 602-29, Jet Propulsion Laboratory, Pasadena, Calif., 1967.

1-16. *Orthographic Atlas of the Moon, Supplement 1 to the Photographic Lunar Atlas*, compiled by D. W. G. Arthur and E. A. Whitaker, Univ. Arizona Press, Tucson, Ariz., 1961.

1-17. *Lunar Charts*, Aeronautical Chart and Information Center, St. Louis, Mo., 1961-68.

1-18. MILLS, G. A.: Absolute Coordinates of Lunar Features. Icarus, vol. 7, 1967, pp. 983-1220, and Icarus, vol. 8, 1968, pp. 90-116.
1-19. ARTHUR, D. W. G.: The Tucson Triangulation, and other related papers, accepted for publication, Communications Lunar and Planetary Lab., University of Arizona, Tucson, Ariz.
1-20. SPRADLEY, L. H.; STEINBACHER, R.; GROLIER, M.; AND BYRNE, C.: Surveyor I: Location and Identification. Science, vol. 157, 1967, pp. 681-683.

ACKNOWLEDGMENT

We extend appreciation to S. E. DWORNIK, NASA, and Dr. T. VREBALOVICH and E. M. CHRISTENSEN, JPL, who were responsible for the planning and conducting of major portions of the Surveyor science effort; to A. L. FILICE, D. WILLINGHAM, J. N. STRAND, and D. L. SMYTHE, JPL, for evaluations of potential landing sites; and to S. Z. GUNTER, J. N. STRAND, Dr. S. L. GROTCH, and R. A. BIDEAUX, JPL, who were responsible for science data handling.

2. Principal Scientific Results From the Surveyor Program

L. D. Jaffe (Chairman), C. O. Alley, S. A. Batterson, E. M. Christensen, S. E. Dwornik, D. E. Gault, J. W. Lucas, D. O. Muhleman, R. H. Norton, R. F. Scott, E. M. Shoemaker, R. H. Steinbacher, G. H. Sutton, and A. L. Turkevich

The successful soft landings made by five Surveyors permitted detailed examinations of the lunar surface at four mare sites along an equatorial belt and at one highland site in the southern hemisphere. The aiming areas, selected before launch, were chosen after examination of telescopic and Lunar Orbiter photographs (except for the Surveyor I mission, which preceded the Lunar Orbiter flights). All five spacecraft landed within these selected areas. The landing sites were:

Surveyor I.—Flat surface inside a 100-km crater in Oceanus Procellarum, 1 radius from the edge of a rimless 200-meter crater.

Surveyor III.—Interior of a subdued 200-meter crater, probably of impact origin, in Oceanus Procellarum.

Surveyor V.—Steep, inner slope of a 9- by 12-meter crater, which may be a subsidence feature, in Mare Tranquillitatis.

Surveyor VI.—Flat surface near a mare ridge in Sinus Medii.

Surveyor VII.—Ejecta or flow blanket north of, and less than 1 radius from, the rim of the crater Tycho in the highlands.

At each of the Surveyor landing sites, the lunar surface is covered by a layer of fragmental debris, predominantly fine grained, which is littered with a variety of rock fragments, and spotted with overlapping small craters. The average thickness of the debris layer, or regolith, was determined for each of the mare landing sites from the depth of the smallest craters with blocky rims. The thickest regolith (10 to 20 meters) was found at the Surveyor VI site. The regolith near Surveyor I is 1 to 2 meters thick, and at the Surveyor V site is less than 5 meters. The thickness of the regolith within the 200-meter crater where Surveyor III landed varies from about 1 to 2 meters on the rim to perhaps 10 meters or more at the crater center.

Small Craters

Small craters account for the irregularities of largest relief on the surfaces at the mare landing sites. Most of the small craters at each of the landing sites have a cup shape with walls and floors concave upward and low, subdued rims, but some are nearly rimless. Most of the cup-shaped craters are believed to be of impact origin. Dimple-shaped craters lacking raised rims and crater chains are common at the Surveyor V site and were observed at the Surveyor III site; these may have been formed by drainage of the surficial fragmental debris into subsurface fissures.

Many irregular craters, ranging in size from a few centimeters to several meters in diameter and lined with clods of fine-grained material, were observed at all landing sites. These are inferred to be secondary impact craters formed by cohesive blocks or clods of weakly cohesive fine-grained material ejected from nearby primary craters.

The cumulative size distribution of small craters a few centimeters to several tens of meters in diameter is consistent with a power law having an exponent of -2. This corresponds to that expected for a steady-state population of craters produced by prolonged, repetitive bombardment by meteoroids and by secondary fragments from the Moon. There are fewer craters larger than 8 meters in diameter at the Surveyor VII site than at the mare landing sites; this indicates that the Tycho rim material on which Surveyor VII landed is relatively young.

Resolvable Rock Fragments

Some rock fragments are on the surface and others are partly embedded to various depths in a finer grained matrix. The fragments are scattered somewhat irregularly, but strewn fields of blocks are found around some of the larger craters. The Surveyor VII landing site has the largest number and variety of resolvable rock fragments; the Surveyor V and VI sites have the least number. Most are relatively angular, but some well-rounded fragments are present and appear, in general, to be embedded fairly deeply in the finer surface material. The fragments tend to be equant in shape, but some are distinctly tabular and a few have the form of sharp, narrow wedges.

Most of the resolvable fragments on the surface are brighter, under all observed angles of illumination, than the unresolved fine-grained matrix. A knobby, pitted surface is the most common surface texture developed on the bright, coarse, rounded rock fragments. The pitted texture and the rounding of the fragments are probably produced by impact of small particles. Further evidence that the surfaces of the rocky fragments have been subjected to an erosive, or abrasive, action is evident on one of the rocks overturned by the surface sampler on Surveyor VII. This rock was rounded on the exposed side and angular on the subsurface side. Spotted rocks, observed at the Surveyor I and V sites, are common near Surveyor VII. In many cases, the lighter material forms slight protrusions on the fragment, suggesting that it is more resistant to lunar erosion processes.

The density of one rock near Surveyor VII was found to be in the range of 2.4 to 3.1 g/cm^3, with a most probable value of 2.8 to 2.9 g/cm^3. A similar rock was broken by a moderately strong blow from the surface sampler. Most of the resolvable blocks on the lunar surface appear to be dense, coherent rock, but some appear less dense and porous. Many blocks at the Surveyor VII site, and a few at the Surveyor I site, are distinctly vesicular. Some coarse fragments are clearly aggregates of smaller particles. Some of these aggregates are compact and angular, whereas others appear to be porous and probably are only weakly compacted.

Particle Size

Between 4 and 18 percent of the lunar surface is covered by fragments coarse enough to be resolved by the television camera (coarser than 1 mm). The size-frequency distribution of resolvable fragmental debris at each Surveyor landing site can be represented by a simple power function. At the mare sites, the exponent of the size-distribution function is, in all cases, less than -2; at the Surveyor VII site, on the rim flank of Tycho, the exponent is -1.8. Fragments coarser than 10 cm are 5 to 10 times more abundant on the rim of Tycho than on the maria.

The ability of the finer grained matrix material to conform to smooth surfaces and to preserve fine imprints; its permeability to gases; its cohesion, and its optical properties, before and after disturbance, all suggest that the bulk of the material has a particle size between 2 and 60 microns.

Structure and Mechanical Behavior of the Fine Material

At all landing sites, the fine matrix, or lunar soil, is granular and slightly cohesive; the soil is compressible, at least in its upper few centimeters, as indicated by the footpad and crushable block imprints; and its static bearing strength increases with depth as follows:

(1) In about the upper millimeter: less than 0.1 N/cm^2 (from imprints of small, rolling fragments).

(2) At a depth of 1 to 2 mm: 0.2 N/cm^2 (from imprints of the alpha-scattering-instrument sensor head).

(3) At a depth of about 2 cm: 1.8 N/cm^2 (from imprints of crushable blocks on Surveyors VI and VII).

(4) At a depth of about 5 cm: 5.5 N/cm^2 (from penetrations of footpads on Surveyor I).

The higher rock population at the Surveyor VII site did not, in general, increase the bearing strength of the soil compared with that at the Surveyor I, III, and VI sites.

The estimated soil shear wave velocity is between 15 and 36 m/sec, and the compressional wave velocity between 31 and 91 m/sec. These estimates, based on oscillations in the spacecraft landing leg forces at touchdown of four landings, are lower than those expected for terrestrial soils with other mechanical properties as listed in this report.

Viscous soil erosion (erosion by the entrainment of soil particles as gas flows over the surface) occurred during vernier-engine and attitude-control jet firings. During the Surveyor V vernier-engine firing, soil and rock fragments up to 4.4 cm in diameter were moved by viscous erosion. At engine shutdown, exhaust gas, which had diffused into the soil, erupted, producing a crater 20 cm in diameter and 0.8 to 1.3 cm deep under one engine.

The permeability of the lunar soil at the Surveyor V site to a depth of about 25 cm is 1×10^{-8} to 7×10^{-8} cm^2. This corresponds to the permeability of terrestrial silts. The soil cohesion bounds were determined to be:

(1) 0.007 to 0.12 N/cm^2 (from vernier-engine firings).
(2) 0.05 to 0.17 N/cm^2 (from attitude-control jet firings).

Lunar material thrown against spacecraft surfaces in several cases adhered to the spacecraft components. Adhesion of soil to the Surveyor VII surface-sampler scoop was observed to increase toward the end of the lunar day. The adhesive strength of the lunar material impacting and adhering to the Surveyor VI photometric target is estimated to be between 10^2 and 10^3 dynes/cm^2.

Soil properties, similar to those described, also were indicated during the surface-sampler operations. The bearing-strength values and the soil behavior during all tests are consistent with a granular material possessing a cohesion of 0.035 to 0.05 N/cm^2, an angle of internal friction of 35° to 37°, and a density of about 1.5 g/cm^3. These values apply to soil depths between a few millimeters and about 10 cm. An increase of strength with depth in the soil near Surveyor III was observed to a depth of about 20 cm; this increase was not found at the Surveyor VII location. The soil was also found to be more brittle at the Surveyor III site.

Assuming that the lunar soil is derived from the rock fragments, the above rock and soil densities indicate a soil porosity of 0.4 to 0.5, on the average, from depths of a few millimeters to about 10 cm.

Optical, Thermal, and Radar Characteristics

Television observations with color filters indicate a gray Moon even in disturbed areas. No demonstrable differences in color were observed on any of the coarse blocks so far examined, which are all gray, but lighter than the fine-grained gray matrix of the surface. Photometric measurements at each Surveyor landing site show that the undisturbed fine matrix of the mare surface has a normal luminance factor (normal albedo) that varies from 7.3 to 8.2 percent, and the mare material disturbed by the footpads and by the surface sampler has a normal luminance factor that ranges from 5.5 to 6.1 percent. The normal luminance factor for the fine-grained, undisturbed and disturbed material at the Surveyor VII site is higher than that of the corresponding mare material. The normal luminance factor for the rock fragments ranges from 14 to 22 percent both on the maria and on the rim flank of Tycho. Light scattered from the surfaces of some rock fragments at the Surveyor VII site is as much as 30 percent polarized at phase angles near 120°. This suggests these rocks are crystalline or glassy, and that their surfaces are relatively free of fine particles.

Observations of the fine-grained parts of the lunar surface disturbed by the landing and liftoff of the Surveyor VI spacecraft, and by rolling fragments set in motion by the spacecraft, have shown that lunar material exposed at depths no greater than a few millimeters has a significantly lower normal luminance factor than the undisturbed surface. A similar, abrupt decrease in normal luminance factor at depths of 3 mm or less was observed at the Surveyor III and V landing sites. The occurrence of this rather sharp contact of material with contrasting optical properties at widely separated localities on the Moon suggests that some process, or combination of processes, lightens the material at the lunar surface. If this is true, it may imply that a complementary process of darkening occurs at depths of a few millimeters and deeper, so that the abrupt albedo contact is not destroyed as a result of repetitive turnover of the lunar surface by solid-particle bombardment.

Where the fine-grained matrix of the surface material was compressed and smoothed by the Surveyor footpads or the surface sampler, the photometric properties were changed. The photometric function of the smoothed surfaces is more like that of a Lambertian scatterer than the undisturbed, fine-grained lunar material. This indicates the pore spaces of the fine-grained material tend to be filled in by compression against smooth surfaces on parts of the spacecraft.

In general, lunar surface temperatures derived from spacecraft thermal data taken during the lunar day are in qualitative agreement with Earth-based data. For each mission during which there was an eclipse, the same value of thermal parameter $(k\rho c)^{-1/2}$ was obtained from spacecraft eclipse and postsunset data, whereas the values for Earth-based postsunset data are lower than those from Earth-based eclipse data. On all spacecraft except one, the same values were obtained in the two directions viewed by the pertinent spacecraft sensors; in the case of Surveyor VII, the thermal parameter values in the two directions were different; this discrepancy apparently was caused by some rocks close to the spacecraft. When Surveyors III and V landed in craters, it was observed that the local lunar surface temperatures depended primarily on the Sun elevation angle to the local lunar surface slope. It should be noted that all Earth-based and spacecraft data indicate that the lunar surface material is a very good thermal insulator.

Radar backscatter data from the lunar surface, at 2.5- and 3.2-cm wavelengths, were obtained during the last 3 minutes of the descent for each Surveyor landing. The radar system consisted of four independent radars. Signal strengths from each beam have been interpreted in the form of the radar cross section as a function of the angle of incidence in the range 0° to 60°. The general form of these functions for all landing sites is approximately the same. However, the entire curve is higher by nearly a factor of 2 for the Tycho region than for the mare sites. Because of strong fading on the radar beams near normal incidence, a reliable estimate of the normal-incidence reflectivity cannot be made. The radar reflectivity of the Tycho rim area is about 30 percent higher than the average reflectivity of the Moon as measured from Earth; the mare areas are about 30 percent lower. In particular, it is estimated that the reflectivity of the Surveyor V area (Mare Tranquillitatis) is between 3 and 5 percent. The Surveyor data confirm a rather low value (about 2.5 to 3) for the dielectric constant of lunar surface material, with a clear distinction between the mare and highland regions.

Chemical Composition

Surveyor obtained the first direct information about the chemical nature of the lunar surface material. At two mare sites (Surveyor V and VI missions) and one highland site (Surveyor VII mission), the presence of magnetic material in the lunar soil was demonstrated, and the amounts of the most abundant chemical elements were established.

Analytical data were obtained on six samples of lunar material, three at the mare sites and three at the highland site. The analyses indicate that the most abundant chemical element on the Moon is oxygen (57 ± 5 atomic percent); second in abundance is silicon (20 ± 5 atomic percent); and third is probably aluminum (about 7 atomic percent). These are, in the same order, the most common elements in the Earth's crust. The three samples from the maria are almost identical chemically, implying that the surface material of large fractions of the lunar maria have this composition. The highland sample differs principally in having about half as much of the "iron"-group elements (titanium through copper) as do the samples from the maria.

The amount of oxygen is estimated to be sufficient to form oxides of all of the metals, and so indicates that the bulk of the material is relatively stable chemically (although a small amount of radiation-decomposed material cannot be excluded). The relative abundance of the principal elements on the lunar surface is similar to that of terrestrial basalts, which often have the amount of magnetic material observed in the lunar soil. The chemical composition found is significantly different from the most common meteorites (metallic or stony) falling on the Earth.

These chemical analyses are in strong disagreement with that expected for primordial solar system material, whether this be considered condensed solar atmosphere, terrestrial ultrabasic rocks, or chondritic meteorites. They clearly contradict a lunar origin for most meteorites and are inconsistent with a lunar origin for tektites.

The similarity of the lunar samples with basaltic composition and the morphology of surface features observed in terrestrial and Lunar Orbiter photographs are strong circumstantial evidence that some melting and chemical fractionation of lunar material has occurred in the past. The bulk composition of the Moon, however, remains obscure. The lower abundance of the "iron"-group elements in the highlands, as compared with the maria, provides an explanation for the albedo differences between these two major geologic units and, in addition, suggests a significant difference in rock density consistent with isostasy.

Observations of the Earth and of the Solar Corona

Pictures of an eclipse of the Sun by the Earth were made during the Surveyor III mission. Using the camera color filters, sufficient data were accumulated to produce color pictures of the light transmitted through the Earth's atmosphere during the eclipse. These results indicated that the observed light primarily was the result of refraction by the Earth's atmosphere and that clouds present at the limb occulted the light.

During the Surveyor VII mission, pictures of the Earth were taken with various polarizing filters. The highly polarized component of light appears to be the result of specular reflection from ocean surfaces. Pictures of Earth-based laser beams directed toward the spacecraft's lunar location were also made in a test of the ability of Earth stations to direct very narrow beams to a specific location on the lunar surface, in preparation for a possible laser reflector experiment.

The solar corona was photographed after sunset on the Surveyor I, V, and VII missions, from the innermost K-corona at 2 solar radii to well beyond the known outer F-corona at 60 solar radii. The data will provide information on the previously unobserved region between 15 and 50 solar radii.

3. Television Observations From Surveyor

E. M. Shoemaker (Principal Investigator), E. C. Morris, R. M. Batson, H. E. Holt, K. B. Larson, D. R. Montgomery, J. J. Rennilson, and E. A. Whitaker

Five successful Surveyor spacecraft landed on the Moon between June 1966 and January 1968 and returned over 87 000 pictures from the lunar surface. Surveyors I, III, V, and VI landed on mare surfaces; Surveyor VII landed in the southern highlands on the flank of the crater Tycho, the youngest, large bright-ray crater on the Moon. Table 3-1 lists the day and times of landing of each spacecraft, their location and the Sun elevation at the time of landing.

Surveyors I, III, V, and VI provided pictures of various lunar-mare features, such as the surface in intercrater areas, the inner walls of a large subdued crater, the inside of a small drainage crater, and a close view of a mare ridge.

The terrain around Surveyor VII, in contrast to the terrain of the maria, consists of ridges and valleys superimposed on a broadly undulating surface. The Surveyor VII pictures revealed a great variety of coarse rock fragments probably excavated from the depths of Tycho. Some fragments are vesicular, others appear dense; some fragments are spotted, suggesting differences in crystallinity or composition in the fragments.

The size distribution of craters and fragments; the thickness of the fragmental debris layer, or regolith; and the colorimetric, photometric, and polarimetric properties of various lunar-surface materials were determined from the pictures for each landing site.

Stereoscopic pictures of the lunar surface were obtained from Surveyors VI and VII. Surveyor VI vernier engines were ignited; the spacecraft lifted off the lunar surface and landed about 2.5 meters from its original position, thus providing a base for stereoscopic pictures. Surveyor VII was equipped with a 9- by 24-cm mirror attached to the spacecraft mast; this mirror was oriented to provide a reflected view, as seen from the television camera, of a small area in front of the spacecraft. Stereoscopic pictures were obtained by recording direct images of this area and images reflected from the mirror.

The television cameras of Surveyors I, V, and VII were operated more than one lunar day. Surveyor I transmitted over 800 pictures during the second lunar day of operation; subsequent engineering interrogations were continued through January 1967. After a warmup period of about 147 hours after sunrise

TABLE 3-1. *Surveyor times, locations, and approximate Sun elevations at landing*

Spacecraft	Landing, GMT hr:min:sec	Location, selenographic coordinates		Approximate Sun elevation at landing, deg
		Longitude	Latitude	
Surveyor I	06:17:36 on June 2, 1966	43.22° W	2.45° S	28
Surveyor III	00:04:17 on Apr. 20, 1967	23.34° W	2.97° S	14
Surveyor V	00:46:42 on Sept. 11, 1967	23.20° E	1.42° N	17
Surveyor VI	01:01:04 on Nov. 10, 1967	1.40° W	0.53° N	3
Surveyor VII	01:05:36 on Jan. 10, 1968	11.47° W	40.86° S	13

on the second lunar day. Surveyor V responded to the first turn-on command and subsequently transmitted over 1000 pictures before the second lunar-day sunset. Surveyor V was revived again the fourth lunar day, but only a few pictures were taken. Surveyor VII was revived about 120 hours after sunrise the second lunar day; about 45 pictures were taken in the 200-line mode before suspension of camera operation.

Most pictures transmitted by the five Surveyor spacecraft were received at the Goldstone, Calif., Tracking Station of the Deep Space Network. Pictures were also received at the Canberra, Australia, and Robledo (near Madrid), Spain, Tracking Stations.

Television Camera

D. R. MONTGOMERY AND E. C. MORRIS

Surveyor's 7.31-kilogram (16.1-lb) television camera (fig. 3–1) consisted of a mirror, filters, lens, shutter, vidicon, and attendant electronic circuitry. Each picture, or frame, was imaged through an optical system onto the photoconductive surface of a vidicon, which was scanned by an electron beam. The camera was designed to accommodate scene luminance levels from about 0.008 to 2600 ft-L, employing both electromechanical mode changes and iris control. On the Surveyor I, III, and V missions, frame-by-frame coverage of the lunar surface could be obtained over 360° in azimuth and from 40° above the plane normal to the camera Z-axis to −65° below this plane. On the Surveyor VI and VII missions, the coverage in elevation was increased to +90° above the plane normal to the camera Z-axis. The camera was capable of a resolution of about 1 millimeter at 4 meters and could focus from 1.23 meters to infinity. Camera operation was controlled by commands from Earth. Commandable operation allowed each frame to be taken with a lens setting and mirror azimuth and elevation position appropriate for a given view of the lunar surface.

The edge of the mirror (fig. 3–2) was a 10.5- by 15-cm ellipse, and the mirror was supported

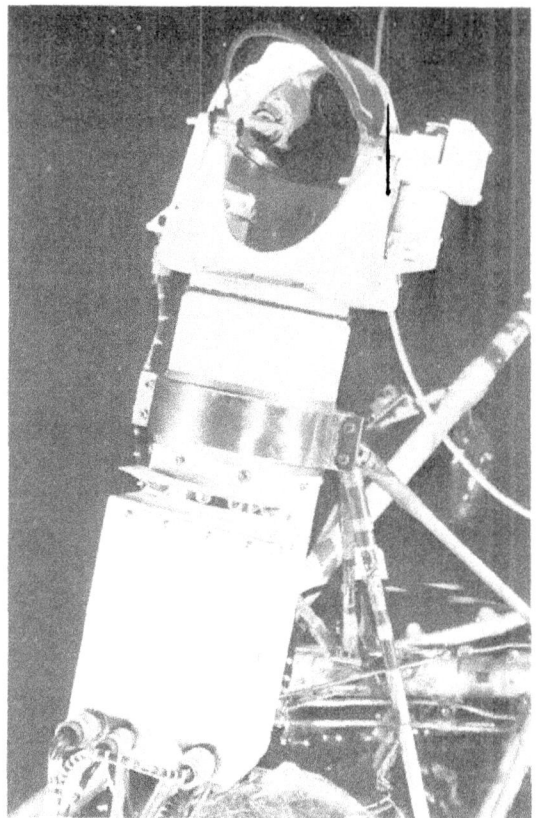

FIGURE 3–1.—Surveyor I television camera.

at its minor axis by trunnions. Its reflecting surface was formed by vacuum deposition of Kanogen on a beryllium blank, followed by a deposition of aluminum and finally by deposition of silicon monoxide. The reflecting surface was flat within one-fourth wavelength at $\lambda = 550$ mμ and had an average specular reflectance of 86 percent. The mirror rotated about two mutually perpendicular axes by means of two drive mechanisms, one for azimuth and the other for elevation; the position of the mirror about each axis was measured by a potentiometer.

Within the mirror housing was a filter wheel (fig. 3–3), which contained three color or polarizing filters, in addition to a fourth section containing a clear element.

The image was formed by means of a variable-focal-length lens (fig. 3–4), placed between the vidicon and the mirror assembly. The focal

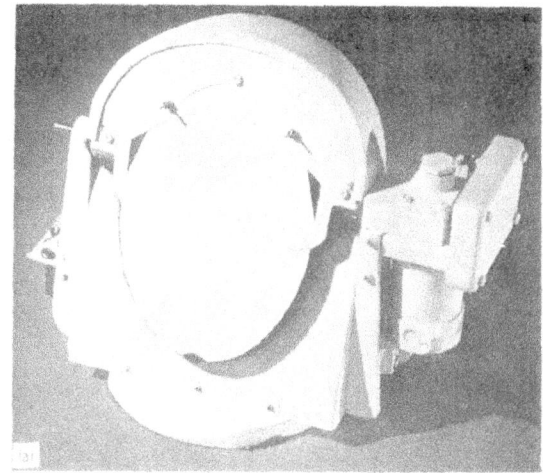

FIGURE 3-2.—Mirror assembly of the Surveyor television camera. (a) Surveyor I television camera hood and mirror assembly. A small visor was added to the top of the hood on the Surveyor III camera. (b) Mirror assembly and redesigned hood used on Surveyors V, VI, and VII.

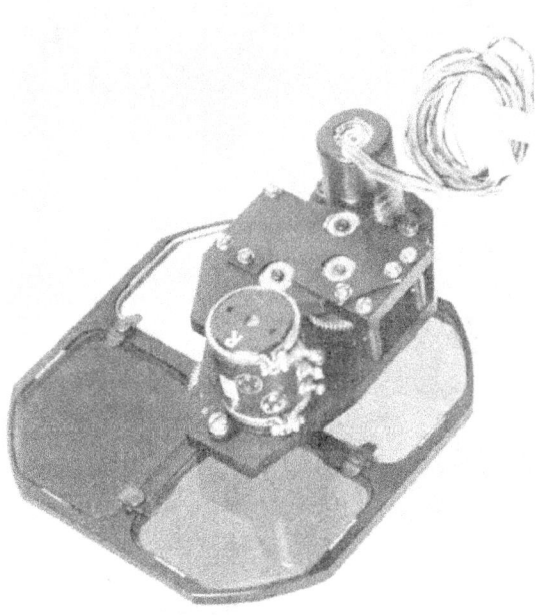

FIGURE 3-3.—Filter wheel of the Surveyor television camera.

length could be varied from 25 to 100 millimeters, resulting in optical fields of view of about 25.3° to 6.43°; however, the camera was always operated at either the 25- or 100-millimeter focal length. Additionally, the lens assembly could be varied in focus by means of a rotating focus cell. An adjustable iris provided effective aperture changes from $f/4$ to $f/22$, in increments which resulted in change of aperture area by a factor of ½. The iris could be controlled by command; also available was a servo-type automatic iris control, which adjusted the aperture area in proportion to the average scene luminance. As in the mirror assembly, potentiometers were geared to the iris, focal length, and focus elements to allow determination of these settings for each picture. A beam splitter on the lens assembly sampled 10 percent of incident light for operation of the automatic iris.

Light energy from object space was converted to an equivalent electrical signal in the image plane by the vidicon tube (fig. 3-5). The size of

FIGURE 3-4.—Lens assembly of the Surveyor television camera. (a) Front view. (b) Back view.

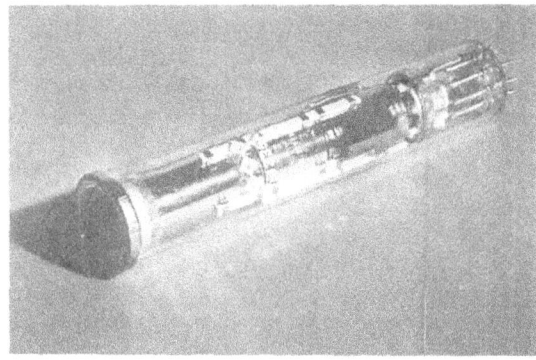

FIGURE 3-5.—Vidicon tube of the Surveyor television camera.

FIGURE 3-6.—Shutter assembly of the Surveyor television camera.

the image frame on the vidicon tube was 11 millimeters square. A reference mark was included in each corner of the scanned frame, which provided, in the video signal, an electronic level representing optical black for the scanned image. In the normal, or 600-line mode of operation, the frame was scanned once each 3.6 seconds. Each frame required nominally 1 second to be read from the vidicon and required 220-kilohertz bandwidth for transmission. In the second mode of operation, one 200-line frame was scanned each 61.8 seconds. Each frame required 20 seconds to complete the video transmission and utilized a bandwidth of 1.2 kilohertz. This 200-line mode was used for omnidirectional antenna transmission from the spacecraft.

A mechanical focal plane shutter, located between the camera lens assembly and the vidicon image sensor (fig. 3-6), could be operated in two modes. In the normal mode of operation, on Earth command, the shutter blades were sequentially driven by rotary solenoids across an aperture in the shutter base plate. The time interval between the initiation of each blade determined the exposure interval, nominally 150 milliseconds. In the other shutter mode, the blades could be positioned to leave the aperture open and the frame scanned every 3.6 seconds during 600-line operation, or every 61.8 seconds during 200-line operation. This open-shutter mode of operation was useful in the imaging of scenes with low luminance levels.

such as the sky with stars, and planets and the lunar surface illuminated by Earth light.

A third operational mode, used for stellar observations and lunar-surface observation under extremely low luminance conditions, is referred to as an integrating mode. This mode also could be applied, by Earth command, to either the 200- or 600-line scan mode. The shutter of the camera was commanded open and the vidicon allowed to accumulate light energy from the scene, after which the shutter was commanded closed and the frame read from the vidicon. Pictures could be taken when scene luminance was as low as 0.008 ft-L with the integrating mode.

Two photometric/colorimetric reference targets were mounted on the spacecraft within view of the camera (fig. 3-7). These targets, one mounted on omnidirectional antenna B and the other on the spacecraft leg adjacent to footpad 3, were oriented so that the line of sight of the camera, when viewing the target, was normal to the plane of the target. Surveyor VII had one additional target mounted on omnidirectional antenna A. Each target was identical and contained a series of 13 gray wedges arranged circumferentially around the target. In addition, three color wedges, whose CIE (Commission Internationale d'Eclairage 1931) chromaticity coordinates are known, were located radially from the target center. A series of radial lines was incorporated in each chart to provide a gross estimate of camera resolution. Finally, the chart contained a centerpost which served as a gnomon, to aid in determining the solar angles after lunar landing.

FIGURE 3-7.—Surveyor V picture of photometric reference target mounted on leg 2 of the spacecraft. The gray steps are indicated by numbers. A small pin protrudes from the center of the target and casts a shadow downward across the target (Sept. 16, 1967, 04:36:25 GMT).

The basic camera design, used on the first three successful Surveyor missions, was later improved for the last two flights. Advances were made primarily in mirror-movement accuracy and, most important, in filter positioning and intermediate iris control.

The sensitivity and dynamic range were different for each Surveyor camera. In general, however, this was not a serious problem during mission operations because of the large range in iris and shutter capability. A comparison of the camera characteristics on each flight is given in table 3-2.

TABLE 3-2.—*Comparison of Surveyor camera characteristics: 600-line mode*

Characteristic	Surveyor I	Surveyor III	Surveyor V	Surveyor VI	Surveyor VII
Dynamic range	13:1	7.9:1	14.4:1	15.4:1	14.1:1
Signal-to-noise ratio, dB	36.4	41	43.9	40.0	43.6
Horizontal relative response at 600 lines (at center of vidicon)	0.17	0.38	0.31	0.20	0.20
Vertical relative response at 600 lines (at center of vidicon)	(a)	0.27	0.40	0.46	0.33
Slope of system transfer characteristic curve	1.4	1.4	1.2	1.1	0.98

a Not tested.

Categories of Pictures

R. M. BATSON AND K. B. LARSON

Surveyor television pictures were taken in sequences called panoramic surveys and in other sequences designed to obtain optimum photographic coverage of areas of special interest. Panoramic surveys cover the area from 65° below the camera horizontal to the horizon. A mechanical stop, 132° counterclockwise and 225° clockwise from the 0° azimuth, prevented taking pictures in a very small sector. During the Surveyor I, III, and V missions, no pictures were taken to the left of the 126° azimuth or to the right of the −213° azimuth because of the possibility that the camera mirror might stick on one of the end stops. Stronger azimuth stepping motors were used on the Surveyor VI and VII cameras, making it possible to take pictures at the end stops without the danger of sticking. Panoramic surveys were taken in both wide- and narrow-angle lens modes.

Narrow-angle panoramas were taken to record surface detail with the highest resolution possible. Between 900 and 1000 pictures were required to record the full panorama in the narrow-angle mode. These pictures were taken every 6° horizontally and every 5° vertically until the entire area visible to the camera was photographed. Successive horizontal rows of pictures were offset 3° from each previous row to avoid gaps in coverage. Iris and focus were set at their optimum values for recording lunar-surface detail. Figure 3–8 (see table 3–3) is a diagrammatic representation of survey panorama sequences in the narrow-angle mode.

TABLE 3–3. — *Calibrated camera elevations (in degrees) for elevation steps* [a] *listed in figs. 3–8 and 3–9*

Elevation step	Surveyor I [b]	Surveyor III	Surveyor V	Surveyor VI	Surveyor VII
0	[c] −68	[c] −66.7	[c] −70.1	[c] −69.97	[c] −69.98
1	[c] −63	[c] −61.7	[c] −65.1	[c] −64.97	[c] −64.98
2	[c] −58	[c] −56.8	[c] −60.1	−59.97	−59.98
3	[c] −53	[c] −51.8	[c] −55.2	−54.97	−54.97
4	[c] −48	[c] −46.9	[c] −50.2	−49.95	−49.96
5	[c] −43	−41.9	[c] −45.3	−44.95	−44.96
6	[c] −38	−37.0	−40.3	−39.96	−39.94
7	[c] −33	−32.0	−35.3	−34.97	−34.94
8	[c] −28	−27.0	−30.3	−29.96	−29.93
9	[c] −23	−22.0	−25.3	−24.96	−24.91
10	[c] −18	−17.1	−20.3	−19.95	−19.94
11	[c] −13	−12.3	−15.5	−14.96	−14.97
12	[c] −8	−7.2	−10.6	−9.96	−9.96
13	[c] −3	−2.2	−5.6	−4.96	−4.98
14	[c] 2	2.9	−.7	.05	.01
15	[c] 7	7.9	4.3	5.06	5.03
16	[c] 12	12.8	9.3	10.05	10.05
17	[c] 17	17.7	14.4	15.04	15.03
18	[c] 22	22.7	19.4	20.04	20.01
19	[c] 27	27.7	24.3	25.05	25.05
20	[c] 32	32.6	29.2	30.05	30.03
21	[c] 37		34.1	35.04	34.99
22				40.04	39.98
23				45.05	44.99

[a] As listed in figs. 3–8 and 3–9.
[b] The sequences shown in figs. 3–8 and 3–9 were not used during the Surveyor I mission, but the nominal Surveyor I elevations are shown here for reference.
[c] Value not specifically calibrated.

TELEVISION OBSERVATIONS

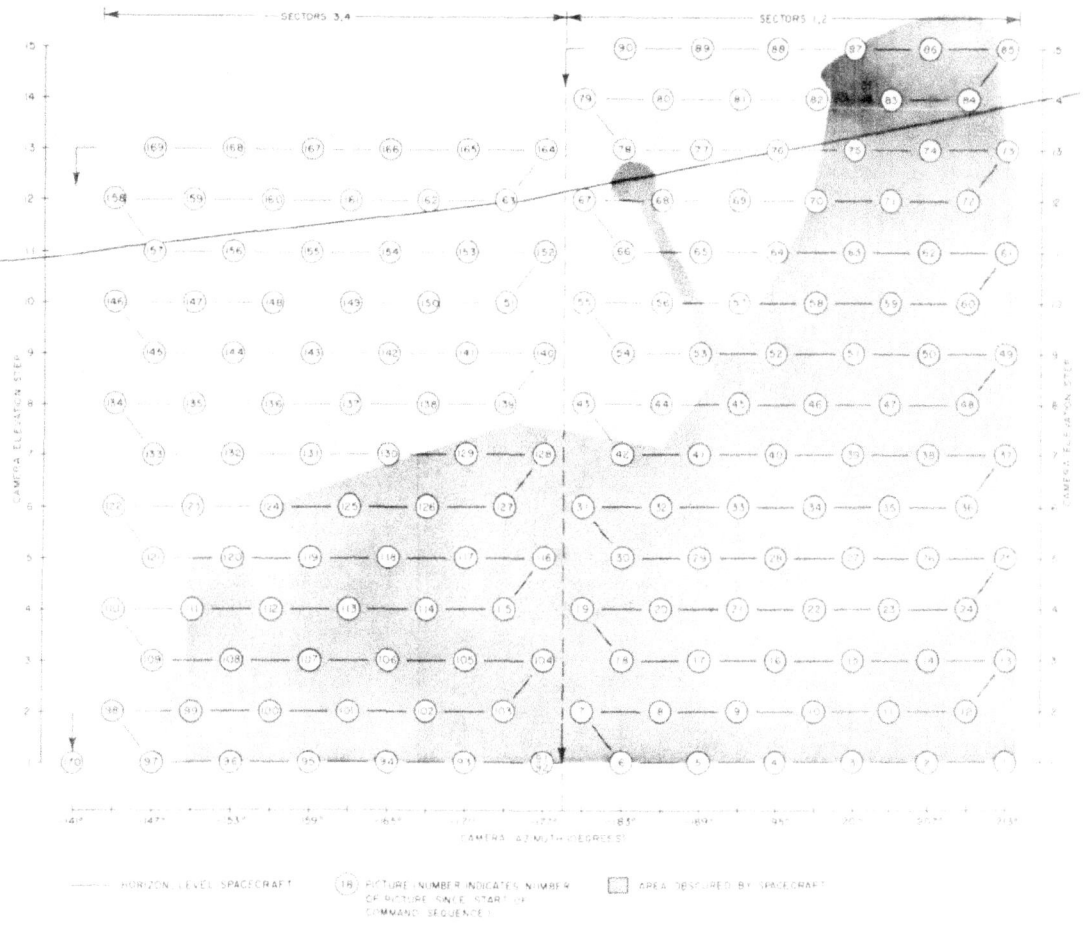

(a)

FIGURE 3-8.—Survey narrow-angle mode panorama sequences. Pictures are taken in the order shown by circled numbers, at camera azimuths and elevations shown. (a) Segment 1.

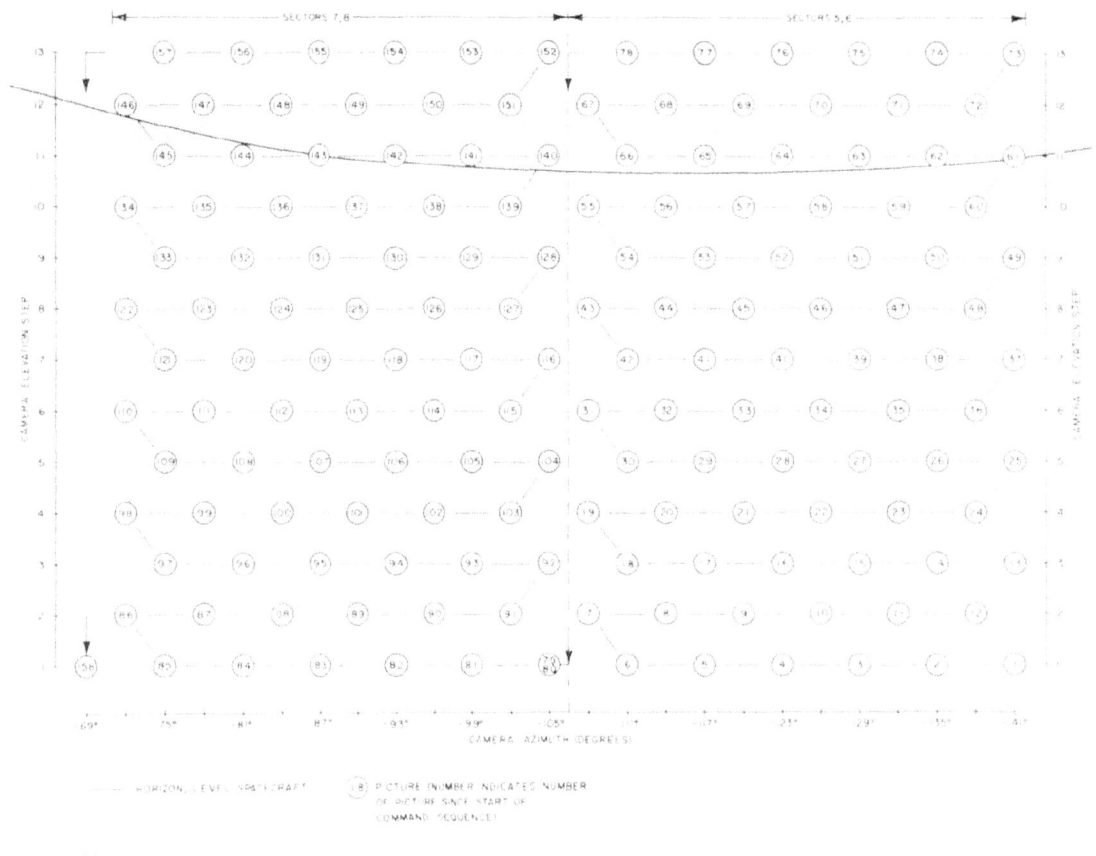

FIGURE 3-8.—Continued. (b) Segment 2.

TELEVISION OBSERVATIONS

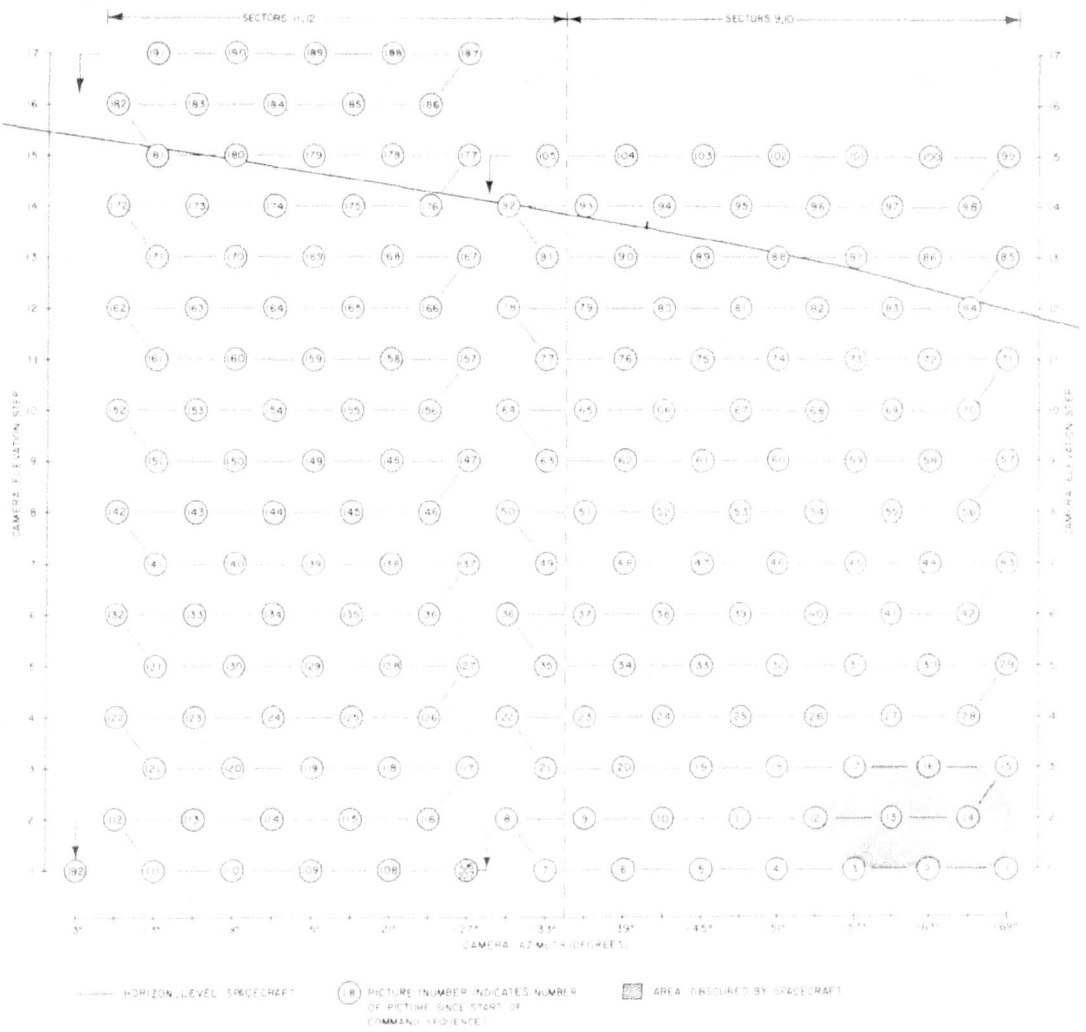

FIGURE 3-8.—Continued. (c) Segment 3.

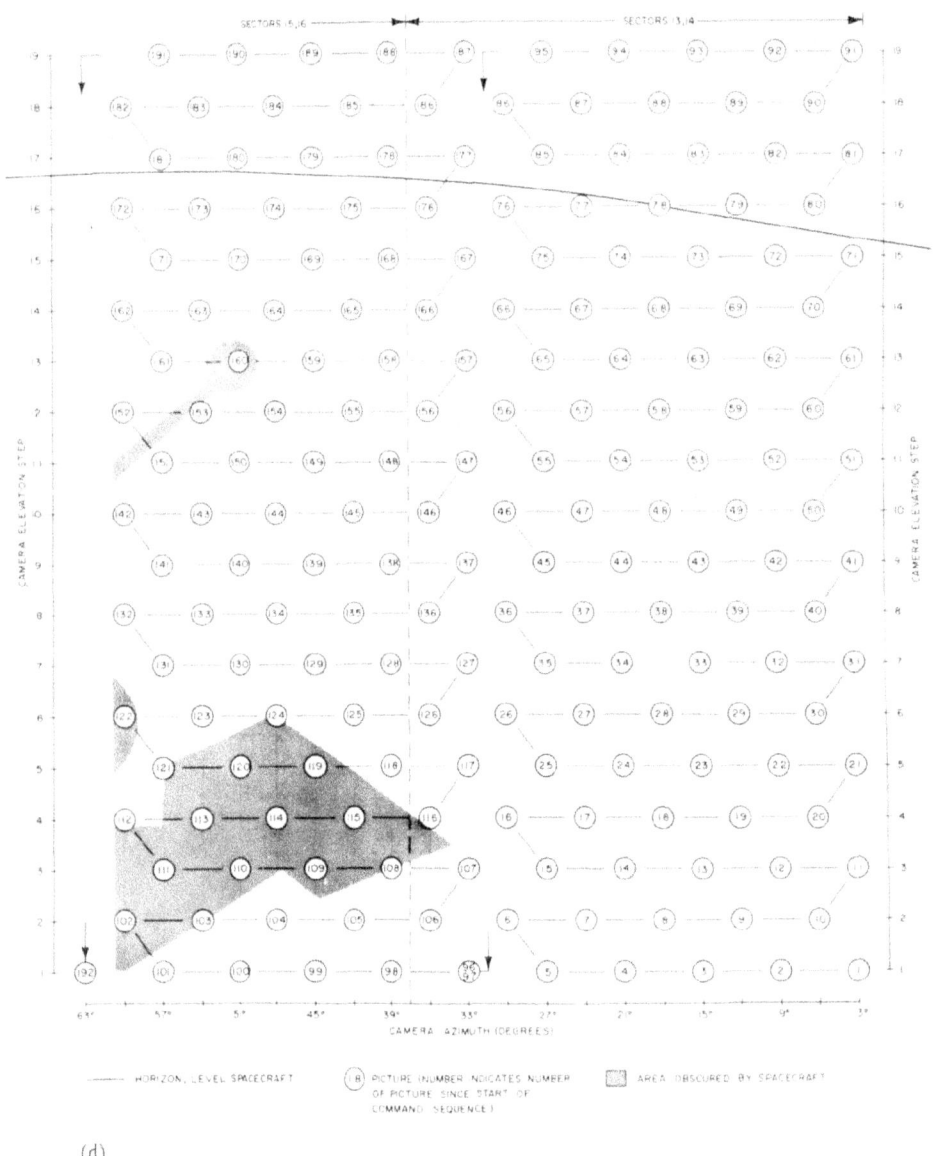

FIGURE 3-8.—Continued. (d) Segment 4.

TELEVISION OBSERVATIONS

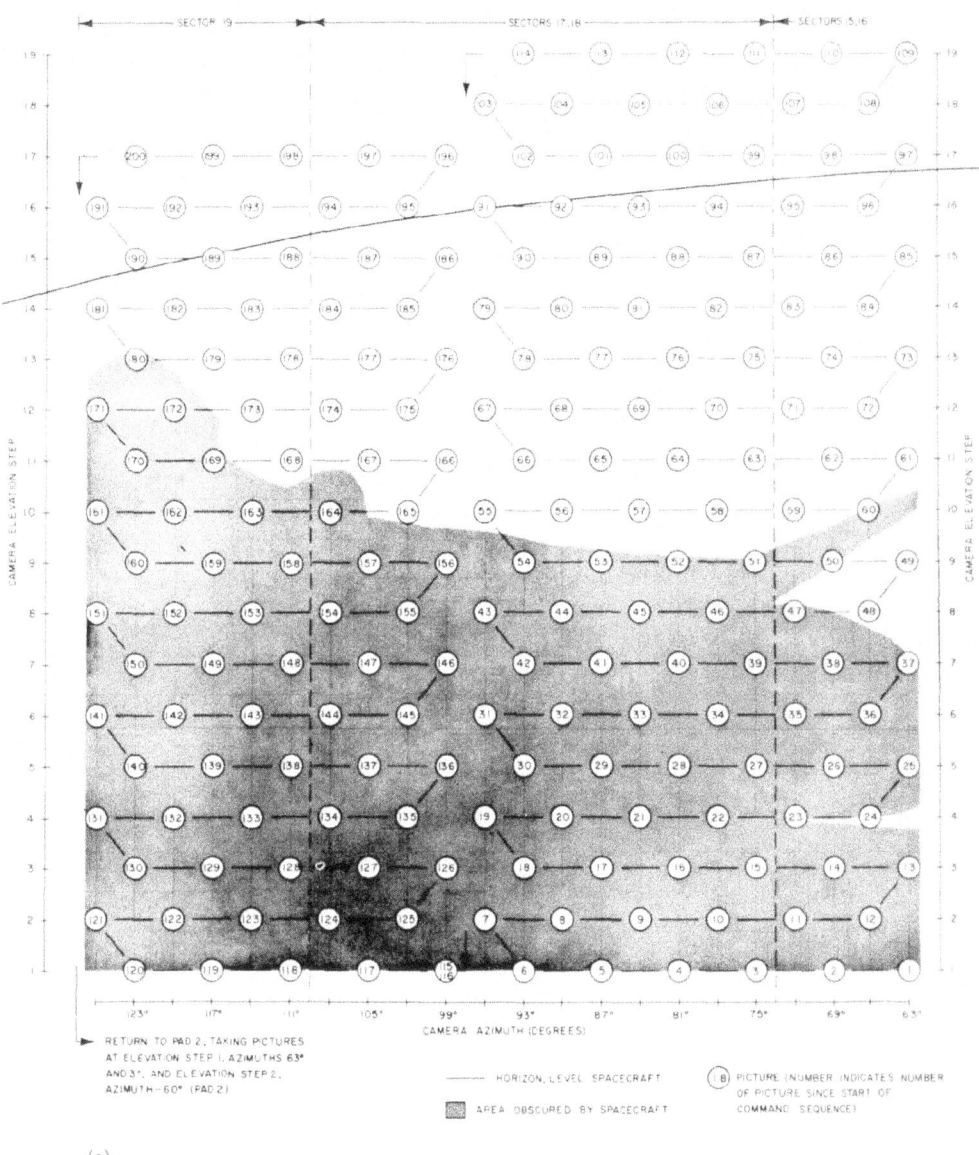

(e)

FIGURE 3-8.—Concluded. (e) Segment 5.

Panoramas could be taken in the wide-angle mode in about one-tenth the time and with about one-tenth the number of pictures required for a panorama of narrow-angle pictures. Although wide-angle Surveyor pictures have only one-fourth the angular resolution of the narrow-angle pictures, wide-angle survey panoramas were useful for reconnaissance examination of the landing sites when high resolution was not essential or when time did not permit the taking of narrow-angle survey panoramas. Figure 3-9 is a diagrammatic representation of the survey panorama sequences for the wide-angle mode.

Colorimetric or polarimetric surveys were taken by repeating surveys at each filter setting. Some were taken according to standard panoramic survey sequences; others were taken of small areas of special interest, and consisted of only a few frames in each filter position.

Photometric surveys were taken to record changes in scene luminance as a function of the angle between the Sun and the surface, and the camera and the surface. Pictures in photometric surveys were taken at selected intervals along lines extending east, west, north, and south from the camera. Special areas of suspected photometric anomalies were also photographed systematically throughout each mission.

Photometric, colorimetric, and polarimetric data were not measured on pictures taken during panoramic surveys because the photometric response of the camera changed nonuniformly with time. It was necessary to stop periodically during photometric, colorimetric, and polarimetric surveys to take control pictures of calibrated color and photometric targets mounted on the spacecraft.

Pictures of stars were taken during each mis-

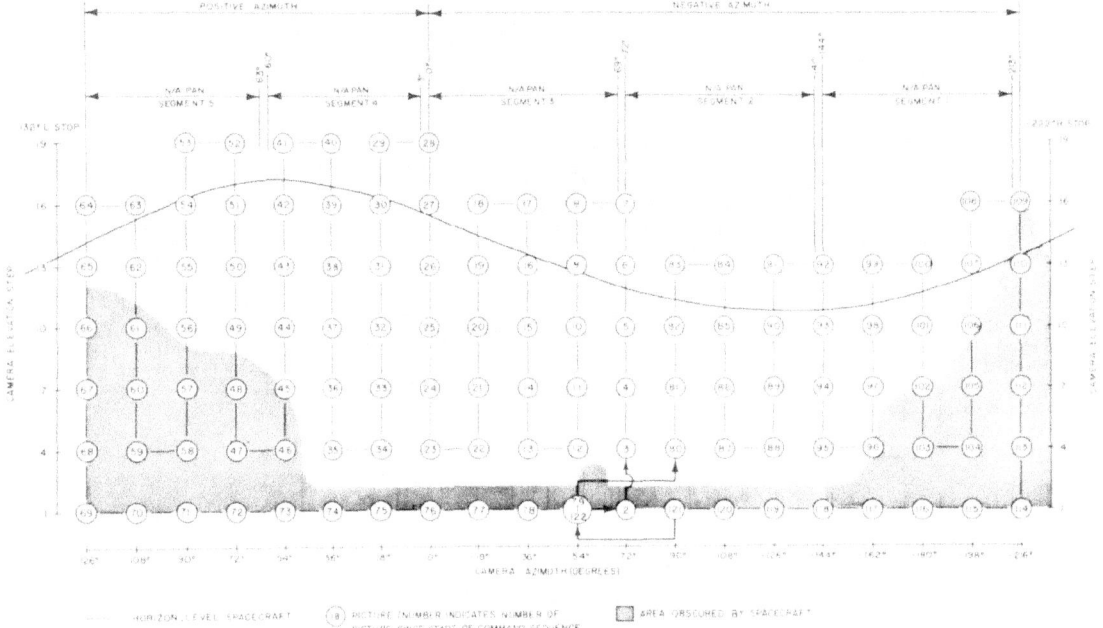

FIGURE 3-9.—Survey panorama sequence used for wide-angle mode operation. Pictures are taken in the order shown by circled numbers at the camera azimuths and elevations shown. See table 3-3 for a listing of camera elevation values in degrees for each elevation step for each Surveyor mission. These sequences and numbering conventions were used during all Surveyor missions except Surveyor I, when vertical sequences were used. Azimuth and elevation values on the pictures were the same for Surveyor I as those shown in the sequences, but the sector-numbering convention was different; for example, Surveyor I, sectors 1 and 2, were equivalent to sectors 9 and 10, on all other missions.

TABLE 3-4. *Categories of pictures taken by the Surveyor cameras* [a]

Category	Surveyor I 1st lunar day	Surveyor I 2d lunar day	Surveyor III	Surveyor V 1st lunar day	Surveyor V 2d lunar day	Surveyor V 4th lunar day	Surveyor VI Prehop	Surveyor VI Posthop	Surveyor VII 1st lunar day	Surveyor VII 2d lunar day	Total
Narrow-angle panoramas	[b]6 459	258	[c]3 385	[b]6 608			[c]6 530	[c]8 336	9 610	[d]25	41 211
Wide-angle panoramas	[c]1 650	106	[c]650	[c]2 509			960	1 297	[c]1 227		8 339
Photometric surveys	90		[b]1 119	[c]2 034			[c]2 266	[c]3 324	[c]2 967		10 800
Focus-ranging surveys	371		372	2 777			810	1 228	1 453		7 091
Stereo mirror survey									299	[d]6	305
Alpha-scattering instrument support				30			75	28	252		385
Surveyor VII alpha-scattering instrument deployment support									253		253
Surface-sampler area surveys			151						800	[d]4	1 045
Surface-sampler operations support			707						1 503		2 210
Special area surveys, magnets, and miscellaneous	1 819	135	653	2 683	1015	[d]63	2 062	1 461	1 258	[d]10	11 159
Earth			[b]64				143		823		930
Stars and planets	57		23	182			91	192	78		623
Shadow progression	202	119	177	895				843	168		2 404
Solar corona	84			74				328	155		641
Total	10 732	618	6301	17 762	1015	63	12 777	17 157	20 916	45	87 386
	11 350		6301	18 876			29 914		20 961		87 386

[a] The number of pictures listed for each mission represents the total number of picture identification data entries received by USGS from JPL. Calibration frames, spurious entries, and duplicate entries such as those for the same pictures recorded at 2 Deep Space Stations, have not been included in the count. Differences between the totals given in this table and those given in ch. 1 of this report probably are a result of the differences in counting methods.

[b] Includes pictures taken through color filters.
[c] Includes pictures taken through polarizing filters.
[d] 200-line television pictures.
[e] Includes 200-line pictures and pictures taken through color filters.
[f] Includes 200-line pictures and pictures taken through polarizing filters.
[g] Includes 200-line pictures.
[h] Includes colorimetric and photometric surveys.
[i] Includes polarimetric and photometric surveys.

sion to measure the orientation of the camera with respect to the lunar surface. During the Surveyor III mission, color pictures were taken of the Earth during its first quarter and during a solar eclipse. During the Surveyor VII mission, pictures of the Earth were taken through polarizing filters throughout the first lunar day. The Earth pictures were used for attitude determination as well as for color and polarization experiments.

Focus-ranging surveys were taken to measure the topography of the near field at the landing sites. Nine or more pictures were taken at successive focus steps at each elevation along given azimuths to determine the points of best focus for each focus setting. Focus-ranging surveys contain an average of 50 to 100 pictures per azimuth.

Topographic computations were also made by measuring the images of shadows on pictures taken at different times during the lunar day. Shadow progression surveys were taken at regular intervals during each Surveyor mission. These surveys usually consisted of 1 to 24 wide- and narrow-angle pictures of the shadow of the spacecraft as it moved across the eastern terrain during the lunar afternoon. During the Surveyor VI mission, shadow progression surveys also were taken from the second camera position of the original sites of contact between the lunar surface and the spacecraft in its first location.

Special area surveys consisted of a series of small surveys of spacecraft parts and spacecraft/surface contact areas. These were taken to examine parts of the spacecraft for possible damage and to examine the interaction of spacecraft parts with the lunar surface. When panoramic surveys were taken, the camera iris and focus were set at values appropriate for the lunar surface; during the special area surveys, they were set at values appropriate for the spacecraft parts. The sequences of these surveys were modified slightly on each mission.

The solar corona was photographed after lunar sunset on the Surveyor I, V, VI, and VII missions. These surveys, which consisted of a series of wide- and narrow-angle pictures taken along the western horizon immediately after sunset, were repeated at regular intervals for several hours after sunset.

Special surveys were made of the entire area in which the surface sampler operated, and pictures were also taken during the operation of the instrument.

The alpha-scattering instrument required only minor support from the television camera. Short predeployment surveys and periodic postdeployment surveys were taken to examine the instrument itself for change or damage. The alpha-scattering instrument was turned over during the Surveyor VI hop, and part of its interior was visible to the camera. Several pictures were taken of this area under different illumination.

During the Surveyor VII mission, the alpha-scattering instrument did not deploy normally to the lunar surface, and the surface sampler was used to deploy the instrument. This activity was supported by pictures from the television camera.

Surveyors V, VI, and VII carried small bar magnets attached to the spacecraft footpads; Surveyor VII had an additional magnet attached to the scoop of the surface sampler. These were surveyed under varying illuminations to investigate them for accumulations of magnetic material.

A small mirror, mounted on the mast of the Surveyor VII spacecraft, was used to take stereoscopic pictures of a small part of the area in which the surface sampler operated.

Table 3-4 lists the categories and numbers of pictures taken by each Surveyor camera.

Location of the Surveyor Spacecraft

E. A. WHITAKER

Selenographic Coordinates

From about 1913 to the present day, the basis for all selenographic coordinates has been the well-known catalogs of Franz (ref. 3-1) and Saunder (ref. 3-2). Thus, the Orthographic Atlas (ref. 3-3) and all the lunar maps and charts prepared by the U.S. Air Force Chart and Information Center, including those of the proposed Apollo landing sites, depend on these

TABLE 3-5. *Comparison of the location of 3 lunar craters*

[As listed in the catalogs of Saunder, Franz, Mills, and Arthur]

Crater	Flamsteed E		Pytheas A		Linné A	
Catalog	λ, deg	β, deg	λ, deg	β, deg	λ, deg	β, deg
Saunder	−45.98	−3.69	−21.73	+20.45	+14.35	+28.93
Franz			−21.71	+20.44	+14.39	+28.93
Mills	−45.99	−3.68	−21.70	+20.45	+14.40	+28.97
Arthur	−45.99	−3.68	−21.69	+20.46	+14.36	+28.95

measures. More recently, these measures were checked and combined into a single catalog by Arthur (ref. 3-4).

New measurements have been made over the last few years by U.S. Army Map Service, U.S. Air Force Chart and Information Center, Baldwin (ref. 3-5), Mills (ref. 3-6), Arthur (ref. 3-7), and others. At the time of writing, the most recent catalog is that of Mills, which uses Schrutka-Rechtenstamm's reduction of Franz's measures (ref. 3-8) as control. A somewhat more extensive catalog is that by Arthur (ref. 3-7). This catalog uses the raw measures made on several sets of plates taken at various observatories and again uses the Schrutka control net; however, the absolute orientation was determined from star trails impressed on a number of Yerkes Observatory 40-in. refractor plates. Typical differences among the various catalogs are shown in table 3-5.

Of the four catalogs, that of Franz probably contains the largest errors, as it was based largely on early Lick Observatory plates of mediocre quality. That of Saunder is generally considered to be more accurate, being based on Observatory of Paris and Yerkes Observatory plates of good quality. The Mills catalog is based on measurements made from 80 films from the observatory at Pic du Midi, and may be superior to the Saunder catalog, but it contains far fewer points. The Arthur catalog may prove to be the most accurate of all, as it employs an improved determination of the position of the Moon's axis with respect to the surface features.

Location of Surveyor I

The Surveyor I spacecraft landed at a position estimated from prelanding tracking data to be 2.49° S, 43.32° W. An early attempt to correlate bright hills on the horizon, visible in the Surveyor I panoramas, with topographic features on ACIC chart LAC 75 was inconclusive, and led to a location situated well outside the 2σ error ellipse of the tracking data (refs. 3-9 and 3-10). Whitaker repeated the attempt, however, and obtained good correlation by using a low sunrise Earth-based photograph of the region, secured a few months earlier with the University of Arizona 61-in. NASA reflector. This yielded coordinates of 2.57° S, 43.34° W, well within the 2σ ellipse (ref. 3-11).

The completion of a narrow-angle mosaic of Surveyor I pictures of a part of the horizon containing both the bright hills and the shadow of the spacecraft at sunset made it possible to determine the true selenographic azimuths of the hills without making reference to camera coordinates. Lunar Orbiter I medium-resolution photographs of the area (the high-resolution photographs were not usable) showed that the correlation between the hills, as viewed by Surveyor I and Lunar Orbiter I, was excellent. This narrowed the area of search for the location of Surveyor I to an area 3 km by 1 km (fig. 3-10). However, the task of correlating craters and other features visible in the Surveyor I panoramas with similar features in the Lunar Orbiter photographs proved difficult, primarily because of the inability to judge distances, and hence dimensions, from the Surveyor panoramas.

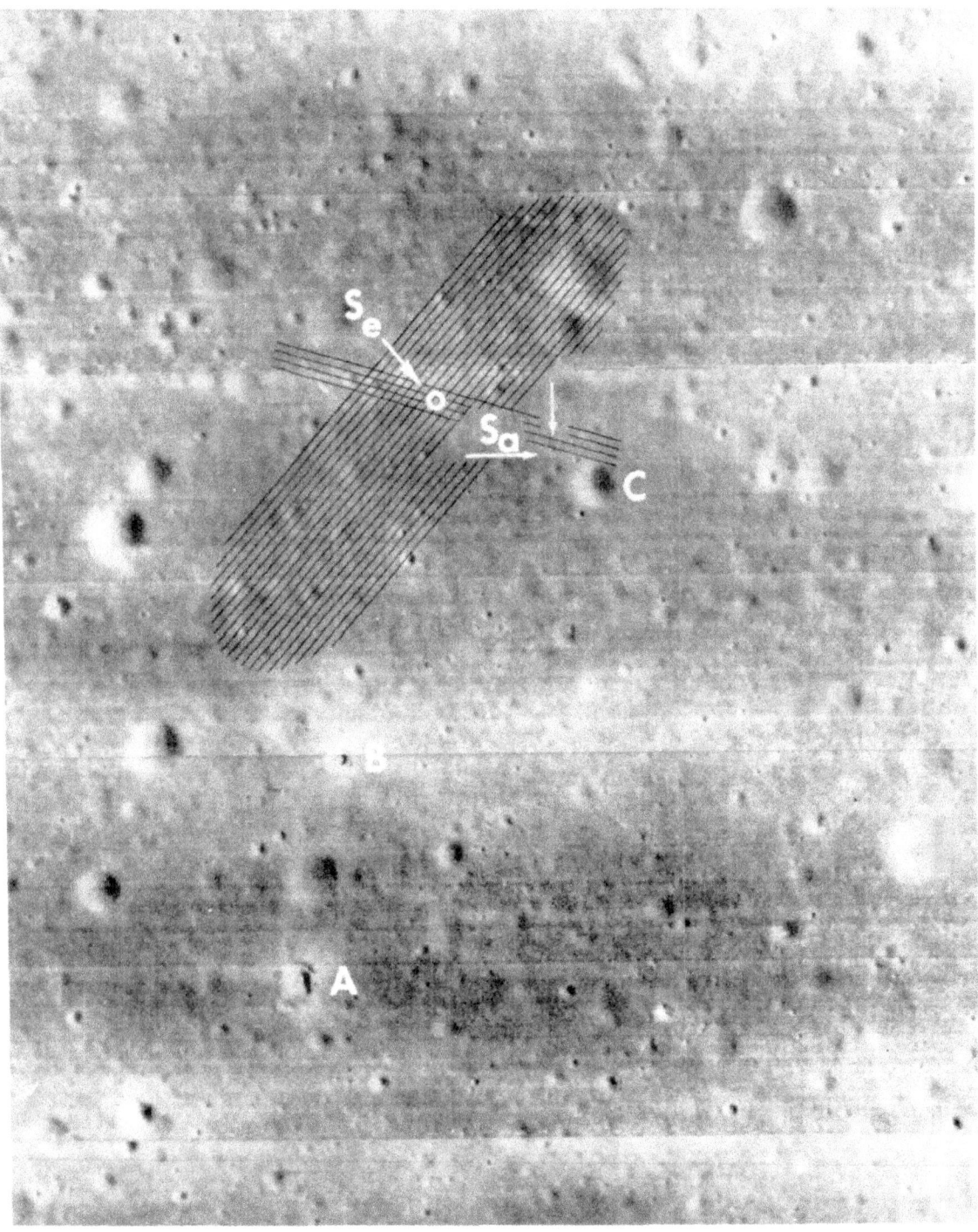

FIGURE 3-10.—Part of Lunar Orbiter I photograph M-210 showing the expected (S_e) and actual (S_a) positions of Surveyor I. The spacecraft is visible as a diffuse bright spot. A and B are rock-strewn craters and C is a crater 180 meters in diameter.

TABLE 3-6. *Selenographic coordinates of 13 craters used to determine the location of Surveyor I*

Crater	Mills		Arthur	
	λ, deg	β, deg	λ, deg	β, deg
Encke E			−40.11	+0.36
Maestlin G	−42.03	−2.01		
Suess F			−44.62	+1.17
Suess FB	−45.57	+.23		
Flamsteed			−44.23	−4.46
Flamsteed F			−44.06	−4.71
Flamsteed FA			−40.75	−3.44
Flamsteed FB			−40.55	−2.32
Flamsteed K	−43.62	−3.09	−43.62	−3.09
Flamsteed D			−44.82	−3.16
Flamsteed E	−45.99	−3.68	−45.99	−3.68
Flamsteed C			−46.20	−5.49
37003			−45.04	−1.94

Subsequently, reflection data from the onboard radar doppler sensors of the Surveyor I spacecraft were made available (ref. 3-12). Two distinct enhancements in reflectivity recorded by the radar doppler receiver 1 were interpreted by Whitaker as having been caused by the beam sweeping across two craters of anomalous appearance (A and B, fig. 3-10), as they were of the correct orientation and spacing.

Others working on this problem assumed that the radar enhancements were caused by the opposite walls of a comparatively large (>1 km) crater (ref. 3-12). This gave a location that could not be reconciled with either the tracking data or the hills on the horizon.

The problem was not finally solved until the general area was photographed in both medium and high resolution by Lunar Orbiter III. The location proposed by members of JPL and ACIC was searched without success. That proposed by Whitaker also was searched unsuccessfully, although a bright, rocklike object casting a long thin shadow was detected close by. It was not possible to reconcile the Surveyor and Lunar Orbiter data until further work by Whitaker et al. established the identification of the bright object casting the long thin shadow as the Surveyor I spacecraft. The anomalous craters A and B (fig. 3-10) were found to be strewn with rocks, undoubtedly the cause of the enhancements in radar reflectivity.

A reexamination of the Lunar Orbiter I photographs showed that Surveyor I was visible as a bright spot on almost all frames that included the landed area; it can be seen in figure 3-10. An investigation similar to Whitaker's was conducted by Spradley et al. (ref. 3-13), who arrived at the same results.

The landed location of Surveyor I was carefully pinpointed on Lunar Orbiter IV photograph H-143. The centers of 13 craters with known coordinates situated around this location were also pinpointed, using a small transparent overlay provided with concentric circles for accurate centering of the craters. All 14 points were then transferred to a plastic overlay, and the positions of the 13 craters, taken from catalogs of selenographic positions (refs. 3-2, 3-4, and 3-6), were noted at the appropriate places. These craters and the sources of the coordinates are shown in table 3-6.

It will be seen that two craters common to both catalogs have identical coordinates, giving some confidence in the accuracy of the measures. A comparison between the coordinates given by Saunder (ref. 3-2) and those listed in the tabulation, where duplicated, shows a definite systematic difference in both λ and β. The mean values are λ (Arthur) − λ (Saunder) = +0.01°; β (Arthur) − β (Saunder) = 0.01°. The coordinates for point 37003 were corrected by these amounts to bring them in line with the others.

A network of lines was next drawn, predominantly in the general directions east-west and north-south, by joining the centers of appropriate craters. By simple proportion, points were then marked on these lines giving the intersects of each full degree line of latitude or longitude as appropriate. Because the Lunar Orbiter photograph was taken nearly vertically and because of the small size and equatorial location of the area photographed, it was expected that the grid lines in the region of interest would depart less than 0.01° from linearity. This was indeed found to be the case. Local divergences in linearity of some of the latitude lines were directly attributable to one crater (37003), whose location was less certain than the other craters. These intercepts were therefore ignored when drawing the grid. The coordinates of Surveyor I were determined by simple interpolation from the enclosing latitude-longitude lines. They are:

$$43.22° \text{ W} \pm 0.01°$$
$$2.45° \text{ S} \pm 0.01°$$

On the traditional Saunder-Franz system, these would be 43.23° W, 2.46° S. The probable errors were not deduced mathematically, but were based on the departure from linearity of the grid intersects.

Location of Surveyor III

The landed location of the Surveyor III spacecraft was pinpointed (ref. 3-14) on Lunar Orbiter IV photograph H-125, as were the following eight measured points: Mills 308, Gambart R, Fra Mauro B, Saunder 233, Lansberg β, Euclides K, Arthur 3402511, and Lansberg N. Adopting a similar procedure to that used for Surveyor I, the following coordinates were determined for Surveyor III:

$$23.34° \text{ W} \pm 0.01°$$
$$2.97° \text{ S} \pm 0.01°$$

On the Saunder-Franz system, these would be 23.34° W, 2.99° S.

The Location of Surveyor V

The landing point of Surveyor V was situated some distance away from the nominal aiming point. The coordinates of the landed position, obtained from the prelanding tracking data, were 23.19° E, 1.50° N. This location, according to the ACIC charts, is situated at the western extremity of Lunar Orbiter V photograph H-78, the only high-resolution frame of the entire Lunar Orbiter series to include it.

Because it landed in a small crater, Surveyor V obtained a foreshortened view of the surrounding terrain. Although many terrain features seen in the Surveyor V panorama appear to be too small to be seen in Lunar Orbiter photograph H-78, this frame was searched extensively but unsuccessfully in the appropriate area for points of correlation (ref. 3-15).

The best postlanding tracking coordinates available at the time of writing were 23.20° E ± 0.03°, 1.42° N ± 0.01°.[1] To note this position with respect to the western border of Lunar Orbiter V photograph H-78, Lunar Orbiter IV photograph H-85 was used as a base. A local grid was constructed on it, as described for Surveyor I, using the five control points.

Sabine B, C, D, E, and Arago CA. The tracked location was then pinpointed on this frame and, by noting its position in relation to nearby topographic features, transferred to Lunar Orbiter V photograph M-74, the best quality medium-resolution frame of the area. The probable error ellipse was added, also the boundaries of Lunar Orbiter V photograph H-78 (shown in fig. 3-11). All topographic details visible in the Surveyor V panoramas are too small to be visible in this photograph. This process revealed a systematic error amounting to 0.1° in longitude on the ACIC charts of the region (RLC 7 and 8).

Location of Surveyor VI

The landed location of Surveyor VI was pinpointed on Lunar Orbiter IV frame H-108, as were the following measured craters: Pallas D, Flammarion A, Réaumur X, and Oppolzer A. The Surveyor VI landing site fell close to the edge of Lunar Orbiter photograph H-108, which prevented the inclusion of control points on the west side; it was therefore necessary to use the overlapping frame H-102 on the west. A number of fiducial points in the overlapping portion was provided by pinpointing the craters Oppolzer A, Bruce, Mills 449, Réaumur Y, Blagg, Réaumur D, and Seeliger A on frame H-102 and constructing an accurate coordinate grid between these points and a number of grid intersects pinpointed on H-108, using the local topography for positioning. Using these intersects and the four craters already noted, it was possible to construct an accurate grid enclosing the Surveyor VI location. The coordinates were measured to be:

$$1.40° \text{ W} \pm 0.01°$$
$$0.53° \text{ N} \pm 0.01°$$

On the Saunder-Franz system, these would be 1.39° W, 0.51° N.

---

[1] F. B. Winn private communication, 1968.

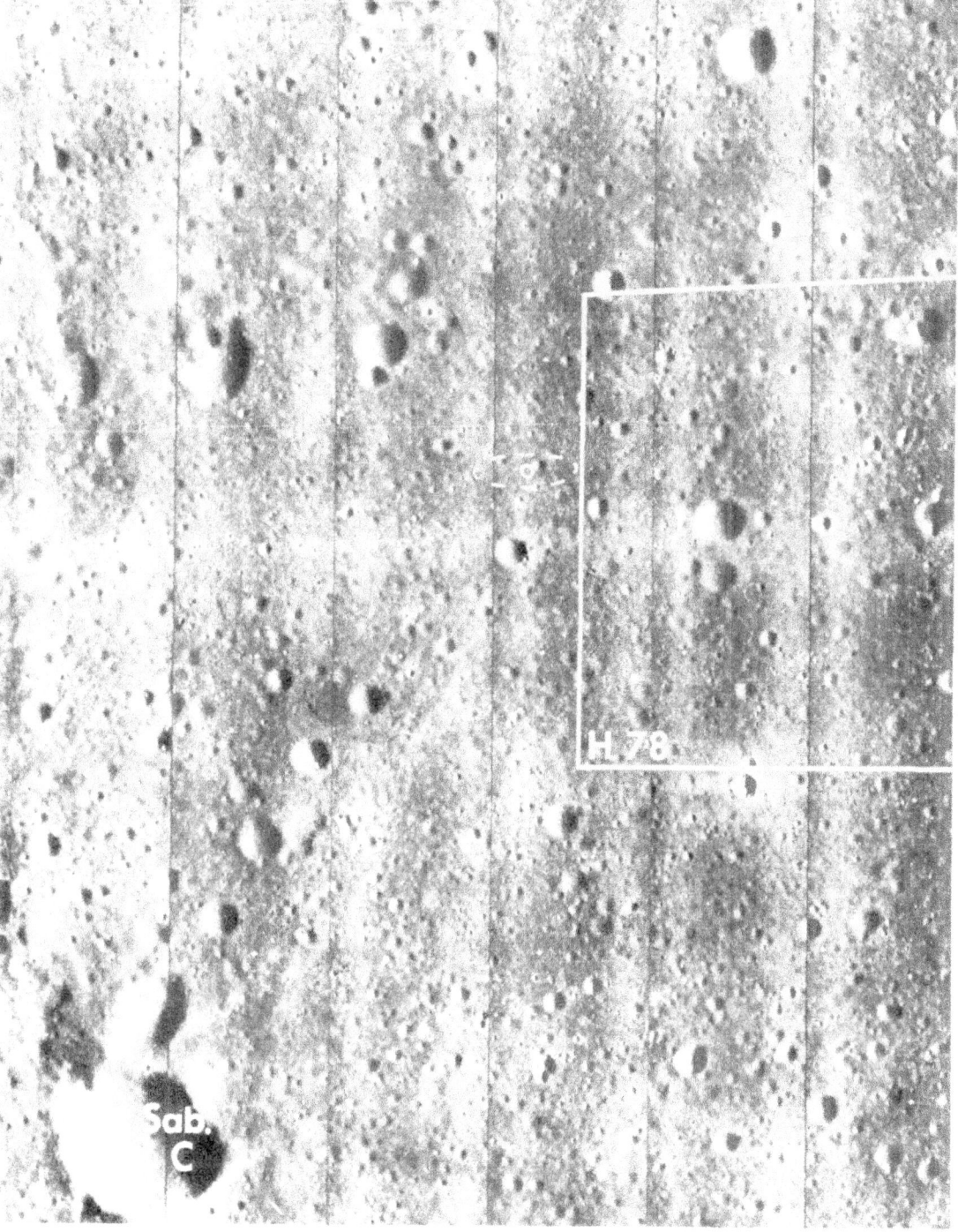

FIGURE 3-11.—Part of Lunar Orbiter V photograph M-74 showing the estimated position of Surveyor V and error ellipse obtained from postlanding tracking data. The boundary of Lunar Orbiter V photograph H-78 is also indicated.

Location of Surveyor VII

The methods used for obtaining the positions of Surveyors I, III, and VI could not be used for determining the location of Surveyor VII because: The selenographic grid is not selenodetic; that is, it assumes the Moon is spherical. For points elevated above the mean sphere and situated some distance from the center of the moon's disk, measured coordinates are greater than the true coordinates; and because, at the latitude of Tycho, latitude circles are sufficiently curved to cause further difficulties in attempting to draw a grid.

The following method was therefore adopted for determining the location of Surveyor VII. A careful comparison was made with respect to the local terrain between Lunar Orbiter V frame M-128 and a photograph (123) taken with the 100-in. telescope at Mount Wilson, which enabled the transference of the pinpointed Surveyor VII location from the former to the latter. As this photograph was used as a basis for sheet D7a in the Orthographic Atlas (ref. 3-3), it was then possible to transfer the point accurately to this sheet. The coordinates of the point, as read directly from the xi-eta grid, were xi, −0.1500, and eta, −0.6550.

Next, it was necessary to check the accuracy of the grid in the general area. The centers of 13 features of known position were marked, and their apparent positions read from the grid. These readings were then compared with the Saunder catalog positions (ref. 3-2) and the mean systematic errors noted. These amounted to −0.00023 in xi, and +0.00007 in eta. Thus, the corrected position was −0.1502, −0.6549, or 11.46° W, 40.91° S.

If the Surveyor location is now assumed to lie 1 km above the measured craters, then these figures reduce to 11.45° W, 40.88° S (Saunder-Franz system). Allowing for the systematic difference between this system and that of Arthur in this region, the final result is: 11.47° W ± 0.02° longitude, 40.86° S ± 0.03° latitude.

Summary

Table 3-1 lists the derived selenographic coordinates for Surveyors I, III, VI, and VII. As Surveyor V has not been located on Lunar Orbiter photographs, the coordinates from tracking data represent a best estimate of its location. The derived coordinates of the four located Surveyors are repeated here:

Surveyor I: 43.22° W ± 0.01° longitude; 2.45° S ± 0.01° latitude

Surveyor III: 23.34° W ± 0.01° longitude; 2.97° S ± 0.01° latitude

Surveyor VI: 1.40° W ± 0.01° longitude; 0.53° N ± 0.01° latitude

Surveyor VII: 11.47° W ± 0.02° longitude; 40.86° S ± 0.03° latitude

The probable errors refer to the precision of the determinations with respect to the Arthur and Mills triangulations, which show small systematic differences from the traditional Saunder-Franz network.

Orientation of the Spacecraft and Television Cameras

J. J. RENNILSON AND R. M. BATSON

The spacecraft attitude, referenced to a local selenographic system, can be determined from three independent kinds of data telemetered from the spacecraft: (1) gyro-error signals and strain-gage deflections on the spacecraft legs, (2) gimbal angles of the solar panel and planar array antenna, and (3) star and planet sightings with the television camera (ref. 3-16). In addition, the attitude of the camera and the spacecraft can be evaluated from observations of the horizon; the east-west component of the camera and spacecraft attitude can be accurately determined from measurement of the position of the western horizon in camera coordinates and the time of sunset.

Gyro-Error Determination of Spacecraft Attitude

The gyro-error signals are initiated by the rotational changes from the spacecraft inertial coordinate system to the final landed attitude. The reference coordinate system before these changes occur is the "attitude hold" orientation of the spacecraft during descent. This reference is established when the spacecraft is about 13 meters above the surface. At this altitude, the spacecraft's Z-axis may not coincide with the local vertical; however, the maximum misalinement is seldom larger than

1.45°. Once on the surface, the leg strain gages offer further corrections to this attitude determination.

Antenna/Solar Panel Positioner (A/SPP) Determination of Spacecraft Attitude

Another method of attitude determination of the Surveyor spacecraft is through the use of the solar panel and planar array gimbal positions. Measured Sun and Earth sightings of the A/SPP are transformed to agree with the local selenographic Sun and Earth vectors. The attitude of the spacecraft is obtained from the difference between the Sun and Earth vectors measured by the A/SPP and the true selenographic Sun and Earth vectors.

In practice, the measured gimbal angles of the A/SPP must be corrected for many errors. These errors arise from mechanical and electrical imperfections. They cause the angle between the Earth and the Sun, as determined by the A/SPP gimbal positions (the vector dot product), to differ from the true selenographic angle. In computing the rotational matrix that defines the attitude of the spacecraft, the contribution of these errors must be minimized. Statistical methods were applied to determine the magnitude of the errors involved and their effect on the determination of the spacecraft attitude. These methods made use of preflight calibration of the errors in the gimbal angle positions and population studies of the errors in the rotational matrices used in calculation of the tilt of the spacecraft. From the many Sun/Earth sightings made during a Surveyor mission, a final matrix was obtained which minimized the errors in the determination of the spacecraft attitude.

Television Camera Determination of Spacecraft Attitude From Star and Planetary Observations

Star and planetary observations with the television camera will yield an attitude matrix relating camera coordinates to selenographic coordinates. If the rotational matrix for camera coordinates to spacecraft coordinates is known, the attitude of the spacecraft can be found.

Errors inherent in the television observations of the stars and planet are numerous. Some errors can be determined accurately and thus a full correction can be made; other errors can be only estimated. Those errors for which the Surveyor cameras were calibrated are (1) image nonlinearity, (2) departure of actual focal length from nominal focal length, (3) misalinement and rotation of the vidicon, and (4) mirror-pointing inaccuracies. Errors for which incomplete or no calibration was made include optical axis/mechanical axis misalinement, lens distortion, and variations of all these errors as a function of temperature. Corrections for errors are usually performed in the order given in the following paragraphs.

Image nonlinearity. The reseau, consisting of a 5 by 5 matrix of dots deposited on the vidicon faceplate, corrects for nonlinearity. The basic assumption made in this correction is that the reseau is distorted linearly in the same manner as the image data. Programs used to correct the image data consist of solving the equations formed by the cross-ratios of the stellar image coordinates and those of the surrounding four closest reseau marks.

Lens distortion. Only one prototype lens of the Surveyor cameras was measured for detailed distortion characteristics; however, flight acceptance curves were obtained for radial or sagittal distortion on each flight lens. These curves can be accurately described by a fourth-order polynomial equation.

Departure of actual focal length from nominal focal length. The calibrated focal length determines the angular scale of the picture. The image coordinates are represented as angular displacements in azimuth and elevation from the coordinates of the center reseau.

Misalinement and rotation of the vidicon. Misalinement errors of the camera optical axis with the center reseau of the vidicon and mirror azimuth axis are known to exist. Because they were minimal, they are not considered in the preliminary data reduction. The rotation of the vidicon as well as the mirror-pointing angles determine the stellar vector in camera coordinates.

Mirror-pointing inaccuracies. Preflight photogrammetric tests of the Surveyor cameras provided estimates of the mirror-pointing inaccuracy of each Surveyor camera except that of Surveyor I. The errors of mirror pointing were sufficiently small as to have a minor effect on the attitude determination and therefore were not considered.

Once the stellar vectors are determined in camera coordinates, the camera coordinate to spacecraft coordinate rotational matrix transforms these vectors to spacecraft coordinates. From this point, the reduction follows the same procedure as the A/SPP attitude determinations.

The rotational matrix of camera coordinates to spacecraft coordinates is one of the largest uncertainties in determining the attitude of the spacecraft. Preflight calibration of this matrix were performed only on Surveyors V, VI, and VII. Stationary points on the spacecraft were observed by the camera, and corrections to the camera/spacecraft matrix were made after landing.

Some variation in the standard deviation of the best-fit attitude matrices have been attributed to slight changes in the attitude of the spacecraft on the lunar surface. Operation of the camera mirror and A/SPP stepping motors caused large oscillations to be imparted to the spacecraft. Flexibility of the legs of the spacecraft, particularly in those cases when the shock absorbers of the legs were not locked, also caused movements of the spacecraft. Monitoring of the horizon by the Surveyor VII camera showed variations in position of objects on the horizon up to 0.2° during the lunar day.

Table 3-7 lists the camera attitude in selenographic coordinates for each successful mission. The attitude was derived from observations of the stars and planets. Table 3-8 lists the combined spacecraft attitudes for each mission from the three sources of data, gyro error, A/SPP position, and television camera observations. The relationship of the angles of spacecraft geometry are shown in the sketch.

Spacecraft attitudes were verified by an independent method of computing the position of the horizon in camera coordinates. If the horizon were a circle with the camera in the center, it could be plotted on a cylindrical projection in camera coordinates as a sinusoidal curve, the amplitude of which (above or below the 0° camera elevation) is equal to the magnitude of camera tilt. The camera coordinates of two or more observed points on the horizon (differing in azimuth by 15° to 165°) would be sufficient to determine the camera tilt and the azimuth of camera tilt. The orientation of the camera parameters can also be determined in the case of a camera displaced from the center of the horizon circle, if the magnitude and direction of the displacement are known.

Errors inherent in computing the attitude of the spacecraft by horizon measurement are the same as those listed for reduction of stellar observations. In addition, no horizon viewed by Surveyor was perfectly smooth, and departure of the observed horizon from the theoretical horizon may also lead to errors in attitude determination. At most Surveyor landing sites, however, the observed horizon is sufficiently close to the theoretical horizon so that when large numbers of points on the horizon were used in several combinations, the standard deviations in computed camera attitude were small.

Most of the observed horizon points used in this kind of attitude computation are selected by subjective interpretation, but the location of the theoretical horizon in places can be determined more rigorously. For example, several distant ridges of known height were visible along the northern horizon at the Surveyor I landing site. The vertical angles between the theoretical horizon and the summits of these ridges, therefore, could be computed as a function of the height and distance of the ridges relative to the camera. The accuracy of this computation is primarily a function of the accuracy with which the heights of the ridges are known; in this case, it was about ±0.1°. Several data points were added to the determinations of horizon position for Surveyor I by this computation.

A single point was added to the data set for each Surveyor by computing the location of the theoretical western horizon as a function of the time of sunset on the camera. Because of topographic prominences west of the spacecraft, sunset came earlier to most Surveyors than that predicted from ephemeris data. The angular height of these prominences above the theoretical horizon could thus be computed as a function of the time differential between actual and predicted sunsets at each landing site. The selenodetic landing site locations and the actual

TELEVISION OBSERVATIONS

TABLE 3-7. *Camera attitudes in selenographic coordinates*

Parameter	ϕ_1, deg [a]	ϕ_2, deg [b]	ϕ_3, deg [c]
Surveyor I	−58.0±0.7	17.1±0.5	56.7±0.7
Surveyor III	228.5±0.9	22.4±0.7	84.0±0.9
Surveyor V	263.0±0.7	33.5±0.5	21.6±0.6
Surveyor VI:			
Before hop	−19.5±0.2	15.3±0.2	49.5±0.2
After hop	−32.3±0.2	12.2±0.2	58.3±0.2
Surveyor VII during most of day	196.5±0.5	14.0±0.3	51.6±0.5

[a] ϕ_1 = angle to the camera tilt vector measured in the horizontal plane from lunar east, negative counterclockwise and positive clockwise.
[b] ϕ_2 = vertical angle of the camera tilt vector measured from local vertical.
[c] ϕ_3 = horizontal angle from the camera tilt vector to the camera 0° azimuth position.

TABLE 3-8. *Spacecraft attitudes*

Spacecraft	Tilt magnitude, β, deg [a]	Tilt direction, α, deg [b]	Roll, θ, deg [c]	Remarks
Surveyor I	0.4	233	89.5	No error analysis performed
Surveyor III	12.0	142.8	44.2	No error analysis performed
Surveyor V	19.5±0.8	262.4±0.8	24.3±2.0	Before static firing
Surveyor V	19.7±1.5	260.1±1.5	25.2±3.0	After static firing
Surveyor VI	1.9±0.7	307.1±0.7	118.6±1.0	Before hop
Surveyor VI	4.0±2.3	41.5±2.3	113.6±1.8	After hop
Surveyor VII	3.2±0.4	279.0±0.4	339.8±0.6	

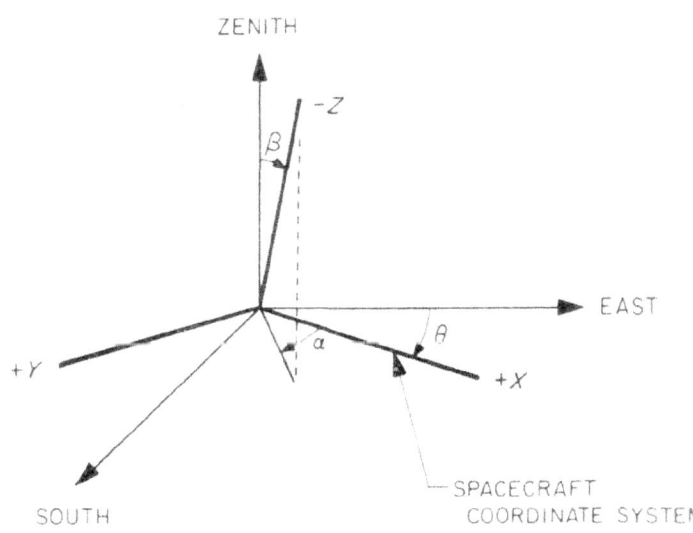

[a] α = azimuth of tilt direction measured positive clockwise from +X axis.
[b] β = magnitude of tilt measured positive clockwise from −Z axis.
[c] θ = roll angle of +X axis from east measured positive clockwise in the spacecraft's x-y plane.

SURVEYOR: PROGRAM RESULTS

TABLE 3-9. *Camera attitudes from horizon measurements*

Parameter	ϕ_1, deg	ϕ_2, deg	ϕ_3, deg
Surveyor I[a]	-63 ± 3	16.1 ± 0.5	52 ± 3
Surveyor III	227 ± 2	23.5 ± 0.2	83 ± 2
Surveyor V (before Sept. 24, 1967)	271 ± 2	31.1 ± 0.2	30 ± 2
Surveyor V (Sept. 24, 1967)	271 ± 2	33.0 ± 0.2	30 ± 2
Surveyor VI before hop (before Nov. 17, 1967, 10:32 GMT)	-22 ± 3	15.4 ± 0.3	47 ± 3
Surveyor VI after hop (after Nov. 17, 1967, 10:32 GMT)	-32 ± 3	11.6 ± 0.3	59 ± 3
Surveyor VII[b] (before Jan. 19, 1968)	196.5 ± 0.5	14.0 ± 0.3	51.6 ± 0.5
Surveyor VII (Jan. 19, 1968)	194.2 ± 0.5	13.8 ± 0.3	49.3 ± 0.5

[a] An angular misalinement of approximately 0.25° in the direction of camera tilt apparently existed in the optical train of the Surveyor I camera.

[b] The horizon at the Surveyor VII landing site was too irregular to make accurate measurements. Therefore, the attitude of the camera as listed in table 3-7 is repeated here. The location of the true western horizon, as determined from the time of sunset, provided an attitude determination consistent with the values shown. The second orientation, relative to the first orientation, was determined by measurements of the shift in the position of the true horizon, as determined from the attitudes listed in table 3-7.

sunset times for each Surveyor mission are known accurately enough so that the location of the western horizon can be computed within approximately $\pm 0.15°$.

Table 3-9 shows the camera attitudes determined from horizon measurements. Most of the measurements agree within probable limits of error with those made by the ASPP and star and planet sightings (table 3-7). The large discrepancy in the different solutions for the Surveyor V orientation is in a north-south direction and cannot be resolved by using the sunset data.

Topographic Mapping Methods

R. M. BATSON

Topographic and planimetric maps have been made from Surveyor television pictures by four basic techniques: (1) focus ranging; (2) stereoscopic photogrammetry; (3) shadow measurement; and (4) photographic trigonometry.

Focus ranging is a near-field (that is, within 10 meters of the camera) mapping method based on the limited depth of field in pictures taken in the narrow-angle mode of the Surveyor lens. Pictures are taken at nine or more focus settings at each camera elevation position along a given azimuth. Small areas in best focus in each picture are located on a mosaic of focus-ranging pictures taken at specific focus settings. The camera azimuth and elevation of the center of the area of best focus for that focus setting is determined by graphical measurement. The location of the point of best focus on the lunar surface is then computed from azimuth, elevation, and calibrated focus distance. A point 10 meters below the intersection of the camera azimuth and elevation rotation axes is used as the origin of the coordinate system, and the x-y plane of the system is oriented parallel to the lunar level plane. Contour lines are drawn by interpolation between control points. Planimetric features, such as craters and rock fragments, are plotted by reference to a grid system on the x-y plane consisting of lines of equal camera azimuth and lines of equal camera elevation, as determined by focus ranging (figs. 3-12 and 3-13).

Topographic maps can be made by focus ranging only if a sufficient number of control points is available. One focus-ranging survey (i.e., a survey along one camera azimuth) results in 50 to 180 pictures, taken at 5 to 20 camera elevation settings and 15 to 25 focus settings. For each survey, 15 to 25 control points are computed. A minimum of 10 focus-ranging surveys is required to provide a sufficient density

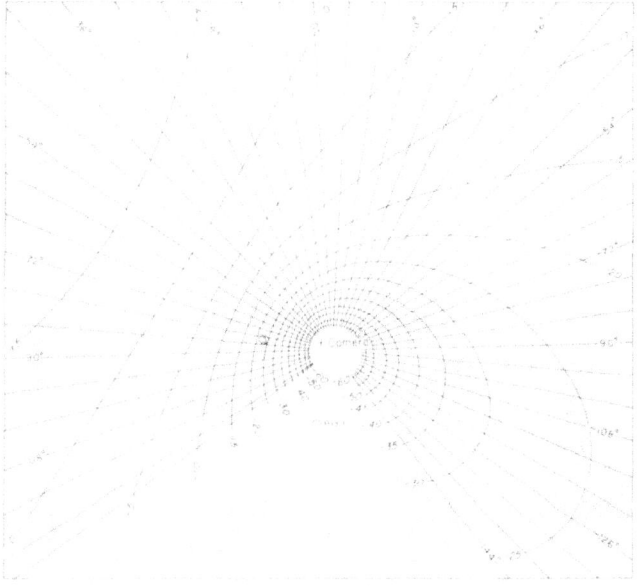

FIGURE 3-12.—Grid on an x–y plane consisting of lines of equal camera azimuth (radiating lines) and lines of equal camera elevation (curved lines). This grid was used to make preliminary planimetric maps of the Surveyor I landing site.

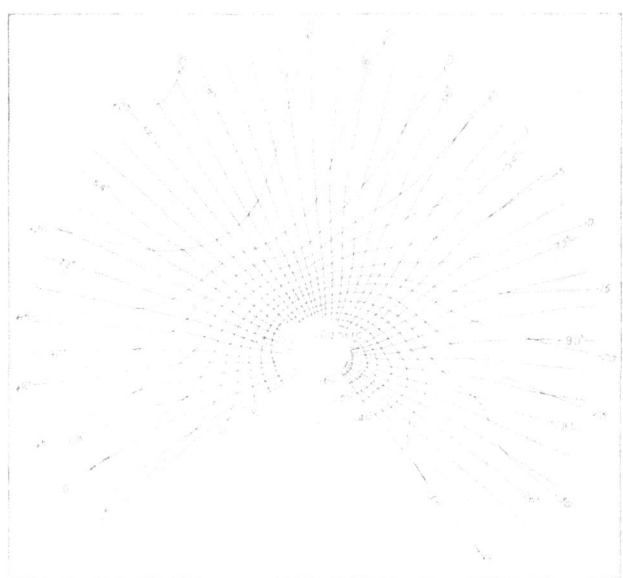

FIGURE 3-13.—Trace of lines of equal camera azimuth and lines of equal camera elevation on a natural lunar surface. Grids like this are used to plot positions of planimetric features (blocks, craters, etc.) at Surveyor landing sites. This grid was made from focus-ranging data taken at the Surveyor V landing site. Deviation of lines from mathematical symmetry is caused by uneven topography.

of control points to map the near field at a Surveyor landing site.

Stereoscopic photogrammetry is the most accurate method for mapping the near field of Surveyor landing sites. Stereopsis is present in pictures of the same surface taken from two different locations. If the relative positions of the two camera stations and the relative camera orientations at the time the pictures were taken are all accurately known, then the locations of the features in the pictures can be computed by triangulation.

Stereoplotting instruments provide a method of continuous analog triangulation. These instruments are used to re-create an optical, three-dimensional model of the surface to be mapped, which can be viewed and measured by the stereoplotter operator. This is done with an index mark in the form of a point of light or a black dot, which appears to float in the three-

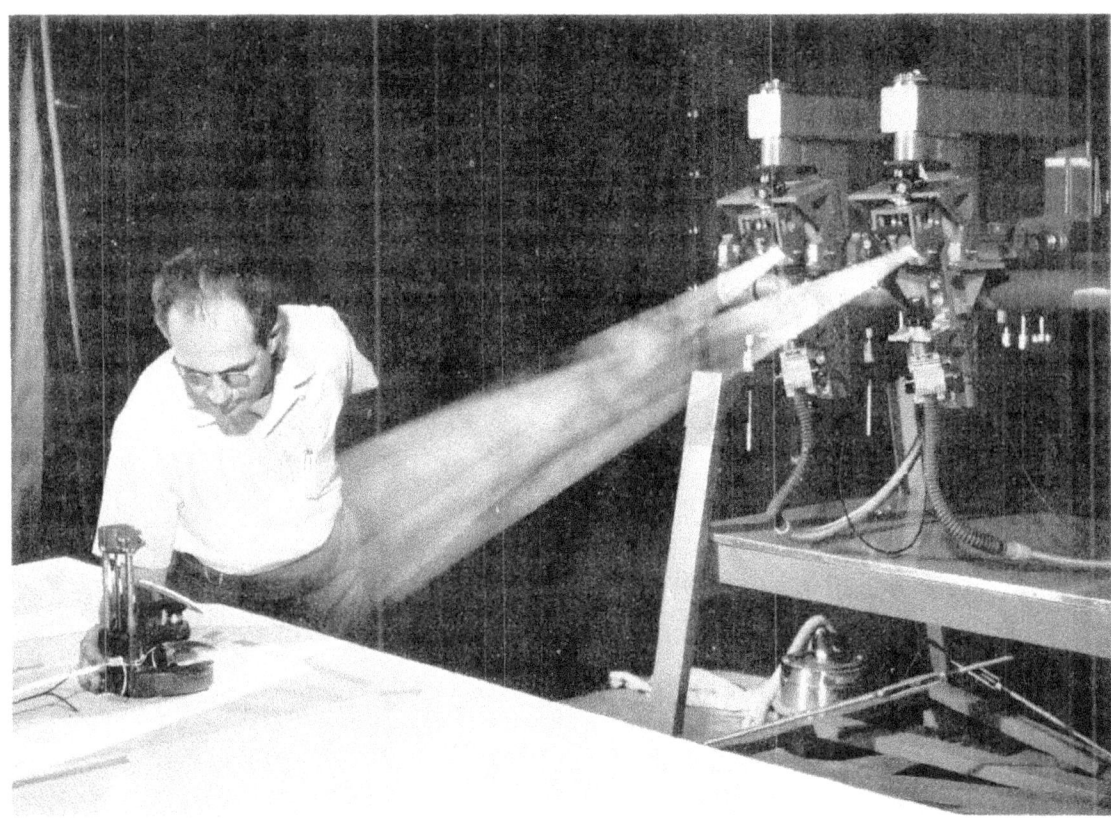

FIGURE 3-14.—Surveyor photorestitutor. This instrument was originally designed to rectify monoscopic Surveyor pictures, but has been modified for use as an anaglyphic stereoplotter. Each projector is a geometric analog of the television camera. Glass transparencies, or "diapositives," made from stereoscopic pairs of pictures are used at each camera position. The projectors and projector mirrors are set in the same relative orientations as the camera at the time each picture was taken. One picture is projected with red light; the other is projected with blue light onto a white screen, or platen. The operator wears spectacles with one red lens and one blue lens, enabling him to see the overlapping images as a single three-dimensional picture or model. As the platen is moved vertically or horizontally, the point of light at its center appears to float in the stereoscopic model. The horizontal position of the point of light is plotted on the map manuscript over which the platen carriage, or tracing table, moves. The vertical position of the point of light can be read from a counter on the tracing table.

dimensional model and which can be moved horizontally and vertically in measurable amounts by the operator. The mark is coupled by a mechanical linkage to a pencil or stylus so that maps or profiles can be traced directly from the stereoscopic models.

Stereoscopic mapping instruments have been in use for many years for terrestrial mapping, but none of them were designed to meet the unique requirements imposed by stereoscopic pictures taken by Surveyor television cameras. Measurements with these pictures must be made analytically, one point at a time, or with a stereoplotter specially designed for use with Surveyor television pictures (fig. 3-14).

Shadow measurement is basically a technique for making a profile along an east-west line through the spacecraft. Data from shadow measurements do not have sufficient density for contour mapping, but are valuable supplements to other data from which the maps are compiled. The method utilizes the known size and orientation of spacecraft parts to compute the size of the shadows of these parts on the lunar surface. From the size of the images of the shadows in the television pictures, their distance from the camera is computed. Camera azimuth, elevation, and distance to the shadow is then used to compute the location of the shadow on the lunar surface, with respect to the camera. The shadow of the television camera is used in the near field, because it is relatively small. The shadow of the solar panel is used for distances between 5 and 100 meters. This shadow, which is approximately 1 meter wide, is too large to make any but gross topographic measurements in the near field. It has been found to be an accurate (± 5 percent) way to measure distances as great as 100 meters from the camera.

Small-scale maps of the far field of Surveyor landing sites are made by correlating features visible on high-resolution Lunar Orbiter pictures with features recorded by the Surveyor television pictures.

We refer to this technique as photographic trigonometry, or "photo-trig." The method is based on the locations of the Surveyor spacecraft on Lunar Orbiter photographs. Heights of features visible on Lunar Orbiter photographs and Surveyor pictures are computed as a function of the distance of the feature from the Surveyor camera and its angular elevation with respect to the Surveyor camera and the lunar level. The former is measured on the Lunar Orbiter photographs, the latter on Surveyor pictures. From the set of heights thus computed, contour lines are plotted by interpolation. The accuracy of these contour lines is a function of distance from the spacecraft and density of control points in the immediate vicinity of the contour lines.

It is difficult to correlate features east or west of the spacecraft. During the lunar morning, features east of the spacecraft are difficult to see, because pictures are taken directly into the Sun, and glare obliterates much of the image area. As the Sun rises higher in the sky, contrast drops and topographic features become difficult to identify. In the lunar afternoon, features to the east of the spacecraft cannot be easily identified because the phase angle in this area is low. Surface detail cannot be seen because the contrast generally is low at low phase angles unless shadows are observed. A complementary set of conditions applies to the area west of the spacecraft.

Landing Site Maps

R. M. BATSON

A topographic map of the landing site of Surveyor 1, the only Surveyor photographed by a Lunar Orbiter spacecraft (fig. 3-15 and ref. 3-13), is being made by the photographic-trigonometry method. The Air Force Chart and Information Center (ACIC) has compiled a map of the site by stereophotogrammetric methods (ref. 3-17) from the Lunar Orbiter photographs. The ACIC maps are being used in conjunction with the Surveyor pictures and the Lunar Orbiter photographs to make accurate correlations and topographic measurements of the Surveyor I landing site.

Only four focus-ranging surveys were taken during the Surveyor 1 mission; consequently, features near the Surveyor 1 spacecraft can be located only approximately by assuming a datum and plotting the location of the feature as a function of its vertical angle from the Surveyor camera and the height of the camera

FIGURE 3-15.— Part of a Lunar Orbiter III frame H-183. The arrow points to the Surveyor I spacecraft.

above the datum. Figure 3-16 is a planimetric map made by this method. Focus-ranging data were not incorporated into this map because these data indicated that the surface was level within the limits of error of the method. Figure 3-17 shows the profiles computed from focus-ranging measurements made during the Surveyor I mission. A shadow progression profile was also made along a line to the east of the spacecraft (fig. 3-18).

The photographic-trigonometry method was first used to make a preliminary topographic map of the Surveyor III landing site (fig. 3-19). Whitaker (ref. 3-14) located the spacecraft by resection on Lunar Orbiter IV photograph H-125 which was taken before the Surveyor III landing. The spacecraft landed in a large crater 200 meters in diameter, and no features farther than about 150 meters from the camera were visible. The topographic measurements for the map in figure 3-19 were based on a camera tilt of 23.5° along the 83° camera azimuth (ref. 3-14). Focus-ranging profiles were not taken during the Surveyor III mission, and no large-scale, near-field maps of the site have been prepared.

The landing site of Surveyor V was the first on which the focus-ranging method was used extensively for detailed mapping (fig. 3-20). Shadow measurements were also made; these

TELEVISION OBSERVATIONS

FIGURE 3-16.—Planimetric sketch of the near field of the Surveyor 1 landing site. This sketch was based on the assumption that the landing site was a plane. Focus-ranging and shadow profiles were taken, but did not consist of a sufficient density of points for accurate mapping.

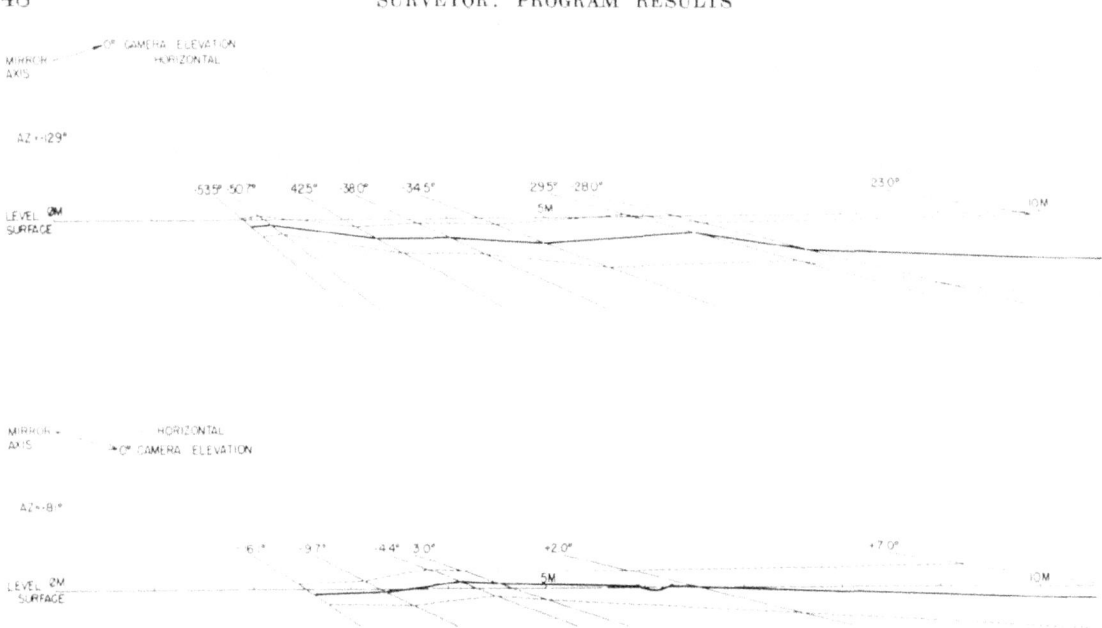

FIGURE 3-17.—Focus-ranging profiles taken during the Surveyor I mission. Horizontal-to-vertical scale ratio is 1:1.

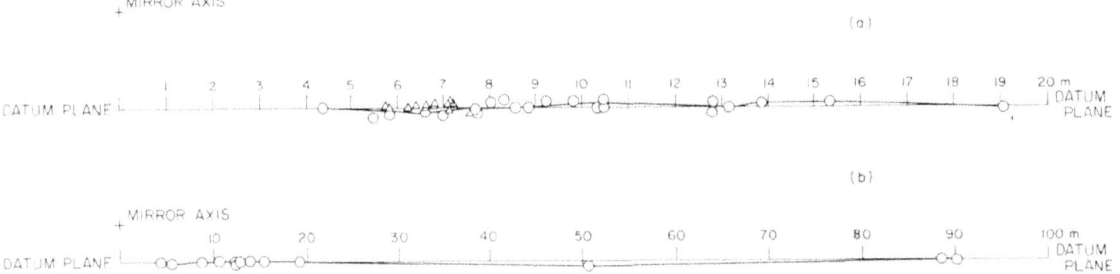

FIGURE 3-18.—Profiles made from shadow progression surveys taken during the Surveyor I mission. Horizontal-to-vertical scale ratio is 1:1. (a) Profile made from the shadow of the television camera and solar panel as it moved eastward from the spacecraft during the midafternoon. (b) Profile made from the shadow of the solar panel as it moved eastward across the lunar surface during the mid and late afternoon.

TELEVISION OBSERVATIONS

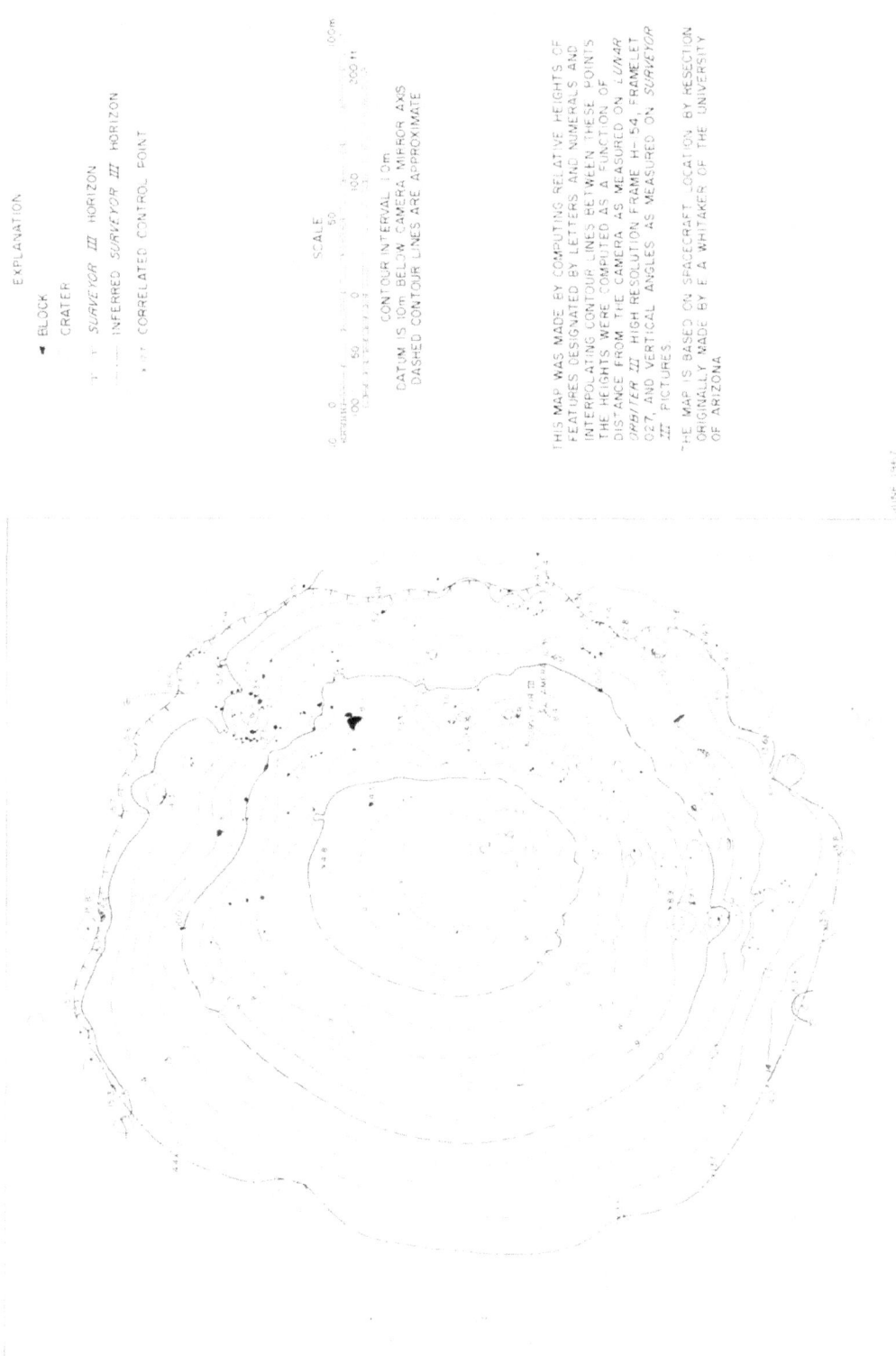

FIGURE 3-19.—Contour map of the Surveyor III landing site. Contour lines were drawn by interpolation between control points computed by the photographic trigonometry method. Probable vertical accuracy is ±0.5 meter.

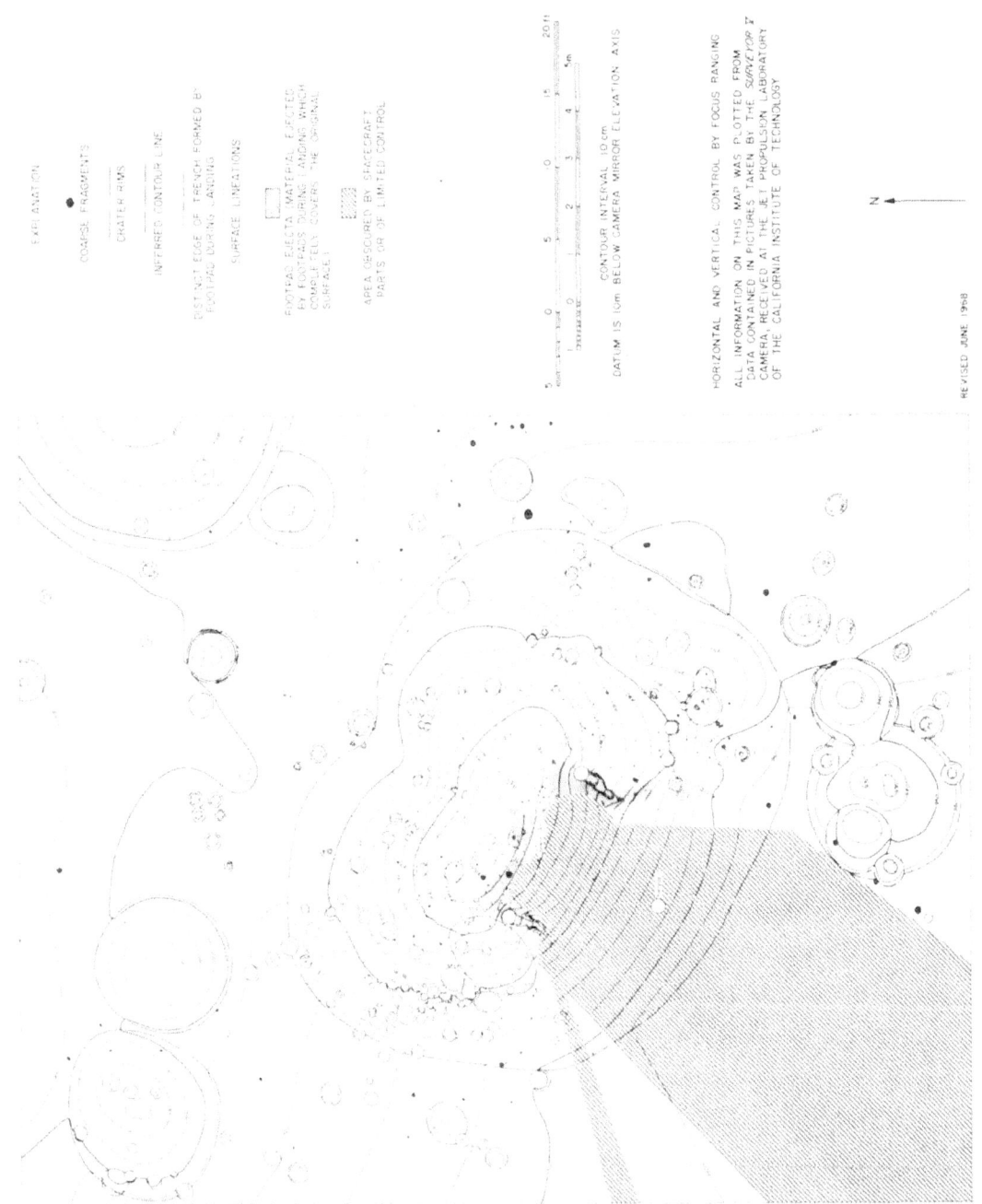

FIGURE 3-20.—Contour map of the near field of the Surveyor V landing site, prepared from focus-ranging data.

FIGURE 3-21.—Contour map of the Surveyor VI landing site. Contour lines were drawn by interpolation between control points computed by photographic trigonometry. Areas known to be obscured to the Surveyor VI camera have been eliminated from the Lunar Orbiter photograph. Depression contours are not shown. Probable vertical accuracy is ±3 meters.

measurements agreed with the focus-ranging measurements and were incorporated into the map as control points.

Stereoscopic measurement of part of the Surveyor V landing site may be possible because of a shift in camera position near the end of the first lunar day. The shock absorbers on legs 2 and 3 contracted and the legs compressed. Footpad 1 rotated about 1.4°, resulting in a vertical separation of camera position of approximately 7.45 cm. This baseline should be sufficient to make measurements of the near field with about the same accuracy as that of focus ranging.

Because Surveyor V has not been located on Lunar Orbiter photographs, mapping by photographic trigonometry is not possible.

All four mapping methods are being used to map the Surveyor VI landing site. A thorough correlation of features on Lunar Orbiter photograph H–121, framelets 264 through 272, has been made with the Surveyor VI pictures. Many features are as far away as 800 meters from the Surveyor VI camera. At this distance, an angular measurement error of ±0.2° results in an error in height measurement of ±2.8 meters. For this reason and because of a relatively low density of control points, a 5-meter contour interval was used (fig. 3–21). No features, with the exception of the small crater designated "B–1," could be located and identified east of the spacecraft; therefore, no contour lines were drawn in this area. Areas obscured to the Surveyor camera have been removed from the superimposed Lunar Orbiter mosaic. Most of the features in the Lunar Orbiter photographs should be visible in Surveyor pictures, but they may not be identifiable because of Surveyor's foreshortened view of the surface. The defiladed areas probably are much more extensive than shown on this illustration. Figure 3–22 is a diagrammatic representation of the location of these features in the Surveyor panorama.

The hop of Surveyor VI displaced the spacecraft 2.54 meters horizontally,[2] −0.12 meter vertically, and rotated it counterclockwise 4.25±0.25°. The displacement was determined

---

[2] An independent measurement made by Christensen et al. was obtained by a different method. (See ch. 4 of this report.)

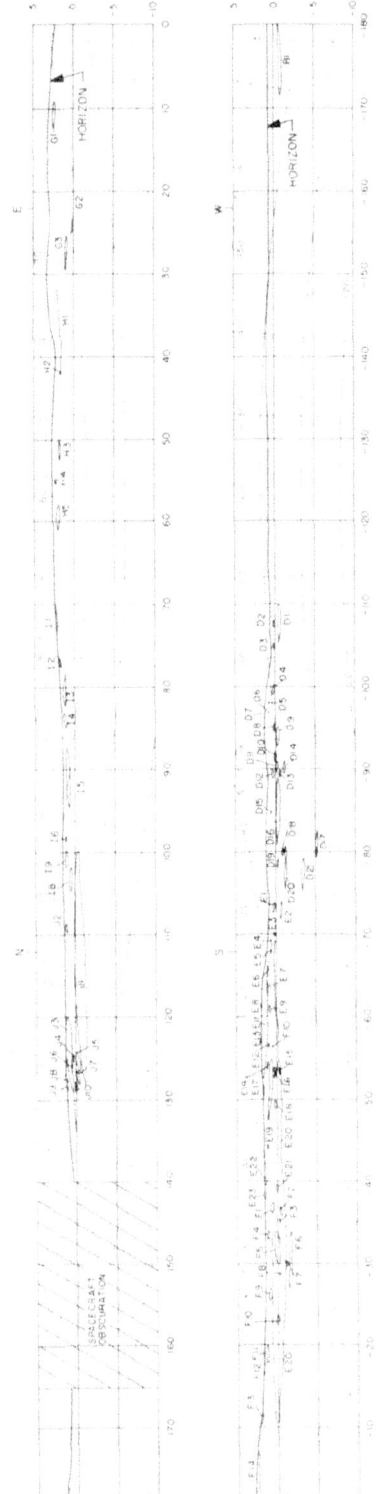

FIGURE 3–22.—Diagram of the Surveyor VI panorama, showing control correlations used to make the map in figure 3–21. The screened lines are the camera coordinates of the panorama.

result is shown in figure 3-23. When vertical heights were computed, it was found that much better correlation between prehop and posthop figures was possible if 12 cm were subtracted from all height measurements made before the hop. Figure 3-24 is the completed focus-ranging map made from surveys taken before and after the hop. Both spacecraft positions are shown and features identified on both sets of data are distinguished from those identifiable on only one.

The displacement in camera position produced by the hop permitted the taking of stereoscopic pictures. These are being used to map the landing site in the area between 5 and 100 meters from the spacecraft. The stereoscopic effect is difficult to use because the hop took place very near lunar noon, when the Sun was high and the contrast was poor in prehop and posthop pictures. For this reason, the stereoscopic mapping requires a combination of point-by-point analytical measurement in conjunction with measurements made with the stereoplotter shown in figure 3-14.

The Surveyor VII landing site is being mapped by combinations of all four mapping methods. Figure 3-25 is the map made by focus ranging. The only stereoscopic effect in the Surveyor VII pictures was produced over a small area about 0.25 m², through the use of a mirror mounted on the spacecraft mast.

A contour map of a small rock fragment (fig. 3-26) was made by stereoscopic measurements of pictures taken directly and through the mirror. This fragment was placed in the area covered by the stereomirror by the surface sampler. Figure 3-27 is a stereogram of the rock showing the measured control points used for compilation of the contour lines. The map was made entirely by the analytical method, one point at a time, through use of a computer program developed at the Jet Propulsion Laboratory. The entire area of stereoscopic coverage on the Surveyor VII landing site is being mapped by the same method and through use of the same computer program. A small-scale map of the Surveyor VII site is also being compiled by photographic trigonometry.

FIGURE 3-23.—Surveyor VI relative orientations. Features shown with screened lines were located by focus-ranging surveys taken before the hop, and features shown with solid lines were located by focus-ranging surveys taken after the hop. The average height of features, measured by focus-ranging before and after the hop, indicates that the second camera position is 12 cm higher than the first. Horizontal displacement was 2.54 meters. Measurements of azimuths of features near the horizon (too far from the spacecraft to be affected by parallax) indicate that the spacecraft rotated counterclockwise 4.25° during the hop.

by focus ranging. Because of the change in illumination angles, both before and after the hop, not all features could be identified on panoramas of pictures taken before and after the hop. A significant number could be located, however, and planimetric maps were made from prehop and posthop focus-ranging surveys. The rotation of the spacecraft during the hop was measured by comparing azimuth angles to horizon features on the assumption that these would not be affected by the parallax between the two camera positions. Focus-ranging planimetric maps of the landing site before and after the hops were superimposed, with a 4.25° rotation, and manipulated laterally until a best fit of features was obtained. The

FIGURE 3-24.—Contour map (by Raymond Jordan) of the near field of the Surveyor VI landing site. The map was made from focus-ranging surveys taken both before and after the hop. Features that could be identified on one set of pictures only (taken either before or after the hop) are shown with shaded lines, as are the locations of spacecraft parts before the hop. All other features are shown with solid lines.

FIGURE 3-25.—Contour map (by Raymond Jordan) of the near field of the Surveyor VII landing site. The map was made from focus-ranging data.

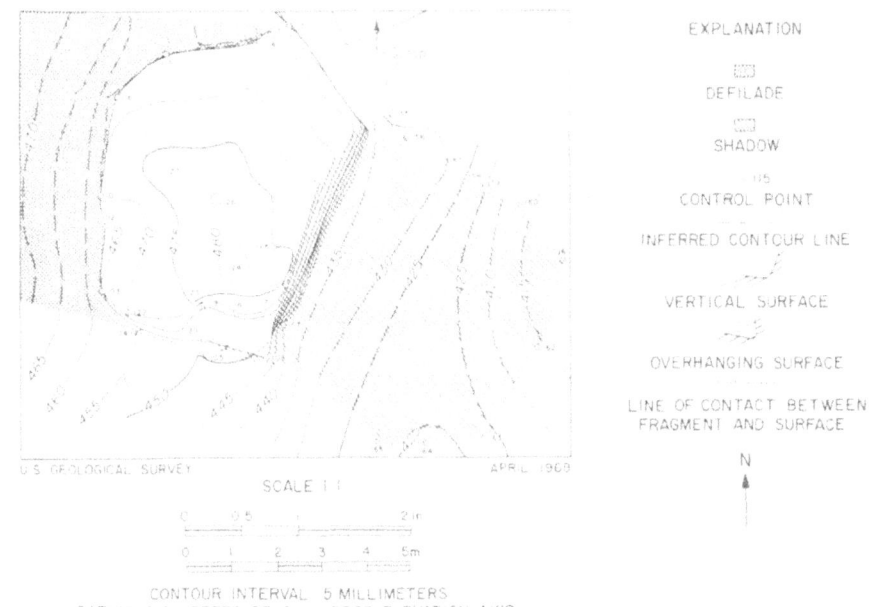

FIGURE 3-26.—Contour map of a small rock fragment at the Surveyor VII landing site. The map was made by photogrammetric measurement of stereoscopically observed control points in the stereomirror area.

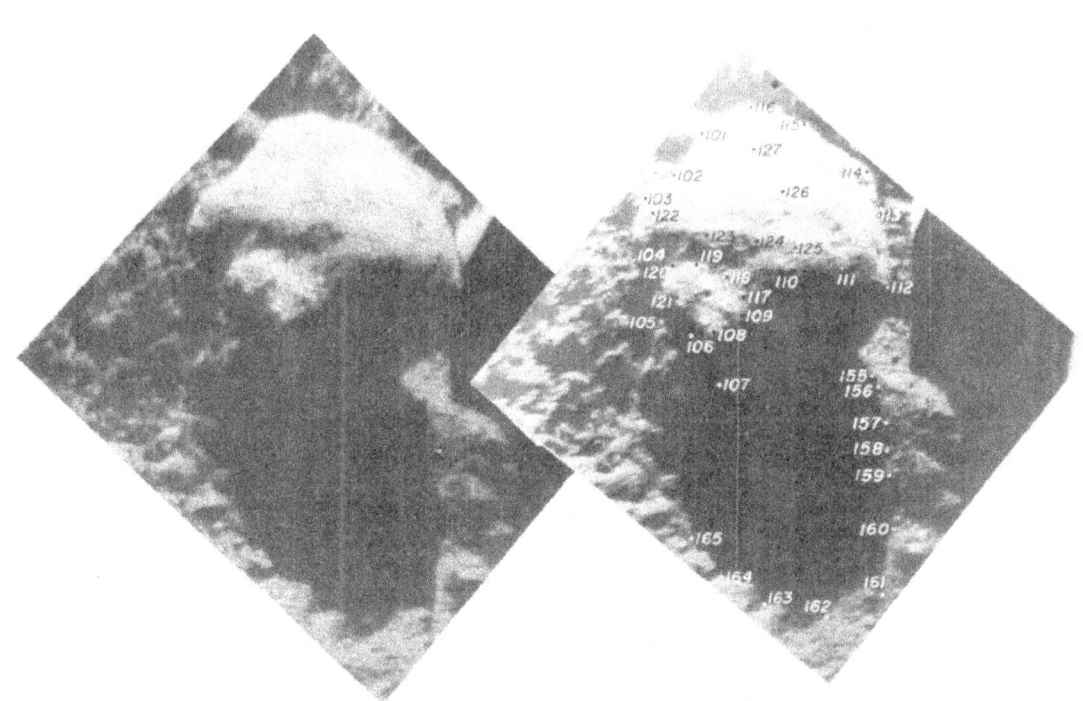

FIGURE 3-27.—Stereogram of the fragment mapped in figure 3-26. The numbered points were used to control the contour lines of figure 3-26.

Geology

Regional Setting

E. C. MORRIS

The first four successful Surveyor spacecraft landed on broad mare plains. Surveyor I landed on a gently undulating mare surface partly enclosed by the rim of a large, nearly buried crater about 100 km in diameter (fig. 3-28) in the southern part of Oceanus Procellarum. The mare material in the eastern half of this crater is considerably darker than the more typical mare material of Oceanus Procellarum, and the Surveyor I site is on the dark mare material. No large rays are found in full-Moon photographs of this area.

Surveyor III landed about halfway up the inner northwest-facing slope of a 200-meter-

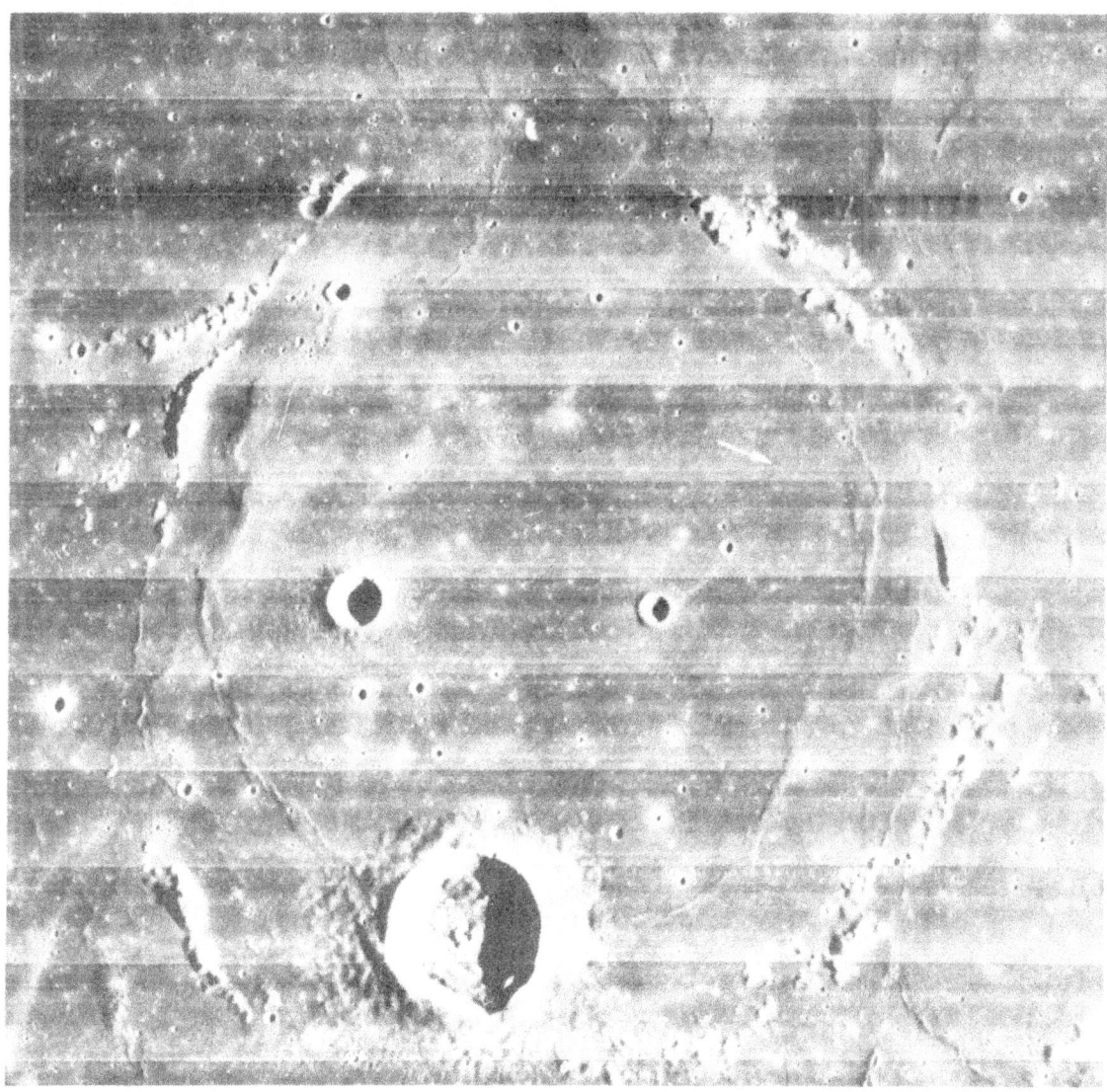

FIGURE 3-28.—Lunar Orbiter IV photograph H-143 of the Surveyor I landing site. Hills north of the landing site, which are part of the rim of the buried crater 100 km in diameter, can be seen in Surveyor panoramas.

FIGURE 3-29.—Small part of Lunar Orbiter III photograph H-154 showing location of Surveyor III in a 200-meter-diameter crater. The small triangle represents the true size and orientation of the spacecraft.

wide subdued crater (fig. 3-29) located in the eastern part of Oceanus Procellarum about 120 km southeast of the crater Lansberg (fig. 3-30(a)). The Surveyor III site is crossed by faint rays from the large crater Copernicus (fig. 3-30(b)) 400 km to the north. About 20 km west of the site, the mare surface is broken by rough, hummocky terrain and numerous isolated hills of the Fra Mauro formation of Imbrian age (ref. 3-18). Low mare ridges and hummocky terrain form the eastern boundary of the smooth patch of mare material on which Surveyor III rests.

Surveyor V landed in the southwestern part of Mare Tranquillitatis in the eastern part of the Moon (fig. 3-31), about 70 km north of the southern boundary of the mare. The region is crossed by faint rays associated with the large crater Theophilus 350 km to the south. The highlands to the west of Mare Tranquillitatis are characterized by prominent northwest-trending ridges and valleys which are part of a system of ridges and valleys known as the Imbrian sculpture. High-resolution Lunar Orbiter photographs of an area near the Surveyor V site reveal many craters about 10 meters across which are also alined in a northwest direction. Some individual craters are markedly

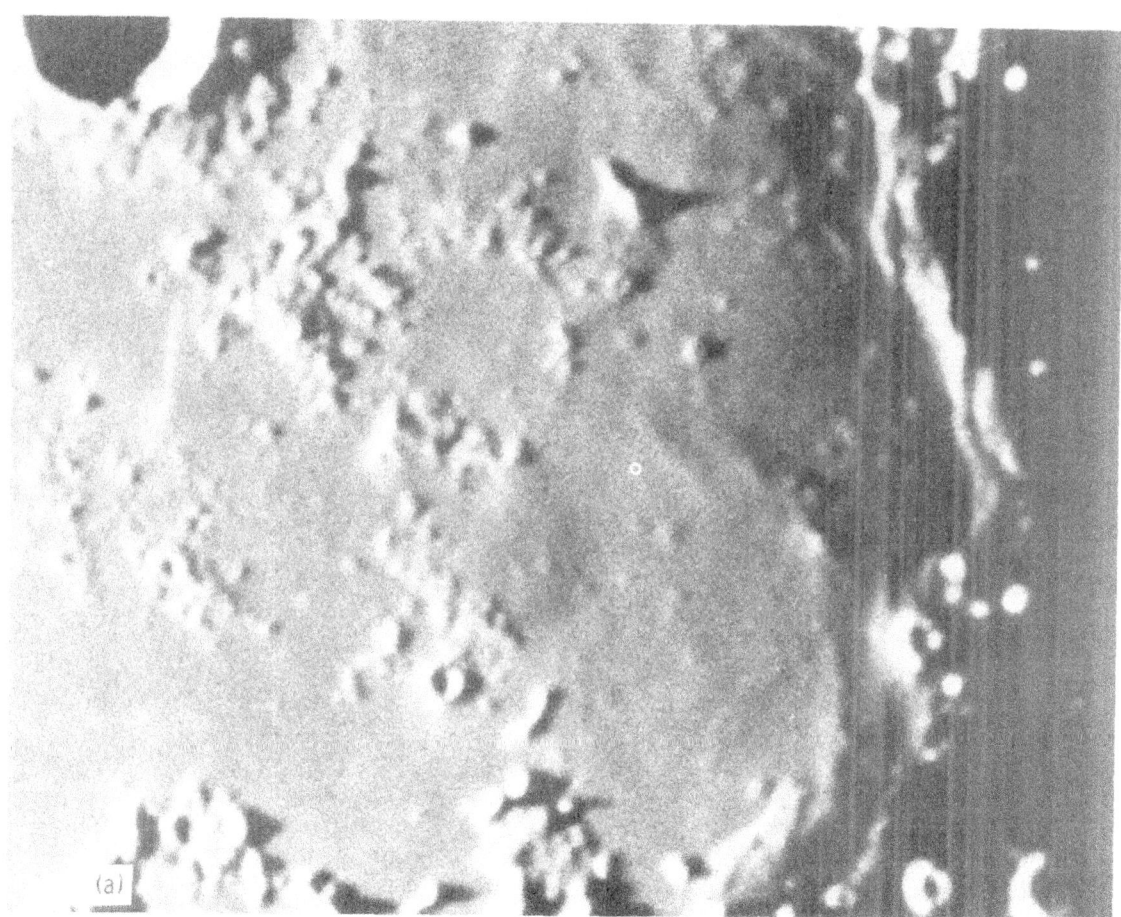

FIGURE 3-30.—The Surveyor III landing site. (a) The crater Lansberg and region to the southwest showing the Surveyor III location (small circle). The low evening illumination emphasizes the low mare ridge segments northeast of the landing site and the rough, hummocky topography northwest of the landing site. (Photograph taken with the 82-in. reflector at McDonald Observatory.) See next page for figure 3-30(b).

elongate in this same direction. These craters may be drainage or collapse craters that are structurally controlled by subsurface fissures and fractures that are related to the Imbrian sculpture. Surveyor V came to rest in one of these small, elongate, rimless craters. The Surveyor V crater is about 9 meters wide, 12 meters long, and about 1 meter deep.

Surveyor VI landed in Sinus Medii, an isolated patch of mare material near the center of the subearth side of the Moon (fig. 3-32). The surface of Sinus Medii has a higher average albedo than most of the maria. The detailed shapes and trends of mare ridges, crater chains, and small shallow trenches on Sinus Medii reflect the dominant structural patterns that occur in the highlands that surround the maria. The most conspicuous structural pattern in the highlands is the northwest-trending Imbrian sculpture. The second most prominent set of linear structures is a northeast-trending system of scarps and ridges.

The most prominent topographic feature in the vicinity of the landed Surveyor VI is a mare ridge about 40 km long that follows a zigzag pattern and trends generally east-west (fig. 3-33). Individual elements of the ridge trend northwest and northeast. The ridge is somewhat smaller than ridges that have been studied through Earth-based telescopes, but it is the

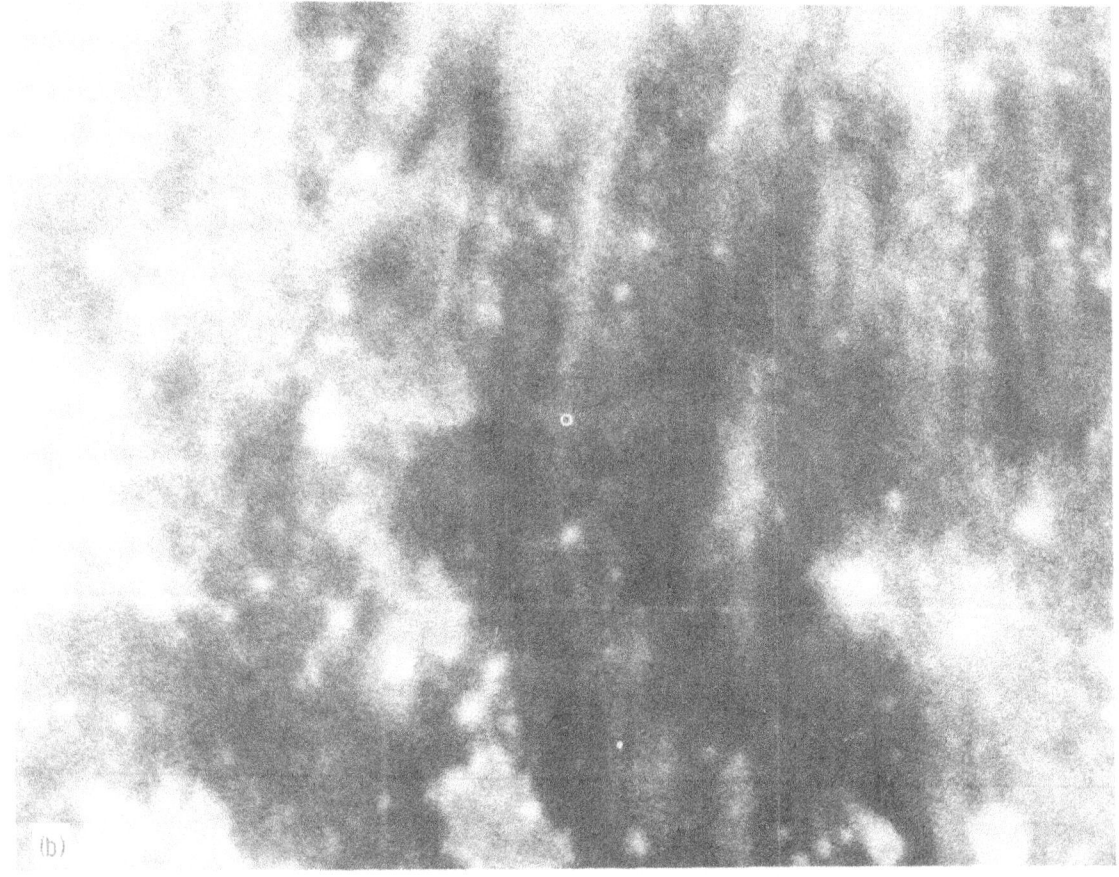

FIGURE 3-30.—Continued. (b) Same region as figure 3-30(a) with full-Moon illumination showing relationship between Surveyor III location (small circle) and Copernicus rays. (Photograph taken with a 40-in. refractor, Yerkes Observatory.)

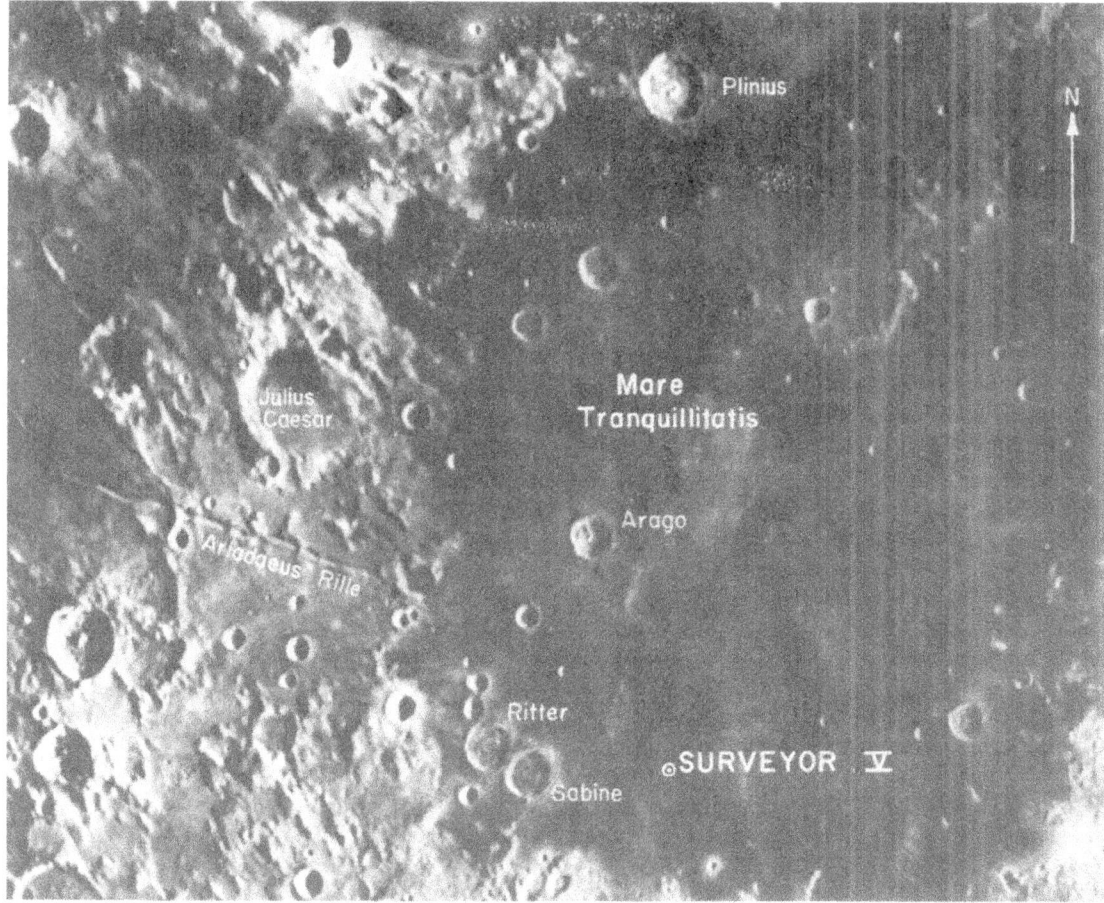

FIGURE 3-31.—Earth-based, telescopic photograph of Mare Tranquillitatis and the highlands to the west. Prominent northwest-trending ridges and valleys in the highlands are part of Imbrian sculpture. (Photograph taken through 36-in. refractor telescope at Lick Observatory.)

FIGURE 3-32.—Earth-based, telescopic photograph of the central region of the Moon. The white circle indicates the location of Surveyor VI. (Photograph taken through the 100-in. telescope at Mount Wilson Observatory.)

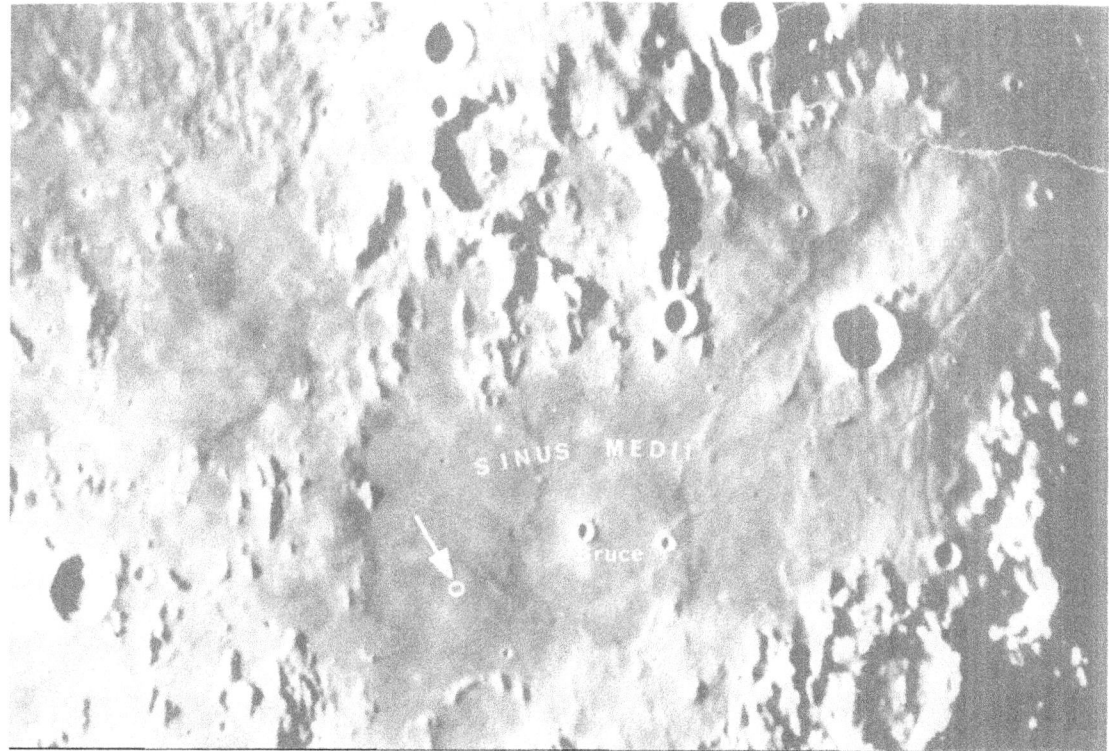

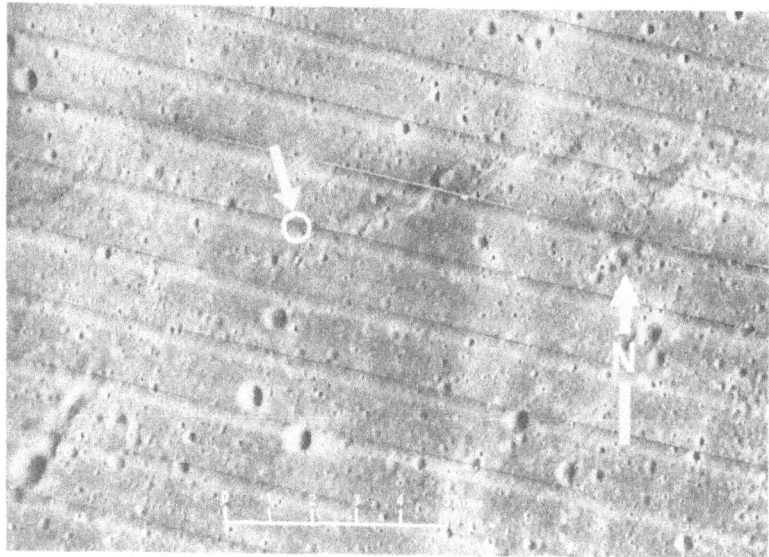

FIGURE 3-33.—Lunar Orbiter II photograph M-113 of Sinus Medii. The Surveyor VI landing site is indicated by the white circle. A mare ridge, which passes just south of the Surveyor VI landing site, can be seen extending across the center of the picture.

FIGURE 3-34.—Lunar Orbiter V photograph M-127 of the crater Tycho and its northern flank. The arrow points to the Surveyor VII landing site.

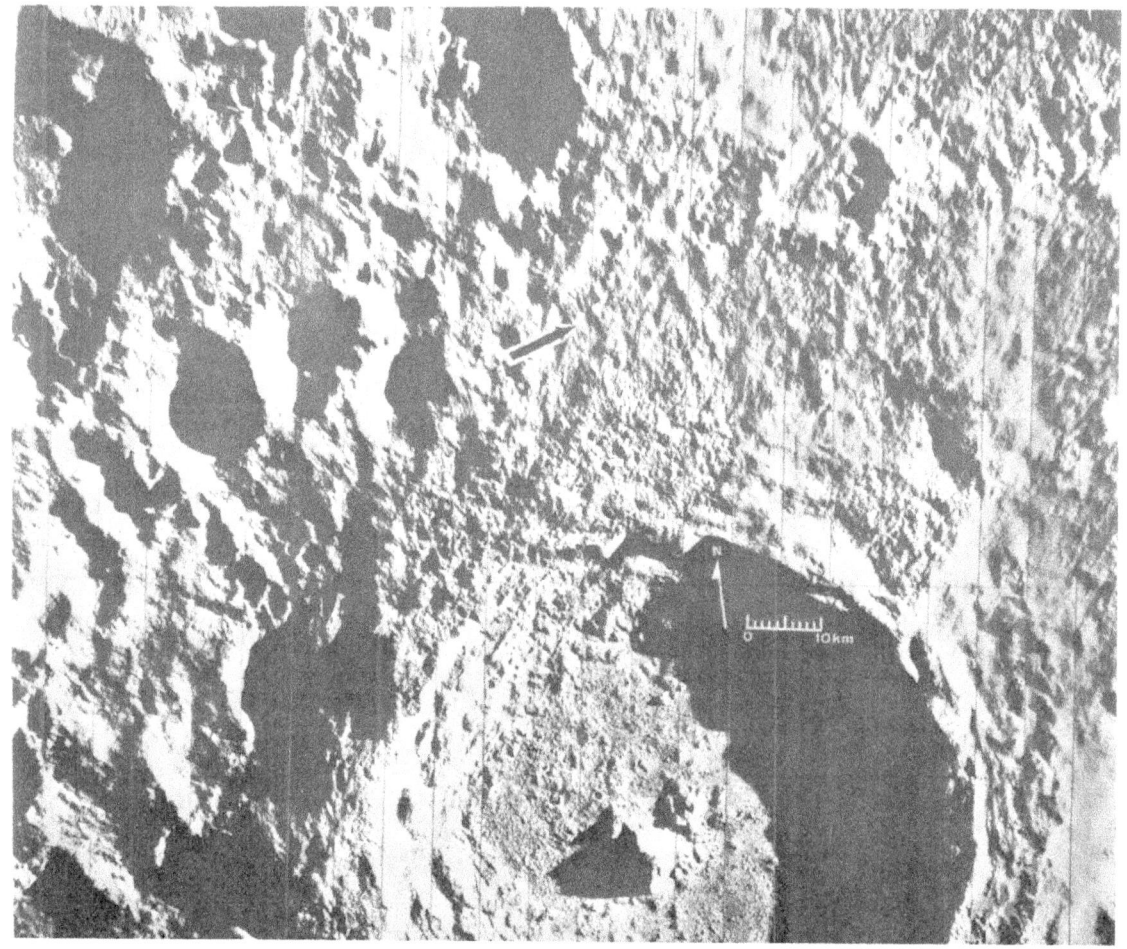

first such feature to be studied from a landed spacecraft.

Surveyor VII, the only spacecraft sent to a highland area, landed on the outer flank of the rim of the crater Tycho, one of the most prominent and well-known features in the lunar highlands (fig. 3-34). The crater is surrounded by the most conspicuous and extensive system of bright rays on the subearth side of the Moon. From the crest extending outward a distance of 10 to 15 km, the rim of Tycho is composed of irregular hills and intervening depressions. From 15 km to a radial distance of about 35 to 40 km, the surface is marked by numerous subradial ridges and valleys, typically 2 to 5 km in length and ½ to 1 km in width, superimposed on a broadly undulating surface. Surveyor VII landed about 30 km north of the rim crest of Tycho on the part of the rim flank marked by these linear ridges.

The rim of Tycho is composed of debris, probably ejected from the crater during its formation, and a sequence of flows that form mappable geologic units. The flows range in form from hummocky, steep-fronted flows to smooth-surfaced flows without marginal scarps. The differences in form are attributed to differences in viscosity of the flows at the time they came to rest. Surveyor VII landing site is on one of the flows whose surface is composed of irregular, low hills and depressions ranging from 100 meters to several hundred meters across with scattered blocks, small craters, and swarms of north-trending fissures that occur on the flow's crest. The Surveyor VII landing site is geologically more complex and contains a greater variety of rock fragments than any of the Surveyor landing sites on the maria.

Craters

E. C. Morris and E. M. Shoemaker

Small craters are the most abundant of the topographic features observed on the lunar surface and account for the irregularities of largest relief on the surface of the landing sites in the maria. Several types of small craters can be recognized: (1) shallow, cup-shaped craters with subdued rims; (2) cup-shaped craters with sharp, raised rims; (3) rimless craters; and (4) irregular or asymmetric craters. Most small craters in the diameter from 10 cm to several meters are cup shaped with concave floors and subdued convex rims. They are difficult to observe under high angles of solar illumination, but are conspicuous at low-illumination angles. A few percent of the craters observed in the Surveyor pictures are cup shaped, with sharp, raised rims. Most of the cup-shaped craters probably are of impact origin.

Rimless craters are prominent at the Surveyor V site; Surveyor V came to rest with two of its footpads in one of these craters. The rimless craters commonly are alined in crater chains (fig. 3-35), and they are inferred to have been formed by drainage of surficial debris into subsurface fissures.

Irregular craters generally are lined with clods of fine-grained material, and some are nearly filled with clods or angular rubble. These craters (fig. 3-36) are interpreted to be secondary craters formed by low-velocity impact of cohesive blocks or clods of weakly cohesive, fine-grained material ejected from nearby primary craters. An irregular crater (fig. 3-37), near Surveyor VII about 3 meters in diameter and filled with coarse blocks up to 60 cm across, is probably a secondary impact crater formed by a large block of rock ejected from a nearby primary crater.

The size-frequency distribution of small craters, a few centimeters to several meters in diameter, was determined from the Surveyor pictures for each of the Surveyor landing sites. Pictures taken during low Sun illumination were used for identification of the craters because small craters are most easily identified and measured under this condition of illumination. Craters smaller than a few centimeters, however, are difficult to recognize at low Sun-elevation angles because the shadows cast by fragments and protuberances tend to hide them. Another factor that tends to lower the observed number of very small craters is the difficulty of recognizing small craters in an oblique view of the lunar surface, such as afforded by the Surveyor pictures. The scale and the ground resolution of the pictures change from the foreground to the background making it difficult

FIGURE 3-35.—Wide-angle picture of the northwest wall of the Surveyor V crater. Chain of small rimless craters 20 to 40 cm in diameter extends from the center to the bottom of the picture (Sept. 23, 1967, 11:26:28 GMT).

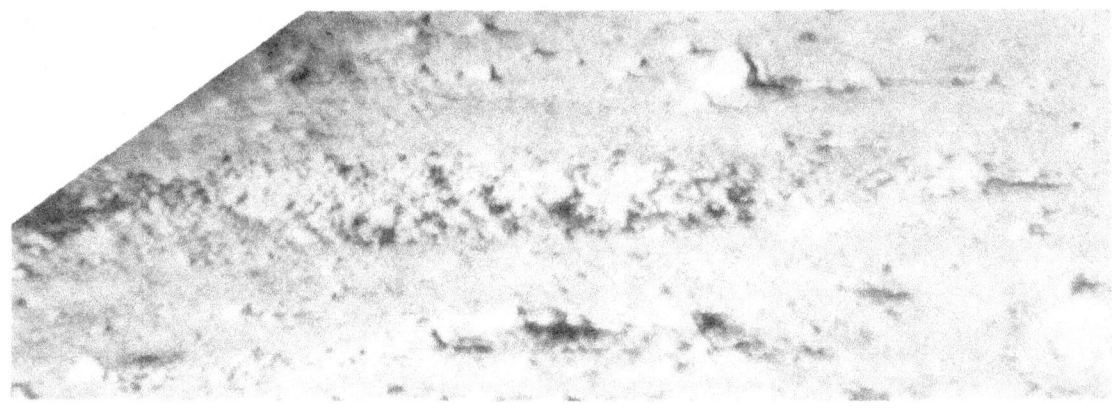

FIGURE 3-36.—Secondary-impact craters. (a) Irregularly shaped crater at the Surveyor I site formed by low-velocity impact of relatively soft clots of material derived from a small primary-impact crater nearby (June 10, 1966, 15:30:18 GMT). (b) Secondary impact crater in moist sand, formed by a clot of weakly cohesive silty clay, ejected from a missile impact crater at White Sands Missile Range, New Mexico. (H. J. Moore, personal communication; photograph by courtesy of U.S. Army.)

FIGURE 3-57.—Secondary-impact crater, about 3 meters in diameter near Surveyor VII, was formed by a large block of fairly strong rock ejected from a nearby large primary crater. The blocky material in the crater and strewn to the left of the crater was probably formed by the breakup of the large block on impact. Fragments in the crater range up to 60 cm across (Catalog 7-SE-22).

to recognize small craters of a size that are easily seen in the near field.

The size-frequency distribution of craters 2 cm to 4 meters in diameter is shown in figure 3-38 for each of the Surveyor landing sites and is compared with a generalized Ranger VII, VIII, and IX curve (ref. 3-19) for craters on the mare plains extrapolated to small sizes. At all landing sites, except Surveyor V, the distribution of the small craters lies close to the extrapolated Ranger VII, VIII, and IX curve. The low frequency of small craters observed at the Surveyor V site may be because of the incompleteness of the observational data. The Surveyor V camera was inclined toward the floor and far wall of the crater in which the spacecraft landed; consequently, more than 80 percent of the field of view below the horizon was occupied by parts of the lunar surface that were not more than 6 meters from the camera. The low oblique view of the lunar surface outside of the Surveyor V crater and unfavorable illumination of the walls of the Surveyor V crater during the lunar day made recognition of the small craters difficult.

If most craters observed on the Moon are of impact origin, the size-frequency distribution of craters a few meters in diameter and smaller should correspond to that expected for a steady-state population of craters produced by prolonged repetitive bombardment by meteoroids and by fragments of the Moon itself (ref. 3-20). The general distribution for small craters on the lunar plains, determined by Trask from Ranger VII, VIII, and IX pictures, is considered to be the steady-state distribution for level surfaces (ref. 3-19). This distribution is a simple function of the form $F = \Phi c^a$, where F is the cumulative number of craters with a diameter equal to and larger than c, and c is the diameter of the craters. The exponent a for craters from 1 meter to several hundred meters in diameter was found to be -2, a value predicted by the steady-state model (ref. 3-21), and the constant Φ was found by

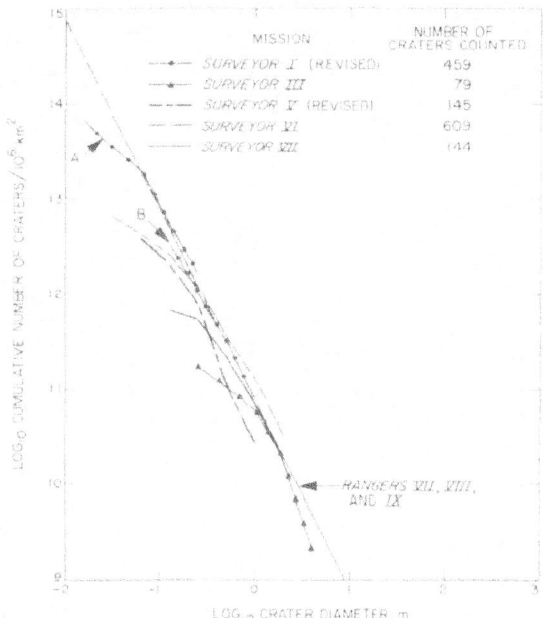

FIGURE 3-38.—Size-frequency distribution of small craters on the lunar surface at Surveyor I, III, V, VI, and VII landing sites compared with the size-frequency distribution of craters on the lunar planes determined from Ranger VII, VIII, and IX pictures and extrapolated to small sizes.

Trask to be $10^{10.9}/m^{-2} \, 10^6 \, km^2$. The size distributions of craters a few centimeters to several meters in diameter, determined from Surveyor pictures, appear to fit this function closely; they show that the function may be extended from craters of 1 meter in diameter, observed by Rangers VII and IX, to craters smaller than 10 cm in diameter.

The upper limiting crater diameter for the steady-state distribution of craters observed on a given surface is a function of the number of large craters on the surface and, by inference, a function of the age of the surface. Above a certain crater diameter, the size-frequency distributions of the craters at the Surveyor landing sites are no longer adequately represented by $F = \Phi c^\mu$, where Φ and μ have the steady-state values but can be represented by other functions, $F = \chi c^\lambda$, where $\lambda < \mu$. The intersection of $F = \chi c^\lambda$ with $F = \Phi c^\mu$ is the upper limiting crater diameter for the steady-state distribution, here designated c_s. In general, the functions $F = \chi c^\lambda$ can be found that fit the crater distributions observed on Lunar Orbiter photographs very closely between c_s and crater diameters of 1 km. The magnitude of c_s increases with increasing cumulative number of craters with diameters greater than c_s; both c_s and the cumulative number of craters at any diameter greater than c_s can be used as a measure of the relative age of the surface. At crater diameters less than c_s, the cumulative number of craters is the same for all surfaces and is, therefore, independent of the age of the surface.

TABLE 3-10. *Constants and exponents of functions of the form* $F = \chi c^\lambda$, $c_s \leq c \leq 1$ km, *fitted to the size-frequency distributions of craters observed from Lunar Orbiter or Ranger photographs of the Surveyor I, V, VI, and VII landing sites*

Landing site	χ, $\dfrac{m^{-\lambda}}{10^6 \, km^2}$	λ [b]	c_s, m [b]
Surveyor I	$10^{12.92}$	-3.15	$10^{0.75}$
Surveyor V [c]	$10^{12.29}$	-2.58	$10^{2.27}$
Surveyor VI	$10^{13.08}$	-3.10	$10^{2.00}$
Surveyor VII (patterned flow)	$10^{11.24}$	-2.93	$10^{0.44}$

[a] $F = \chi c^\lambda$, $c_s \leq c \leq 1$ km

where F is the cumulative number of craters with diameter equal to, or larger than, c per $10^6 \, km^2$, and c is the diameter of craters in meters.

[b] $c_s = \left(\dfrac{\Phi}{\chi}\right)^{1/(\lambda-\mu)}$ $\Phi = 10^{10.9}/m^\mu \, 10^6 \, km^2$ $\mu = -2$

[c] Function fitted to size-frequency distribution of craters observed from Ranger VIII photographs by N. H. Trask (ref. 3-19).

Functions $F = \chi c^\lambda$ and the values of c_s are shown in table 3-10 for the Surveyor I, V, VI, and VII landing sites. The functions and values of c_s listed for Surveyor I, VI, and VII are based on crater distributions, shown in figures 3-39 to 3-41, measured from Lunar Orbiter photographs; the function listed for the Surveyor V site is based on the crater distribution measured by Trask from Ranger VIII pictures

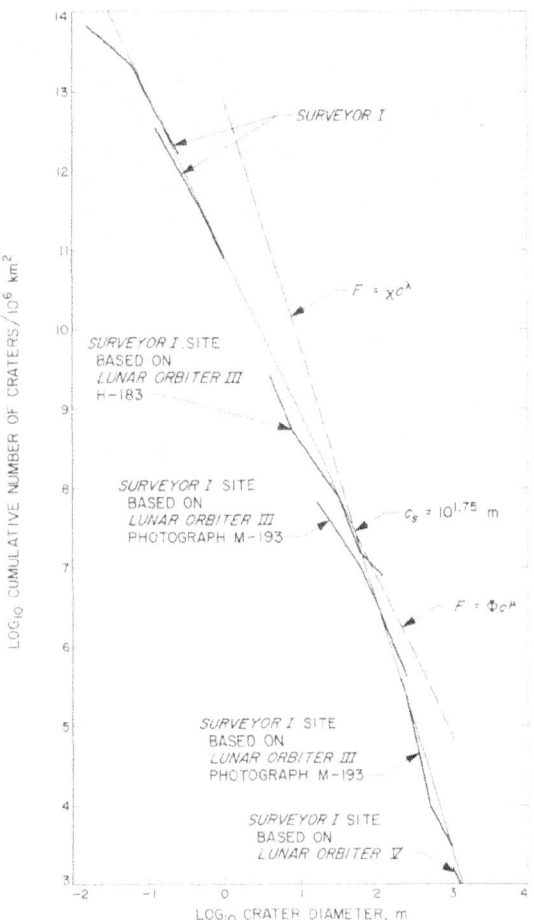

FIGURE 3-39.—Cumulative size-frequency distribution of small craters on the lunar surface in the vicinity of Surveyor I, determined from Surveyor I pictures and Lunar Orbiter III and V photographs.

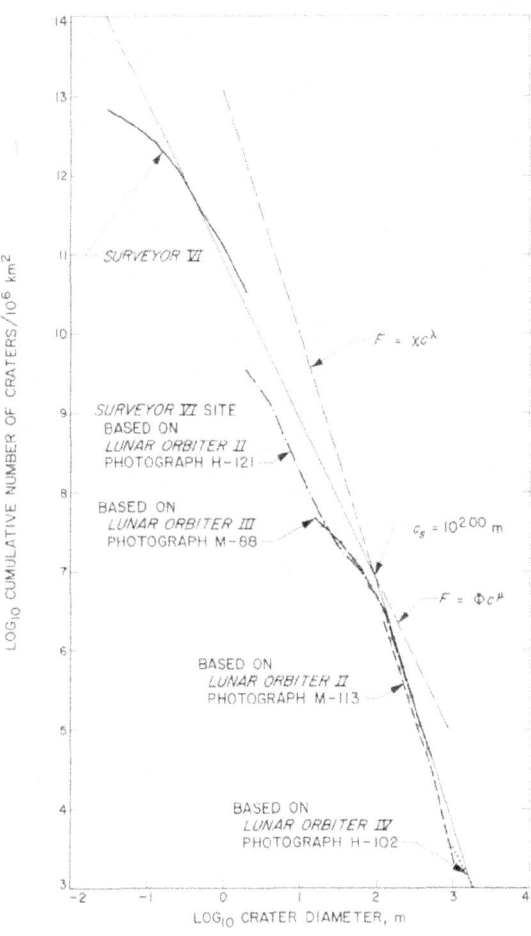

FIGURE 3-40.—Cumulative size-frequency distribution of small craters on the lunar surface in the vicinity of Surveyor VI, determined from Surveyor VI pictures and photographs from Lunar Orbiter II, III, and IV.

of Mare Tranquillitatis. The crater distribution at the Surveyor III landing site, although not listed, is similar to that at Surveyor I. On the basis of the observed distributions, the sequence of ages of the surfaces at the landing sites listed from the oldest to the youngest is (1) Surveyor V, $c_s = 186$ meters; (2) Surveyor VI, $c_s = 100$ meters; (3) Surveyor I, $c_s = 56$ meters; and (4) Surveyor VII, $c_s = 2.7$ meters. In terms of cratering history, the surface on which Surveyor VII landed on the rim of Tycho is much younger than the surfaces at any of the mare landing sites.

Fragmental Debris

E. C. MORRIS AND E. M. SHOEMAKER

The surface on which the five successful Surveyor spacecraft landed consists of a fragmental debris layer, or regolith, composed of poorly sorted or well-graded fragments that range in size from large blocks to fine particles too small to be resolved by the Surveyor camera (< 0.5 mm). The number of resolvable particles per unit area varies from site to site; 3 to 18 percent of the surface was found to be occupied by fragments larger than 1 milli-

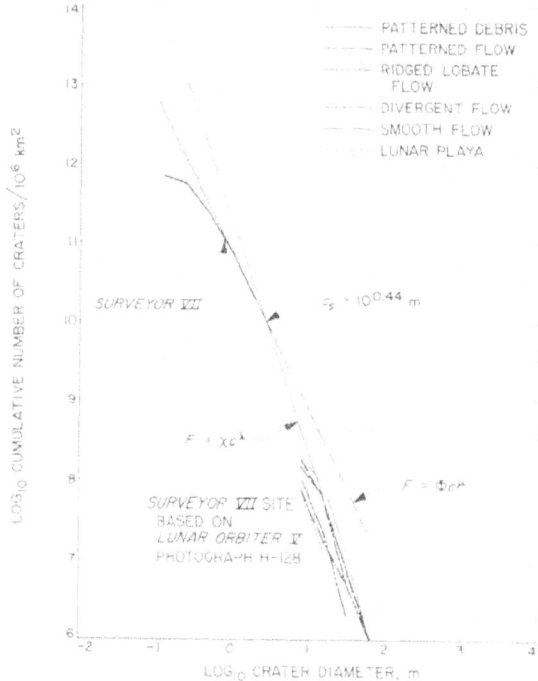

FIGURE 3-41.—Cumulative size-frequency distribution of small craters on the lunar surface in the vicinity of Surveyor VII, determined from Surveyor VII pictures and Lunar Orbiter V photograph H-128.

meter. The greatest number of large fragments was observed at the Surveyor VII site where fragments larger than 1 millimeter in diameter occupy about 18 percent of the surface and fragments coarser than 10 cm in diameter occupy about 10 percent of the surface. Coarse fragments are an order of magnitude more abundant at the Surveyor VII site than fragments of similar size at the Surveyor V and VI sites, which have the least number of coarse fragments.

Many of the larger fragments observed at each landing site are partly, or entirely, surrounded by a fillet or embankment of fine-grained material (fig. 3-42). In general, the best developed fillets occur around the largest blocks. These fillets probably are formed by the ballistic trapping of small particles sprayed out from nearby parts of the lunar surface by multiple small-impact events. An unusually large and well-developed fillet occurs around block G at the Surveyor VII landing site on a smooth patch of material north of the spacecraft. Block G is about 5 meters in diameter and has nearly vertical sides more than 2 meters high. The fillet surrounding the block is about 20 meters in diameter at the base and about 1 meter high at the contact with the block (fig. 3-43). It is possible this large fillet has a different origin than most of the fillets observed.

Most fragments larger than 1 meter in diameter on the lunar surface are relatively rounded, except in a few cases where they are part of strewn fields of blocks associated with sharp-rim craters (fig. 3-44). For the most part, the large, rounded fragments seem to be fairly deeply embedded in the surface on which they are found (ref. 3-14). Some smaller fragments, however, are resting on the fine-grained matrix of the surface without being significantly embedded in this material (fig. 3-45).

Because of the relatively large size and number of blocks in strewn fields of blocks at the Surveyor III landing site, it has been possible to conduct a preliminary statistical investigation of the roundness and degree of burial of these blocks and the relationship of roundness and burial to the characteristics of the principal crater associated with each strewn field.

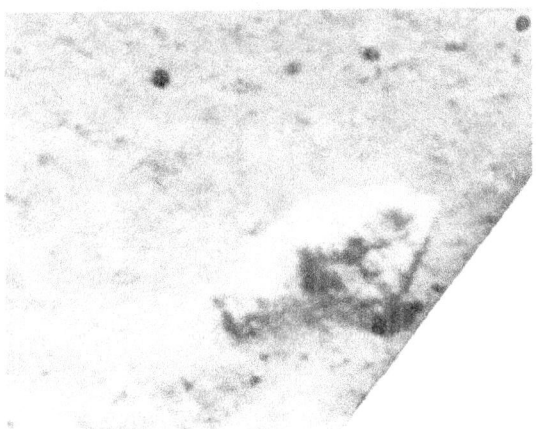

FIGURE 3-42.—Part of a narrow-angle Surveyor III picture, showing an angular block close to the spacecraft and fine-grained material banked against the side of the block facing the camera. The block is 7 cm across. Notice small particles 1 to 5 mm in diameter that can be resolved in the fine-grained debris (Apr. 28, 1967, 14:30:51 GMT).

FIGURE 3-43.—Mosaic of narrow-angle Surveyor VII pictures of an area 350 meters northeast of the spacecraft showing a large block about 5 meters across (block G) and about 2 meters high. The fillet surrounding the block is about 20 meters in diameter and about 1 meter high at the contact with the block (Jan. 10, 1968, 06:54:33 and 06:58:12 GMT; Jan. 11, 1968, 12:55:16 GMT).

FIGURE 3-44.—Angular blocks, up to 2 meters in diameter, which form part of a strewn field of blocks that surround a sharp-rim crater 13 meters in diameter, near Surveyor III (Apr. 21, 1967, 05:59:43 GMT; computer processed).

The coarsest blocks scattered about the surface of the Surveyor III site occur primarily in two distinct strewn fields. One field (area B, fig. 3-46) is associated with a sharp, raised-rim crater about 13 meters across on the northeast rim of the main crater in which Surveyor III landed; the other field (area A, fig. 3-47) is associated with two adjacent subdued craters high on the southwest wall of the main crater.

FIGURE 3-45.—Part of a wide-angle Surveyor III picture, showing rounded fragment 20 cm across lying on top of the lunar surface (Apr. 26, 1967, 09:07:06 GMT).

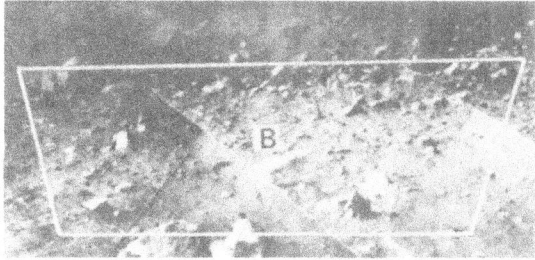

FIGURE 3-46.—Mosaic of narrow-angle Surveyor III pictures, showing a crater 13 meters across and associated strewn field of blocks on the northeast wall of the main crater, in which the spacecraft is located. The outline shows the boundary of area B, in which roundness factor and burial factor of blocks were measured (Catalog 88-SI).

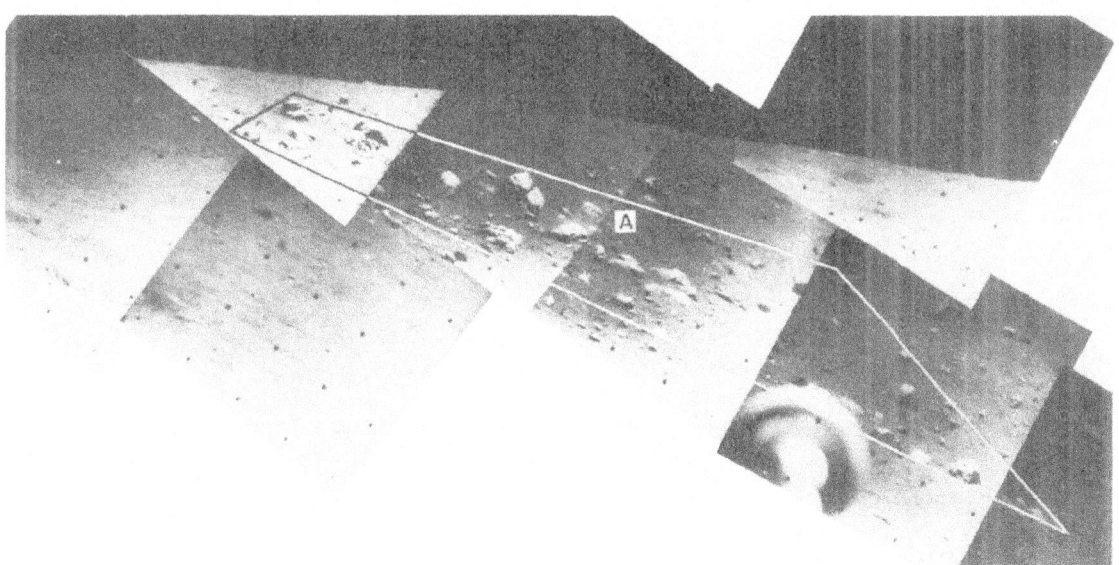

FIGURE 3-47.—Mosaic of narrow-angle Surveyor III pictures, showing part of southwest wall of main crater and strewn field of blocks. Outline shows boundary of area A, in which roundness factor and burial factor of blocks were measured. Two subdued rim craters with which the blocks are associated are present, but difficult to discern in these pictures because of high Sun illumination (Catalog 87-SI).

Most of the blocks in the strewn field associated with the crater to the northeast of the Surveyor III spacecraft (area B, fig. 3-46) are clearly related to the crater because there is a rapid increase in spatial density of blocks toward the crater. The crater is also occupied by blocks. The blocks outside are inferred to have been ejected from this crater and to have been derived from material that underlies the surface at depths of only 2 or 3 meters. The observed blocks are strikingly angular and range from a few centimeters (the limit of resolution at this distance from the camera) to more than 2 meters across. Blocks associated with the two subdued craters to the southwest (area A, fig. 3-47) show a similar range in size but are more rounded. The larger of these two craters is about 15 meters in diameter; it is inferred that most of the blocks were ejected from the larger crater.

To obtain a measure of roundness that could be used for statistical studies, a descriptive parameter that may be obtained from pictures, here called the roundness factor, was devised as follows. Circles are fitted to all the corners or curved parts of the outline of each block silhouetted against the more distant lunar scene (fig. 3-48). The geometric mean of the radii of these circles is then divided by the radius of the circle that just encloses the outline of the block. This ratio is the roundness factor and, for blocks that are not deeply buried in the surface, it will vary between the limits of 0 and 1. For very round fragments whose tops are just exposed above the surface, it is possible to obtain values for the roundness factor larger than 1, although no values this high were observed for the blocks measured in the strewn fields.

The roundness factor was measured for 25 blocks located within a confined area in each strewn field (fig. 3-49). Blocks associated with the sharply formed crater to the northeast (area B) exhibit a mean roundness of 0.17 with

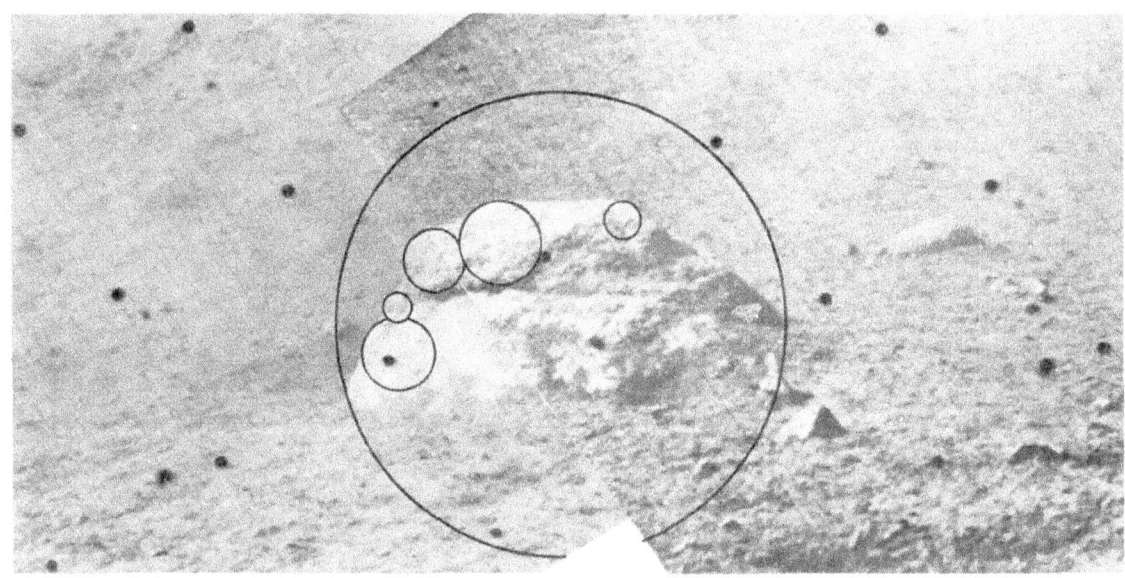

FIGURE 3-48.—Mosaic of two narrow-angle Surveyor III pictures, showing block about 0.5 meter across close to the spacecraft and position and size of circles used in measuring roundness factor. The largest circle encompasses the entire block. Smaller circles are fitted to corners and rounded parts of the outline of block that occults the distant lunar scene. The geometric mean of the radii of the small circles divided by the radius of the large circle is defined as the roundness factor (Apr. 30, 1967, 14:54:23 and 14:52:22 GMT).

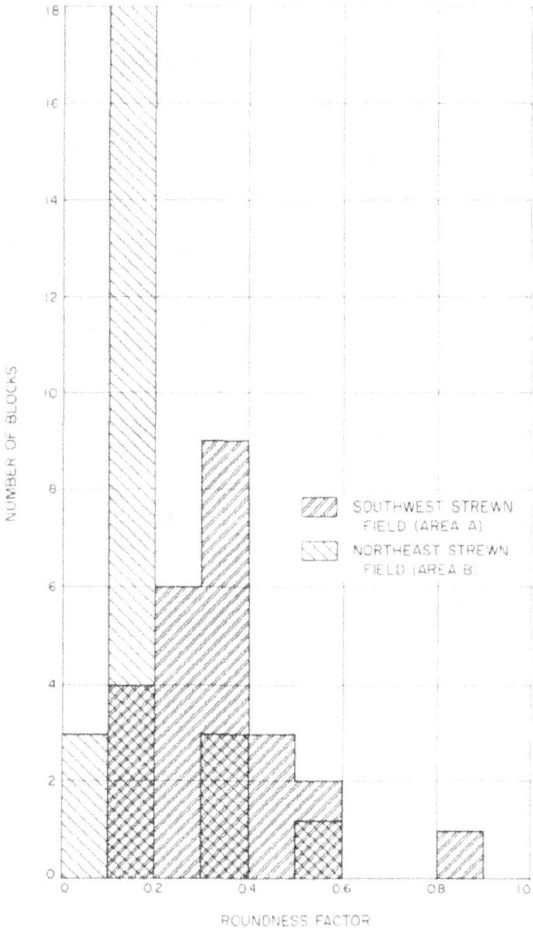

FIGURE 3-49.—Histograms showing frequency distribution of roundness factors for 25 blocks in area A and 25 blocks in area B. Blocks in area A, associated with subdued rim craters, are significantly more rounded than blocks in area B, associated with a sharp-rim crater.

a standard deviation of roundness of 0.11. The blocks associated with the more subdued, rounded-rim crater to the southwest exhibit a mean roundness of 0.33 and a standard deviation of roundness of 0.17. The difference in roundness between these two samples of blocks is significant by Student's t-test at the 0.999-probability level.

A measurement of degree of burial of blocks in the lunar surface was obtained by the following method. The angle between a line parallel to the horizon that meets the block where its outline against the more distant lunar scene intersects the surface, and the tangent to the outline of the block at this point was measured on each side of each block (fig. 3-50). The sum of these two angles for each block, divided by 2π radians, is here defined as the burial factor; values of this parameter can vary between 0 and 1. Rounded fragments whose tops just barely show above the surface have burial factors that approach 1, whereas rocks that sit on the surface and exhibit overhanging sides have burial factors that approach 0.

Measurement (fig. 3-51) of the burial factor for the same 25 blocks in each strewn field that were studied for roundness gave the following results: The mean burial factor of blocks associated with the sharp-rim crater to the northeast (area B) is 0.62 with a standard deviation of burial factor of 0.09. The blocks associated with the more subdued, rounded-rim crater to the southwest (area A) have a mean burial factor of 0.69 with a standard deviation of burial factor of 0.07. The difference between these means is significant at the 0.995-probability level by Student's t-test.

No significant correlation was found between roundness and burial of individual blocks within each strewn field. The linear correlation coefficient between the roundness factor and burial factor for the blocks in the strewn field around the northwest crater is -0.07; for the blocks in the strewn field associated with the southwest crater, it is -0.16. Both these coefficients are well below the 95-percent confidence level. If the blocks in both the strewn fields are examined as a single sample, the linear correlation coefficient between roundness factor and burial factor is $+0.13$, which is also below the level of significance. Examination of the scatter diagram (fig. 3-52) of burial factor versus roundness factor shows that, although there is no significant linear correlation, relatively few blocks in the strewn fields tend to have both high roundness and a low burial factor.

Although there is no significant linear correlation between roundness and burial for blocks in a given strewn field presumably of one age, there should be a correlation between the roundness and degree of burial for fragments generally mixed together in the debris layer, or

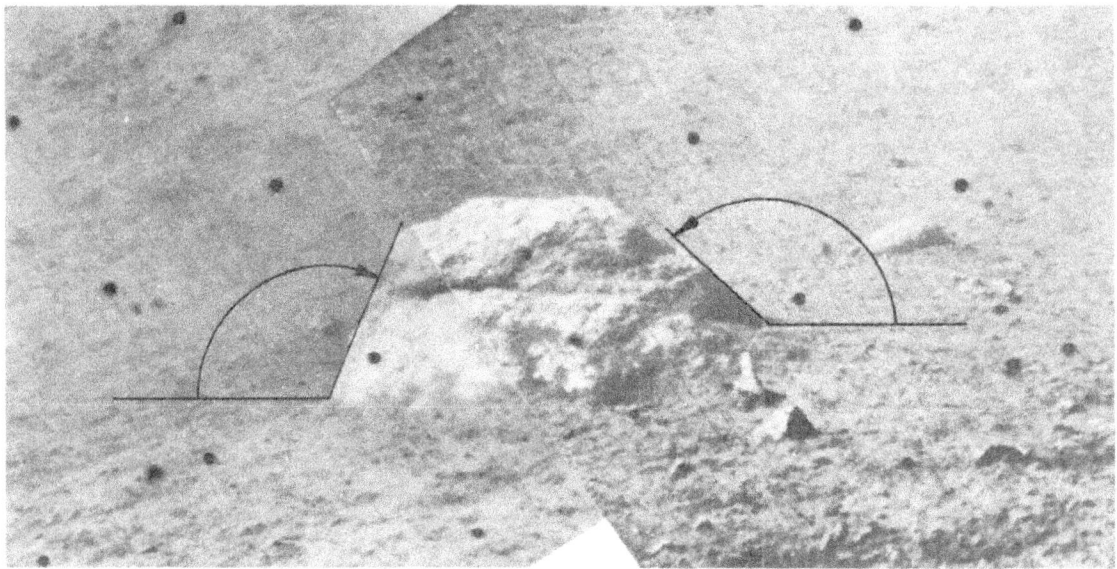

FIGURE 3-50.—Mosaic of two narrow-angle Surveyor III pictures, showing block about 0.5 meter across close to spacecraft and angles measured to determine burial factor. Angles are measured between lines parallel with the horizon and the tangents to the outline of the block, where the outline of the block against the more distant lunar scene meets the surface. The sum of the two angles divided by 2π radians is defined as the burial factor (Apr. 30, 1967, 14:54:23 and 14:52:22 GMT).

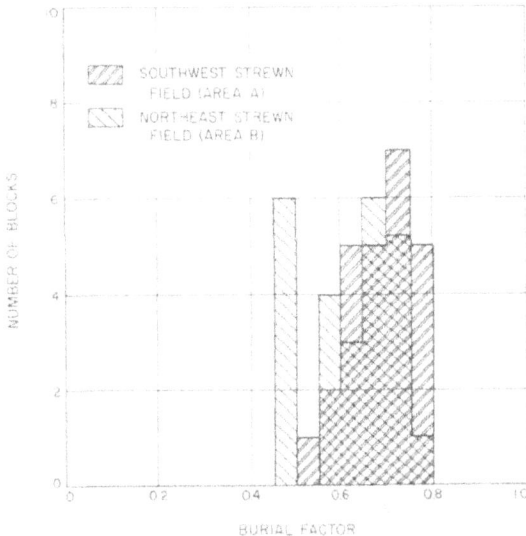

FIGURE 3-51.—Histograms showing frequency distribution of burial factors for 25 blocks in area A and 25 blocks in area B. Blocks in area A, associated with subdued rim craters, are significantly more deeply buried in the surface than the blocks in area B, associated with a sharp-rim crater.

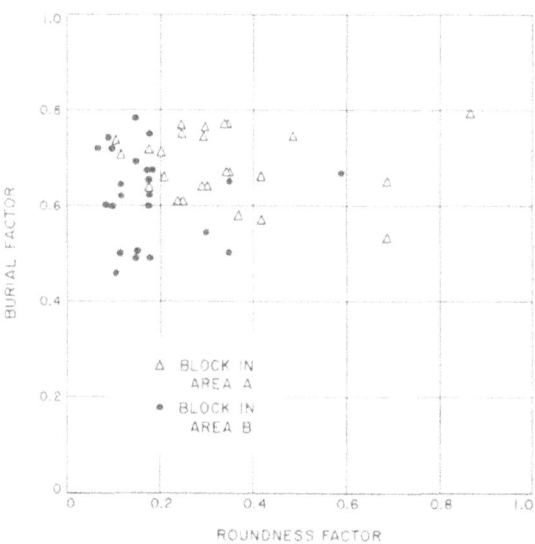

FIGURE 3-52.—Scatter diagram of roundness factor versus burial factor for 50 blocks in areas A and B. The roundness factor and burial factor have no significant linear correlations; relatively few blocks, however, exhibit both high roundness and low burial.

FIGURE 3-53.—Part of narrow-angle Surveyor III picture, showing blocky fragments on north wall of crater in which the spacecraft is located. Some of the largest blocks shown are tabular in shape and appear to be laminated (Apr. 30, 1967, 14:52:35 GMT).

regolith. Further studies will be required to confirm this hypothesis.

In summary, the blocks associated with the more subdued craters have twice as high a mean roundness factor as the blocks associated with the crater with a sharp raised rim, and the blocks around the subdued crater are significantly more buried in the lunar surface than the blocks around the crater with the sharp raised rim. These results suggest that blocks freshly exposed on the lunar surface tend in time to be rounded off by solid particle bombardment and possibly by evaporation of material by the solar wind or other high-energy radiation. Initially, the ejected blocks tend to be shallowly embedded in the lunar surface; but in time they may become partly or completely covered by ejecta arriving from other parts of the surface. Progressive burial of blocks may occur also as a result of downslope movement of the debris layer.

Fragments less than 10 or a few tens of centimeters across exhibit a wide range of shapes, and many are conspicuously angular, especially at the Surveyor I, III, and VII sites. Some of the smaller fragments seem to have been broken along joint planes and tend to have planar surfaces with rectangular outlines, but others are highly irregular in shape. Some fragments exhibit fresh-appearing conchoidal spall or fracture surfaces. Tabular or platy fragments were observed at some of the mare landing sites (fig. 3-53); they resemble rock slabs derived from flow-banded lavas.

The fragments exhibit a wide variety of surface textures and structures, but those at the Surveyor VII landing site exhibit a far greater variety than the fragments observed on any single mare site. Some fragments are plain, but others are spotted. Some fragments appear to be massive, but others exhibit well-developed linear structures on their surfaces, which probably correspond to internal planar or linear structures. Most fragments appear to be relatively dense, but some are clearly vesicular. Most of the fragments probably are pieces of coherent rock, and the variety of observable characteristics suggests a variety of lithology.

Nearly all bright, rounded fragments on the lunar surface have a knobby, pitted surface texture (fig. 3-54). The pitted texture is not present on highly angular, faceted blocks; it is inferred to be produced by some of the processes that produce the rounding. The pits probably

FIGURE 3-54.—Broken block 30 to 40 cm across about 2½ meters from Surveyor VII camera. Knobby, pitted surface texture is common on most rounded blocks on the lunar surface. Note the large cracks on the block. Separation of the pieces of the block may be a result of thermal expansion and contraction or to seismic events (Jan. 15, 1968, 11:51:36 GMT).

are produced by the impact of small particles.

Light spots, which occur on a large number of fragments at the Surveyor VII site, are most easily observed on relatively smooth, clean fracture surfaces (fig. 3-55). In most cases, the spots on a given fragment have irregular, diffuse margins and vary widely in size. In many cases, the light material forms slight bumps, or protrusions, from the surfaces of the fragments; the raised relief of the light material suggests it is more resistant to processes of erosion occurring on the Moon's surface. A large, angular block near Surveyor I was also spotted, or mottled, and the light material formed marked bumps or protrusions (fig. 3-56). The block at the Surveyor I site has a pronounced set of fractures that appear to intersect. These fractures resemble cleavage planes produced during plastic flow of rock under moderately high shock pressure. The block lies in a swarm of similar smaller fragments that are strewn in the direction of the long axis block. It appears that the main piece has broken, perhaps on impact with the surface, and that it has relatively low shear strength. Spotted, or mottled, rocks were also observed at the Surveyor V site.

A densely spotted fragment (fig. 3-57) which lies about 2 meters from the camera at the Surveyor VII site has spots ranging in size from less than 1 millimeter to about 30 millimeters. The spots occupy about 30 percent of the surface of the fragment. The size-frequency distribution of these spots (fig. 3-58) suggests they may be fragments, possibly xenoliths, which were partly assimilated in the dark matrix material. The slope of the integral size-frequency function, however, is somewhat steeper than that expected for most fragmentation processes. A more likely explanation for the light spots is that they represent parts of the fragment that differ from the matrix in crystallinity, or in composition or size of constituent crystals. Somewhat similar spots occur in partially crystallized volcanic rocks and a variety of metamorphic rocks on Earth.

Small, elongate spots, ranging from 1 to 10 millimeters in length, were observed on a conchoidal fracture on one fragment close to Surveyor VII (fig. 3-59). They occupy a few percent of the surface of the fragment, and the long axes of the spots tend to be oriented parallel with one another. Their orientation

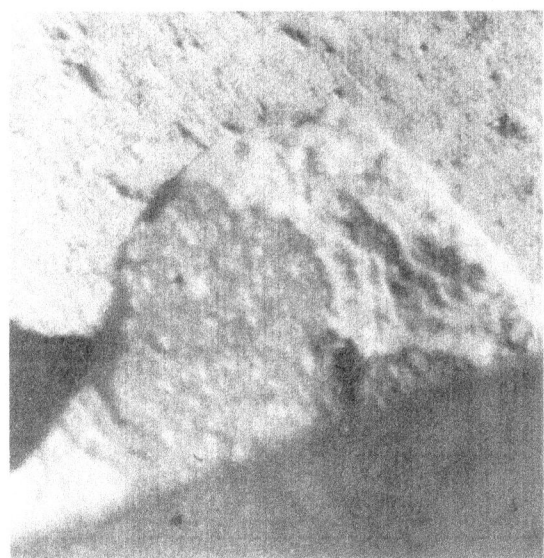

FIGURE 3-55.—Spotted fragment about 1½ meters from Surveyor VII camera. Bright spots have indistinct boundaries and vary from less than 1 mm to about 8 mm across (Jan. 15, 1968, 11:51:05 GMT).

FIGURE 3-57.—Spotted rock 25 cm across, about 2 meters from Surveyor VII camera. The spots range from less than 1 mm to 3 cm in size. Note the indistinct boundaries and irregular shapes of most spots (Jan. 11, 1968, 06:29:29 GMT; computer processed).

FIGURE 3-56.—Mottled rock about 50 cm long near Surveyor I. Lighter areas, which range from a few millimeters up to 3 cm across, appear to stand out as knobs, perhaps as a result of differential erosion. Note intersecting fractures and swarm of smaller fragments to the left of the block (June 3, 1966, 09:33:59 and 09:33:07 GMT).

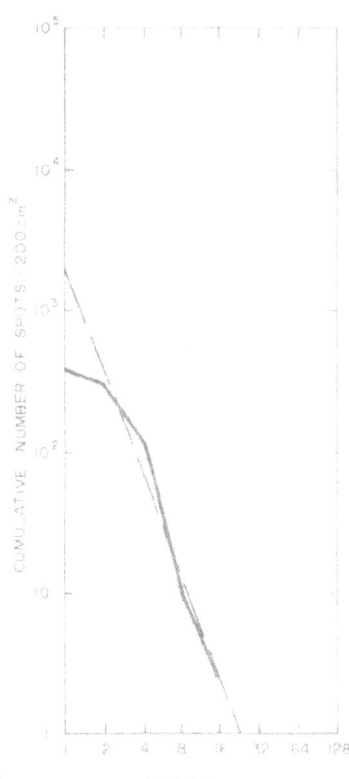

FIGURE 3-58.—Cumulative size-frequency distribution of bright spots on spotted rock shown in figure 3-57. The dashed line represents the mean distribution of the bright spots and is the plot of the function $N = 2 \times 10^3 \, D^{-1.8}$, where N is the cumulative number of spots and D is the diameter of spots in millimeters.

FIGURE 3-59.—Angular block about 18 cm across, about 2½ meters from Surveyor VII camera. Block has a conchoidal fracture surface and bright elongate spots that are roughly parallel and range from 1 to 2 mm wide and up to 10 mm long (Jan. 13, 1968, 10:31:04 GMT).

suggests a flow lineation or flow foliation fabric; their relatively high albedo suggests they may may be rich in feldspar. This suggestion is consistent with chemical analyses of both the fine-grained matrix of the regolith and an individual rock at the Surveyor VII landing site. These analyses indicate the rocks at the Surveyor VII landing site are rich in the elemental constituents of plagioclase feldspar.

Most angular fragments scattered over the lunar surface are conspicuously brighter than the fine-grained matrix of the regolith at nearly all angles of solar illumination. A few angular fragments, on the other hand, are nearly as dark as the fine-grained material of the surface.

These dark, angular fragments appear to be pieces of rock and not aggregates of fine-grained material; one small, dark fragment was attracted to magnets on the surface sampler at the Surveyor VII site. It is possible that most of the dark, angular fragments are rock rich in magnetite or other minerals of high magnetic susceptibility, or that they are mineralogically different in other respects from most of the other rock fragments on the surface.

Some fragments scattered about the Surveyor sites exhibit one or more sets of linear ridges and grooves on their surfaces. Many of the same fragments also have nearly planar surfaces with rectangular outlines (fig. 3-60). The low

FIGURE 3-60.—Angular block about 10 cm across, about 9 meters from Surveyor VII, with two sets of structures on the surface facing the camera. One set consists of ridges parallel with the top edge of the block; the other set consists of ridges and grooves that intersect the first set at an angle of about 70° (Jan. 11, 1968, 23:56:00 GMT).

FIGURE 3-61.—Angular block 10 cm across, about 3½ meters from Surveyor VII camera. Two sets of intersecting structures form a pattern on the surface of the block. One set consists of ridges parallel with the edge of one side of the block; the other set consists of short, deep grooves that intersect the first set at an angle estimated to be about 45° (Jan. 11, 1968, 10:11:26 GMT).

FIGURE 3-62.—Vesicular fragment about 35 cm across, about 7 meters from Surveyor VII camera. Note the slight banding caused by subdued ridges and grooves that extend from the upper right to the lower left of the block. The long axes of the vesicles are oriented at an angle of about 70° to the banding (Jan. 15, 1968, 12:15:11 GMT).

ridges and grooves tend to be parallel with the edges of some of the larger planar surfaces (fig. 3-60). These ridges and grooves probably were developed by differential erosion along the exposed edges of planar structures within the blocks. The planar structures may be flow banding, joints, rhythmic layering, or other primary structures commonly found in igneous rocks. On the surfaces of some fragments, intersecting sets of linear structures are visible (figs. 3-61 and 3-62). The presence of intersecting sets of structures suggests these fragments have been dynamically metamorphosed. One set of structures probably corresponds to an original or primary structure; the other set may correspond to a secondary structure produced by metamorphism such as slaty cleavage.

Many fragments in the vicinity of the Surveyor VII spacecraft and some fragments near

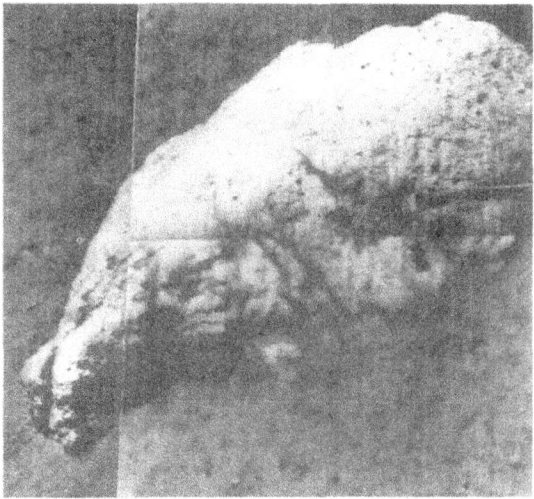

FIGURE 3-63.—Vesicular rock approximately 50 cm long near Surveyor I. Elongate vesicles (dark elliptical spots) range up to 8 mm long and have their major axes roughly parallel.

FIGURE 3-65.—Rounded, vesicular fragment 10 cm across, about 3 meters from Surveyor VII camera. The large vesicle near the bottom is about 1 cm across. The small vesicles are up to 10 mm long, but are only a few millimeters across (Jan. 11, 1968, 23:50:47 GMT).

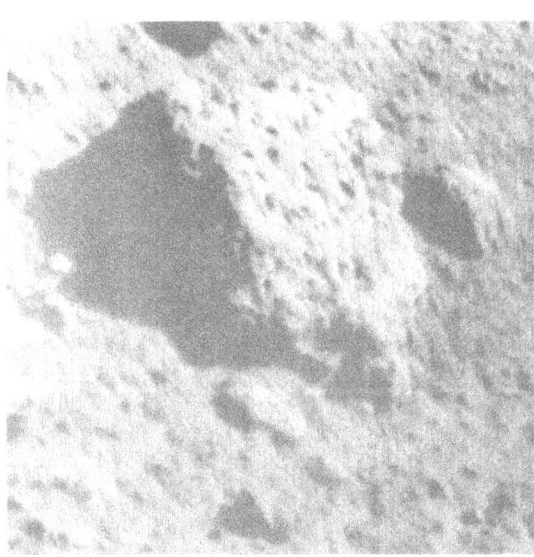

FIGURE 3-64.—Vesicular fragment about 10 cm across, about 2½ meters from Surveyor VII camera. Vesicles are 3 to 5 mm across and up to 10 mm long. Most of the vesicles are elongate with the long axes oriented approximately parallel to one another (Jan. 18, 1968, 13:56:36 GMT).

FIGURE 3-66.—Vesicular block about 50 cm long and 15 cm thick, about 4½ meters from Surveyor VII camera. Vesicles are up to 1 cm across and 2 cm long. Note smooth, undulating, slightly concave surface on the bottom side of the block and the parallel band of elongate vesicles whose long axes intersect the edge of the smooth surface at a large angle (Jan. 11, 1968, 14:04:59 GMT).

Surveyors I and III have deep pits on their surfaces; these pits range from a fraction of a millimeter to a centimeter across. They are almost certainly vesicles produced by exsolution of a volatile phase at the time the material was molten. Examples of vesicular fragments are shown in figures 3-63 to 3-66. The vesicles on these fragments are from 2 to 10 millimeters long, with their long axes oriented parallel or approximately parallel to one another. Parallel orientation of the vesicles is a common feature of the observed vesicular fragments. In some cases, the vesicles are extremely elongate, as shown in a fragment illustrated in figure 3-65. This fragment has fairly large, nearly equidimensional vesicles about a centimeter across and smaller vesicles as much as 1 cm long, but only 1 to 2 millimeters wide.

One of the most interesting fragments observed at the Surveyor VII landing site just south of the spacecraft seems to be a member of a pile of fragments partially obscured by the spacecraft (fig. 3-66). This fragment has two kinds of surfaces: One side is a smooth, undulating, slightly concave surface; the rest of the surface is relatively rough or porous in texture and is partly occupied by vesicles. A row of vesicles parallel with the edges of the smooth surface occurs along the side of the fragment facing the camera. Some of the other vesicles observed on this side of the fragment are elongate and oriented parallel to one another; the orientation of their long axes intersects the row of vesicles and the edge of the smooth surface at a fairly large angle. Thus, in this fragment, there is evidence both of melting and of intersecting structures. Its structure is similar to that of the shock-metamorphosed fragment from Meteor Crater, Arizona, shown in figure 3-67. The smooth, undulatory surface may be a chilled margin, as found on the surfaces of shock-melted ejecta from impact craters.

FIGURE 3-67.—Vesicular, shock-melted Coconino sandstone fragment ejected from Meteor Crater, Arizona. Elongate vesicles have formed along the relict slaty cleavage planes. Note relict bedding that parallels the top of the fragment. Slaty cleavage produced in the sandstone during plastic flow under high shock pressures intersects the relict bedding at an angle of about 80°.

Size-Frequency Distribution of Fragmental Debris

E. M. Shoemaker and E. C. Morris

The size-frequency distribution of the resolvable fragments on the lunar surface was studied at each landing site by choosing sample areas near the spacecraft so that the resolution and area covered would provide particle counts spanning different but overlapping parts of the particle-size range. Studies were made of the parts of the surface undisturbed by the spacecraft at each site. Sample areas were selected that appeared to be representative of the areas surrounding the spacecraft; areas that appeared to have anomalously high or anomalously low particle abundances were avoided. The fragmental material, disturbed by the footpads of the spacecraft during landing also was studied at some sites, and special studies were made at the Surveyor III landing site of the size distribution of the fragmental debris in the strewn fields of blocks surrounding craters with raised rims.

About 1000 to 2800 particles, ranging in size from 1 or 2 millimeters to about 10 cm to 1 meter, were counted on the undisturbed parts of the surface at each landing site. The counts were made from pictures taken at high Sun angles. In these pictures, the low-relief features of the surface, such as small craters and lumps or aggregates of fine particles, are not generally observable; only features of sharp, or abrupt, relief can be detected. Thus, the particles counted were selected on the basis of sharpness of relief. In addition, nearly all the particles counted were significantly brighter than the unresolved fine-grained matrix. Sharpness of relief and brightness are the criteria used to distinguish objects judged to be individual rocky chips of fragments from weakly coherent aggregates of fine-grained particles. The weakly coherent aggregates generally exhibit low relief and are photometrically indistinguishable from the unresolved fine-grained matrix of the regolith. The size distribution of weakly coherent, dark aggregates in the footpad ejecta was studied at the Surveyor I and V landing sites (refs. 3–10 and 3–15), but these data are not discussed here.

Larger rocky fragments on the lunar surface, in general, are easily distinguished and measured. The data for these coarser fragments probably contain relatively little error or investigator bias attributable to difficulties of fragment recognition. Significant investigator bias probably enters into the count of small particles, however, particularly where the size of the particles counted approaches the limit of resolution of the television pictures. Although the effective ground resolution in pictures taken near the spacecraft is about 0.5 millimeter in optimum cases, a practical cutoff in particle counts at each landing site was about 1-millimeter particle diameter. All counts were made by one investigator (Morris); thus, the investigator bias for small particles probably is consistent for the data from each landing site.

Variations exist in the accuracy of the counts at different landing sites. These variations are caused primarily by variation in quality of the pictures and, in part, to the completeness of coverage of the surface at Sun angles appropriate for particle studies. The data for the Surveyor III landing site are the least accurate from the standpoint of both picture quality and picture coverage.

The particle counts from individual sample areas from each landing site show an approxi-

TABLE 3–11. *Constants and exponents of functions of the form $N = KD^\gamma$, $1\ millimeter \leq D \leq K^{-1/\gamma}$, fitted to observed size-frequency distributions of surface particles at the Surveyor landing sites*

Landing site	K, mm$^{-\gamma}$ 100 m^{-2}	γ [a]	Number of particles measured
Surveyor I	5.0×10^5	-2.1_1	2192
Surveyor III	3.3×10^5	-2.5_6	1068
Surveyor V	$1.2_5 \times 10^6$	-2.6_5	2803
Surveyor VI	$1.9_1 \times 10^6$	-2.5_1	1766
Surveyor VII	7.9×10^5	-1.8_2	2077
Total			9906

[a] $N = KD^\gamma$, $1\ \text{mm} \leq D \leq K^{-1/\gamma}$, where N is the cumulative number of particles with diameter equal to, or larger than, D per 100 m^2, and D is the diameter of particles in millimeters.

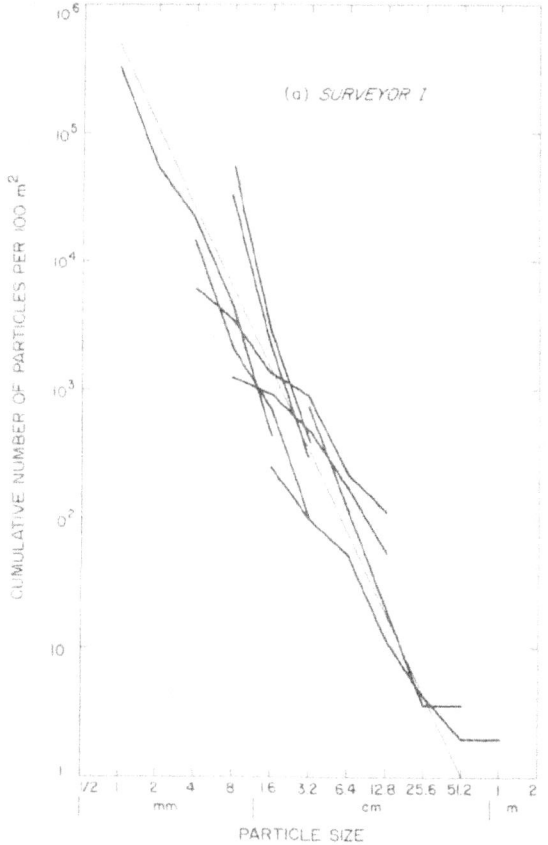
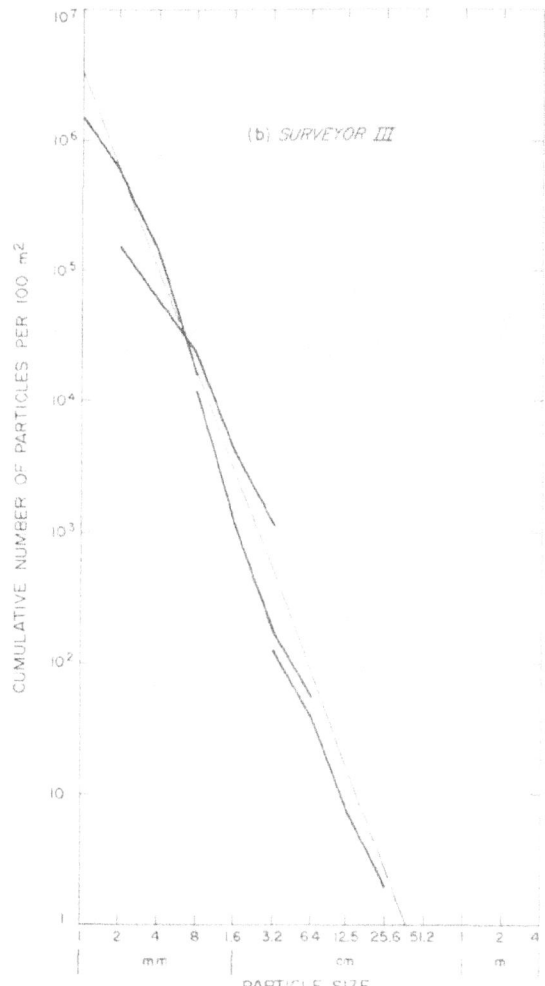

FIGURE 3-68.—Cumulative size-frequency distributions of surface particles at the Surveyor landing sites. Solid lines are distributions observed in individual sample areas. Dashed lines are power functions fitted to the observed size-frequency distributions at each landing site: (a) Surveyor I, (b) Surveyor III, (c) Surveyor V, (d) Surveyor VI, and (e) Surveyor VII.

mately linear relationship, between the log of the cumulative particle counts and the log of the particle size (fig. 3-68). The general trend of the particle counts at each landing site, for all sample areas combined, shows a similar linear relationship. A power function, therefore, was fitted to the data from each landing site to represent the general form of the particle-size distribution. A best fit was made by eye; no attempt was made to fit the observations to a given function by least-squares or other statistical methods, inasmuch as the data are preliminary and do not warrant more sophisticated analysis. Final studies of particle sizes will be based on a much more comprehensive count of the particles observed at each site. Constants and the exponents of functions of the form $N = KD^2$, $1 \text{ mm} \leq D \leq K^{-1/2}$, where N is the cumulative number of particles with diameter equal to or larger than D per 100 m^2, and D is the diameter of particles in millimeters, fitted by eye to the data at each landing site, are shown in table 3-11; the functions are shown in figures 3-69 and 3-70.

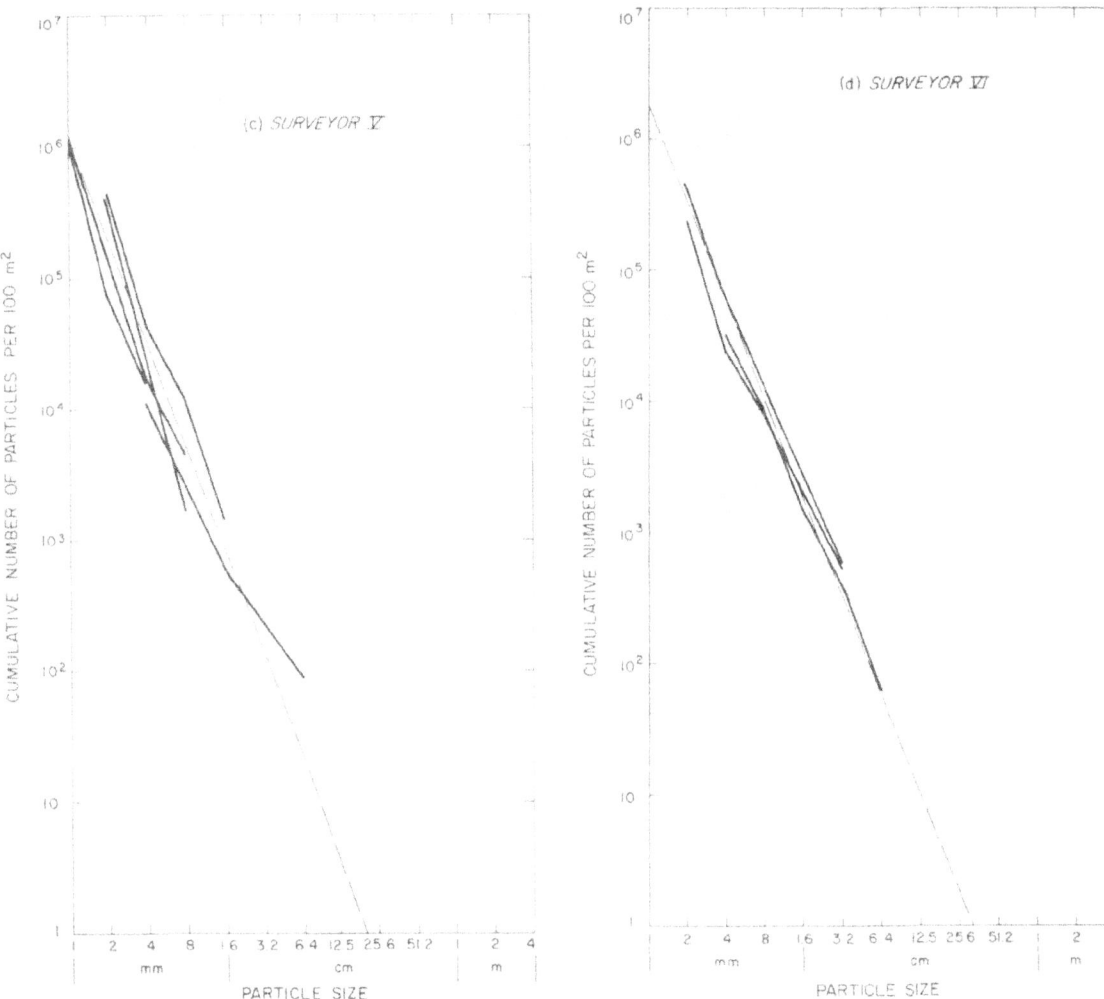

FIGURE 3-68.—Continued. (c) Surveyor V.

FIGURE 3-68.—Continued. (d) Surveyor VI.

As shown in figure 3-69, there is considerable scatter of the data about the fitted functions. For the function fitted to the first set of data acquired for the Surveyor I landing site, $K=3\times10^5$ and $\gamma=-1.7$. These values, published in references 3-9 and 3-22, were based on a count of about 825 particles. A later fit gave $K=5.0\times10^5$ and $\gamma=-2.1$, based on the particle count of 2192. The difference in the two fitted distributions illustrates the possible errors in estimation for the fitted functions. This difference is caused primarily by considerable variation of particle distribution from one sample area to another. Thus, there are fairly large statistical errors in the values of K and γ at each landing site as well as investigator bias. Precision of estimation is improved by increasing the number of particles counted. We estimate that the probable error for the values of K at each landing site (table 3-11) is approximately $\times 2$ and the probable error for the values of γ is about ± 0.2.

If the estimates of precision are correct, it can be seen from table 3-11 that although the precision of estimation is low, some real differences exist in the size-frequency distribution of particles between some of the landing sites. Fairly conspicuous differences in size distributions among the different landing sites are apparent simply by visual inspection of the

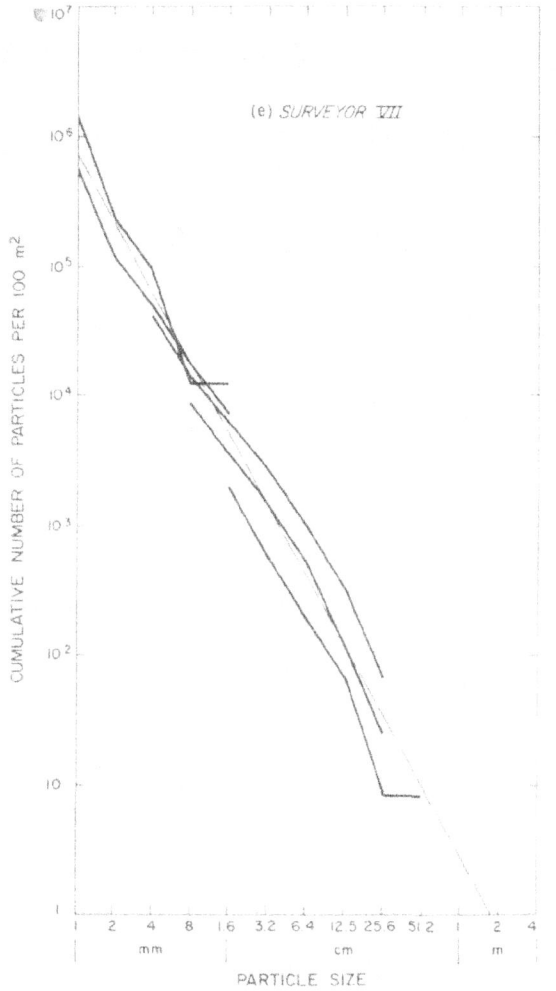

FIGURE 3-68.—Concluded. (e) Surveyor VII.

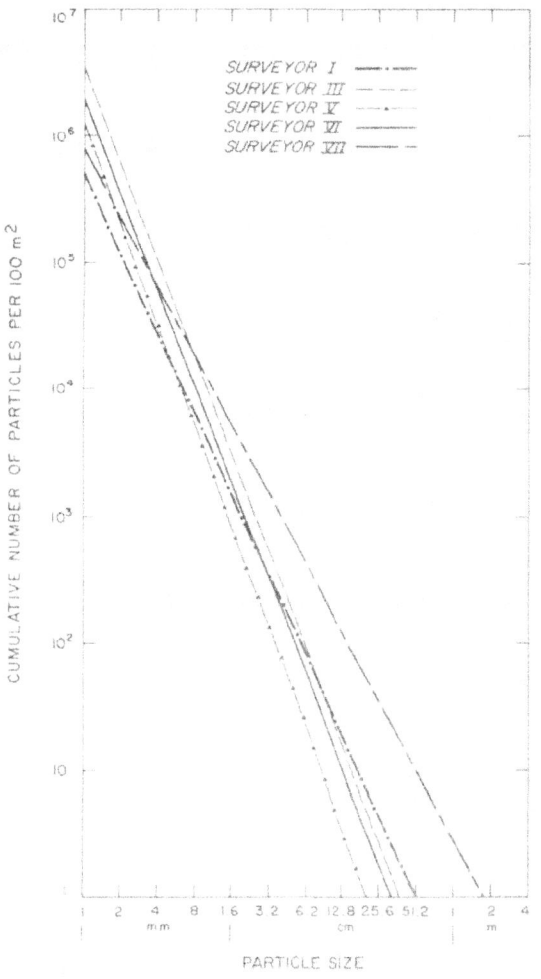

FIGURE 3-69.—Power functions fitted to observed size-frequency distributions of surface particles at the Surveyor landing sites.

pictures. Particle-size distributions at the Surveyor III, V, and VI landing sites are essentially the same within the errors of estimation. The observed distributions at these three sites, however, are significantly different from the distributions observed at the Surveyor I site and at the Surveyor VII site. The distributions observed at the Surveyor I and Surveyor VII sites not only differ from those observed at the other localities but probably also differ from one another. The particle-size distribution at the Surveyor VII site, which is based on a large number of fairly coarse particles, probably is the most accurate of the observed distributions. Fragments 10 cm and larger in diameter occur more frequently by an order of magnitude at the Surveyor VII landing site than at the Surveyor III, V, and VI landing sites. Similarly, there are more coarse fragments at the Surveyor I landing site than at the other mare landing sites, although the fragments are not nearly as abundant at the Surveyor I landing site as at Surveyor VII.

Although the observed particle counts are useful for comparing one landing site with another, the primary interest in these data is their use in estimating particle-size distribution by volume of the lunar regolith. Particle counts on the surface may be used to estimate the particle-size distribution in the regolith, if it is assumed that the particles exposed at the surface are representative of the particles in the subsurface. In this case, the Rosiwal principle (ref. 3-23) may be applied: the ratio of the area occupied by the particles of a given size to the total area studied is equal to the ratio of the volume of the particles of this size to the total volume.

In a rigorous application of Rosiwal analysis to estimation of grain-size distribution, it must be taken into account that the full size of most particles is not revealed along a given surface or cross section. The size of about half the particles on the lunar surface will be underestimated, and a range of sizes will be observed for grains that actually have the same mean diameter. Corrections required for this effect (refs. 3-24 and 3-25) are large if the actual range of particle size is small, but the correction is small if the actual range of grain size is very large. The corrections will be ignored in the analysis presented here.

We have tested in the laboratory the application of the Rosiwal principle to surface particle counts of the type made from the Surveyor pictures by comparing the counts of particles from pictures of the surface of a layer of fragmental debris with volumetric particle analysis of the same layer carried out by standard sieving methods. These tests show that if there is no vertical variation in grain-size distribution within the layer or vertical variation in porosity, the surface particle counts can be used to estimate the volumetric particle-size-frequency distribution with good precision. The assumption that the particle-size distribution of the lunar regolith does not vary vertically has been challenged by Scott and Roberson (ref. 3-26). We believe that there are minor vertical variations in the grain-size distribution, which will cause small errors in the application of the Rosiwal principle, but that there are no major differences between the particles on the surface and the particles in the subsurface. We shall return to this question in more detail at a later point.

Of equal concern in the application of the Rosiwal principle to estimating the volumetric particle distribution of the lunar regolith is vertical variation in porosity. Results obtained from studies of the mechanical properties of the lunar surface (refs. 3-27 to 3-30) suggest that the lunar regolith is more porous in the upper few millimeters than at greater depths. With such a vertical variation in porosity, coarse fragments will tend to be more abundant at the surface in proportion to the fine-grained matrix than they are at depth. A qualitative assessment of the importance of porosity will be made by calculating volumetric particle-size distribution on the basis of different assumed porosities.

To apply the Rosiwal principle, the cumulative areas occupied by the particles, as a function of particle size, must be calculated from the cumulative particle-frequency distribution. The cumulative number of particles per unit area at each landing site has been represented by the function

$$N = KD^\gamma \qquad (1)$$

differentiation of equation (1) gives

$$\frac{dN}{dD} = \gamma K D^{\gamma-1} \qquad (2)$$

The area of individual fragments on the surface can be estimated by

$$a = \frac{\pi}{4} D^2 \qquad (3)$$

where a is the cross-sectional area of individual fragments. If A is the cumulative area of fragments of diameter equal to or larger than D, then

$$\frac{dA}{dD} = a \frac{dN}{dD} \qquad (4)$$

Combining equations (2), (3), and (4),

$$\frac{dA}{dD} = \frac{\gamma \pi K}{4} D^{\gamma+1} \qquad (5)$$

Integration of equation (5) between the limits of the maximum-size fragment, $K^{-1/\gamma}$ and D yields

$$A = \int_{K^{-1/\gamma}}^{D} \frac{\gamma \pi K}{4} D^{\gamma+1} dD = \alpha D^{\gamma+2} - \rho \quad (6)$$

where $\alpha = \gamma \pi K/4(\gamma+2)$ and $\rho = \gamma \pi K^{-2/\gamma}/4(\gamma+2)$.

A choice must now be made of the reference area with which A is to be compared. This choice is a function of porosity. For 0 percent porosity, the reference area is equal to the unit area to which N is referred. In the case of the distributions given in table 3-11 and shown in figures 3-69 and 3-70, the reference area is 100 m² or 10^8 mm². For values of porosity equal to P, only $100(1-P)$ percent of the surface can be occupied by particles. For 50 percent porosity, for example, the reference area is 50 m² or 0.5×10^8 mm².

The volumetric particle-size-frequency distribution, expressed as the percentage of cumulative volume as a function of D, is given by

$$V = \frac{100A}{A_r} = \frac{100(\alpha D^{\gamma+2} - \rho)}{A_r} \quad (7)$$

where V is the percentage of cumulative volume of particles with diameter equal to or larger than D, and A_r is the reference area $= 10^8(1-P)$ mm². Solutions for the constants in equation (7), using the surface particle-size distributions shown in table 3-11, are listed in table 3-12; the volumetric particle-size distributions are shown in figures 3-70, 3-71, and 3-72. Two different values of porosity, 0 and 50 percent, have been adopted for the distributions shown. Observations of the mechanical properties of the upper part of the lunar regolith suggest it has about 50 percent porosity near the surface (refs. 3-27 to 3-30). The regolith, in most places, probably has a significantly lower porosity at depths greater than 5 to 10 cm.

It may be seen from figures 3-70, 3-71, and 3-72 that a relatively small fraction of the total volume of the lunar regolith is composed of particles coarse enough to be resolved by the Surveyor cameras. On the basis of the 0-percent porosity model, between 3.7 and 18 percent of the regolith is composed of particles coarser than 1 millimeter; on the basis of the 50-percent porosity model, 7.3 to 36 percent of the regolith

TABLE 3-12. *Constants and exponents of volumetric particle-size-frequency distribution functions of the lunar regolith based on observed surface particle-size distributions*

Landing site	α, mm$^{-\gamma}$ [a]	ρ, mm³ [a]	$\gamma + 2$ [a]
Surveyor I	$7.5_3 \times 10^6$	$3.8_5 \times 10^6$	-0.11
Surveyor III	$1.1_8 \times 10^7$	$4.3_8 \times 10^6$	$-.56$
Surveyor V	$4.0_1 \times 10^6$	$1.2_7 \times 10^6$	$-.65$
Surveyor VI	$7.3_9 \times 10^6$	$3.8_7 \times 10^6$	$-.51$
Surveyor VII	$-6.2_8 \times 10^6$	$-2.1_6 \times 10^7$	$+.18$

[a] $V = \dfrac{100(\alpha D^{\gamma+2} - \rho)}{A_r}$ $\quad 1 \text{ mm} \geq D \geq K^{-1/\gamma}$

where V is the percentage of cumulative volume of particles with diameter equal to or larger than D, D is the diameter of particles in millimeters, and A_r is the reference area

$= \begin{cases} 10^8 \text{ mm}^2, \text{ for 0 percent porosity model.} \\ 0.5 \times 10^8 \text{ mm}^2, \text{ for 50 percent porosity model.} \end{cases}$

is composed of particles coarser than 1 millimeter. The largest proportion, by volume, of resolvable particles was observed at the Surveyor VII landing site and the smallest at the Surveyor I landing site. A large and conspicuous difference among the fragment-size distribution at the Surveyor VII landing site on the rim of Tycho and the distributions observed at the landing sites on the maria are shown in figures 3-70, 3-71, and 3-72.

It is of interest to inquire whether the observed distribution of resolvable particles can be used to infer or predict the particle-size distribution of the unresolved matrix. The functions given in table 3-12 have been extrapolated to 1 micron or to 100 percent volume in figures 3-70 through 3-72. As shown in these figures, these extrapolations lead to two quite different results, depending on the values of γ observed for the distributions of resolvable surface particles. At the Surveyor III, V, and VI landing sites, the cumulative volume goes to 100 percent in the particle-size range between 5 and 100 microns. At these sites, γ varies between -2.5_1 and -2.6_5. At the Surveyor I site, however, where $\gamma = 2.1_1$, the cumulative volume does not reach 100 percent unless the size-distribution function is extrapolated to particle

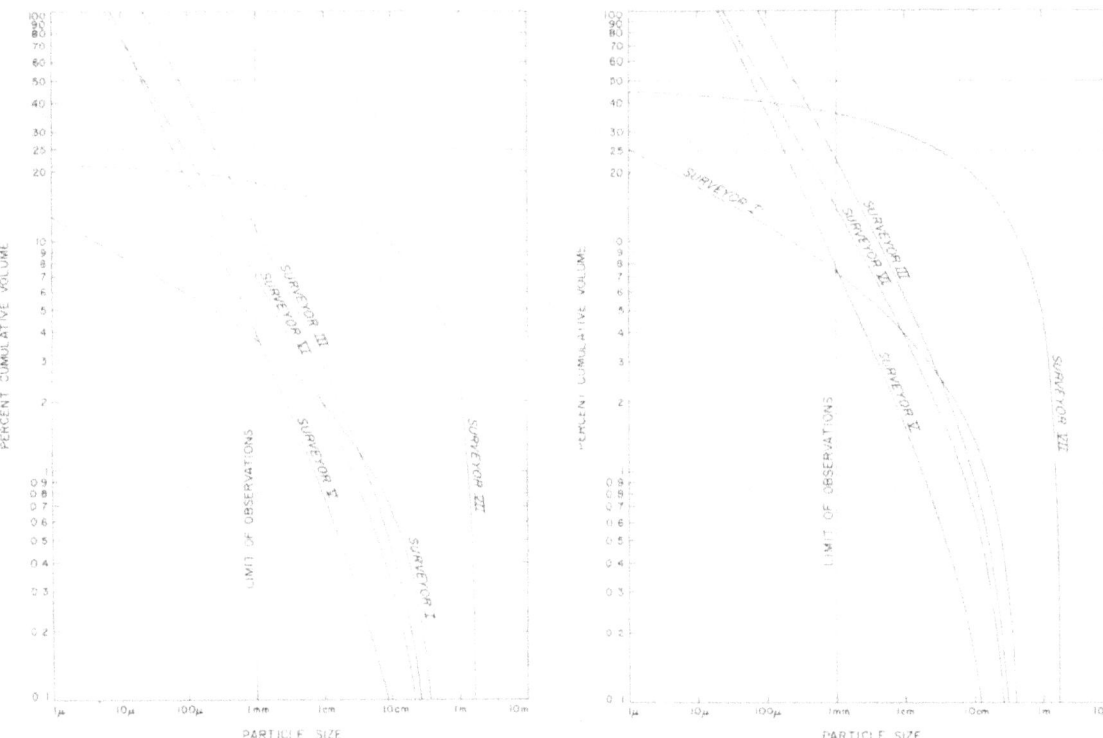

FIGURE 3-70.—Volumetric particle-size-frequency distributions of the lunar regolith for 0 percent porosity, based on power functions fitted to observed size-frequency distributions of surface particles. Dashed lines represent extrapolations below observed particle size of functions listed in table 3-11.

FIGURE 3-71.—Volumetric particle-size-frequency distributions of the lunar regolith for 50 percent porosity, based on power functions fitted to observed size-frequency distributions of surface particles. Dashed lines represent extrapolation below observed particle size of functions listed in table 3-12.

sizes in the subatomic range. At the Surveyor VII site, the maximum volume of particles, based on the extrapolation of the observed size distribution to infinitesimally small particles, is only 24.1 percent of the total volume on the 0-percent porosity model and 48.2 percent on the 50-percent porosity model. Particles smaller than 1 millimeter, therefore, must be represented by a different distribution function from that fitted to the resolvable particles at both the Surveyor I and Surveyor VII sites.

For the three observed particle distributions that give plausible results when extrapolated into the submillimeter particle-size range (Surveyors III, V, and VI), the bulk of the volume consists of particles less than a few hundred microns in diameter (table 3-13). The median particle diameter ranges from 23 to 75 microns on the 0-percent porosity model and from 60 to 255 microns on the 50-percent porosity model. While these results are at least physically plausible, they are not consistent with other indirect lines of evidence about the grain-size distribution of the unresolved matrix. Photometric and polarimetric observations of the lunar surface suggest that most of the particles are finer than 20 microns. Thus, the size distribution of the unresolved matrix at these sites may also be somewhat different from that indicated by the extrapolation of the observed particle-size distribution.

It should not be surprising that the observed particle-size distribution of the resolvable fragments cannot be extrapolated without corrections to very small particle sizes. As shown in fig. 3-69, the large bulk of the lunar

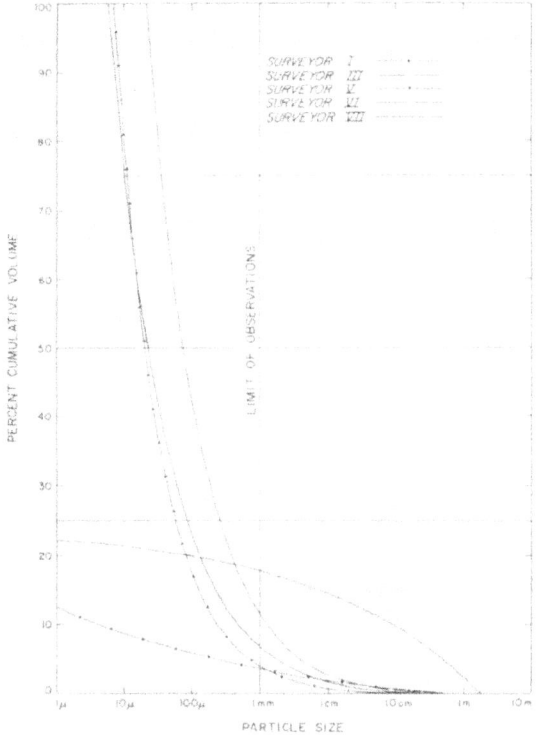

FIGURE 3-72.—Volumetric particle-size-frequency distributions of the lunar regolith for 0 percent porosity. Distributions are the same as shown in figure 3-71; however, V=percentage of cumulative volume is shown with arithmetic instead of logarithmic scale in order to illustrate the forms of the distributions. Below D 1 mm, the distributions are extrapolations of the functions listed in table 3-12.

regolith at all of the Surveyor landing sites consists of submillimeter particles; only a very small part of the distribution of particles, in terms of percent volume, is in the size range of resolvable particles. Small errors in the estimated values of γ for the observed parts of the particle-size distributions have a strong effect on the results obtained from extrapolation of the observed distributions to small particle sizes. By selecting revised values of γ, however, new power functions can be found for the surface particle distributions which correspond to volumetric particle distributions that go to 100 percent volume for any arbitrarily defined lower limiting particle size.

Several lines of evidence suggest that a significant fraction of the particles in the lunar regolith is near 1 micron in diameter, but that only a relatively small fraction of the particles is much finer than 1 micron. The general polarimetric and photometric properties of the lunar surface can be simulated closely in the laboratory with certain irradiated rock powders, if all of the powdered material is ground so as to pass a 20-micron sieve (ref. 3-31). The low degree of polarization of light scattered from the lunar surface indicates that many particles have diameters not much above the limits set by diffraction; that is, about 1 micron (ref. 3-32). Size analyses of fragmental material produced by impact and explosive cratering show that particles as small as 1 micron are

TABLE 3-13. *Volumetric particle-size-frequency distribution of lunar regolith by quartiles, based on functions listed in table 3-12 extrapolated to 100 percent volume*

Landing site	0 percent porosity model, particle diameter, μ				50 percent porosity model, particle diameter, μ			
	At 25 percent volume [a]	At 50 percent volume [a]	At 75 percent volume [a]	At 100 percent volume [a]	At 25 percent volume [a]	At 50 percent volume [a]	At 75 percent volume [a]	At 100 percent volume [a]
Surveyor I	5×10^{-3}	2×10^{-5}	5×10^{-7}	4×10^{-8}	1.1	5×10^{-3}	2×10^{-4}	2×10^{-5}
Surveyor III	255	75	37	22	870	255	125	75
Surveyor V	60	20	11	7	180	60	31	20
Surveyor VI	89	23	11	6	340	89	40	23
Surveyor VII	[b]	[b]	[b]	[b]	3.0 cm	[b]	[b]	[b]

[a] Percent volume is cumulative from largest particle to smallest.
[b] Maximum cumulative volume of particles based on extrapolation of observed size distribution to infinitesimally small particles is 24.1 percent of total volume, for 0 percent porosity model, and 48.2 percent for 50 percent porosity model.

produced.[3] Most rock-forming silicate minerals do not crush readily to particles much smaller than 1 micron, however.

The number of interplanetary particles that strike the Moon's surface, and which are here assumed responsible for most of the fragmentation, increases rapidly with decreasing size, but there is a definite lower limit to their size. Particles with densities such as those of ordinary rock-forming silicates are not stable in the solar system if their diameters are much smaller than about 1 micron. Particles much smaller than 1 micron are ejected from the solar system by radiation pressure (ref. 3-33). Abundant particles in the size range of 1 to 10 microns tend to be produced by impact of small interplanetary particles, but submicron particles probably are produced only in small quantity.

If 1 micron is adopted as a lower limit of size for the particle-size-frequency distributions of the lunar regolith, truncated power functions can be found to represent the particle-size distribution between 1 micron and 1 millimeter that will permit the cumulative volume to go to 100 percent at 1-micron particle diameter at each Surveyor landing site. The total size-frequency distribution of particles at each site can then be represented by two power functions: the function $N = KD^\gamma$, fitted to the observed distribution in the particle-size range 1 mm $\leq D \leq K^{-1/\gamma}$, and a second function $N = KD^{\gamma'}$, in the particle-size range 1 micron $\leq D \leq 1$ millimeter. The value of γ' will be uniquely determined under the conditions:

(1) That the total area of particles between 1 micron and 1 millimeter diameter is equal to $A_r - A_1$, where A_1 is the cumulative area of particles equal to and larger than 1 millimeter (obtained from the observed distribution).

(2) That N is identical for both $N = KD^\gamma$ and $N = KD^{\gamma'}$ at $D = 1$ millimeter, or, in other words, that K is identical for both functions.

By analogy with equation (6), the cumulative area, $A_{D<1}$ of particles of diameter equal to or larger than D, but less than 1 millimeter, can be expressed by

$$A_{D<1} = \int_{D=1}^{D} \frac{\gamma' \pi K D^{\gamma'+1}}{4} dD = \alpha'(D^{\gamma'+2} - 1) \quad (8)$$

[3] E. M. Shoemaker, unpublished data.

where $\alpha' = \gamma' \pi K / 4(\gamma' + 2)$.

The solution for γ' is facilitated by reorganizing equation (8) in the form

$$\gamma' = \frac{\log_{10}\left(\frac{A_{D<1}}{\alpha'} + 1\right)}{\log_{10} D} - 2 \quad (9)$$

Introducing the boundary conditions,

$$A_{D<1} = A_r - A_1 \quad \text{at} \quad D = 10^{-3} \text{ millimeters}$$

equation (9) becomes

$$\gamma' = \frac{\log_{10}\left(\frac{A_r - A_1}{\alpha'} + 1\right)}{-3} - 2 \quad (10)$$

which is solvable for $\gamma' < -2$. The area A_1 at each Surveyor landing site is found by solving for $D = 1$ millimeter in equation (6). If specific values of the reference area A_r are adopted, corresponding to specific values of porosity P, then equation (10) can be solved explicitly for γ' by successive approximation. Solutions for γ' at each landing site, for 0 and 50 percent porosity, are listed in table 3-14.

The derived total size-frequency distribution of surface particles at each Surveyor landing site is shown in figure 3-73 for the 0-percent porosity model of the lunar regolith, and in figure 3-74 for the 50-percent porosity model. At each site and for each model of porosity, the derived frequency distribution comprises the two functions $N = KD^\gamma$, 1 mm $\leq D \leq K^{-1/\gamma}$, listed in table 3-11, and $N = KD^{\gamma'}$, 1 $\mu \leq D \leq 1$ mm, listed in table 3-14.

It may be seen from figures 3-73 and 3-74 that the general trend of the derived particle-size distributions below 1 millimeter is similar to the general trend of the observed distributions above 1 millimeter in both the 0- and 50-percent porosity models. If only mare sites are considered, the mean value of γ' for the 0-percent porosity model is -2.3_9, as compared with a mean value for γ of -2.4_8; for the 50-percent porosity model, the mean value of γ' is -2.2_3. The mean of all the values of γ and γ' for the mare sites is -2.4_4 for the 0-percent porosity model and -2.3_6 for the 50-percent porosity model. Functions $N = KD^\gamma$, fitted by eye to

TABLE 3-14. *Constants and exponents of truncated power functions* $(N=KD)^{\gamma'}$, $1~\mu \leq D \leq 1$ mm) *derived from observed size-frequency distributions of surface particles at the Surveyor landing sites; cumulative volume of particles = 100 percent at* $D=1$ *micron*

Landing site	0 percent porosity model		50 percent porosity model	
	K, $\frac{mm^{-\gamma'}}{100~m^2}$ [a]	γ' [a]	K, $\frac{mm^{-\gamma'}}{100~m^2}$ [a]	γ' [a]
Surveyor I	5.0×10^5	-2.58	5.0×10^5	-2.45
Surveyor III	3.3×10^6	-2.21	3.3×10^6	-2.01_7
Surveyor V	1.25×10^6	-2.42	1.25×10^6	-2.27
Surveyor VI	1.91×10^6	-2.33	1.91×10^6	-2.17
Surveyor VII	7.9×10^5	-2.47	7.9×10^5	-2.29

[a] $N=KD^{\gamma'}$, $1~\mu \leq D \leq 1$ mm, where N is the cumulative number of particles with diameter equal to or larger than D, and D is the diameter of particles in millimeters.

represent the average total particle distributions from 1 micron to $K^{-1/\gamma}$ at the mare sites, have $K=1.1 \times 10^5$, $\gamma = -2.4_2$, for the 0-percent porosity model, and $K=7 \times 10^5$, $\gamma = -2.3_5$, for the 50-percent porosity model. Both of these fitted functions lie generally within the envelope of individual functions $N=KD^\gamma$, 1-mm $\leq D \leq K^{-1/\gamma}$, and $N=KD^{\gamma'}$, $1~\mu \leq D \leq 1$ mm, for the mare sites. In other words, for the mare sites, the mean derived frequency distribution of surface particles smaller than 1 millimeter for both the 0- and 50-percent porosity models is nearly an extension to small sizes of the mean observed distribution of particles larger than 1 millimeter. The fit is closest for the 0-percent porosity model.

In the case of the Surveyor VII landing site, on the rim of Tycho, the derived distributions of particles smaller than 1 millimeter, for both the 0- and 50-percent porosity models, are within the envelope of derived distributions for the mare sites. There is a marked difference, however, between the derived distribution of particles smaller than 1 millimeter at the Surveyor VII site and the distribution of the observed larger particles. The value of γ' is -2.4_7 on the 0-percent porosity model and -2.2_9 for 50 percent porosity, whereas γ is -1.8_2. Thus, the fine particles at the Surveyor VII site probably are similar in distribution to those on the maria but the observed coarse particles follow a conspicuously different distribution.

Total volumetric particle-size-frequency distributions that correspond to the surface particle distributions shown in figures 3-73 and 3-74 can be derived for the lunar regolith. For particle sizes equal to or greater than 1 millimeter, the volumetric distributions are the same as those given in table 3-12 and shown in figures 3-70 and 3-71. For particle sizes smaller than 1 millimeter, the volumetric particle-size-frequency distribution, expressed as the percentage of cumulative volume, is given by

$$V = \frac{100(A_{D<1}+A_1)}{A_r} \quad (11)$$

on the basis of the Rosiwal principle. Combining equations (8) and (11),

$$V = \frac{100[\alpha'(D^{\gamma'+2}-1)+A_1]}{A_r} \quad 1~\mu \leq D \leq 1~\text{mm} \quad (12)$$

Solutions for the constants of equation (12), based on the constants and exponents of the truncated power functions given in table 3-14, are listed in table 3-15. The volumetric particle-size-frequency distributions corresponding to the combined functions listed in tables 3-12 and 3-15 are shown in figure 3-75 for the 0-percent porosity model, and in figure 3-76 for the 50-percent porosity model. Solutions for these functions at the quartiles are listed in table 3-16.

On the basis of the derived volumetric particle-size-frequency distributions shown in figures 3-75 and 3-76, the estimated median

TABLE 3-15. *Constants and exponents of truncated volumetric particle-size-frequency distribution functions derived from observed size-frequency distributions of surface particles at the Surveyor landing sites; cumulative volume of particles = 100 percent at D = 1 micron*

Landing site	0 percent porosity model		50 percent porosity model		A_1, mm² [a]
	α', mm$^{-\gamma'}$ [a]	$\gamma'+2$ [a]	α', mm$^{-\gamma'}$ [a]	$\gamma'+2$ [a]	
Surveyor I	$1.7_5 \times 10^6$	−0.58	$2.1_4 \times 10^6$	−0.45	3.67×10^6
Surveyor III	$2.7_3 \times 10^7$	−.21	$3.0_8 \times 10^8$	−.017	1.14×10^7
Surveyor V	$5.6_9 \times 10^6$	−.42	$8.2_7 \times 10^6$	−.27	3.89×10^6
Surveyor VI	$1.0_7 \times 10^7$	−.33	$1.91_9 \times 10^7$	−.17	7.00×10^6
Surveyor VII	$3.2_6 \times 10^6$	−.47	$4.8_4 \times 10^6$	−.29	1.78×10^7

[a]
$$V = \frac{100[\alpha'(D^{\gamma'+2} - 1 \text{ mm}^{\gamma'+2}) + A_1]}{A_r} \quad 1\,\mu \leq D \leq 1 \text{ mm}$$

where V is the percentage of cumulative volume of particles with diameter equal to or larger than D, D is the diameter of particles in millimeters, A_1 is the cumulative area of surface occupied by particles with $D \geq 1$ mm, and A_r is the reference area

$$= \begin{cases} 10^8 \text{ mm}^2, \text{ for 0 percent porosity} \\ 0.5 \times 10^8 \text{ mm}^2, \text{ for 50 percent porosity} \end{cases}$$

TABLE 3-16. *Volumetric particle-size-frequency distribution of lunar regolith by quartiles, based on observed distributions in the particle-size range above 1 mm and truncated power functions derived from the observed distributions in the size range 1 micron to 1 mm; cumulative volume of particles = 100 percent at D = 1 micron*

Landing site	0 percent porosity model, particle diameter, μ				50 percent porosity model, particle diameter, μ			
	At 25 percent volume [a]	At 50 percent volume [a]	At 75 percent volume [a]	At 100 percent volume [a]	At 25 percent volume [a]	At 50 percent volume [a]	At 75 percent volume [a]	At 100 percent volume [a]
Surveyor I	12	3.3	1.6	1.0	26	4.9	1.8	1.0
Surveyor III	126	15	3.1	1.0	850	88	6.3	1.0
Surveyor V	25	5.2	1.9	1.0	75	9.1	2.5	1.0
Surveyor VI	48	7.6	2.3	1.0	226	21	3.6	1.0
Surveyor VII	83	6.2	2.0	1.0	3.0 cm	45	4.1	1.0

[a] Percent volume is cumulative from largest particle to smallest.

particle diameter of the lunar regolith ranges from 3.3 to 15 microns and averages 7.5 microns for the 0-percent porosity model. For the 50-percent porosity model, the estimated median particle diameter ranges from 4.9 to 88 microns and averages 33 microns. These estimates for median particle diameter may be compared with the conclusion of Christensen et al. (ref. 3-27) that "a significant number of particles are in the silt-sized range (that is, smaller than 0.06 millimeter)," based on comparison of the Surveyor III footpad imprints with simulated footprints in granular materials of various size distributions. At the Surveyor III site, our estimate of the median particle diameter is 15 microns for the 0-percent porosity model, and 88 microns for the 50-percent porosity model. Jaffe et al. (ref. 3-34) found that, on the assumption that a bright band of light observed along the horizon after sunset at the

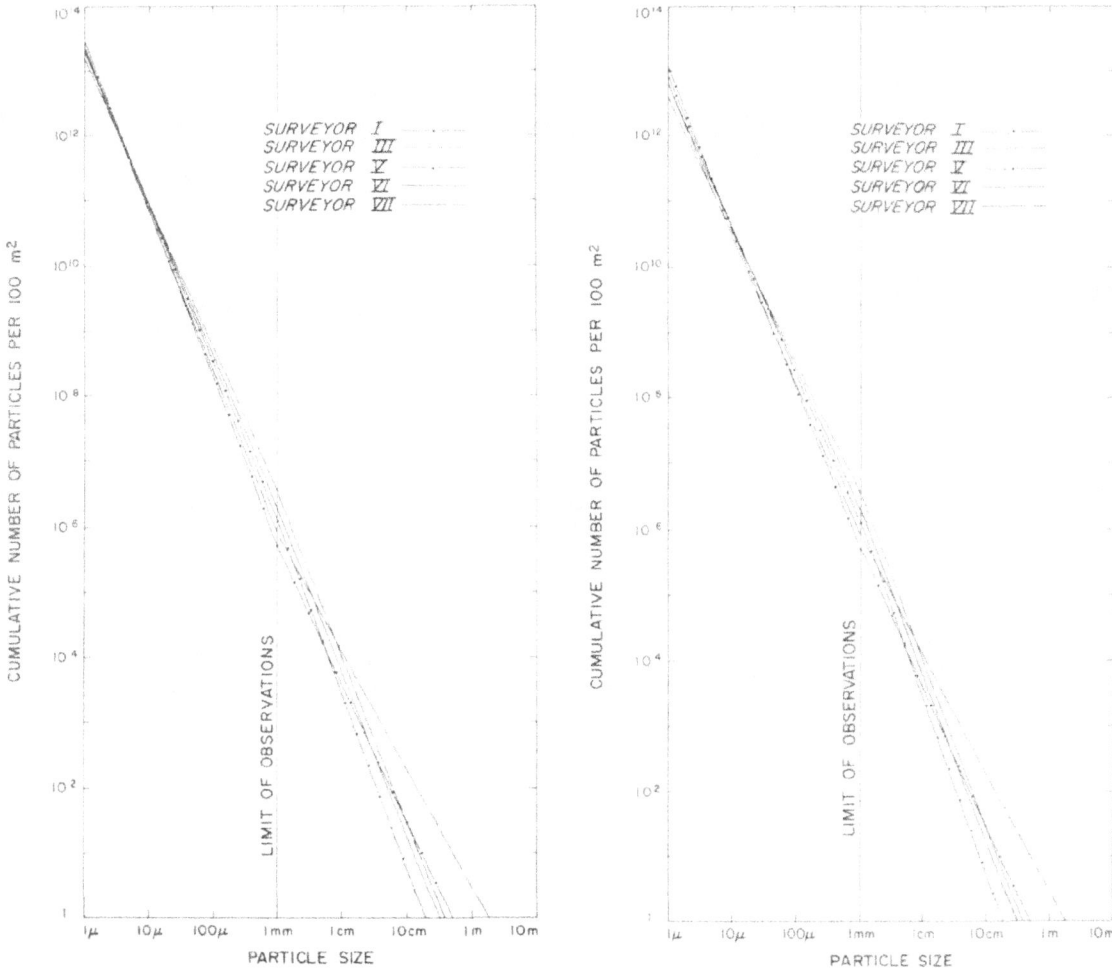

FIGURE 3-73.—Power functions derived from observed size-frequency distributions of surface particles at the Surveyor landing sites for the 0-percent porosity model of the lunar regolith. The derived particle distribution at each site is represented by two functions, $N=KD^\gamma$, where $1 \text{ mm} \leq D \leq K^{1/\gamma}$, and $N=KD^{\gamma'}$, where $1\mu \leq D \leq 1$ mm, which are listed in tables 3-11 and 3-14. Cumulative volume of particles in the regolith at each landing site goes to 100 percent at 1-micron particle diameter for the surface particle distributions illustrated.

FIGURE 3-74.—Power functions derived from observed size-frequency distributions of surface particles at the Surveyor landing sites for the 50-percent porosity model of the lunar regolith. The derived particle distribution at each site is represented by two functions, $n=KD^\gamma$, where $1 \text{ mm} \leq D \leq K^{1/\gamma}$, and $N=KD^{\gamma'}$, where 1 micron $\leq D \leq 1$ mm, which are listed in tables 3-11 and 3-14. Cumulative volume of particles in the regolith at each landing site goes to 100 percent at 1-micron particle diameter for the surface particle distributions illustrated.

Surveyor VI site was caused by diffraction by small particles, the observations indicate a mean particle size less than 10 microns. We estimate the median particle diameter at the Surveyor VI site is 7.6 microns for the 0-percent porosity model and 21 microns for the 50-percent porosity model.

On the basis of a gaseous diffusion-caused eruption crater produced by firing the vernier engines of the Surveyor V spacecraft after

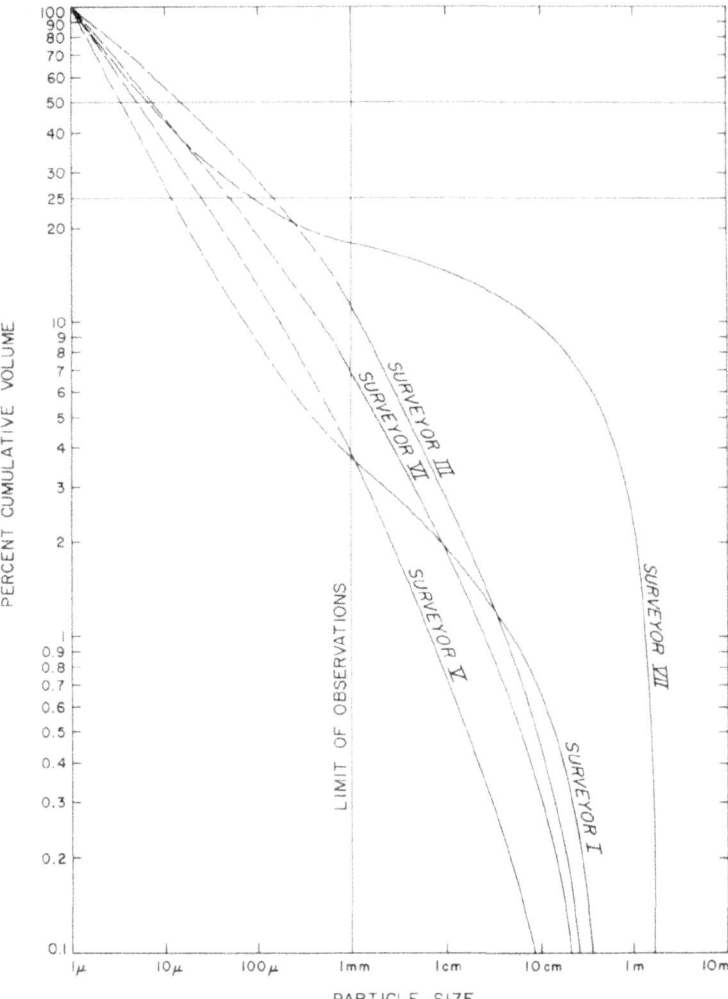

FIGURE 3-75.—Volumetric particle-size-frequency distributions of the lunar regolith for 0 percent porosity, constrained so that cumulative volume goes to 100 percent 1-micron particle diameter. Solid lines, based on power functions fitted to observed size-frequency distribution of surface particles 1 mm and larger in diameter, are listed in table 3-12. Dashed lines, based on power functions for particles smaller than 1 mm in diameter, are listed in table 3-15.

landing, Christensen et al. (ref. 3-28) state that "the estimated lunar permeability indicates most of the particles are in the 2 to 60μ size range." We find in the average case that the central 50 percent of the regolith by volume (the distribution by volume between the first and third quartile, table 3-16) lies between 2.2 and 59 microns for the 0-percent porosity model and between 3.7 and 835 microns for the 50-percent porosity model. At the Surveyor V site, the central 50 percent of the particles by volume lies between 1.9 and 25 microns for the 0-percent porosity model and between 2.5 and 75 microns for the 50-percent porosity model. This agreement between two completely independent estimates of grain-size distribution is better than should be expected, considering the uncertainties and approximations used in both methods of estimation. We conclude that our extrapolations of the observed surface particle-size distribution into the 1-micron to 1-millimeter-size range, as illustrated in figures 3-73 and 3-74, are supported by the studies of the mechanical properties of the lunar regolith.

It is appropriate to return at this point to the question of whether the particles observed on the surface are representative of the subsurface material in the lunar regolith. The general smoothness of the walls of trenches dug by the surface sampler at the Surveyor III and VII landing sites and the smooth forms of the imprints made by Surveyor footpads

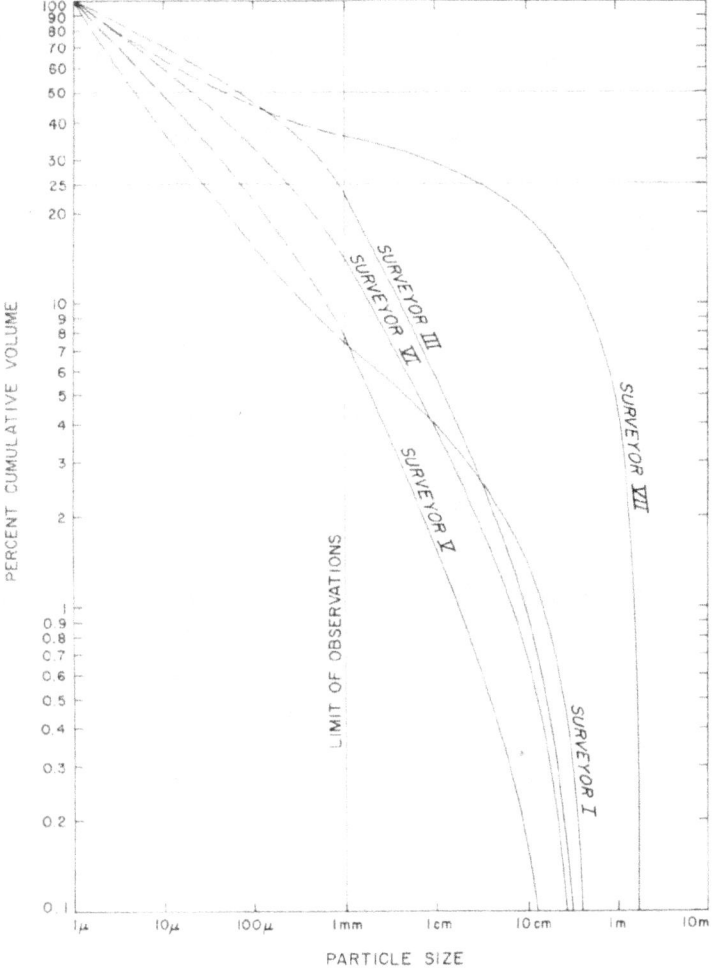

FIGURE 3-76.—Volumetric particle-size-frequency distributions of the lunar regolith for 50 percent porosity, constrained so that cumulative volume goes to 100 percent at 1-micron particle diameter. Solid lines, based on power functions fitted to observed size-frequency distribution of surface particles 1 mm and larger in diameter, are listed in table 3-12. Dashed lines, based on power functions for particles smaller than 1 mm in diameter, are listed in table 3-15.

have given the impression to some observers that coarse particles of the size found on the surface are much less abundant or are absent in the subsurface. Scott and Roberson (ref. 3-26), from their observations at the Surveyor VII landing site, state: "A distinct impression is gained from the surface-sampler work that the surface rocks lie on a relatively fine-grained granular material, and that this material does not contain rock fragments of a size comparable to the fragments on the surface." Gault et al. (ref. 3-35), on the basis of scintillation of striations on the wall of a trench dug by the surface sampler at the Surveyor VII site, concluded that grains as coarse as 200 microns might be present, but that this represented the maximum possible grain size of the material exposed in the walls of the trench.

Our interpretation of the features shown in the wall of the trench described by Gault et al. and shown in figure 3-28 of reference 3-35 is quite different. Inspection of the walls reveals at least two holes from which fragments several millimeters across may have been plucked. In addition, many of the striations in the wall of the trench, attributed by Gault and others to the surface sampler, may have been formed by coarse grains dragged by the surface sampler along the walls. We believe that the walls of the surface-sampler trenches and footpad imprints are relatively smooth because the bulk of the subsurface material is, in fact, very fine grained, as indicated in table 3-16, and it is also somewhat compressible (ref. 3-30). Coarse particles, therefore, tend to be pressed into the fine-grained matrix by the action of the surface sampler and spacecraft footpads and are obscured by the fine-grained material. Simulation of the action of the surface sampler and footpads in a soil model with grain-size distributions like those shown in figures 3-75 and 3-76 would be desirable for further investigation of this problem, but we have not seen convincing evidence that the size distribution of particles in the shallow subsurface is significantly different from that on the surface.

Thickness of the Regolith

E. M. SHOEMAKER AND E. C. MORRIS

The lunar regolith has been defined (ref. 3-36) as a layer of fragmental debris of relatively low cohesion that overlies a more coherent substratum. It covers nearly all parts of the lunar surface observed on the maria by Surveyors I, III, V, and VI; it is inferred to have been derived primarily by a process of repetitive bombardment, which also produced the majority of small craters observed nearly everywhere on the lunar surface. In most places on the maria, the regolith is very fine grained; 90 percent or more of the regolith consists of fragments finer than 1 millimeter.

The thickness of the regolith may be estimated from a variety of observational data. One of the most direct estimates is provided by the observed depths, or the estimated original depths, of craters with blocky rims. These craters have been excavated partly in more coherent or coarser grained material that underlies the regolith; their original depths, therefore, exceed the local thickness of the regolith. The estimated thickness of the regolith at a given Surveyor landing site is bracketed by the original depths of the smallest craters with blocky rims and the original depths of the largest, sharp, raised-rim craters without blocks.

The actual thickness of the regolith probably varies considerably at a given site, and generally only a limited number of craters is observable in the size range of interest. Thus, the estimates of regolith thickness are necessarily approximate. At the Surveyor I landing site, for example, the smallest blocky-rim crater was found to be 9 meters in diameter,[4] whereas the largest crater with a smooth, raised rim was 3.3 meters in diameter. The depth-to-diameter ratio of newly formed craters of this size and type is close to 1:3. The estimated depth of the regolith at the Surveyor I site, therefore, is between 1 and 3 meters.

At the Surveyor III site, the smallest blocky-rim crater observed along the walls of the main crater in which Surveyor III landed is 13 meters in diameter; the largest observed crater with a smooth raised rim is about 3 meters in diameter. The thickness of the regolith along the main crater walls is thus estimated to be between 1 and 4 meters. On the mare plain at the Surveyor VI site, no blocky-rim craters were observable from the Surveyor camera; smooth, raised-rim craters as much as 30 meters in diameter were observed however. Thus, we concluded that on the plain the thickness of the regolith locally exceeds 10 meters (ref. 3-37). On the mare ridge at the Surveyor VI site, the smallest blocky-rim crater observed is 30 meters in diameter and the largest smooth-rim crater is 22 meters in diameter; there the regolith is estimated to be between 8 and 10 meters thick.

Another estimate of the thickness of the regolith may be obtained from the inferred original depth of craters believed to have been formed by drainage of the regolith material into subregolith fissures. Rimless, elongate craters observed in the vicinity of the Surveyor V landing site are believed to have been formed by drainage, and the small crater in which the spacecraft landed appears to be a member of this class. The original depths of the largest craters formed by drainage represent the minimum local thicknesses of the regolith. As the angle of repose on the walls of these craters is about 35°, the original depths of the drainage craters were probably about one-third the

---

[4] Revised estimate based on distance from Surveyor I to this crater observed on Lunar Orbiter III high-resolution photographs.

widths or minor axes of the craters. The minor axis of the Surveyor V crater is 9 meters; the minimum estimated thickness of the regolith is, therefore, 3 meters. The maximum thickness of the regolith derived from one-third the diameter of the smallest blocky-rim crater at the Surveyor V landing site is 5 meters.

A dimple crater with a minor axis of 20 meters was observed in the floor of the main crater at the Surveyor III landing site. If the dimple crater has been formed by drainage, as we believe, it indicates the regolith in the floor of the main crater is about 7 meters or more thick.

At the Surveyor VII landing site, on the rim flank of Tycho, there is an ambiguity both in prior definition and in observational evidence that may be used to interpret the presence, thickness, and characteristics of the regolith. The difficulty arises from the fact that possibly several, and at least one, of the geologic units that make up the rim of the crater are fragmental debris. In the case of the patterned debris, one of the most widespread units on the Tycho rim, the material of this geologic unit also appears to have relatively low cohesion.

We do not intend to apply the term "regolith" to such widespread blankets of fragmental ejecta associated with large, individual craters on the Moon. These units, inferred to be formed by a single event or by a sequence of a small number of events during a well-defined, short interval of time in lunar history, are more appropriately treated as mappable, regional geologic units. They may be expected to have certain internal consistencies of structure and to exhibit systematic lateral variations in grain size, shock metamorphism, and other characteristics.

The regolith, on the other hand, is conceived here as a thin layer of material that forms and progressively evolves over a longer period of time as a result of an extremely large number of individual events, and possibly as the result of interaction of a number of different processes. The regolith is a strictly surficial layer of debris that conceals underlying geologic units in most places on the Moon. Its thickness and other characteristics are a function of total exposure time of the different parts of the lunar surface to a number of surface processes. A new surface freshly formed by a volcanic flow or ash fall or by a deposition of an extensive ejecta blanket around a new crater has no regolith; the process of its development, however, begins almost immediately, and the regolith gradually becomes thicker with the passage of time. In this respect, we consider the regolith as a surficial layer analogous to soils on the Earth.

Where a regolith has developed on a fragmental geologic unit such as a regional ejecta blanket or a debris flow, the practical distinction between the regolith and the underlying fragmental material must be based on differences in grain size and aspects of physical and chemical alteration that can be recognized through the data at hand. In particular, the presence of numerous, small craters; a photometrically observable alteration profile or coatings or alteration rinds on individual fragments; or a grain-size distribution of the surficial material similar to that observed elsewhere on the regolith, can be used as evidence for its presence.

At the Surveyor VII landing site, there is good evidence for the presence of a thin regolith. The most important evidence is the presence of abundant craters smaller than 3 meters in diameter; the size frequency distribution of the small craters corresponds to the steady-state distribution. This suggests that the surface has been repeatedly cratered by very small craters. Material excavated by the surface sampler is, for the most part, very fine grained and is similar in mechanical properties and in general optical properties to the regolith observed on the maria. Furthermore, as on the maria, an abrupt change in albedo was found at shallow depth. We conclude that this material is part of a regolith similar in origin to that observed on the maria. In some places, much more coherent material or much coarser fragmental material was encountered by the surface sampler at depths less than 2.5 cm; elsewhere, fine-grained material of low cohesion was found to extend to a depth of at least 15 cm (ref. 3-26). These depths may represent the approximate range of thickness of the regolith at the Surveyor VII site.

TABLE 3-17. *Apparent thickness of the lunar regolith at the Surveyor landing sites*

Landing site	Diameter of largest crater with smooth, raised rim, m	Diameter of smallest crater with blocky rim, m	Minor axis of largest crater inferred to have been produced by drainage, m	Estimated thickness of regolith, m
Surveyor I	3.3	9		1 to 3
Surveyor III:				
Wall of main crater	3	13		1 to 4
Floor of main crater			20	≥7
Surveyor V		15	9	3 to 5
Surveyor VI:				
Mare ridge	22	30		8 to 10
Mare plain	30	(a)		>10
Surveyor VII	60	3		2 to 15 cm [b]

[a] None observed.
[b] Estimated from the depth of trenches excavated by the surface sampler (ref. 3-26).

The smallest crater with a conspicuously blocky rim observed at the Surveyor VII landing site is about 3 meters in diameter. We believe smaller blocky-rim craters were not observed because there is only a low probability that individual smaller craters intersect very coarse blocks in the fragmental unit that underlies the regolith; blocky crater rims would be expected only where such coarse blocks were encountered (ref. 3-36).

The estimated thickness of the regolith at the five Surveyor landing sites is summarized in table 3-17. The thickest regolith was observed at the Surveyor VI site and the thinnest regolith at the Surveyor VII site.

Disturbances of the Surface

E. M. SHOEMAKER AND E. C. MORRIS

At each of the Surveyor landing sites, wherever the lunar surface was disturbed, dark, fine-grained material was exposed beneath a light surface layer (refs. 3-10, 3-14, 3-15, 3-36, and 3-37). Material ejected by footpad impacts consisted primarily of dark clods or aggregates of fine-grained particles (fig. 3-77). The surface sampler exposed dark material at depths as shallow as a few millimeters.

On the basis of observations at the Surveyor I and III landing sites, the hypothesis was advanced (ref. 3-14) that the subsurface material, exposed by landing of the Surveyor spacecraft and by manipulation of the surface, is dark because the subsurface particles are coated with a dark substance called lunar varnish. Under this hypothesis, the rocky fragments are generally brighter than the fine-grained particles on the surface and conspicuously brighter than the subsurface fine-grained material because they are devoid of varnish. It is supposed that if the varnish at one time had

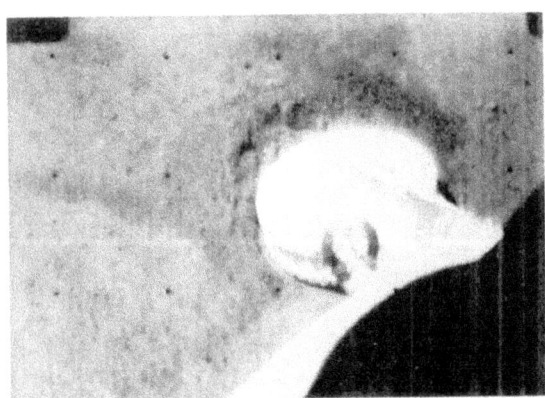

FIGURE 3-77.—Wide-angle picture of footpad 2 of Surveyor I showing raylike deposits of dark subsurface material extending to almost 1 meter from the edge of the footpad. The dark subsurface material was kicked out onto a light surface layer by the footpad during landing of the spacecraft (June 5, 1966, 11:29:49 GMT).

FIGURE 3-78.—Small fragment, about 6 cm across, turned over by surface sampler on Surveyor VII, exposing the dark underside. Part of the dark coating has been scraped away by the surface sampler (Jan. 18, 1968, 06:03:17 GMT).

FIGURE 3-79.—Small, rounded fragment, about 5 cm across, picked up and turned over by Surveyor VII surface sampler. Contact between a dark coating on underside of fragment that has been turned up and the bright top surface that has been turned down can be seen along the front surface of the fragment in shadow, which is partly illuminated by sunlight scattered from the spacecraft. The contact between the bright and dark surfaces can also be seen along the sides of the fragment (Jan. 19, 1968, 06:47:40 GMT).

been deposited on these fragments, it has subsequently been scrubbed off by the same processes of erosion that produce rounding of the fragments. A thin layer of fine particles on the undisturbed parts of the lunar surface is lighter than the subsurface material because these particles also tend to be scrubbed, but the surface layer of fine particles is darker than the exposed surfaces of the rocks because the scrubbing is incomplete, owing to relatively rapid turnover of the particles. Under this hypothesis, the deposition of varnish must take place on the surfaces of fragments at depths as shallow as a few millimeters, or the abrupt decrease of albedo with depth would not persist in the face of repetitive cratering. It may be expected, on the basis of the lunar-varnish hypothesis, that the buried undersides of the coarser fragments are coated with the varnish.

A test of the hypothesis was provided at the Surveyor VII site by the overturning of a number of coarse fragments by the surface sampler. Two of the overturned fragments are shown in figures 3-78 and 3-79; in both cases, the undersides of these objects proved to be dark. On the object shown in figure 3-78, most of the dark material may simply be dark, fine-grained particles adhering to the rock. The coating was partly scraped away by the surface sampler which exposed material of much higher albedo beneath the coating. On the rounded, rocklike object shown in figure 3-79, the coating evidently is very thin; however, it proved to be resistant to abrasion and scraping by the surface sampler. This coating may be the postulated layer of varnish.

Origin of the Lunar Regolith

E. M. SHOEMAKER AND E. C. MORRIS

A simple ballistic model for the origin of the lunar regolith is presented here to account for (1) the size-frequency distribution and variation in distribution of craters on the surface of the regolith, (2) the thickness and variation of thickness of the regolith, and (3) the size-frequency distribution and variation of distribution of the fragmental debris of which the regolith is composed. We recognize that high-energy radiation, mass wasting, and probably other processes have played a role in producing the observed features of the regolith, but we believe the effects of these other processes are subordinate to the effects of solid-particle bombardment of the lunar surface. For simplicity of statement, the subordinate processes will not be considered in this discussion.

The large majority of craters observed on the lunar surface in all size classes is here interpreted to be of impact origin. At present, it is not possible to determine the ratio of primary to secondary impact craters. Some very low-velocity, secondary-impact craters appear to be recognizable in Surveyor pictures from the presence of projectile fragments lining the craters. Higher velocity, secondary-impact craters probably cannot be distinguished from primary-impact craters. On the basis of the probable rate of production of secondary-impact craters during primary cratering events (ref. 3-20), we believe that most craters, in the size range observed in the Surveyor pictures, probably are secondary craters.

Regardless of whether the observed craters are predominantly primary or secondary in origin, the size distribution of small craters appears to be a time invariant or steady-state distribution. Below a certain size limit, the size-frequency distribution of craters is the same at all Surveyor landing sites despite differences in the abundance of larger craters and probable differences in age of the surfaces. This is shown most dramatically in the case of the Surveyor VII landing site on the rim of Tycho, where the observed distribution of craters smaller than 3 meters in diameter is closely similar to the distribution observed on the maria, even though the abundance of craters larger than 100 meters in diameter is more than an order of magnitude smaller on the rim of Tycho than it is on the maria. The upper crater size limit for the steady-state distribution of craters varies as a function of the number of large craters and the age of the surface. It ranges from about 3 meters at the Surveyor VII landing site to about 50 to 200 meters at the mare landing sites.

The range in thickness of the regolith for a surface of a given age can be predicted from the observed size-frequency distribution of the craters on the basis of the following simplified model. Functions $F = \chi c^\lambda$, fitted to the observed size distribution of craters larger than c_s (the steady-state limit), are assumed to represent the size distribution of all the small craters that have actually been formed. At any crater diameter below c_s, the difference between $F = \chi c^\lambda$ and $F = \Phi c^\mu$ represents the number of craters lost by erosion or burial as a result of repetitive cratering and other processes of crater destruction. The diameter of the largest crater that has been lost or has become unrecognizable is c_s. The loss of craters with diameters close to c_s occurs partly by erosion of the rim of the crater and partly by filling of the crater by debris derived from other craters both near and far. In terms of crater depth, most of the change in these craters occurs as a result of filling, and the thickness of the deposit filling an old crater of diameter c_s, which has just disappeared, is given by the original depth of the crater minus the original rim height. This thickness is about one-fourth c_s and represents the maximum thickness of the regolith in a sample area of 10^6 km².

The minimum thickness of the regolith is the depth of the smallest crater, now filled, whose floor forms a local part of the base of the regolith. The approximate size of this crater can be obtained from the solution for the cumulative area of all the filled craters, the floors of which, when connected together, form the entire base or lower contact of the regolith. If

$$F = \chi c^\lambda \quad (13)$$

represents the cumulative frequency distribution of all the craters smaller than c_s that have been formed on a given surface, and

$$a_c = \frac{\pi}{4} c^2 \quad (14)$$

represents the area of craters of diameter c, then

$$\frac{dA_c}{dc} = a_c \frac{dF}{dc} \quad (15)$$

where A_c is the cumulative area of craters

equal to, or larger than, c whose floors constitute part of the base of the regolith and

$$A_c = \int_{c_s}^{c} \frac{\lambda \pi \chi}{4} c^{\lambda+1} \, dc = \frac{\lambda \pi \chi}{4(\lambda+2)} (c^{\lambda+2} - c_s^{\lambda+2}) \quad (16)$$

In addition, if the craters were uniformly spread over the surface so that at $A_c = 10^6$ km² the surface was completely occupied by craters equal in diameter to, or larger than, the corresponding value of c, and if at $A_c = 2 \times 10^6$ km², every place on the surface had been occupied twice by craters equal in diameter to, or larger than, the value of c corresponding to $A_c = 2 \times 10^6$ km², then a critical value $c_{\min}$ would exist such that the floors of craters smaller than $c_{\min}$ could not be part of the base of the regolith, but must lie above the base of the regolith. The depth of the craters with diameter equal to this critical value of c is taken as the minimum thickness of the regolith. For the spatial distribution of craters that would give the least variance in the thickness of the regolith, the critical value of c is reached at $A_c = 2 \times 10^6$ km², or about twice the reference area for the distribution $F = \chi c^\lambda$. In other words, after the surface has been covered about twice over with craters, the floors of smaller craters that may be considered in the integration indicated in equation (16) do not contribute significantly to the base of the regolith. On the basis of equation (16), we may write

$$c_{\min} = \left[c_s^{\lambda+2} + \frac{4(\lambda+2) A_m}{\lambda \pi \chi} \right]^{1/(\lambda+2)} \quad (17)$$

where

$c_{\min}$ = the critical value of c or the diameter of the smallest crater whose floor is part of the regolith

A_m = total area of craters whose connected floors constitute the base of the regolith
$= 2 \times 10^6$ km²

The floors of actual craters are concave upward and those parts of the floors of intersecting craters that form the base of the regolith are each a part of a concave surface. For simplicity, however, let each concave element composing the base of the regolith be approximately represented by a flat surface of a characteristic depth h, where

$$h = qc \quad (18)$$

and q is a constant of proportionality between the diameter of a crater and the thickness of the regolith that subsequently develops where this crater was once formed. The constant q is somewhat smaller than the depth-to-diameter ratio of the original craters and is close to 1/4. On the basis of equations (16) and (18), an approximate frequency distribution for the depth of the regolith may be written in the form

$$H \approx \frac{100 A_c}{A_m} = \frac{100 \lambda \pi \chi (1/q)^{\lambda+2}}{4(\lambda+2) A_m}$$
$$(h^{\lambda+2} - h_s^{\lambda+2}) \qquad h_{\min} \leq h \leq h_s \quad (19)$$

where

H = cumulative percentage of the surface underlain by a regolith of thickness equal to, or greater than, h
$h_{\min} = q c_{\min}$ = minimum thickness of the regolith
$h_s = q c_s$ = maximum thickness of the regolith

The median thickness of the regolith, h_{med}, is obtained from the solution for equation (19) at $H = 50$ percent.

Solutions for $h_{\min}$, h_s, and h_{med} for the Surveyor I, V, VI, and VII landing sites are listed in table 3–18, together with our best estimates of the thickness of the regolith at these four sites. The solutions shown are based on the observed values of χ, λ, and c_s at each of the four sites listed in table 3–10. No data are listed for the Surveyor III landing site because all the observations made from Surveyor III pertain to the inside of a 200-meter crater; the observed thicknesses of the regolith there should be compared with a more complex model than the one we have described.

It may be seen from table 3–18 that our best estimates of the thickness of the regolith, in general, lie within the range of $h_{\min}$ to h_s and are close to h_{med}. At the Surveyor VI landing site, our best estimate of the thickness

TABLE 3-18. *Predicted thicknesses of the lunar regolith, based on eqs. (17) to (19) and the functions $F=\chi c^\lambda$ listed in table 3-10, compared with best estimates of the thicknesses of the lunar regolith*

Landing site	h_{min}, m [a]	h_i, m [a]	h_{med}, m [a]	Best estimate [b] of thickness of the regolith, m
Surveyor I	1.6	14	3.1	1 to 3
Surveyor V	1.2	47	3.2	3 to 5
Surveyor VI	2.7	25	4.6	8 to 10 (mare ridge) >10 (mare plain).
Surveyor VII	.05	.7	.09	0.02 to 0.15

[a] h_{min} = predicted minimum thickness of regolith.
h_a = predicted maximum thickness of regolith.
h_{med} = predicted median thickness of regolith.
[b] Best estimates based on observational data listed in table 3-17.

of the regolith is twice to more than twice as great as h_{med}. Although a difference this large in one out of four cases is consistent with our model of thickness distribution, it is also possible that the thickness of the regolith at the Surveyor VI site is truly anomalous. An anomaly of this type could be caused by the presence on the mare surface of an initial layer of fragmental material, such as a fragmental flow top or a layer of volcanic ash, or to a layer of pyroclastic material deposited at some later time. It is of interest that the order of the Surveyor landing sites by median thickness of the regolith predicted from the observed crater distribution is the same as the order on the basis of the estimates of thickness derived from other observational data. This strengthens our confidence in the approximate validity or predictive value of the proposed model.

It remains now to see whether the size-frequency distribution of the fragments composing the lunar regolith can be accounted for by means of the ballistic model. To examine this question, it is convenient to express the size-frequency distribution in terms of cumulative number of particles per unit volume of the regolith as a function of the mass of the particles. If the cumulative number of particles per unit area is given by $N=KD^\gamma$, as defined above, then

$$N_v = K^{1-(1/\gamma)} D^{\gamma-1} \quad (20)$$

where

N_v = cumulative number of particles equal to or larger than D per $10^8/(2) K^{-1/\gamma}$ mm^3
D = diameter of particles in millimeters

The mass of a particle m can be estimated by

$$m = \frac{\pi \nu}{6} D^3 \quad (21)$$

where ν is the density of the particle. Combining equations (20) and (21)

$$f = N_v = km^\epsilon \quad (22)$$

where

f = cumulative number of particles equal to or greater in mass than m per $10^8/(2) K^{-1/\gamma}$ mm^3

$$k = K^{1-(1/\gamma)} \left(\frac{6}{\pi \nu}\right)^{(\gamma-1)/3}$$

$$\epsilon = \frac{(\gamma-1)}{3}$$

Taking the functions $N=KD^\gamma$, fitted by eye to represent the average total particle distributions from 1μ to $K^{-1/\gamma}$ at the mare sites, $\gamma = -2.4_2$ for the 0-percent porosity model, and $\gamma = -2.3_5$ for the 50-percent porosity model. The corresponding values of ϵ are -1.1_4 for the 0-percent porosity model, and -1.1_2 for the 50-percent porosity model. We wish to see whether these values could have been predicted. For a given value of ϵ, the constant k of equation (22) and K of equation (1) are determined from the assumed condition that the cumulative volume of the regolith goes to 100 percent at $D=1\mu$.

The regolith will be assumed to be composed primarily of the cumulative ejecta from craters with a size-frequency distribution $F=\chi c^\lambda$, $c \leq c_s$. The diameter c of each impact crater is related to the kinetic energy and mass of the projectile which formed it by the scaling relation

$$c = \theta E^\tau = \theta \left(\frac{v^2}{2}\right)^\tau m_p^\tau \qquad (23)$$

where

$E=$ kinetic energy of the projectile
$v=$ velocity of the projectile
$m_p=$ mass of the projectile
$\theta=$ a scaling constant
$\tau=$ an exponent close to 1/3

Assuming v to be constant, equations (13) and (23) may be combined to give

$$f_p = F = k_1 m_p^\eta \qquad (24)$$

where

$f_p=$ cumulative number of projectiles equal to, or greater in mass than, m_p per 10^6 km^2

$k_1 = \chi \theta^\lambda \left(\frac{v^2}{2}\right)^{\lambda\tau}$

$\eta = \lambda\tau$

The mean for the four estimates of λ listed in table 3-10 is -2.94. For $\tau=1/3$, a good empirical value for small craters formed in coherent rock (ref. 3-38), the mean value of η is -0.98. Most of these projectiles are believed to be secondary fragments of the Moon. For primary projectiles striking the Moon, estimates of ϵ', based on meteor photographs and on masses of meteorites recovered on Earth, range from -0.8 to -1.34 (ref. 3-38).

Gault, Shoemaker, and Moore (ref. 3-38) have shown that the cumulative frequency of particles ejected from craters formed in coherent targets by projectiles with the mass frequency distribution $f_p = k_1 m_p^\eta$ can be expressed in the form

$$f_e = k_1 k_2 m_e^\omega \qquad (25)$$

where

$f_e=$ cumulative number of ejected particles equal to, or greater in mass than, m_e per 10^6 km^2
$m_e=$ mass of ejected particle

and

$$\omega = \frac{\eta+1}{\delta} - 1 \qquad (26)$$

where

$\delta=$ the exponent of the function $m_b = k_3 M_e^\delta$
$m_b=$ mass of the largest particle
$M_e=$ total mass of particles ejected from a crater

From a variety of empirical data, δ is found to range from 1.0 for small craters to about 0.8 for large craters. For $\eta = -0.98$, derived from the $F = \chi c^\lambda$ distributions of craters, $\omega = -.098$; for $-1.34 \leq \eta' \leq -0.8$, estimated for primary projectiles, $-1.44 \leq \omega' \leq -0.75$.

In reality, the ejecta are repeatedly reworked by small craters, and the ejecta from craters of diameter less than c_{min} do not add to the volume of the regolith. Thus, the volume of the regolith is only a very small fraction of the cumulative volume of craters $c_l \leq c \leq c_s$, where c_l is of the order of 10 microns or less. To a first approximation, however, the exponent ω should be close to ϵ. The principal effect of repeated cratering is to grind the regolith finer and to increase the proportion of fine to coarse fragments. It may be seen that ω is slightly greater than ϵ ($\omega = -0.98$, $\epsilon = -1.1_4$, -1.1_2) but that $-1.45 \leq \omega' \leq -0.75$ brackets ϵ. Considering the uncertainties introduced by the approximations used, the agreement is perhaps as good as could be expected.

The size-frequency distribution of resolvable fragments at the Surveyor III, V, and VI landing sites was found to be the same, within the errors of estimation, but at the Surveyor I and VII landing sites, coarse fragments are more numerous. At the Surveyor VII site, the spatial frequency of coarse fragments is close to that observed in the fields of coarse blocks around blocky-rim craters. It is significant that coarse fragments are most abundant at the sites with the thinnest regolith. Considering all five sites, a strong inverse correlation exists between the abundance of coarse blocks and the thickness of the regolith. This correlation may be explained in terms of the probability that the Surveyor touchdown points, or the sample areas studied, are close to blocky-rim craters. Where the regolith is thick, only relatively large craters have blocky rims, and these are widely spaced; where the regolith is thin, smaller, more closely spaced craters have blocky rims, and there is a higher probability that the sample areas studied for particle-size distribution will contain coarse blocks from these craters.

Direct evidence suggests that the occurrence of coarse blocks in anomalous abundance is related to the proximity to blocky-rim craters. At the Surveyor I site, for example, a prominent 27-meter-diameter, blocky-rim crater lies about 60 meters from the spacecraft; many coarse blocks observed nearby may have been derived from this crater. The Surveyor VII site is on the rim of Tycho, and the great number of coarse fragments observed there is probably related to the fact that the regolith has been formed on relatively coarse fragmental debris of the Tycho rim. Most fragments larger than 30 cm (diameters greater than twice the thickness of the regolith) are probably part of the fragmental, patterned flow material that directly underlies the regolith at the Surveyor VII site.

Photometry of the Lunar Regolith, as Observed by Surveyor Cameras

H. E. HOLT AND J. J. RENNILSON

Photometric data provided by the Surveyor pictures have been used to (1) determine the photometric function of the surface in the vicinity of the spacecraft, (2) compare macroscopic textures of the surface with the observable photometric function, (3) test for symmetry or degeneracy of the photometric function, and (4) investigate the effect of scale on the photometric properties of the lunar surface. The photometric characteristics of lunar materials observed around each of the Surveyor spacecraft and their variation or pattern of variation from place to place can be used to distinguish different lunar materials.

Preflight calibrations of each Surveyor television camera system response, combined with frequent calibrations on the lunar surface using photometric targets mounted on the spacecraft and photometric control in the ground recording system, were used in the reduction of lunar photometric data acquired from each landed Surveyor.

Undisturbed Material

At each landing site, the photometric properties of the undisturbed, fine-grained material are remarkably similar and closely correspond to the photometric properties of local areas observed telescopically. It was possible to make measurements of the surface luminance at the mare landing sites along the Equator over a much wider range of photometric geometry than could be obtained from the Earth-based observations, which are restricted to east-west phase planes (Earth-Moon-Sun plane). Surveyors I, V, and VI viewed rather flat, but undulating, cratered plains, and as the Sun passed almost directly over the spacecraft, relatively complete goniophotometric measurements were obtained. Surveyor III landed within, and on the east side of, a subdued, 200-meter-diameter crater, which restricted the photometric geometry; a dust-coated camera mirror compromised the quality of the photometric data. Surveyor VII viewed a rolling-to-hilly terrain within the lunar highlands, where the Sun did not rise more than 48° above the northern horizon; this restricted the photometric geometry.

Terrain areas were selected for photometric measurement on the basis of surface flatness, homogeneity, and relative direction from the camera. Photometric data were collected, when possible, along lunar azimuths of 0°, 45°, 135°, 180°, 225°, 270°, and 315° to test the symmetry of the local photometric function. The photometric sample areas ranged from 2 meters to approximately 100 meters from the camera. Uniform-appearing areas near the camera did not contain unresolved fragments larger than a few millimeters, while uniform-appearing areas 75 to 100 meters away could contain fragments up to several centimeters in diameter. Surface flatness was judged by studying pictures taken at low Sun elevations; areas that exhibited little relief under grazing illumination were selected for photometric study.

Photometric data were reduced by both digitizing analog magnetic-tape recordings of the picture data and measuring the film record by a densitometer. Photometric control between pictures was maintained by an eight-step gray scale recorded with each picture. The film densities of the selected areas were first compared to the gray scale of that frame, and the luminances were calculated from the transfer characteristic determined from the picture closest in time of the photometric target on the spacecraft. The luminances from the gray steps of the photometric target were computed from the preflight goniophotometric calibration, the lunar photometric angles existing at that time, and an assumed solar illuminance of 13 000 lumens. Corrections for iris and filter differences were applied to the luminance values computed for the selected areas.

Surveyor photometric data reveal that the luminance of the fine-grained surface increases gradually with decreasing phase angle, and a rapid increase occurs between 10° and 1° phase angles. Among the various sample areas of fine-grained material at any given phase angle, only a small variation in luminance as a result of the change in angle between the surface normal projected into the phase plane and the reflected, or emergent, light rays was observed. The surface luminance gradually increases as this angle increases when the projected surface normal occurs on the Sun side of the reflected light vector (negative α). A rapid decrease in luminance occurs as the angle increases when the projected surface normal appears on the reflected ray side of the phase plane (positive α). These angles are shown in figure 3-80.

The general photometric properties of the uniform, fine-grained material viewed by all Surveyor cameras did not reveal any variations dependent on lunar azimuth. The photometric function of the fine-grained material appears symmetrical and similar to the terrestrially measured lunar photometric function which was based on measurements restricted to east-west phase planes (Earth-Moon-Sun plane). The Surveyor data also indicate that the lunar function is essentially scale independent down to resolutions of 1 cm; topographic irregularities larger than this appear to have little effect on

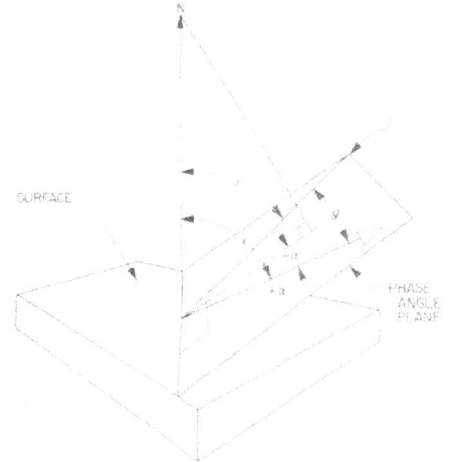

FIGURE 3-80.—Lunar photometric geometry as viewed from Surveyor camera on the lunar surface. Symbols: ⊙, Sun; ⚹, camera line of sight; g, phase angle; i, angle of incidence; N, surface normal; α, luminance longitude (negative toward Sun side of camera line of sight); ϵ, angle of emittance.

the telescopically measured function. The lunar photometric function of N. N. Sytinskaya and V. V. Sharonov (ref. 3-39), compiled and improved by D. Willingham (ref. 3-40), can be used in photometric data analysis of images from Ranger, Lunar Orbiter, and Surveyor with resolutions as small as a few millimeters (fig. 3-81) as well as for telescopic resolutions of one-half km or more. Local concentrations of rock fragments larger than a few centimeters, however, can cause local variations of the photometric function because the rocks have a different photometric function.

Luminance measurements made at phase angles of less than 5° were extrapolated to the peak luminance of 0° phase angle. Homogeneous lunar-surface areas exhibit their maximum luminance at 0° phase angle, regardless of the emittance angle and orientation of the surface element. The peak luminance divided by the solar illuminance is called the normal luminance factor, or normal albedo. Estimates of the normal albedo were made of the undisturbed, fine-grained material at each landing site. These estimates ranged from 7.3 to 8.5 percent at the mare sites; the estimated normal albedo was 13.4 percent at the Tycho site in the lunar highlands (table 3-19).

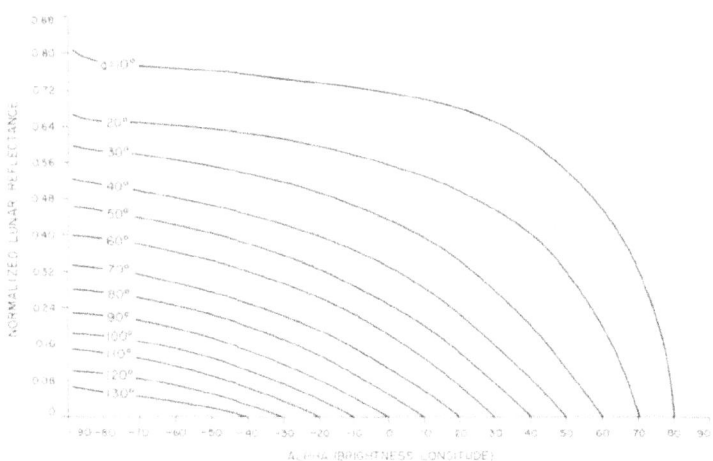

FIGURE 3-81.—Normalized lunar reflectance versus luminance (brightness) longitude based on terrestrial observations by Sytinskaya and Sharonov (ref. 3-39). Surveyor photometric measurements from +20° to −89° phase angles were similar to the Earth-based telescope measurements, which indicates that the lunar photometric functions were essentially similar to scale independent to resolutions of a centimeter.

TABLE 3-19. *Albedo of Surveyor landing sites*

Landing site	Normal albedo (normal luminance factor)		L_U/L_E
	L_U [a]	L_E [b]	
Surveyor I	7.3	5.5	1.33
Surveyor III	8.5	7.6	1.09
Surveyor V	7.9 to 8.4	7.5	1.06
Surveyor VI	8.2	6.1	1.35
Surveyor VII	13.4	9.6	1.39

[a] Normal albedo of undisturbed material.
[b] Normal luminance of footpad ejecta.

Variations in the albedo of the undisturbed, fine-grained surface were not evident from small phase-angle luminance measurements except at the Surveyor V landing site. The near-zero-phase luminances measured from the walls and floor of the Surveyor V crater indicated a nearly uniform normal albedo of 7.9 percent, but the more level surface outside the crater exhibited a normal albedo of 8.4 percent. Slight variations in the luminance of fine-grained material were observed at larger phase angles (60° to 90°) in the middle to distant areas (50 to 200 meters) at the Surveyor I, V, and VII landing sites. Some of these areas occurred near blocky-rim craters from which many small fragments could have been ejected. The variations could be produced by swarms, or strewn fields, of small fragments (1- to 5-cm diameter) that cannot be resolved by the camera.

Disturbed Material

Photometric properties of the fine-grained material were different where the fine-grained material was dislocated or overturned by the spacecraft footpads, surface sampler, or by rolling fragments. The disturbed material is 5 to 28 percent darker than adjacent undisturbed material (table 3-19). Before the Surveyor observations, it was not known that darker material occurred immediately beneath the surface.

The lighter surface layer is extremely thin; rolling fragments smaller than 2 cm across exposed darker material along their tracks, which were at most a few millimeters deep. Vertical exposures along the walls of trenches dug by the surface sampler did not reveal any visible truncated edge of a light surface layer. The lighter surface layer was always destroyed whenever the surface was scraped by the surface sampler. It is concluded that the light surface layer is no more than 1 millimeter in thickness and may be less than ½ millimeter thick.

The measured part of the photometric function of the disturbed material extends from 20° to 90° in phase angle and does not differ appreciably in form from the function of the undisturbed material. Ratios of the luminances of the disturbed and undisturbed material show more accurately the differences in the photometric function of the disturbed material with reference to the more precisely measured function of the undisturbed, fine-grained material (fig. 3-82). The derived photometric function appears to be very similar to that of the undisturbed material. Lack of available photometric data at phase angles from 0° to 20° does not allow determination of the complete nature of this function for undisturbed material, however.

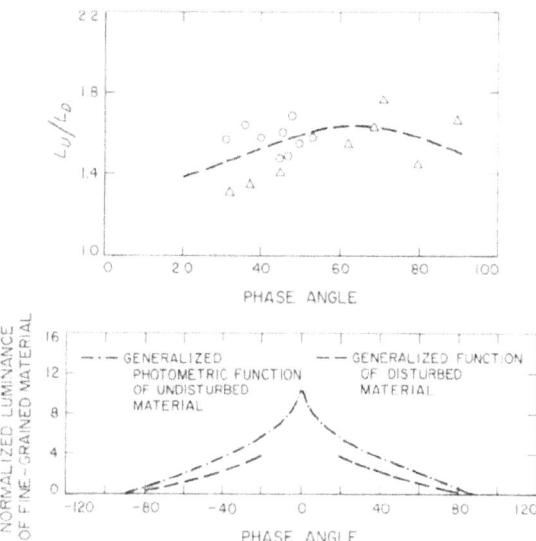

FIGURE 3-82.—Ratios of luminances observed from adjacent undisturbed and disturbed fine-grained, lunar-surface materials are shown on the upper diagram; the dashed line delineates the average ratio. In the lower diagram the reflectance function of the disturbed material is plotted with respect to the function of the undisturbed fine-grained material. The lack of available data from 0° to 20° phase angle does not permit determination of the complete function for the disturbed material.

The occurrence of a very thin, easily destroyed, light surface layer was one of the most surprising observations made with the Surveyor television cameras. The ratio of the normal albedo of undisturbed surface material to the normal albedo of disturbed subsurface material at the various Surveyor sites (table 3-19) may provide a clue to processes affecting the light surface layer. At the Surveyor I, VI, and VII landing sites, about the same ratio was found for nearly flat areas. At the Surveyor III and V landing sites, the ratio of albedos was much smaller; there the measured surfaces slope 14° or more. Pictures from Surveyor III provide evidence that mass wasting or surface creep is actively occurring on lunar slopes (ref. 3-14). The creep movement may be caused by seismic shocks from meteoroid impacts or internal readjustments or by thermal expansion and contraction. This movement would tend to mix the lighter surface material with darker subsurface material and there the ratio would be low as a consequence of the high rate of mixing.

The lowest ratio, or contrast between surface and subsurface materials, was observed at the Surveyor V crater where the steepest slopes were observed. This crater probably was formed by drainage of the fragmental material of the regolith into a subsurface cavity, or fissure, which would cause mixing of material along the crater wall. Continued creep inside the crater may prevent the formation of a light surface layer equal in albedo to that formed outside the crater. The estimates of normal albedo support this concept; the level surface near the crater exhibits an albedo of 8.4 percent; inside the crater, the albedo is 7.9 percent.

Rock Fragments and Blocks

All coarse rock fragments and blocks that protrude above the general level of the surface are brighter at low angles of solar illumination and large phase angles than the fine-grained material. At small phase angles, most fragments and blocks are brighter than the fine-grained material, but the contrast is much less, and a few rocks are darker than the matrix material. The normal albedo of the larger measured fragments ranges from 9 to 22 percent. The photometric function of the measured rocks differs markedly from that of the fine-grained material; it resembles more closely that of a Lambertian scattering surface than the function for fine-grained material.

A few rock fragments at the Surveyor V and VII landing sites showed small, bright areas, which varied in position as a function of illumination angle. These transient bright spots may be caused by specular reflections from crystals or glassy material.

Most rock surfaces are essentially dust free; a coating of dust, even as little as 0.1 millimeter thick, would significantly alter their photometric function. The fact that some rocks exhibit spots and other forms of albedo differences, as well as exhibiting specular reflections, is indicative of nearly clean rock surfaces. Several rocks have irregular surfaces, or pockets, where fine-grained material has accumulated; this is easily recognized by the contrasting photometric properties of the fine-grained material. Also, the polarimetric function of the larger blocks is markedly different from the adjacent fine-grained surface material, another indication of essentially dust-free surfaces.

Measurements of rock luminances by the Surveyor cameras were made under limited geometric conditions. The calculated absolute luminances showed greater variations than measurements for the fine-grained material, partly because of less accurate photometric control, but chiefly because of the effects of diverse orientation of the surfaces. Ratios of the luminances of the rocks to the luminances of the adjacent fine-grained material more accurately illustrate differences between the photometric function of the rocks and that of the fine-grained material (fig. 3–83). The derived function for the rocks appears nearly Lambertian; a small distinct peak may occur on the photometric function around zero-phase angle.

Colorimetric Observations of the Lunar Surface

J. J. RENNILSON

Color filters were incorporated in the Surveyor I, III, and V television cameras in anticipation that color might be an aid in discriminating between lunar-surface materials. Because only a limited number of filters could be used in the television camera, three-color colorimetry was selected as the best method for measuring and describing the colors observed with the camera (ref. 3–41).

The color filters in the camera-filter wheel for Surveyor I were selected so that the overall camera-filter spectral response (fig. 3–84) would duplicate the standard color-matching functions of colorimetry as well as possibly using

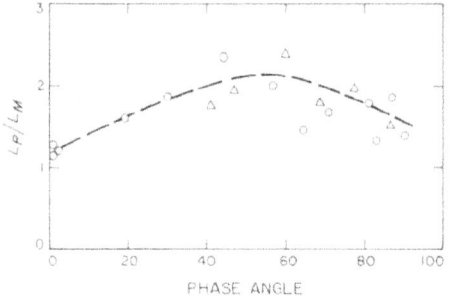
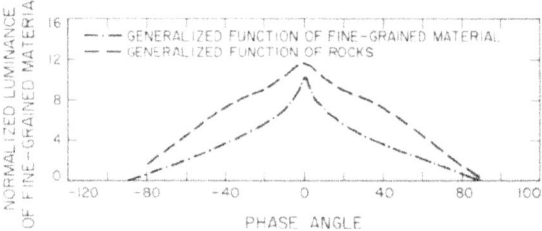

FIGURE 3–83.—Ratios of the luminances from rocks and adjacent fine-grained material are shown in the upper diagram, the dashed line representing the average ratio. In the lower diagram the generalized reflectance function of rocks is plotted with respect to the function of the undisturbed fine-grained surface material. The reflectance curve for rocks appears nearly Lambertian with a small backscatter peak superimposed near the zero-phase angle.

single filters. Figure 3–85 is a graph of the overall camera-filter spectral response showing the fit to the standard Commission Internationale d'Eclairage 1931 (CIE) color-matching functions.

A technique suggested by Davis and Wyszecki (ref. 3–41) was used for selecting color filters to fit the response of the Surveyor III and V television cameras approximately to the color-matching functions. Two filters were used in series in the optical train of the camera. The filter glass components had to be 1 millimeter or more in thickness to withstand the rigors of space flight. Because of weight constraints, the filter pairs were limited to a total thickness of 3.0 millimeters. A special computer program was used for determining the ideal thicknesses and combinations of filters required to fit the Surveyor III and V camera systems' spectral response to the CIE color-matching functions; the fit obtained was fairly good (figs. 3–86 and 3–87). The filters were coated

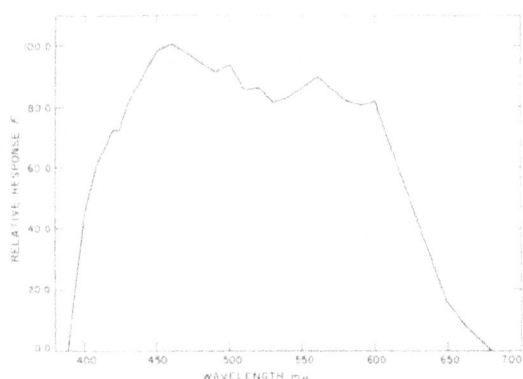

FIGURE 3-84.—Spectral-response curve of the Surveyor I television camera at the clear position of the filter wheel.

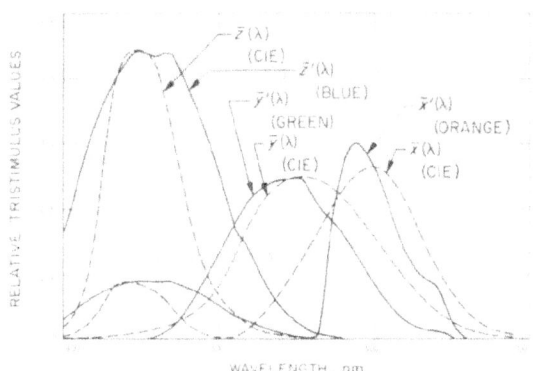

FIGURE 3-85.—Overall camera-filter spectral response of the Surveyor I camera using single filters in the optical path. The solid lines are camera-filter response curves; the dashed lines are the CIE color-matching functions. The second maximum in the $\bar{x}'$ curve is obtained from a reduced value of the $\bar{z}'$ function added to the original camera-filter response.

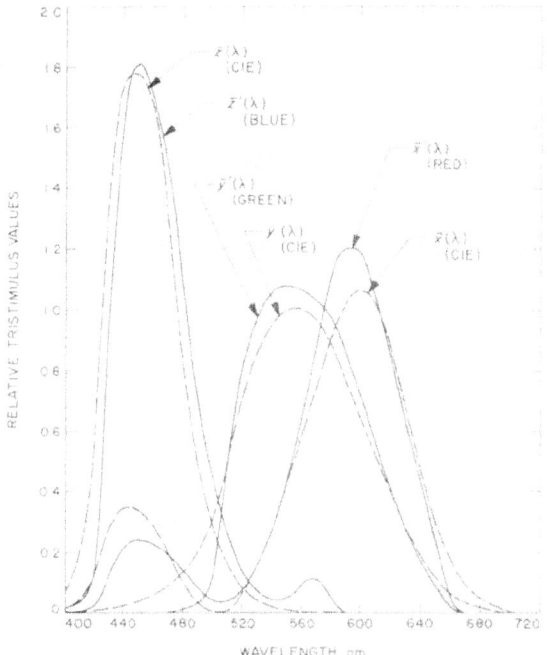

FIGURE 3-86.—Camera-filter spectral response functions of the Surveyor III camera compared with CIE color-matching functions. Camera-filter spectral-response functions are shown with solid lines and the CIE color-matching functions shown with dashed lines.

with a neutral density deposit of Inconel so that without varying the aperture approximately equal video signals would be produced by exposure to a daylight source.

In order to measure color from the television pictures, it is necessary to determine the camera tristimulus values, which are proportional to the video voltage. The proportionality factors may be determined by measuring the video signal when the camera is exposed to object colors of known spectral radiance. Tristimulus values for the Surveyor III and V cameras were determined by observing a 3×3 matrix of filter/source combinations before launch; the proportionality factors were obtained by least-squares solution. The chromaticity coordinates of these nine filter/source combinations were calibrated with a spectroradiometer.

The spectral response of the vidicon tubes used in the Surveyor television cameras is sensitive to temperature. Variation of the spectral response of the Surveyor television cameras with temperature was not calibrated before flight; thus, tristimulus values obtained from measurements of pictures of calibrated color targets, taken on the lunar surface at the operating temperature of the camera, must be used for accurate calculation of color. The color targets were provided as parts of the two photometric targets mounted on the spacecraft. Three colors were present on each target, and they were calibrated before launch with a spectroradiometer while irradiated by a known spectral source at various angles of incidence.

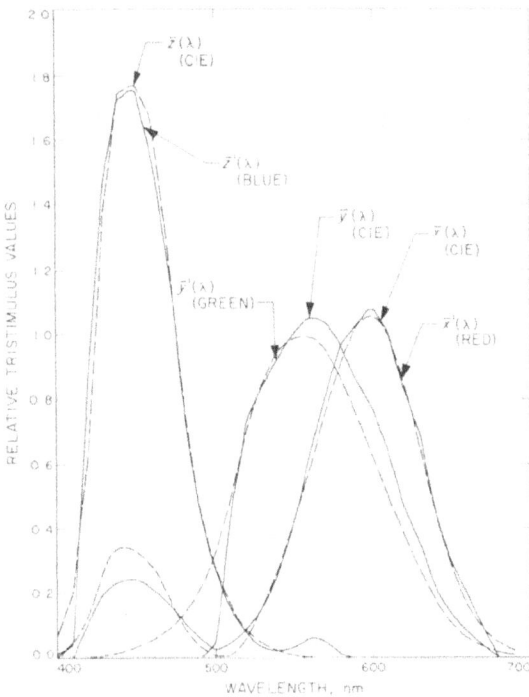

FIGURE 3-87.—Camera-filter spectral-response functions of the Surveyor V camera compared with CIE color-matching functions. Camera-filter spectral-response functions are shown with solid lines and the CIE color-matching functions shown with dashed lines.

Sets of pictures for colorimetric measurement were taken of various parts of the lunar surface at the Surveyor I, III, and V landing sites. Some of the larger blocks and the material disturbed by the surface sampler on Surveyor III were of special interest. Preliminary attempts to look for color differences have been made by preparing color pictures, by color reconstitution methods, using the preflight calibration for control. The preliminary results show the disturbed and undisturbed surface material at the Surveyor I and III sites to be relatively uniform dark gray. However, the preliminary colorimetric reduction of the Surveyor V data indicates the lunar surface at the Surveyor V site is slightly olive gray. At none of the sites were demonstrable differences in color observed on any of the coarse blocks so far examined, which are all gray, but lighter gray than the fine-grained matrix of the surface.

Polarimetric Observations of the Lunar Regolith

H. E. HOLT

Polarizing filters were installed in the Surveyor VI and VII television cameras to serve as analyzers for detection and measurement of the linearly polarized component of light reflected from the lunar surface. Areas around Surveyors VI and VII were selected for polarimetric study according to their geometric orientation relative to the camera and Sun. Sample areas were selected along the east-west Sun line to compare low-resolution telescopic measurements of polarized moonlight, made of areas more than 100 km^2, to high-resolution Surveyor measurements, covering areas from 1 cm^2 to 1 m^2. Other sample areas were selected northwest, north, northeast, southeast, and south of the spacecraft to provide coverage to test for symmetry in the lunar polarization functions. Additional areas containing large rock fragments, disturbed material, and footpad imprints were selected for special examination. After lunar sunset, pictures were taken of the lunar surface illuminated by earthlight to measure the depolarization of earthlight reflected from the lunar surface.

Method of Polarimetric Measurements

Pictures of the lunar terrain were taken with three polarizing filters rotated sequentially in front of the camera lens, while the aperture and other camera conditions were held constant. Variation in the apparent radiance of the same image element contained in the three pictures is caused by a polarized component in the light incident on the filters. The greater the degree of polarization, or percentage of linearly polarized light in the light scattered from the lunar surface, the greater the variation in apparent radiance of image elements in pictures taken through the three filters. Laboratory tests with a slow-scan television camera and three polarizing filters have shown that as little as 5 percent linearly polarized light can be measured with moderate precision and as little as 3 percent can be detected.

The orientation of the polarizing filters remains fixed with respect to the camera mirror and rotates with respect to the picture format.

The Surveyor VI camera was tilted about 16° from the lunar vertical toward the lunar azimuth 60° before the hop, and 13° toward 56° azimuth after the hop; the Surveyor VII camera was tilted about 16° from the lunar vertical toward the lunar azimuth 290°. Pictures taken in the direction of the camera tilt plane will have the polarization axis of filter 2 parallel to the horizon and the axis of filter 4 normal to the horizon. At other camera viewing positions, the axes of filters 2 and 4 are inclined to the left or right of these positions, reaching the maximum inclination of 16° at viewing positions at right angles to the camera tilt plane.

For a first approximate analysis, the degree of polarization was computed by dividing the difference between the luminances observed through filters 2 and 4 by the sums of the luminances. The percentage of polarized light determined by this preliminary method of analysis includes polarization introduced by the camera mirror. Final corrections for the polarization introduced by the mirror will be based on further tests of mirrors of the type used in the Surveyor camera. Preliminary polarimetric calibrations of Surveyor camera mirrors has shown that a linearly polarized component of light will be partially changed to an elliptically polarized component. Both the aluminum oxide mirror surface and the silicon monoxide overcoat are involved in introducing polarimetric errors; the errors do not appear to exceed a few percent.

Polarimetry of Fine-Grained Materials

Light scattered from the lunar surface is partially plane polarized at most phase angles. The degree of polarization of sunlight scattered from fine-grained areas of the lunar surface was found to depend principally on the phase angle. The orientation of the scattering surface has a negligible effect on the degree of polarization, although the plane of polarization of the linearly polarized light, in most cases, tends to be parallel to the scattering surface.

The degree of polarization of light scattered from the fine-grained lunar surface on both the maria (Surveyor VI) and highlands (Surveyor VII) is similar to that observed telescopically. Below phase angles of 35°, the polarization was low to undetectable; an increasing degree of polarization was observed from 35° to 90° phase angles, and as much as 19 percent polarization between 90° to 110° phase angles. The degree of polarization decreases at larger phase angles, declining to 8 to 11 percent at 155° phase angle. Differences in the degree of polarization at a given phase angle in light scattered from surfaces north and south of the camera were within the scatter of the data; there is no evidence for lack of symmetry in the polarization function.

Light scattered from the fine-grained material observed by Surveyor VII in the lunar highlands was also partially plane polarized, but the maximum degree of polarization was less than half that observed on the mare site. A smooth curve through the data points shows less than 4 percent polarization below 60° phase angle, 4 percent polarization at 60° phase angle, and a maximum of 7 percent at 100° phase angle, followed by gradually decreasing polarization at larger phase angles. The degree of polarization did not vary significantly as a function of the lunar azimuth of observation.

The preliminary polarization measurements of light scattered from the undisturbed, fine-grained lunar surface material at the Surveyor VI and VII sites indicate that the polarimetric functions at resolutions of about 1 cm are similar to the telescopically observed polarimetric functions (fig. 3-88) determined for areas of more than 100 km² of lunar surface. The polarimetric function obtained from Surveyor observations and at the telescope is also similar to that obtained in the laboratory for powders of basic and basaltic rocks (fig. 3-89; refs. 3-42 and 3-43). Thus, the polarimetric results are consistent with elemental analyses from the alpha-scattering experiment conducted on Surveyors V, VI, and VII (ref. 3-44).

Polarimetry of Rock Fragments

Rock surfaces observed on the maria and highlands exhibited a greater variation of polarizing properties than the fine-grained material. A few rock surfaces are similar in their polarizing properties to the adjacent fine-grained material, while others produce a maximum polarization of the scattered light; that is, two or more times greater than that produced

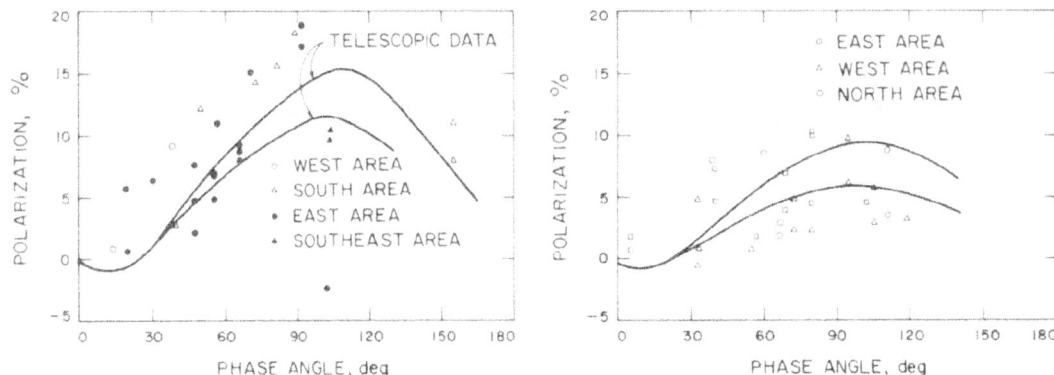

FIGURE 3-88.—Preliminary polarization measurements for undisturbed fine-grained material on the lunar surface near Surveyor VI (maria) at left, and Surveyor VII (highlands) at right, compared with polarimetric functions derived from telescopic measurements. The two curves from telescopic data represent the limits of the range of polarimetric measurements (refs. 3-42 and 3-43).

by the fine-grained surface materials. The degree of polarization of light scattered from some rock surfaces varies mostly as a function of phase angle, and the orientation of the scattering surface has a subordinate effect.

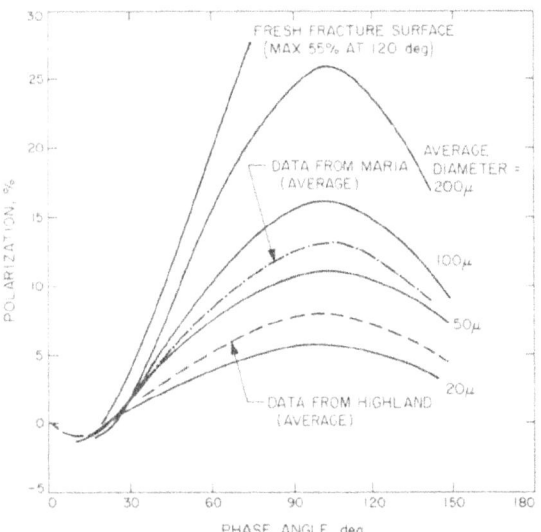

FIGURE 3-89.—Polarization curves for fresh fracture surfaces of olivine basalt and powdered olivine basalt samples (refs. 3-42 and 3-43). For comparison, the average telescopically measured polarization function for fine-grained mare and highland materials are superimposed on the polarization curves for basalt.

The orientation of the plane of polarization of the linearly polarized light component contained in the reflected sunlight tends to be parallel to the viewed rock-surface element.

Only a few rock surfaces were observed from Surveyor VI, mostly to the south and east of the camera, where observations were limited to a small range of phase angles (20° to 95°). The polarization effects of rock surfaces at phase angles less than 60° were undistinguishable from the effects of the fine-grained materials. Between 70° and 95° phase angles, the degree of polarization of light scattered from the rocks increased at a greater rate than the polarization of light scattered from the fine-grained material (fig. 3-90). Light scattered from two rocks exhibited 25 and 26 percent polarization at 95° phase angles. These rocks have an estimated normal albedo (normal luminance factor) that is at least two times higher than the normal albedo of the adjacent fine-grained material. Other rock fragments with lower albedos produce lower degrees of polarization. The brighter rock fragments caused as much as 50 percent more polarization than the fine-grained material.

More rock fragments were viewed from Surveyor VII than from Surveyor VI, and polarization measurements were obtained over a greater range of phase angles (5° to 135°).

The rock surfaces again exhibited a more irregular polarization function than the fine-grained material of the lunar highlands. The light scattered from some rocks showed only slightly more polarization than that of the fine-grained materials; other rocks caused three to four times greater polarization (fig. 3-90). The light scattered from some rocks exhibits a rapid increase in degree of polarization, starting at 6 to 9 percent and going to 25 to 34 percent, over the interval from 70° to 125° phase angles. Maximum degree of polarization caused by individual rocks occurred between phase angles of 114° to 128°. The measurements were obtained only at 8° to 13° phase-angle intervals, however, and the peak degree of polarization from rocks causing strong polarization of light scattered occurs between 120° to 122° phase angles. The rocks producing high-peak polarization are among the brightest rocks observed on the lunar surface and have an estimated normal albedo nearly twice that of the fine-grained material.

Rocks causing less polarization also have a lower albedo, and peak polarization occurs at a phase angle nearer that observed for the fine-grained material. The combination of lower albedo and an intermediate photometric function suggests that some of these rocks may consist of shock-altered rock or shock-lithified, fine-grained material.

Depolarization of Earthlight Reflected From the Lunar Surface

During the lunar day, polarimetric measurements were made of an area on the fine-grained material of the highland surface about 3.1 meters northeast (44° selenographic azimuth) of the Surveyor VII spacecraft. The highland area exhibited an average polarimetric function with about 6½ percent polarization at a phase angle of 95°. Pictures were also taken through the polarizing filters of this fine-grained material about 12 hours after sunset, when the lunar surface was illuminated by earthlight. The phase angle of the polarimetric observations under earthlight was about 92°. The degree of polarization of the incident-integrated earthlight was estimated to be 15±2 percent from a series of pictures of the Earth taken through polarizing filters within a few minutes of the pictures taken of the lunar surface.

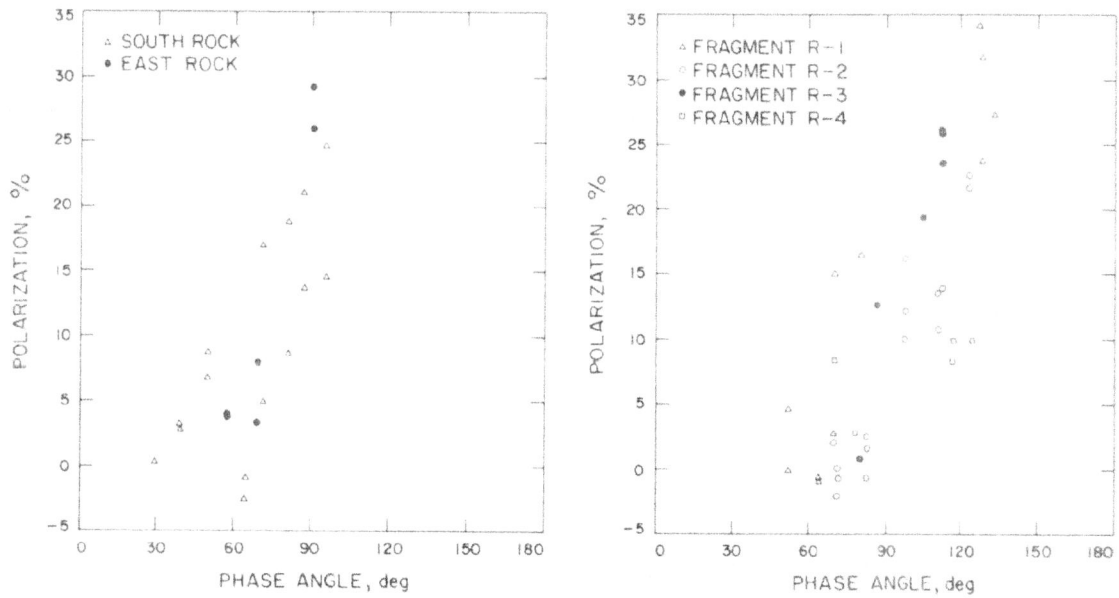

FIGURE 3-90.—Preliminary polarization measurements of light scattered from surfaces of rock fragments near Surveyor VI (maria) left and Surveyor VII (highlands) right.

Preliminary reduction of the polarimetric data indicates that the polarization of earthlight scattered from the lunar surface is about 10 ± 2 percent at a phase angle of 92°. The observed depolarization of the scattered earthlight is probably caused by multiple scattering from the surfaces of grains composing the very porous lunar surface.

Summary of Polarimetric Measurements

Preliminary reduction of the polarization measurements from Surveyor VI and VII pictures indicates that the polarimetric function of the fine-grained part of the regolith reaches a maximum of 7 to 9 percent in the highlands and 16 to 19 percent on the maria. These polarimetric functions, measured for areas of 1 cm² to 1 m², are similar to those measured at the telescope for areas of more than 100 km²; thus, the polarimetric functions are nearly independent of scale. The polarization function of fine-grained lunar material closely matches the polarization function of basalt and gabbro powders (fig. 3-89). The polarization function for the mare sites more closely matches that of basalt powders about 75 microns in mean size; the polarization function for the highland site is similar to basalt powders ground to about 30 microns in diameter (fig. 3-89).

The polarimetric function of the lunar rocks varies from nearly the same as that of the fine-grained surface material to functions with maxima more than twice the maxima observed for the fine-grained surface material. The rocks producing strong polarization effects also have high estimated albedos. Rocks that produce polarization somewhat similar to the fine-grained material and have an intermediate albedo may consist of shock-lithified, fine-grained material, or strongly shock-altered, coarse crystalline rock.

The overall observed polarimetric function of lunar rocks is more like that of terrestrial basalts and gabbros than any other common rock type; it is intermediate between the functions of fresh fracture surfaces; and basalt crushed to grains several hundred microns in diameter (figs. 3-89 and 3-90). As most observed lunar rock fragments have been disturbed by cratering events, the fragments have undergone varying degrees of shock alteration and comminution of crystals. The polarization function of these rock fragments might be expected, therefore, to be intermediate between the function of essentially uncrushed crystalline material and the function of finely fractured material.

Sunset Observations

J. J. RENNILSON

During observations of the western horizon shortly after sunset, an unexpected phenomenon was noticed in the Surveyor missions. A bright line with several bright beads was observed along the western horizon as late as 1 hour after the upper limb of the Sun had set. The beads disappeared by groups as the Sun dropped lower behind the local horizon. Although first recorded by Surveyor I, the phenomenon was first recognized during the Surveyor VI mission. It was again observed during the Surveyor VII mission, but polarimetric measurements of the phenomenon were taken only with Surveyor VII. To date, only the Surveyor VI data have been reduced to photometric units.

The Surveyor VI data consisted of a total of seven narrow-angle frames, spanning about 30 minutes in time. Frames numbered 1, 2, 4, 5, and 6 were reduced using the mission analog magnetic tapes and the data processed for absolute luminances (ref. 3-37); frames 3 and 7 were omitted from the final reduction because of inconsistencies in their exposure data. Figure 3-91 shows the location of the photometric data in the processed frames. All significant picture elements were plotted in terms of their television line and position along that line. The heavy line in figure 3-91 represents the apparent path of the Sun as it traversed the western horizon. Dots on this line are positions of the Sun's center in each of the usable frames. The dots, designated 18 and 19, are reseau marks on the vidicon. The angular separation of these dots is 1.383°. Notice that the distribution of picture elements in this illustration is for the first frame only.

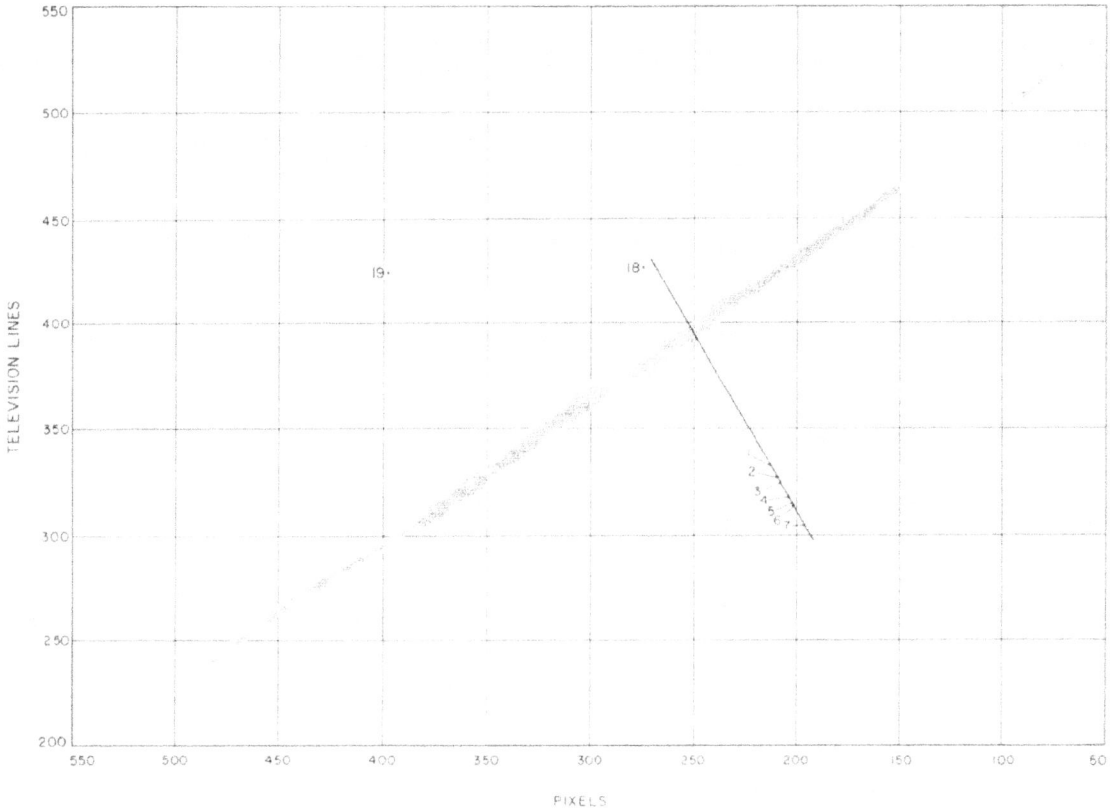

FIGURE 3-91.—Location of photometric data used to determine the luminance of the bright line on the western horizon after sunset at the Surveyor VI site on Nov. 24, 1967, 14:15:26 GMT. All significant picture elements (pixels) are plotted according to position on the television scan lines. The heavy line is the apparent path of the Sun; the numbered positions 1 to 7 are the centers of the Sun's disk in the pictures.

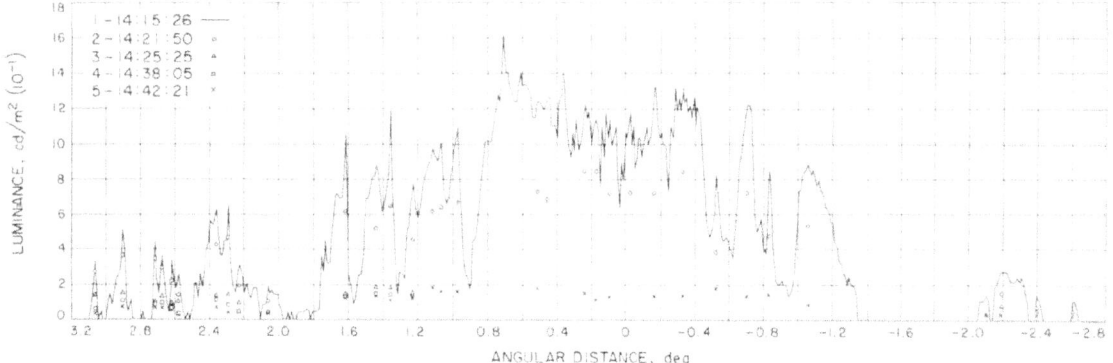

FIGURE 3-92.—Luminance variation of the bright line along the western horizon after sunset at the Surveyor VI site. Angular distance is measured right or left of the intersection of the Sun's path with the horizon. The solid line represents the data from the first picture only; maximum values of the other pictures are shown with identifying symbols.

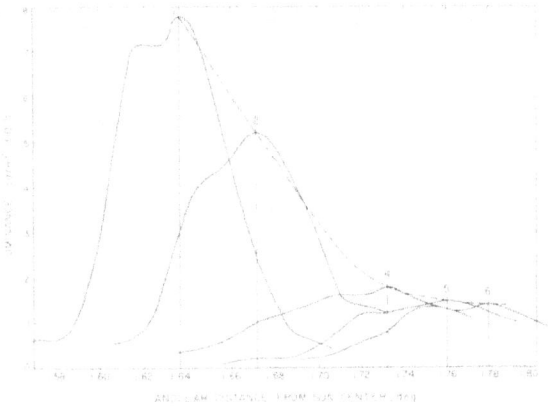

FIGURE 3-93.—Luminance of a bead (line 320, element 360) in the bright line on the horizon after sunset at the Surveyor VI site. Luminance is plotted as function of angular distance from the center of the Sun's disk for six pictures.

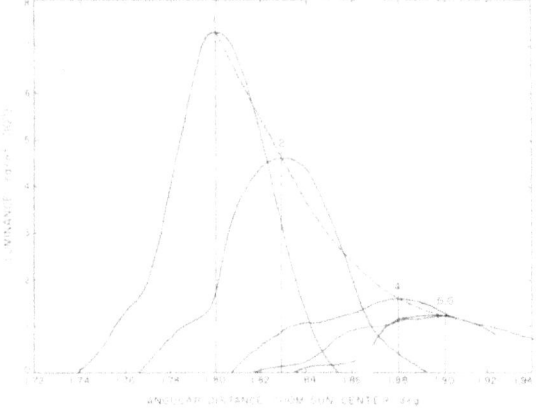

FIGURE 3-94.—Luminance of a bead (line 310, element 375) in the bright line on the horizon after sunset at the Surveyor VI site. Luminance is plotted as a function of angular distance from the center of the Sun's disk for six pictures.

Preflight calibration was used to establish the correct luminance value for each picture element. Television frames obtained during this calibration consisted of pictures of a uniform area light source. Variation in the luminance of this source caused proportionate changes in the camera video signals. Computer processing of these signals at the same iris position resulted in a transfer characteristic for each picture element. Thus, effects such as camera shading and nonlinearity of camera response were considered.

Figure 3-92 shows the variation of luminance along the horizon, through the center of the bright area, and relates luminance to angular distance along the horizon. For convenience, this angular distance is measured right and left from the intersection of the Sun's path and the horizon. Negative angles are measured from the left of 0°. For clarity, the solid line represents the data for the first frame only. The maximum values of the other frames are plotted with identifying symbols. If video signals were saturated, the data points were not included in these frames. It may be seen in figure 3-92 that positions of the beads do not change as the Sun sets. No new beads appeared and the brightness of each bead decreased with time.

Figures 3-93 and 3-94 show the luminance profiles of two beads. The centers of these beads are marked by the line and element coordinates shown in figure 3-91. The angular distance refers to the angle between the center of the bead and the center of the Sun. As the Sun's angular distance increased, the position of the beads' luminance maxima shifted. The dotted line indicates the trace of the maxima from each frame. The luminance of the beads rapidly decreases with time, reaching a minimum value about 20 minutes after their appearance. The exact time at which the beads completely disappeared is not known.

Photometric and Polarimetric Observations of the Earth

H. E. Holt

Throughout the lunar day, a series of pictures was taken intermittently by the Surveyor VII camera through the polarizing and clear filters to obtain polarization measurements of earthlight (fig. 3-95). These pictures are being used to study the integral photometric and polarimetric functions of the Earth, as viewed from the Moon during the January 1968 lunation. A sequence of Earth pictures was taken every

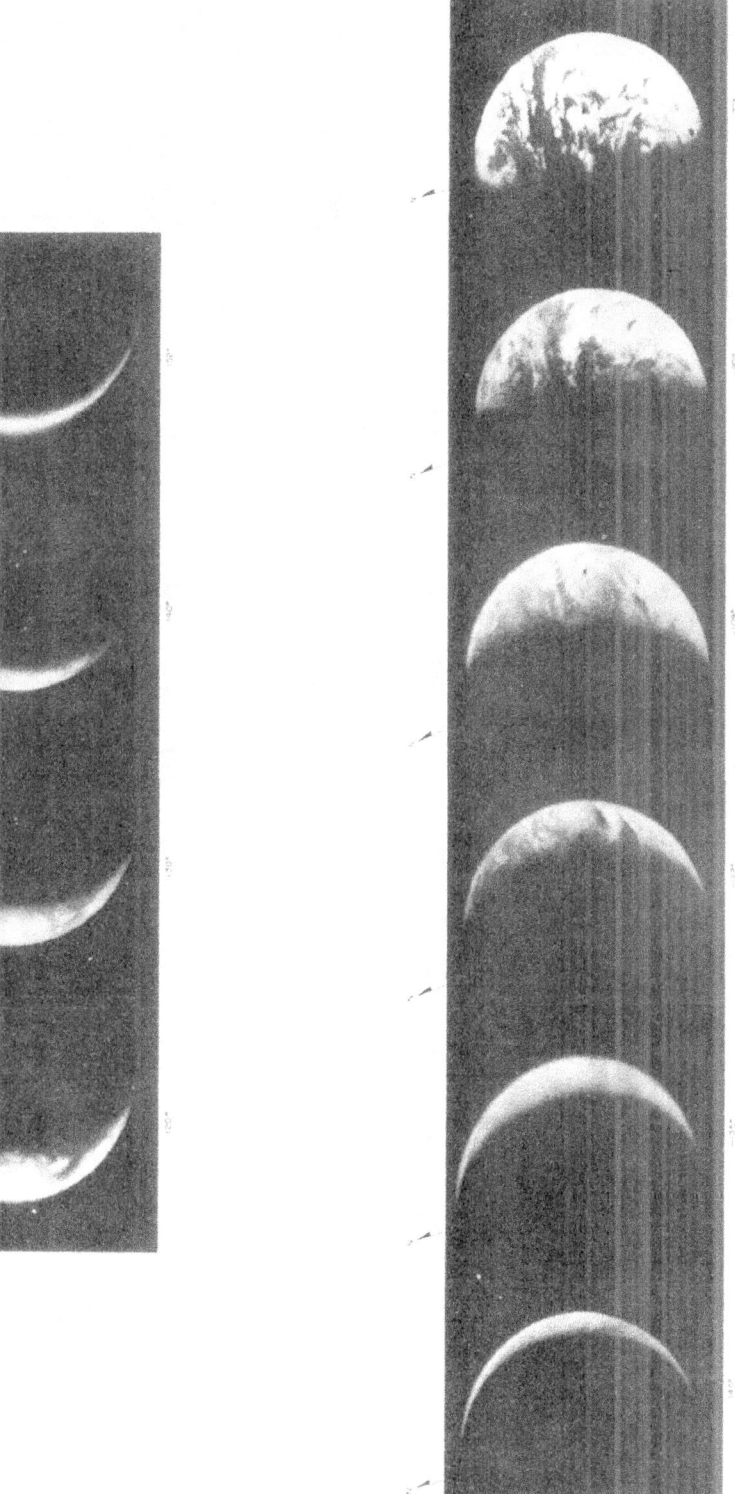

FIGURE 3-95.—A series of pictures of the Earth taken by Surveyor VII camera during the first lunar day after landing. The first four pictures show a waning Earth, while the later pictures show a waxing Earth. The phase angle is shown below each picture.

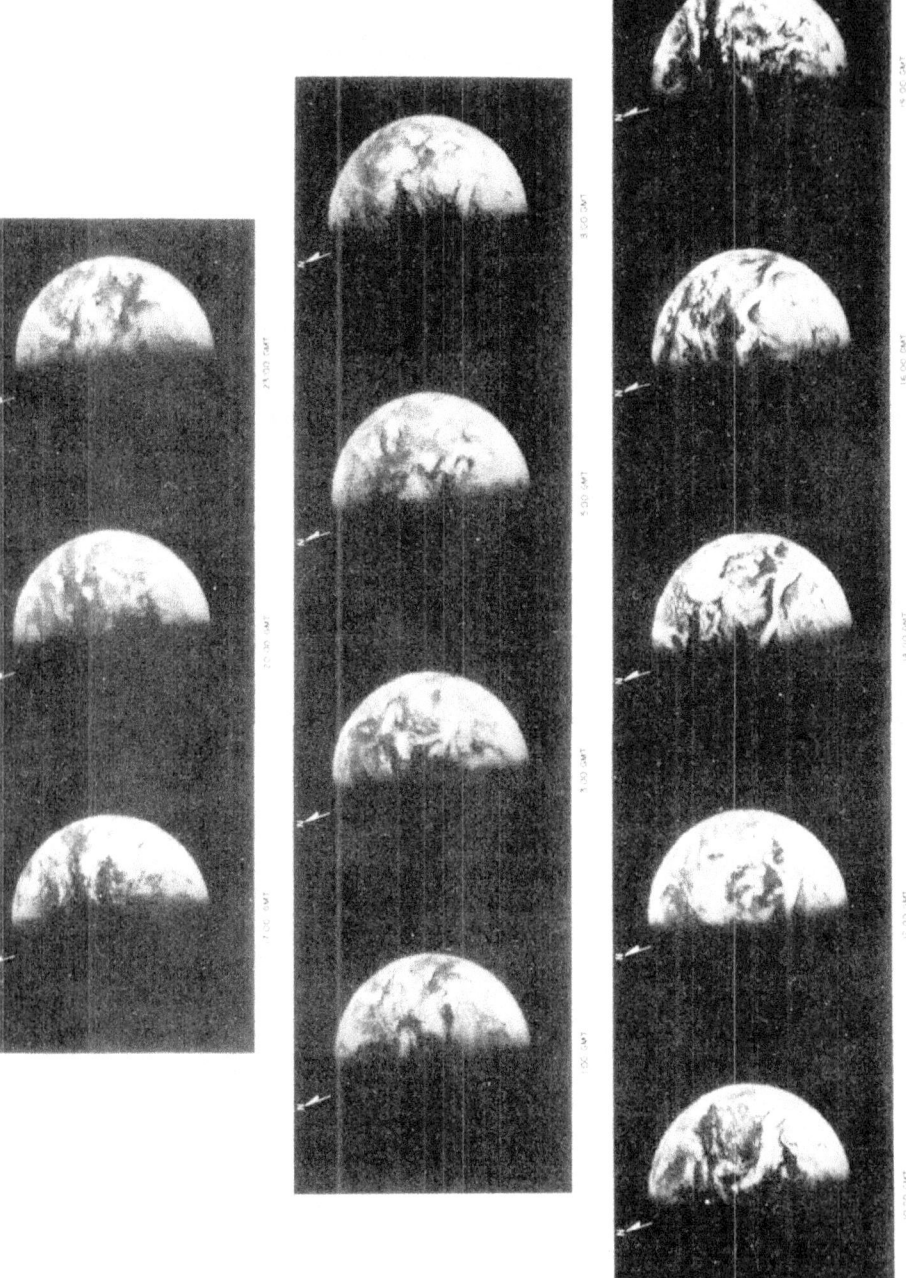

FIGURE 3-96.—Twelve pictures, covering a 26-hour period on Jan. 22 and 23, 1968, of the partially illuminated Earth taken by the Surveyor VII camera. The Earth rotated 180° from left to right during this 26-hour viewing period, or about 30° between successive pictures. Sunrise occurs on eastern Australia at about 20:00 GMT. At 03:00 GMT, sunrise occurs along the east coast of Africa; most of the Indian Ocean is covered by clouds.

2 or 3 hours (fig. 3-96) starting at 17:11 GMT on January 22, 1968, and terminating at 19:37 GMT on January 23, 1968. The pictures taken during the 26-hour period provide the information for studying the variation in reflectance of Earth as a function of rotation and changing cloud distribution during a single day. The Earth's reflectance observed from the Surveyor VII pictures was about 15 to 20 percent higher than expected, and variation in reflectance of 11 ± 3 percent occurred between 10:11 and 12:13 GMT on January 23.

Preliminary polarization measurements indicate that the polarized component of the earthlight varies as a function of cloud cover and the changing patterns of oceans and continents during rotation. Specular polarization appears to occur over an area of about 2×10^6 km² in the approximate geometric center of the Earth's illuminated crescent. The degree of polarization of earthlight from the specular reflection area varied from 26 to 30 percent over clear parts of the oceans, 12 to 16 percent over clear parts of the continents, and 4 to 8 percent over clouds. Thus, the cloud distribution over the area of specular reflection has a strong effect on the degree of polarization of earthlight. The degree of polarization of earthlight is much less from areas beyond the zone of specular reflection. Digital data-processing procedures are being conducted to determine more accurately the polarimetric function of the Earth.

Eclipse of Sun by Earth, as Seen From Surveyor III

E. M. SHOEMAKER, J. J. RENNILSON, AND
E. A. WHITAKER

In late morning of the first lunar day of the Surveyor III mission, an unusual opportunity occurred to observe an eclipse of the Sun by the Earth; this eclipse took place on April 24, 1967. Were it not for the fact that the spacecraft was tilted as much as 14.7° to the west and was oriented favorably with respect to azimuth, it would not have been possible to observe the Earth from a landing site at 23° W longitude because of the limited range of elevation angles through which the mirror can be stepped. To observe the Earth, the mirror was pointed upward and positioned at its highest permissible elevation step, and wide-angle pictures of the eclipse were obtained. The image of the Earth was reflected from very near the upper edge of the mirror. During the eclipse, two series of pictures (20 pictures total) were obtained through the color filters. The first series of pictures was obtained at approximately 11:24 GMT; the second set was obtained approximately 37 minutes later. The pictures were taken at two iris positions, and multiple pictures were taken through each filter.

During the eclipse, the Sun passed behind the Earth along a path that brought the position of the center of the Sun, as seen from the Moon, to within 15 minutes of the sublunar point on the Earth (fig. 3-97). At the time the Sun was most nearly centered behind the Earth, the projected center of the Sun lay northeast of the sublunar point. The sublunar point was at about 172° W longitude and 12.5° S latitude at the time the first series of pictures was taken, and at about 179° E longitude and 12.5° S latitude at the time the second series of pictures was taken. These positions are in the southwest Pacific. The limb of the Earth lay along western North America, the eastern Pacific, eastern Antarctica, the central Indian Ocean, southeast Asia, central China, eastern Siberia, and a short arc across the western Arctic Ocean.

In the first series of eclipse pictures, the Earth is partly surrounded by a halo of refracted light that varies greatly in brightness from one position to another along the limb (fig. 3-98). A very bright region, approximately 60° in arc length, lies along the northern part of the limb, nearest the position of the Sun. In the majority of pictures taken, parts of the image of the halo in this region are saturated. On either side of this bright region, the halo has a beaded appearance; small bright areas of short arc length are separated from other bright areas by sectors of the halo that are relatively faint. Most of these bright areas or beads are only a few degrees in length, but one relatively bright sector, about 20° long, is present that cannot be resolved into separate beads. At least 12 beads can be distinguished in the halo.

A gap ranging from about 50° to more than 90° is present in the images of the halo along

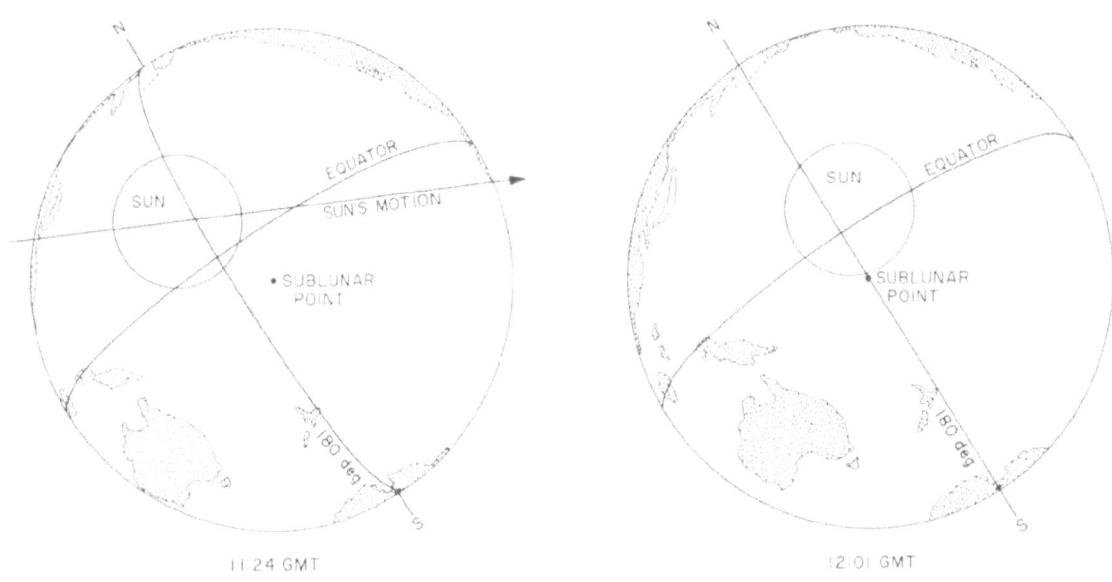

FIGURE 3-97.—Diagrams showing orientation of Earth and position of the Sun, as seen from the Moon on Apr. 26, 1967, at 11:24 GMT and 12:01 GMT. A series of pictures of the eclipse of the Sun by the Earth was taken by the Surveyor III television camera at approximately each of the times illustrated.

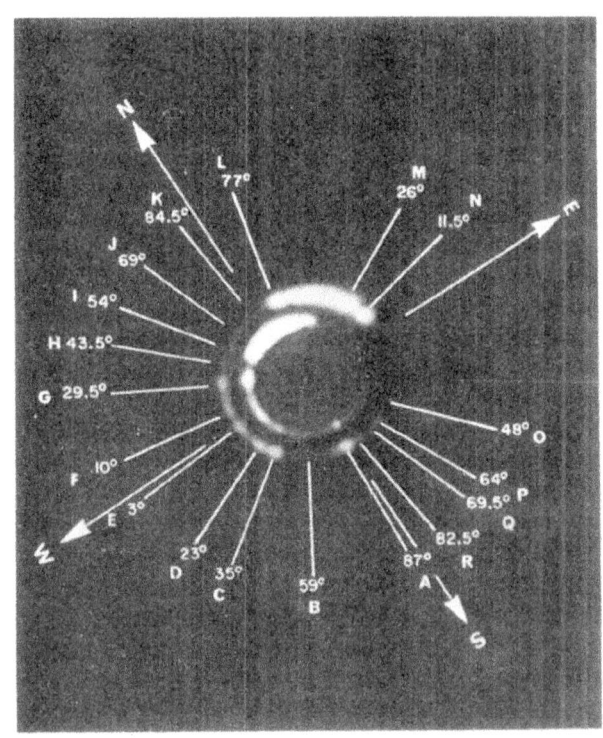

FIGURE 3-98.—Superimposed Surveyor III pictures showing distribution of light in the refraction halo of the Earth at the time the first series of eclipse pictures was taken and at the time the second series was taken. An eclipse image taken from the first series of pictures has been reduced in size and is shown nested within an eclipse image taken from the second series of eclipse pictures. The angular orientation of both images is the same. Line marked N-S shows projection of Earth's axis on plane of pictures. Eighteen beads, identified by letters, can be distinguished. Note that the angular position of beads in the refraction halo tends to remain the same; the bright region nearest the Sun changes position between the time of the first series of pictures and the time of the second, following the Sun.

the eastern limb of the Earth. Over most of the arc length of the gap, the halo is too faint to be detected with the exposures used, but over a short sector of the gap, the image of the Earth may have been cut off by the edge of the camera mirror.

In the second series of eclipse pictures, the very bright region in the halo shifts to the northeastern part of the limb, following the Sun (fig. 3-98). More of the eastern limb was bright enough to be detected in the second series of television pictures than in the first series, and the gap was reduced to an arc length no greater than 40°. At least 18 beads can be distinguished in the halo in the best exposed pictures. Many of these beads occur at nearly the same angular position, relative to the projection of the Earth's axis, as the beads observed in the first series of pictures. The beads are clearly related to features in the Earth's atmosphere, in contrast to the brightest region in the halo, which is related to the position of the Sun.

To identify the atmospheric features controlling the distribution of the beads, each bead's position in the second series of pictures was measured relative to the projection of the Earth's axis. These positions were plotted on the trace of the limb on stereographic projections of the Northern and Southern Hemispheres of the Earth. The plotted positions of the beads were then compared directly with stereographic mosaics of ESSA 3 pictures of the Earth taken on the day before the eclipse (fig. 3-99(a) and (b)). Even though there was some shift in cloud patterns between the time the ESSA 3 pictures were taken and the time of the eclipse, it can be seen that the beads occur predominantly over clear or largely clear regions between the clouds. Clouds tend to occult the refracted rays of the Sun, most of which pass through the low atmosphere at the limb; the beads occur at depressions in the optical silhouette of the Earth.

Preliminary reduction of the colorimetric information contained in the pictures has begun. Six pictures, one taken through each of the three color filters during each of the two periods of observation (fig. 3-100), were digitized using equipment at the Jet Propulsion Laboratory. The video voltage recorded on magnetic tape was divided into 64 equally spaced levels. For calibration, the preflight recording of the 3×3 filter matrix was also digitized. Equations for computing tristimulus values were derived from the digital printout of the preflight calibration tape.

The digitization procedure adopted for the television pictures generates a larger number of digital picture elements along a scan line than there are scan lines in the picture; the digital picture is a rectangular matrix of 600×684 elements. In the first series of pictures, the image of the refraction halo is 54 lines high and 61 picture elements wide. The total number of picture elements yielding colorimetric data in each of the digitized pictures from this series was 644. Chromaticity coordinates for selected picture elements were calculated, by means of the tristimulus value equations, from the digital voltages of corresponding elements in pictures taken through each of the three color filters. Because of present uncertainties about the preflight calibration tapes and because of possible jitter or other displacement of image points in corresponding pictures, the calculated chromaticity coordinates may have an error of as much as 0.03 in x and y.

Chromaticity coordinates were calculated for 18 points on the images in the first series of eclipse pictures and were plotted on a chromaticity diagram (fig. 3-101). The location of these colorimetric measurements with respect to the eclipse images is shown in figure 3-102. Also plotted on the chromaticity diagram are the locus of color temperatures for a body that obeys Planck's law (a blackbody) and the locus of color temperatures for natural daylight as far as 4800° K. Loci of correlated color temperature (ref. 3-45) are shown crossing the Planckian locus in figure 3-101.

Most colorimetric measurements were made in the bright region of the halo controlled by the position of the Sun. Most measurements in this region have a correlated color temperature close to 4800° K. Beads in the halo exhibit lower correlated color temperatures. The center of bead G, close to the bright region, has a correlated color temperature of about 4000° K, and the center of bead A, which lies over

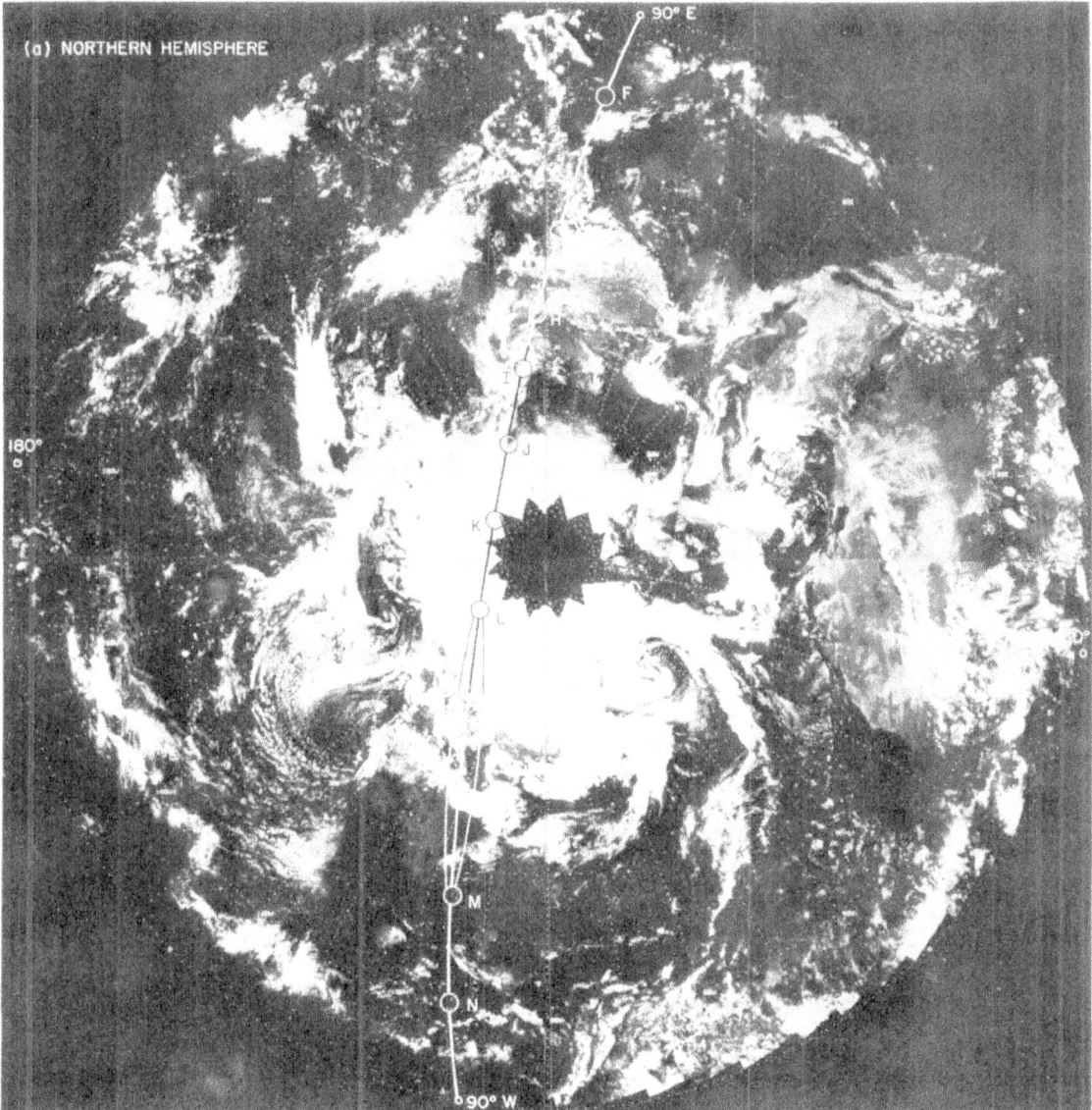

FIGURE 3-99.—Stereographic mosaics of ESSA 3 pictures of the Earth, taken the day before the eclipse. (a) Northern Hemisphere. (b) Southern Hemisphere. Bright areas are clouds and, in mountainous regions and the Arctic, snowfields; dark areas are clear. Note position of African continent which may be seen in the clear areas on right-hand side of (a). Trace of limb of Earth, as seen from the Moon, and positions of beads and the bright region in the refraction halo of the Earth observed in the second series of eclipse pictures taken by Surveyor III are shown by the white circles and lines that extend from 90° W to 90° E. Beads occur in areas that are largely clear. Letters beside symbols for beads correspond to letter identification of beads in figure 3-98 (mosaic provided by Dr. D. S. Johnson, National Environmental Satellite Center, ESSA).

TELEVISION OBSERVATIONS 123

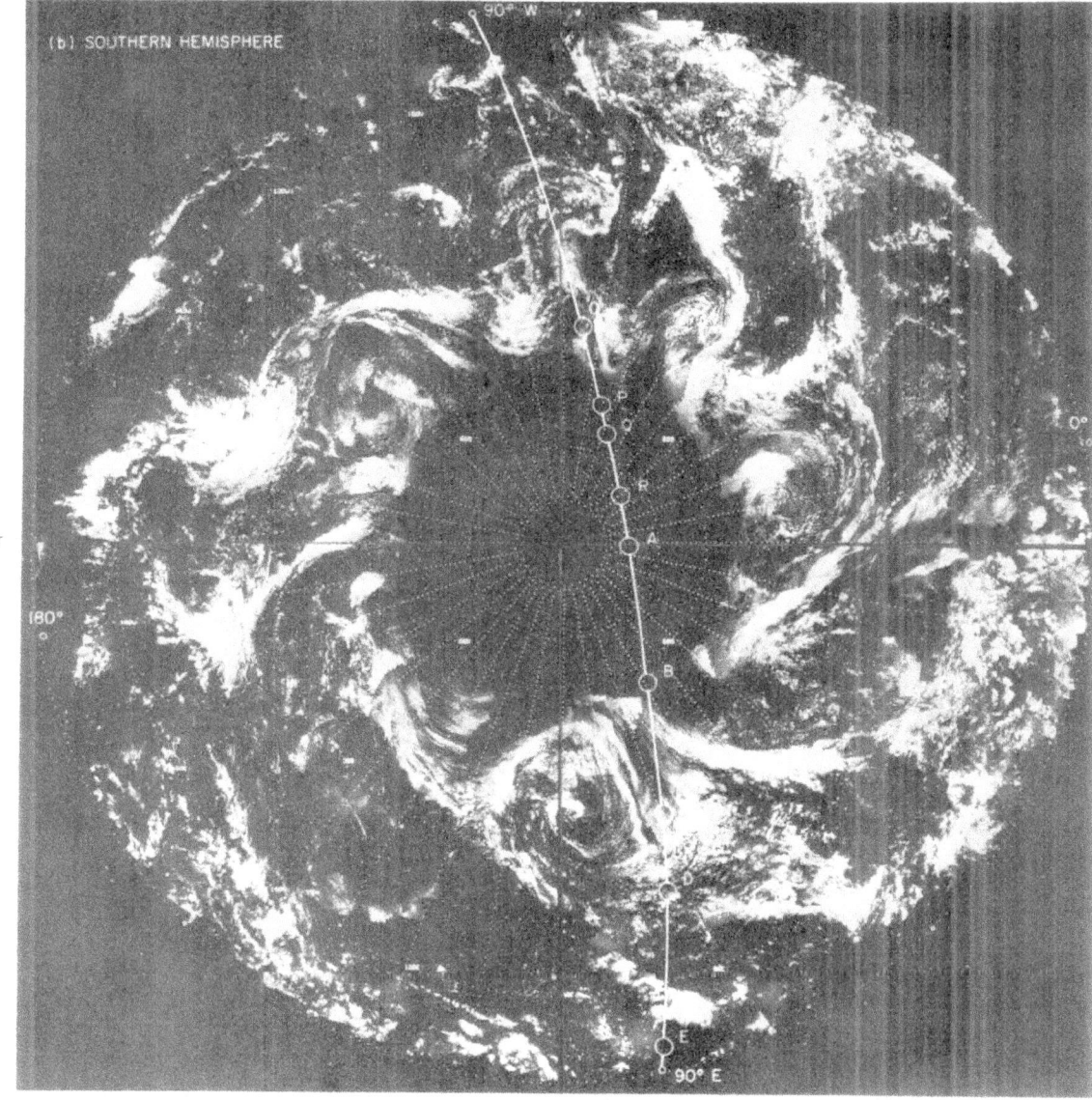

FIGURE 3-99.—Continued. (b) Southern Hemisphere.

Antarctica, farthest from the projected position of the Sun, has a correlated color temperature of approximately 2850° K. The correlated color temperature tends to decrease in directions away from the projected position of the Sun and also tends to decrease toward the inner edge of the halo. As would be expected, the color temperature tends to be lower for light that followed paths of greater atmospheric absorption. Most of the colors present in the images had purities less than 50 percent.

References

3-1. FRANZ, J.: Die Randlandschaften des Mondes. Novo Acta Leopoldina, vol. 99, no. 1, Halle, 1913, pp. 1–96.

3-2. SAUNDER, S. A.: Determination of Selenographic Positions and the Measurement of

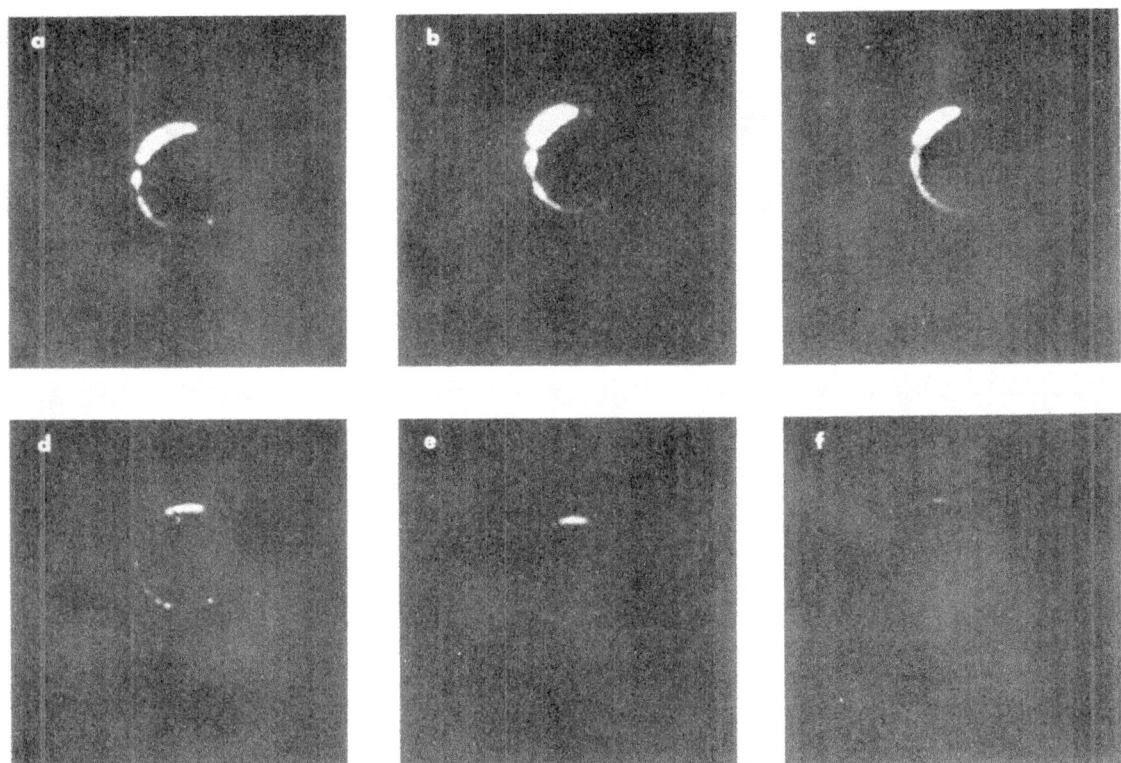

FIGURE 3-100.—Wide-angle Surveyor III pictures showing eclipse of Sun by Earth as observed on Apr. 24, 1967, through the filters indicated. (a) $\bar{x}'$ (red), 11:31:40 GMT; (b) $\bar{y}'$ (green), 11:23:06 GMT; (c) $\bar{z}'$ (blue), 11:24:01 GMT; (d) $\bar{x}'$, 12:02:10 GMT; (e) $\bar{y}'$, 12:03:10 GMT; (f) $\bar{z}'$, 12:02:44 GMT. First set of figures includes parts (a), (b), and (c); second set of figures includes parts (d), (e), and (f).

Lunar Photographs. Mem. Roy. Astron. Soc., vol. LX, pt. 1, 1910.

3-3. Orthographic Atlas of the Moon, Supplement 1 to the Photographic Lunar Atlas, compiled by D. W. G. ARTHUR and E. A. WHITAKER, Univ. of Arizona Press, Tucson, Ariz., 1961.

3-4. ARTHUR, D. W. G.: Consolidated Catalog of Selenographic Positions. Communications Lunar and Planetary Lab., vol. 1, no. 11, Univ. of Arizona, Tucson, Ariz., 1962.

3-5. BALDWIN, R. P.: The Measure of the Moon. Univ. of Chicago Press, 1963.

3-6. MILLS, G. A.: Absolute Coordinates of Lunar Features. Icarus, vol. 7, 1967, pp. 193-200; and Icarus, vol. 8, 1968, pp. 90-116.

3-7. ARTHUR, D. W. G.: The Tucson Triangulation, and other related papers, accepted for publication. Communications Lunar and Planetary Lab., Univ. of Arizona, Tucson, Ariz.

3-8. SCHRUTKA-RECHTENSTAMM, G. V.: Neureduktion der 150 Mondepunkte der Braslauer Messungen von J. Franz. Osterr. Akad. Wiss. Math-Naturw. Kl. Sitzber, pt. IIa, vol. 165, nos. 1-4, 1958, pp. 97-126.

3-9. JAFFE, L. D., ET AL.: Surveyor I: Preliminary Results. Science, vol. 152, 1966, pp. 1737-1750.

3-10. RENNILSON, J. J.; DRAGG, J. L.; MORRIS, E. C.; SHOEMAKER, E. M.; and TURKEVICH, A.: Lunar Surface Topography. Surveyor I Mission Report. Part II: Scientific Data and Results, Tech. Rept. 32-1023, Jet Propulsion Laboratory, Pasadena, Calif., Sept. 10, 1966, pp. 7-44.

3-11. WHITAKER, E. A.: Surveyor I Location. Science, vol. 153, 1966, p. 1550.

3-12. BROWN, W. E., JR.: Lunar Surface Surveyor Radar Response. J. Geophy. Res., vol. 72, 1967, p. 791.

3-13. SPRADLEY, L. H.; STEINBACHER, R.; GROLIER, M.; and BYRNE, C.: Surveyor I: Location and Identification. Science, vol. 157, 1967, pp. 681-683.

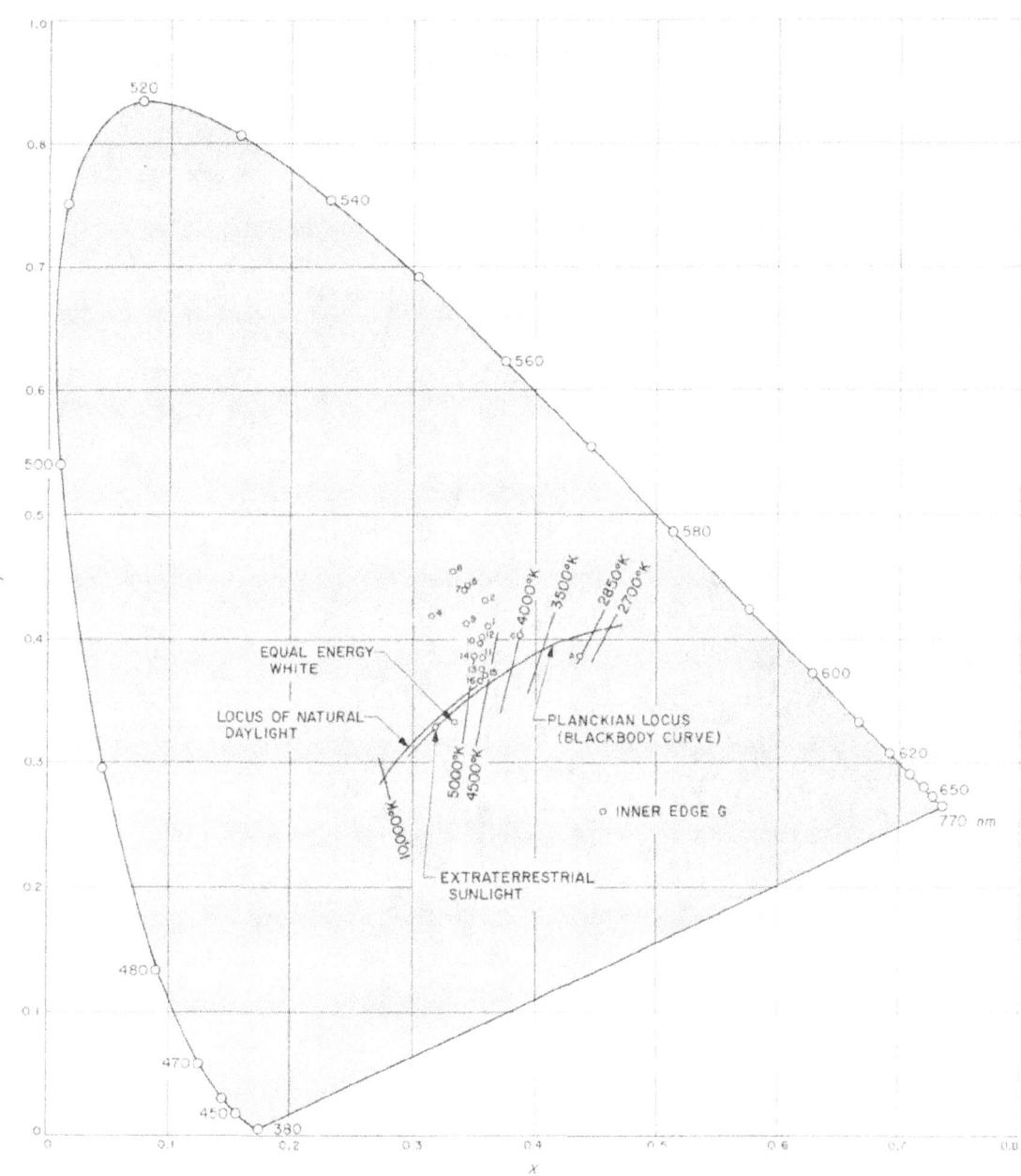

FIGURE 3-101.—Chromaticity coordinates of selected points from the first series of Surveyor eclipse pictures. For purposes of comparison, the Planckian locus with correlated color temperature lines is shown with the locus of natural daylight, as measured by Y. Nayatio and G. Wyszecki (ref. 3-41).

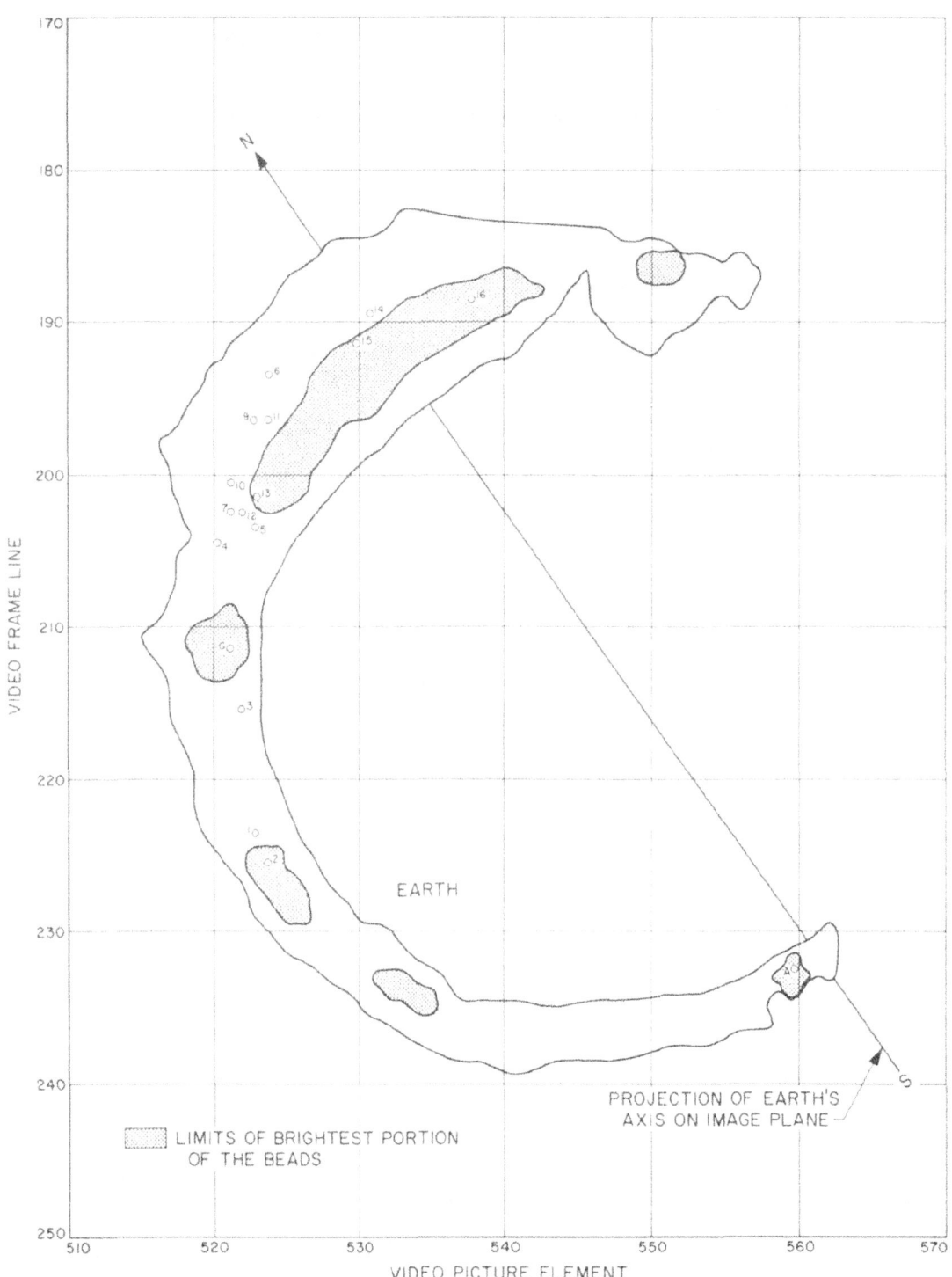

FIGURE 3-102.—Diagram of the solar eclipse showing positions of points on the color composite image that were measured for color. The letters refer to the beads.

3-14. SHOEMAKER, E. M.; BATSON, R. M.; HOLT, H. E.; MORRIS, E. C.; RENNILSON, J. J.; and WHITAKER, E. A.: Television Observations From Surveyor III. J. Geophys. Res., vol. 73, no. 12, June 15, 1968, pp. 3989–4043.

3-15. SHOEMAKER, E. M.; BATSON, R. M.; HOLT, H. E.; MORRIS, E. C.; RENNILSON, J. J.; and WHITAKER, E. A.: Surveyor V: Television Pictures. Science, vol. 158, no. 3801, Nov. 3, 1967, pp. 642–652.

3-16. LOCKMAN, R. L.; HEATHCOTE, W. R.; and VOLANSKY, S. A.: Surveyor Attitude Determination on Lunar Surface. Report SSD 88021R, Hughes Aircraft Co., El Segundo, Calif., Feb. 1968.

3-17. Air Force Chart and Information Center, Lunar Map, Surveyor I Site, 1:2000. First ed., Jan. 1968.

3-18. EGGLETON, R. E.: Geologic Map of Riphaeus Mountains Region of the Moon. U.S. Geological Survey Miscellaneous Geologic Investigations, Map I-458, 1965.

3-19. TRASK, N. J.: Size and Spatial Distribution of Craters Estimated From Ranger Photographs. Ranger VIII and IX. Part II: Experimenters' Analyses and Interpretations, Tech. Rept. 32-800, Jet Propulsion Laboratory, Pasadena, Calif., Mar. 15, 1966, pp. 249–338.

3-20. SHOEMAKER, E. M.: Preliminary Analysis of the Fine Structure of the Lunar Surface in Mare Cognitum, The Nature of the Lunar Surface, The Johns Hopkins Press, Baltimore, 1966, pp. 23–77.

3-21. MARCUS, H.: A Stochiastic Model of the Formation and Survival of Lunar Craters. 5, Approximate Diameter Distribution of Primary and Secondary Craters. Icarus, vol. 5, 1966, p. 590.

3-22. Lunar Surface Topography and Geology. Surveyor I: A Preliminary Report, NASA SP-126, 1966.

3-23. ROSIWAL, A.: Über geometrische Gesteinsanalysen. Verhandl. der K.K., Geol. Reichsanstalt., Wien, 1898, pp. 143–175.

3-24. PACKHAM, G. H.: Volume-, Weight-, and Number-Frequency Analysis of Sediments from Thin-Section Data. J. Geol., vol. 63, 1955, pp. 50–58.

3-25. ROETHLISBERGER, H.: An Adequate Method of Grain-size Determination in Sections. J. Geol., vol. 63, 1955, pp. 579–584.

3-26. SCOTT, R. F.; and ROBERSON, F. I.: Soil Mechanics Surface Sampler. Surveyor VII: A Preliminary Report, NASA SP-173, 1968, pp. 121–161.

3-27. CHRISTENSEN, E. M.; BATTERSON, S. A.; BENSON, H. E.; CHOATE, R.; JAFFE, L. D.; JONES, R. H.; KO, H. Y.; SPENCER, R. L.; SPERLING, F. B.; and SUTTON, G. H.: Lunar Surface Mechanical Properties. Surveyor III: A Preliminary Report, NASA SP-146, 1967, pp. 94–120.

3-28. CHRISTENSEN, E. M.; BATTERSON, S. A.; BENSON, H. E.; CHOATE, R.; HUTTON, R. E.; JAFFE, L. D.; JONES, R. H.; KO, H. Y.; SCHMIDT, F. N.; SCOTT, R. F.; SPENCER, R. L.; and SUTTON, G. H.: Lunar Surface Mechanical Properties. Surveyor V: A Preliminary Report, NASA SP-163, 1967, pp. 43–87.

3-29. CHRISTENSEN, E. M.; BATTERSON, S. A.; BENSON, H. E.; CHOATE, R.; HUTTON, R. E.; JAFFE, L. D.; JONES, R. H.; KO, H. Y.; SCHMIDT, F. N.; SCOTT, R. F.; SPENCER, R. L.; SPERLING, F. B.; and SUTTON, G. H.: Lunar Surface Mechanical Properties. Surveyor VI: A Preliminary Report, NASA SP-166, 1968, pp. 41–95.

3-30. CHOATE, R.; BATTERSON, S. A.; CHRISTENSEN, E. M.; HUTTON, R. E.; JAFFE, L. D.; JONES, R. H.; KO, H. Y.; SPENCER, R. L.; and SPERLING, F. B.: Lunar Surface Mechanical Properties. Surveyor VII: A Preliminary Report, NASA SP-173, 1968, pp. 83–119.

3-31. HAPKE, B.: Lunar Surface: Composition Inferred From Optical Properties. Science, vol. 159, no. 3810, Jan. 5, 1968, pp. 77–79.

3-32. HAPKE, B.; and VAN HORN, H.: Photometric Studies of Complex Surfaces With Applications to the Moon. J. Geophys. Res., vol. 68, no. 15, Aug. 1963, pp. 4545–4570.

3-33. BELTON, M. J. S.: Dynamics of Interplanetary Dust. Science, vol. 151, no. 3706, Jan. 7, 1966, pp. 35–44.

3-34. JAFFE, L. D.; BATTERSON, S. A.; BROWN, W. E., Jr.; CHRISTENSEN, E. M.; DWORNIK, S. E.; GAULT, D. E.; LUCAS, J. W.; NORTON, R. H.; SCOTT, R. F.; SHOEMAKER, E. M.; STEINBACHER, R. H.; SUTTON, G. H.; and TURKEVICH, A. L.: Principal Scientific Results. Surveyor VI: A Preliminary Report, NASA SP-166, 1968, pp. 1–3.

3-35. GAULT, D. E.; ADAMS, J. B.; COLLINS, R. J.; KUIPER, G. P.; MASURSKY, H.; O'KEEFE, J. A.; PHINNEY, R. A.; and SHOEMAKER, E. M.: Lunar Theory and Processes. Surveyor VII: A Preliminary Report, NASA SP-173, 1968, pp. 233–276.

3-36. SHOEMAKER, E. M.; BATSON, R. M.; HOLT, H. E.; MORRIS, E. C.; RENNILSON, J. J.; and WHITAKER, E. A.: Television Observations From Surveyor VII. Surveyor VII: A Preliminary Report, NASA SP-173, 1968, pp. 13–81.

3-37. MORRIS, E. C.; BATSON, R. M.; HOLT, H. E.; RENNILSON, J. J.; SHOEMAKER, E. M.; and WHITAKER, E. A.: Television Observations From Surveyor VI. Surveyor VI: A Preliminary Report, NASA SP-166, 1968, pp. 11–40.

3-38. GAULT, D. E.; SHOEMAKER, E. M.; and MOORE, H. J.: Spray Ejected From the Lunar Surface by Meteoroid Impact. NASA Tech. Note D-1767, 1963.

3-39. SYTINSKAYA, N. N.; and SHARONOV, V. V.: Leningrad Univ., Astron. Observatoria, Uchenye Zapisky, no. 153, pp. 114-154 (1952) (in Russian); Trudy 16, 114-154 (1953). Trans. by Space Technology Lab., Inc., ARTIC STL-TR-61-511-23, May 1961.

3-40. WILLINGHAM, D.: The Lunar Reflectivity Model for Ranger Block III Analysis. Tech. Rept. 32-664, Jet Propulsion Laboratory, Pasadena, Calif., Nov. 1964.

3-41. DAVIES, W. E.; and WYSZECKI, G.: Physical Approximation of Color-Mixture-Functions. J. Opt. Soc. Am., vol. 52, no. 6, June 1962, pp. 679-685.

3-42. DOLLFUS, A.: The Polarization of Moonlight. *In:* Physics and Astronomy of the Moon, Academic Press, 1962, pp. 131-160.

3-43. WRIGHT, F. E.; WRIGHT, F. H.; and WRITE, H.: The Lunar Surface Introduction. *In:* The Moon, Meteorites, and Comets—The Solar System, vol. IV, Univ. Chicago Press, 1962, pp. 1-47.

3-44. FRANZGROTE, E. J.; PATTERSON, J. H.; and TURKEVICH, A. L.: Chemical Analysis of the Moon at the Surveyor VII Landing Site: Preliminary Results. Surveyor VII Mission Report, Part II: Science Results, Tech. Rept. 32-1264, Jet Propulsion Laboratory, Pasadena, Calif., Mar. 15, 1968, pp. 241-265.

3-45. JUDD, D. B.; MACADAM, D. L.; and WYSZECKI, G.: Spectral Distribution of Typical Daylite as a Function of Correlated Color Temperature. J. Opt. Soc. Am., vol. 54, no. 8, 1964, pp. 1030-1040.

4. Lunar Surface Mechanical Properties

R. Choate, S. A. Batterson, E. M. Christensen (Chairman), R. E. Hutton, L. D. Jaffe, R. H. Jones, H. Y. Ko, R. F. Scott, R. L. Spencer, F. B. Sperling, and G. H. Sutton

Significant new knowledge on the mechanical properties of the lunar soil has been provided by the Surveyor series of spacecraft (fig. 4–1). This report summarizes the mechanical properties sections of the Surveyor Mission Reports (refs. 4–1 to 4–5), which contain information on soil cohesion, adhesion, permeability, compressibility, particle-size distribution, bearing strength, elasticity, frictional and light reflectance characteristics. Mechanical property estimates presented are results of interpretations of landing telemetry data and television pictures of footpad and crushable block landing imprints, alpha-scattering-instrument imprints, tracks made by rolling stones, and rock fragments lying on the surface. The surface sampler, flown on Surveyors III and VII (see ch. 5 of this report), also obtained data on mechanical properties.

To study lunar soil erosion effects and to determine soil properties, Surveyor vernier engines and attitude control jets were operated after the landings. Erosion effects were observed when: (1) Surveyor V vernier engines were fired at low thrust, without spacecraft movement; (2) Surveyor VI vernier engines were fired at higher thrust, resulting in a hop of 2.4 meters[1] (ref. 4–4); and (3) Surveyor VI attitude control gas jets were operated.

Soil at the five landing sites was found to be remarkably similar in many of its physical properties, much more similar than would be expected at five widely separated terrestrial sites selected in a manner similar to those of Surveyor. At each site, the soil is granular and slightly cohesive. Although predominately fine grained, the soil contains a small percentage of rock fragments in a wide range of sizes.

Spacecraft Landings

Description

Surveyors I, VI, and VII made nominal landings on relatively flat surfaces; Surveyor V landed at close to nominal conditions, but in a small crater with an inner slope of about 20°. For these spacecraft, vernier-engine shutdown occurred at 4.3 ± 1.4-meter altitude, when the spacecraft were essentially horizontal, with vertical velocities of 1.5 ± 0.5 m/sec and horizontal velocities less than 1 m/sec. Surveyor III made an unplanned triple landing because the vernier engines continued to fire during the first two touchdowns. The second touchdown occurred 24 seconds after the first touchdown; 12 seconds later (½ sec after termination of engine thrust by Earth command) the third and final touchdown occurred. During this landing sequence, the spacecraft progressed down the 10° (average) inner slope of the large crater in which it landed (ref. 4–2). Landing conditions for each Surveyor mission are summarized in table 4–1.

The spacecraft landing system (fig. 4–2) consisted of three legs with crushable footpads (fig. 4–3) and three cylindrical crushable blocks mounted under the frame near the leg attachments. During landing, the legs rotated upward around their hinge axes, and energy was absorbed by compressing the shock absorbers. When leg rotation was sufficiently great, the blocks contacted the surface and provided additional energy-absorption capability. After

---

[1] Values are given in centimeter-gram-second units. To convert to foot-pound-second units, the following factors apply: 1 m = 3.28 ft; 1 cm = 0.394 in.; 1 N (newton) = 10^5 dynes = 0.225 lb; 1 N/cm² = 1.45 lb/in.²; 1 dyne/cm² = 10^{-5} N/cm² = 1.45×10^{-5} lb/in.²; 1 poise = 1 dyne sec/cm² = 2.089×10^{-3} lb sec/ft².

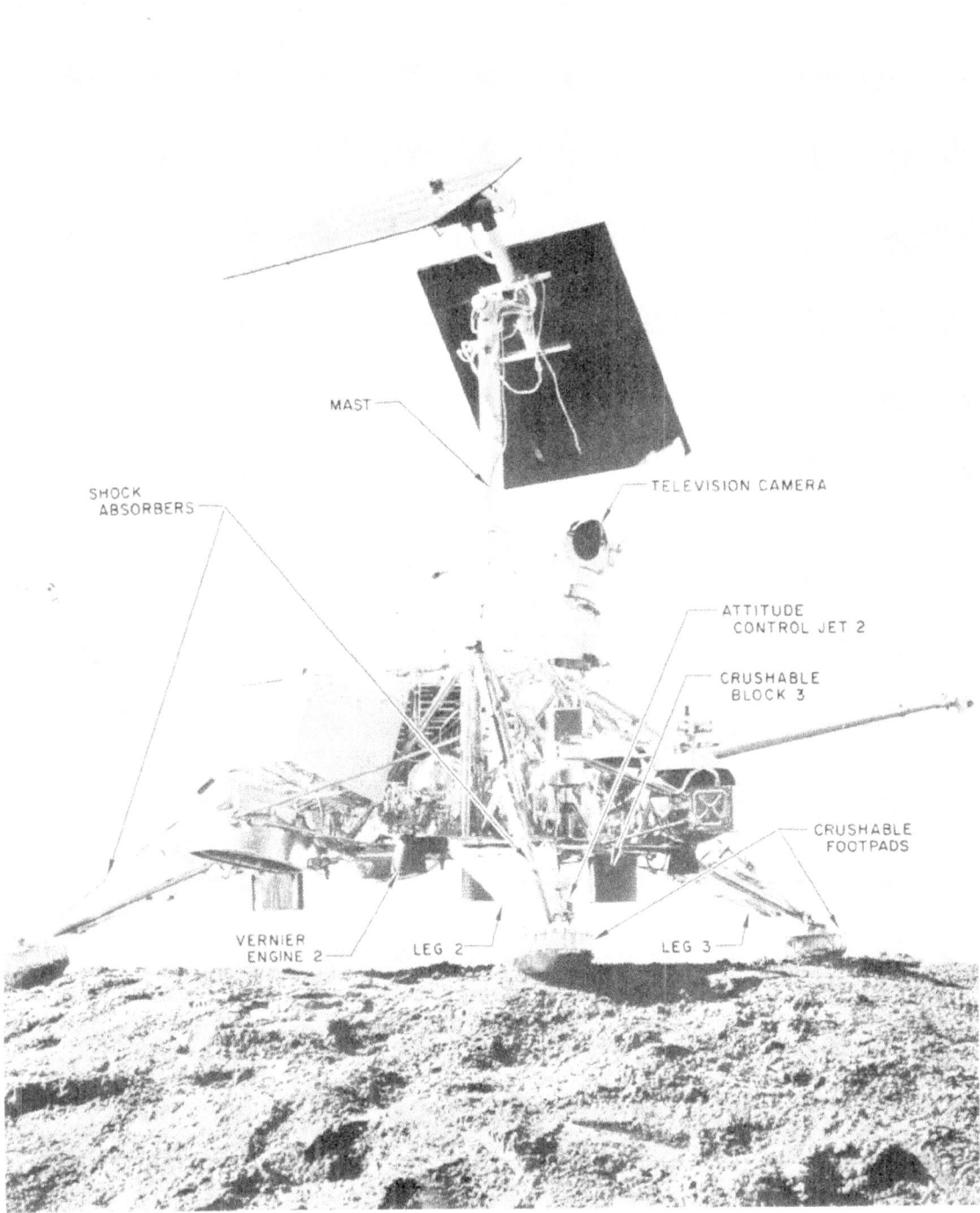

FIGURE 4-1.—Ground level view of a Surveyor spacecraft. Each spacecraft has three symmetrically located footpads, shock absorbers, crushable blocks, vernier engines, and three attitude control jets.

TABLE 4-1. *Summary of landing conditions*

Parameter	Surveyor I	Surveyor III			Surveyor V	Surveyor VI		Surveyor VII	Tolerance
		Touch-down 1	Touch-down 2	Touch-down 3		Initial	Hop		
Vertical landing velocity, m/sec	3.6	1.8	1.4	1.5	4.2	3.4	3.8	3.8	±0.4
Horizontal landing velocity, m/sec	0.3	0.2	0.8	0.9	0.3	0.3	0.5	0.1	±0.2
Angle of surface slope, deg	1.0	11.5	14.0	9.2	19.5	0.9		3.1	±0.5
Maximum axial shock-absorber load (leg 1), N	6190	2970	1420	3860	5620	7000	14 900	7330	±5%
Maximum axial shock-absorber load (leg 2), N	7110	3060	2800	2440	7280	8000	7800	7800	±5%
Maximum axial shock-absorber load (leg 3), N	6220	3680	2350	4120	7300	7000	8600	6540	±5%
Mass of spacecraft at landing, kg	294.5	305.7		299.0	303.7	300.5	299.8	306.0	±0.5
Depth of footpad 2 [a] penetration, cm	3		2	2	4±2	6	6	4	±1
Depth of footpad 3 [a] penetration, cm	2		2	4	4±2	3	3	4	±1
Maximum travel distance of ejecta, cm	50			70	80	50	90	40	±5

[a] The footpad depths listed are the estimated depths of the footpads below the undisturbed surface in their final position; depths of penetration correlating to times at peak shock-absorber loads were generally different than the final depths; see table 4-2.

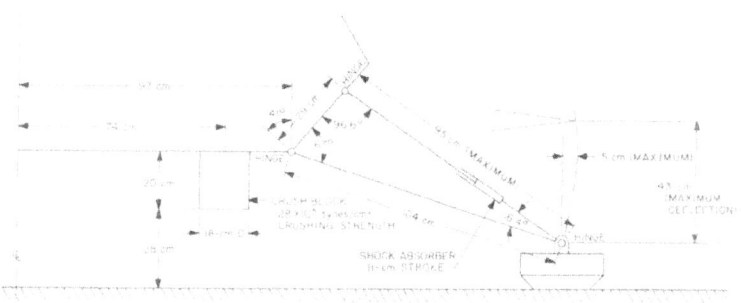

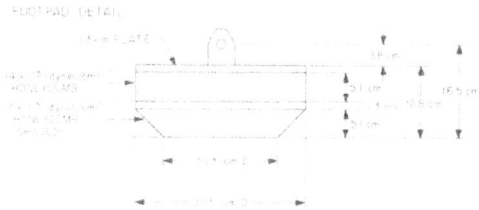

FIGURE 4-2.—Dimensions of landing gear assembly.

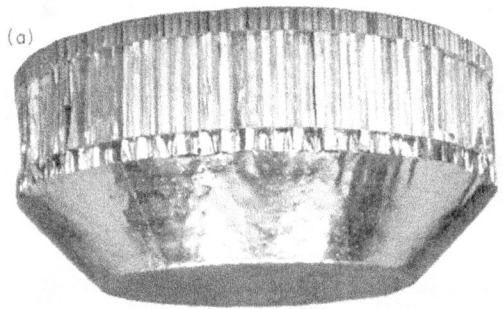

FIGURE 4-3.—Surveyor footpad. (a) Side view. (b) Bottom view, showing waffle pattern in aluminum foil covering formed by internal honeycomb structure. The footpad diameter is 30.5 cm at the top and 20.3 cm at the bottom.

landing, the shock absorbers reextended to their original undeflected lengths.

A temperature-compensated strain-gage bridge was mounted on each shock absorber to monitor its axial load history, which was telemetered to Earth in the form of a continuous analog signal. Initial strain-gage histories for each landing are shown in figure 4-4. Except for the first two landings of Surveyor III, all recordings indicate zero-force level before touchdown because the spacecraft were in a free-fall

FIGURE 4-4.—Telemetry data showing initial shock-absorber axial load histories during Surveyor landings. In most cases, oscillations continue beyond the 2.5 seconds shown. (a) Surveyor I: landing. (b) Surveyor III: first touchdown. (c) Surveyor III: second touchdown. (d) Surveyor III: final landing. (e) Surveyor V: landing. (f) Surveyor VI: initial landing. (g) Surveyor VI: hop landing. (h) Surveyor VII: landing.

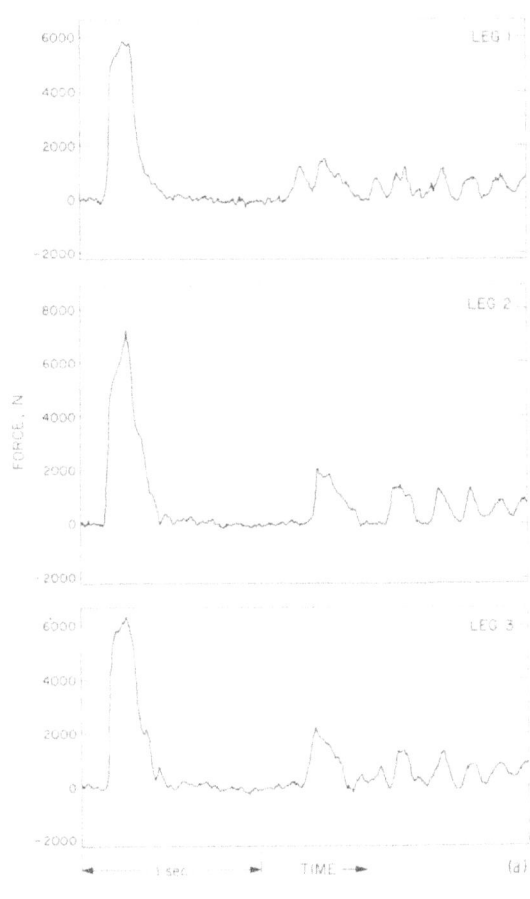
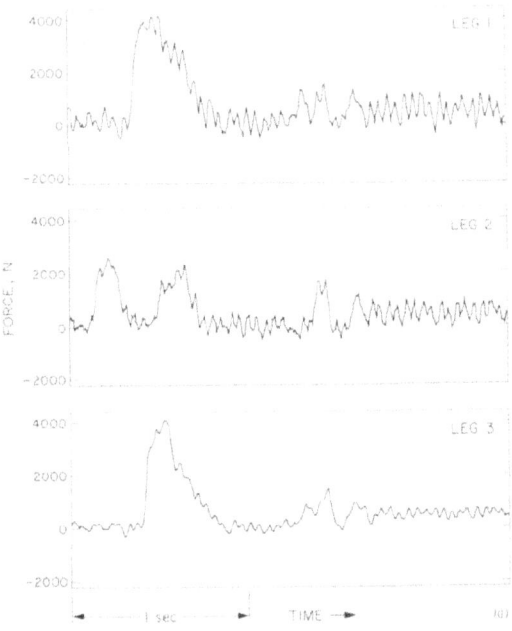

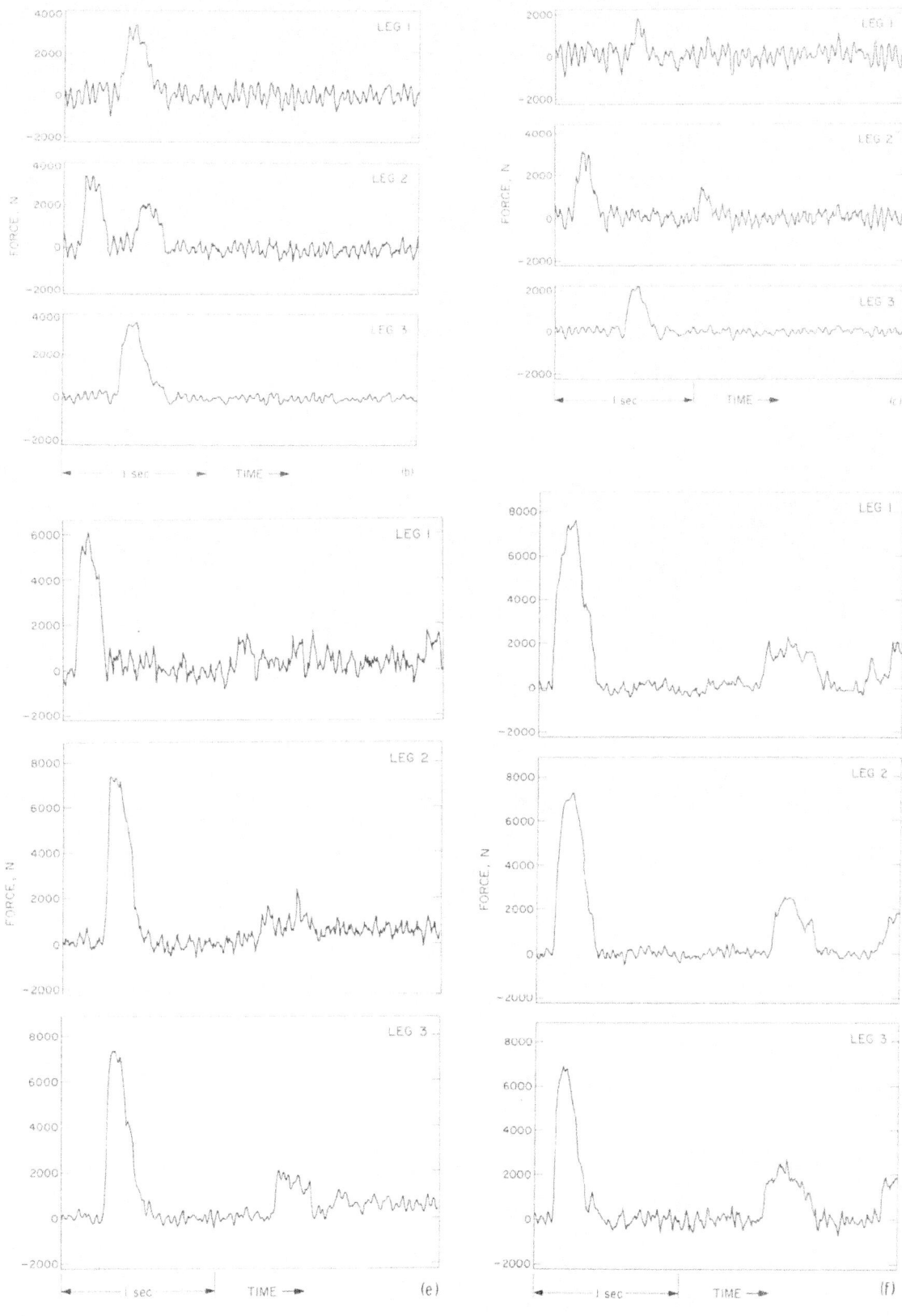

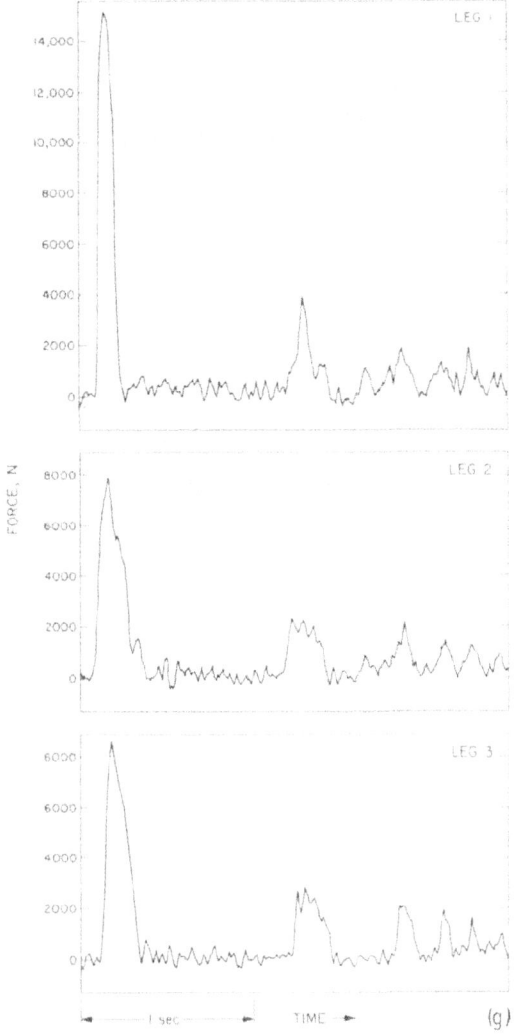
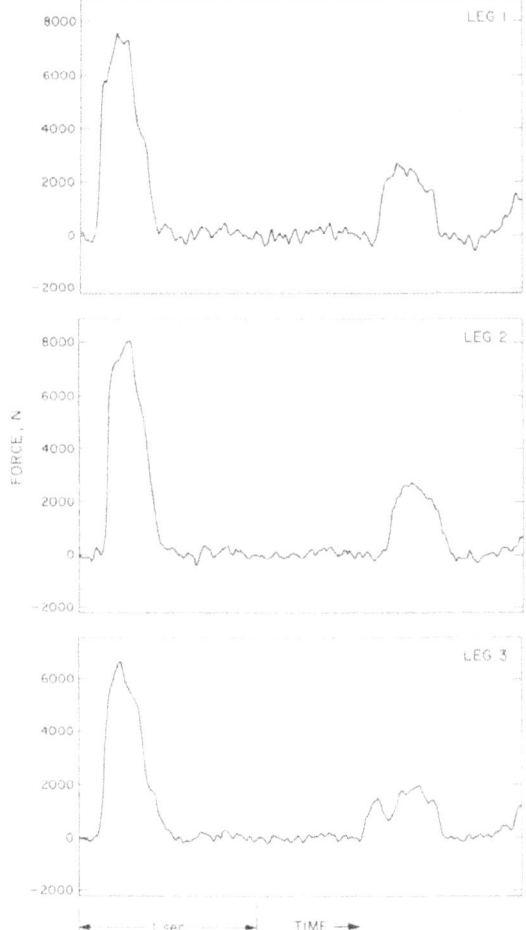

state after vernier-engine cutoff, which occurred about 1.3 seconds before touchdown. However, even for the first two Surveyor III landings, the deviation from zero was insignificant on the scale of the recordings. Footpad impact was indicated by a rapid increase in force which reached its maximum within 0.1 to 0.15 second. As shown in figure 4-4, about 0.3 second after first contact with the lunar surface, the forces returned to zero level, indicating elastic rebound of the spacecraft. Rebound occurred in all landings; it was most pronounced for Surveyors V, VI, and VII because of a higher shock-absorber spring rate. Reimpact, in some cases followed by a second slight rebound, was followed by a ring-out oscillation, after which (beyond the range of the display in fig. 4-4) the force indications settled at values indicating the lunar spacecraft weight.

The second landing of Surveyor VI occurred after the vernier engines had been fired in a test on the lunar surface. This firing caused the spacecraft to lift off and land 2.4 meters from the first landing. Complete strain-gage records for the Surveyor VI initial landing, presented in figure 4-5, contain the most clearly defined ring-out oscillations of any Surveyor landing.

Spacecraft/Soil Interactions

Visible imprints made by the footpads (figs. 4-6 and 4-7[2]) during the landings indicate that

[2] Unless otherwise noted, all individual pictures and mosaics are from narrow-angle television frames.

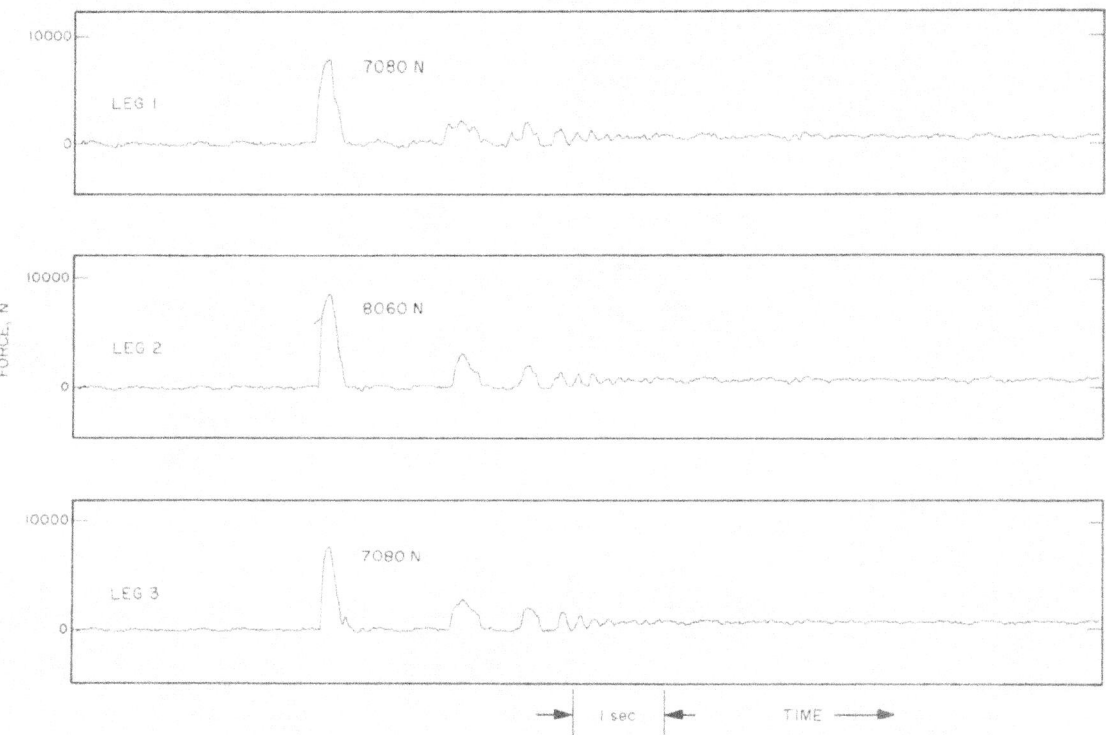

FIGURE 4-5.—Surveyor VI shock-absorber axial load histories which show the complete ring-out oscillations during the initial landing. These late oscillations are the most highly developed of any Surveyor landing.

the lunar soil has been compressed, displaced, and ejected. (Portions of the footpad imprints were obscured by the footpads and by other spacecraft components.) Motion of the footpads during impact pushed the lunar soil downward and outward. The horizontal motions of the footpads, caused both by horizontal spacecraft velocity and by outward movement of the legs during shock-absorber deflection, formed ridges, primarily outboard of the footpads. The footpads are surrounded by ejected soil that appears darker than the undisturbed surface. The ejecta, which extend to distances as great as 90 cm from the footpads, consist of fine and coarse soil thrown out over the surface during impact and subsequent sliding of the footpads.

Figures 4-6 and 4-7 show footpads 2 and 3 in their final positions and the adjacent areas of Surveyors I, III, V, VI (before and after hop), and VII. (Footpad 1 and the adjacent area were not visible to the television camera.) Figure 4-8 shows the imprints made during the second landing event of Surveyor III, as viewed from the spacecraft after the final landing. Figure 4-9 shows the imprints made by footpads 2 and 3 during the initial landing of Surveyor VI, as viewed from the spacecraft after the hop.

Footpad imprint details vary, primarily because of the different landing conditions encountered by each spacecraft. After Surveyor I bounced, the footpads came to rest in the initial imprints (figs. 4-6(a) and 4-7(a)). The imprints made during the first landing event of Surveyor III were not located; however, imprints made during the second landing event and rebound are shown in figure 4-8. Multiple imprints were made by each footpad during the third landing event (figs. 4-6(b) and 4-7(b)). The Surveyor V footpads plowed trenches (figs. 4-6(c) and 4-7(c)) in the soil as the spacecraft slid down the inner slope of a small crater 9 to 12 meters in diameter. The footpads rotated about their hinge axes

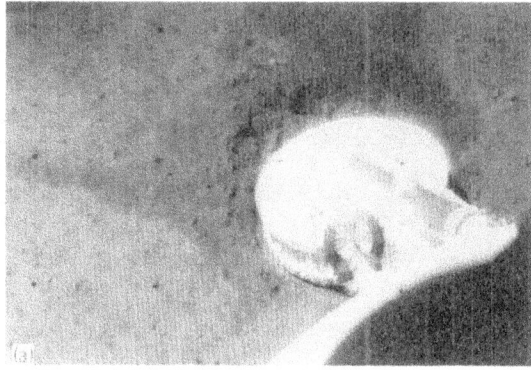

FIGURE 4-6.—Lunar surface areas disturbed during footpad 2 landing impacts. (a) Surveyor I: wide-angle picture showing rayed ejecta pattern (June 5, 1966, 11:29:49 GMT). (b) Surveyor III: mosaic showing high reflectivity of bottom of first two overlapping imprints outboard of footpad. Edge of a third imprint, formed in ejecta from first two impacts, is visible over top of footpad (Apr. 26, 1967, Catalog 3-MP-3). (c) Surveyor V: wide-angle pictures showing trench formed as spacecraft slid down the inner slope of the small crater in which it landed. The deepest penetration was on the uphill end of the trench where initial impact occurred (Sept. 14, 1967, Catalog 5-MP-19). (d) Surveyor VI: after initial landing. Wide-angle frames showing part of imprint made during first impact and symmetrically distributed ejecta formed when the spacecraft had a near-vertical final descent onto an almost horizontal lunar surface (Nov. 15, 1967, Catalog 6-SE-25). (e) Surveyor VI: after hop. Part of mosaic of wide-angle pictures showing portion of imprint made during first impact at right of footpad; spacecraft bounded back toward prehop position after making this imprint. Most soil was ejected in the direction the spacecraft was moving during the hop (Nov. 17, 1967, Catalog 6-SE-9). (f) Surveyor VII: imprint made during first impact is to the lower right of the footpad. The rock beside the footpad is about 20 cm across (Jan. 11, 1968, Catalog 7-SE-18).

MECHANICAL PROPERTIES

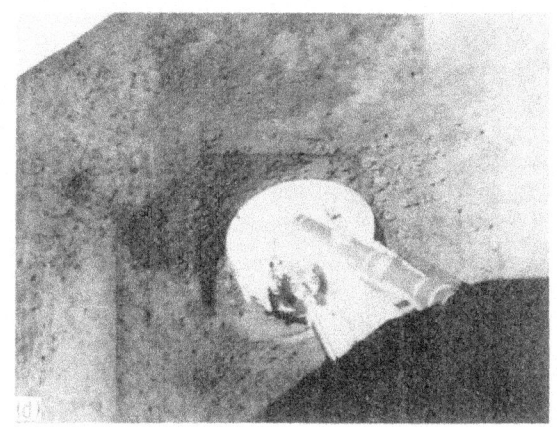

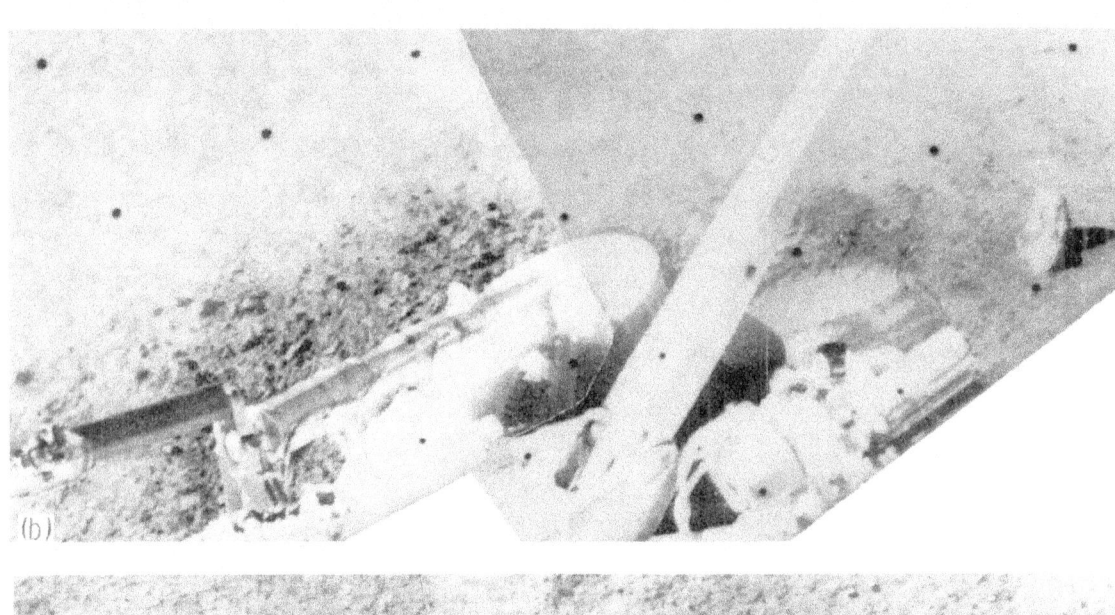

FIGURE 4-7.—Lunar surface areas disturbed during footpad 3 landing impacts. (a) Surveyor I (June 12, 1966, Catalog 1-SE-5). (b) Surveyor III; wide-angle pictures showing imprint below antenna boom made during first impact of the final landing (Apr. 30 and May 1, 1967). (c) Surveyor V; wide-angle pictures showing far wall of trench (below shock absorber) formed as spacecraft slid downslope (Sept. 20, 1967, Catalog 5-MP-33). (d) Surveyor VI; after initial landing (Nov. 17, 1967, Catalog 6-SE-19). (e) Surveyor VI; after hop (Nov. 18, 1967, Catalog 6-SE-24). (f) Surveyor VII; one of the numerous rocks in the landing area made a hole in the crushable footpad behind the magnet bracket (Jan. 17, 1968, Catalog 7-SE-20E).

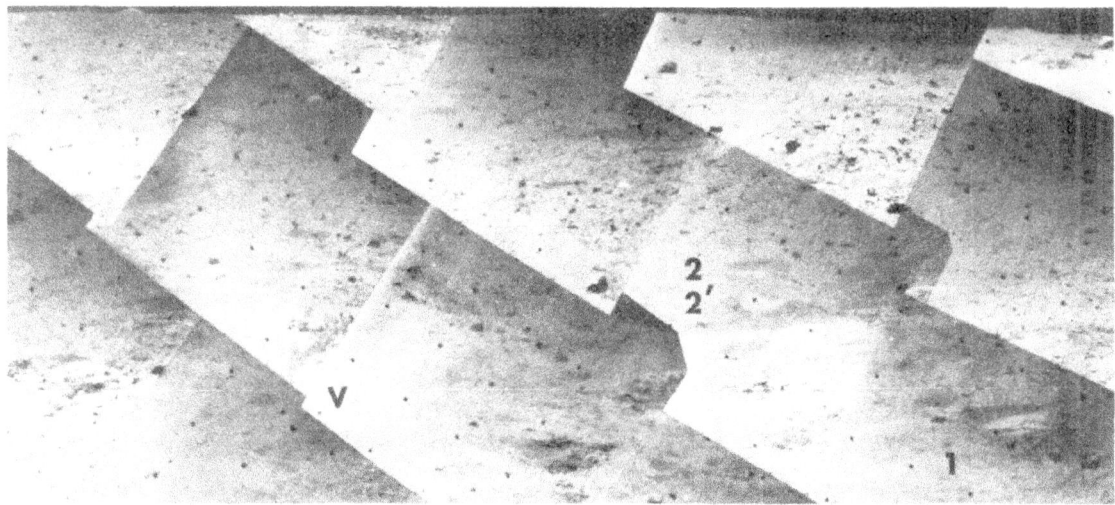

FIGURE 4-8.— Mosaic showing soil disturbance, about 11 meters from the television camera, caused by second landing event of Surveyor III. Footpad imprints are indicated by numerals and vernier engine 3 disturbance by a V. A second footpad 2 imprint formed during a short bounce is indicated by the symbol 2' (Apr. 26, 1967, Catalog 3–MP–1).

during the slide, with their outboard edges tipped downward, and lunar soil was deposited on the footpad tops (fig. 4-6(c)). Footpad imprints in figures 4-6(d) and 4-7(d) were made at first impact during the initial landing of Surveyor VI; other pictures of these same imprints were obtained after the hop (fig. 4-9). Figures 4-6(e) and 4-7(e) show the footpad imprints made visible because of the spacecraft bounce after the hop. Because of the substantial horizontal velocity acquired during the hop, a greater amount of soil was ejected and was thrown further by the footpads than during any other landing except for the downhill slide of Surveyor V (fig. 4-6(d)). Footpad 3 of Surveyor VII came to rest in its initial imprint on top of a small rock (fig. 4-7(f)). A rock, about 18 cm long and at least 10 cm high, near the final position of footpad 2, appears to have been moved during landing (fig. 4-6(f)).

Pictures of the areas disturbed by the footpads show clearly that the lunar surface material is granular. Many fragments are visible in the television pictures. However, as shown in the smoothed walls and bottoms of imprints and trenches made by the footpads, the surface sampler, and the sensor head of the alpha-scattering instrument, most of the soil is composed of particles finer than the resolution of the television camera (approximately 1 mm at footpad distances). Fragments larger than a few millimeters in diameter are more abundant in material ejected by the footpads than on the nearby undisturbed surface. This observation has been confirmed by particle count (see fig. 3-23 of ref. 4-6). Thus, many of the larger fragments in the disturbed material must be aggregates of fine soil displaced during footpad impacts. Most of the fragments in the ridges formed by the footpads are dark; similar dark fragments are visible on the undisturbed surface. Lighter fragments are visible on the undisturbed surface and occasionally on the disturbed surface; they probably are rock fragments and are much more abundant at the Surveyor VII highland site than at the mare sites.

Imprints formed by the crushable blocks mounted under the spaceframe were observed in Surveyor I, VI, and VII pictures (fig. 4-10). On all spacecraft except Surveyor I, flat or curved auxiliary mirrors were attached below the spaceframe to improve the view of the crushable blocks and of the area directly under them. Some soil particles could be seen adhering to the bottom of a Surveyor V crushable block.

Some of the footpad imprint pictures show

MECHANICAL PROPERTIES 141

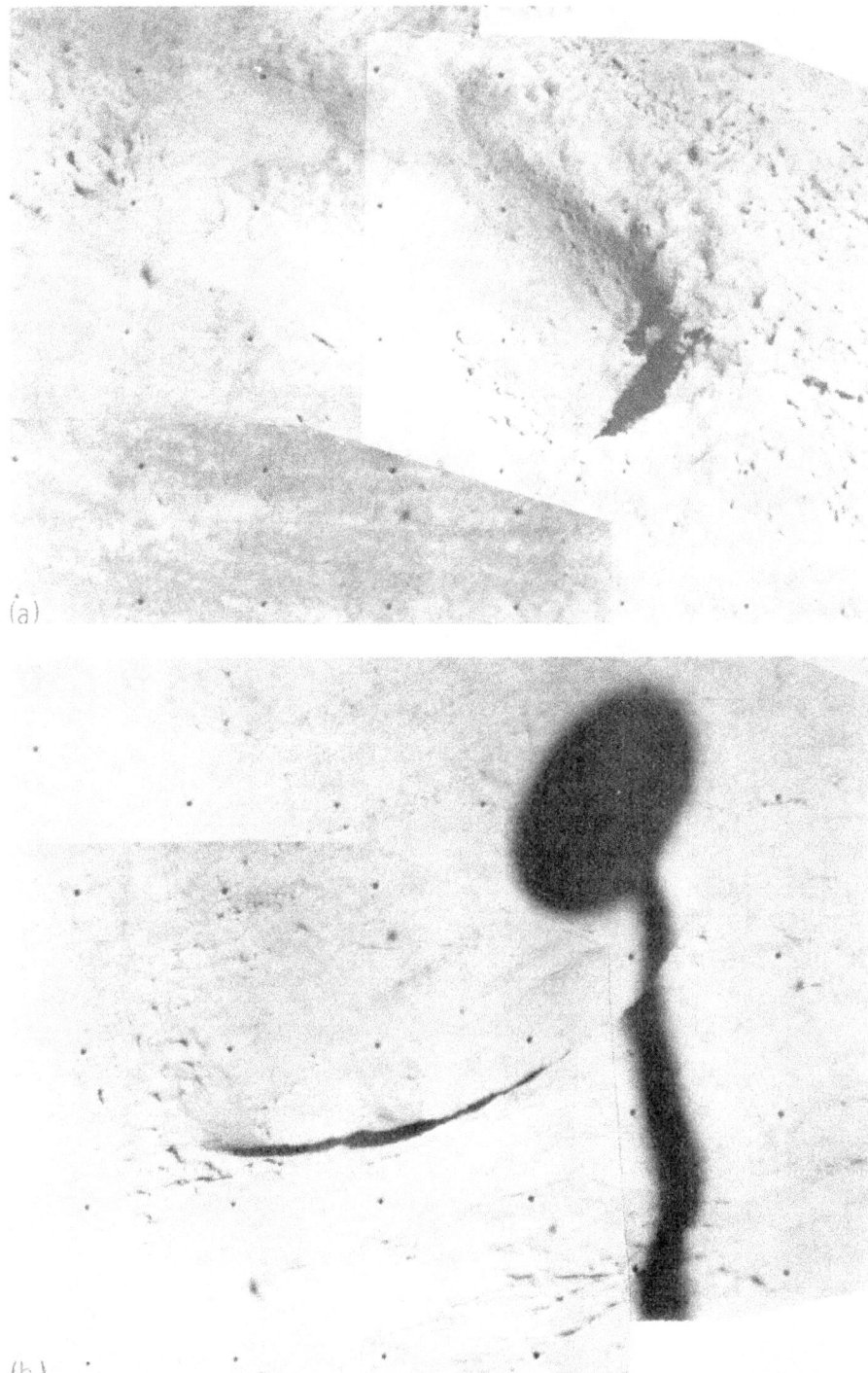

FIGURE 4-9.—Surveyor VI footpad imprints made during initial landing, as seen from posthop position. (a) Footpad 2 imprint (Nov. 17, 1967, Catalog 6-SE-39). (b) Footpad 3 imprint (Nov. 22, 1967, Catalog 6-SE-38).

FIGURE 4-10.—Imprints on lunar surface made by crushable blocks. (a) Surveyor I: crushable block 3 imprint. Imprint walls are almost vertical; depth of penetration is about 2 cm (June 3, 1966, 06:34:09 GMT, computer processed). (b) Surveyor VI: crushable block 3 imprint, made during first landing, as seen through auxiliary mirror. The mound of soil in the center of the imprint was left when the aluminum foil that covers the bottom of the cored block ruptured during landing (Nov. 12, 1967, 07:08:08 GMT). (c) Surveyor VI: crushable block 3 imprint made during first landing, as seen from posthop position. This imprint is the same imprint seen in figure 4-10(b) (Nov. 17-18, 1967, Catalog 6-SE-23). (d) Surveyor VII: crushable block 2 imprint, as seen through auxiliary mirror (Jan. 10, 1968, 05:45:23 GMT, computer processed).

clearly that the soil is inelastically compressible. For example, the second impact of footpads 2 and 3 on the Surveyor VI initial landing, overlapping the flat bottoms of the first imprints, produced sharp indentations without raised rims (fig. 4-9). This indicates that, even after the compaction caused by the first impact, the soil underwent additional anelastic compression. Further evidence of the compressibility of the top few centimeters of the soil is provided by the doughnut-shaped imprints of the crushable blocks (fig. 4-10). In the center of each imprint is a mound of soil, which was formed during landing when the aluminum sheet that covers the bottom of the cored block collapsed. The crushable block imprints have sharp outer edges and show no raised rim or disturbed soil outside the depressions, again indicating that the soil is compressible. Soil clumps, which fell from the crushable block assemblies as they retracted after landing, cover the imprint bottoms.

Evidence of soil cohesion may be seen in many of the Surveyor television pictures (fig. 4-11). A soil clump thrown 30 cm by a Surveyor V footpad had sufficient strength to resist crumbling when it landed. Another soil clump fractured on impact without disintegrating after being ejected by a Surveyor VII footpad (fig. 4-11(i)). Footpad imprints have steep walls with open fractures; soil clumps overhang a Surveyor VI imprint in cantilever fashion (fig. 4-11(f)). Cracks in surface material that had been moderately disturbed were also visible near Surveyor VII crushable block 3 imprint (ref. 4-5). The surface samplers on Surveyors III and VII dug trenches which retained vertical walls 10 to 15 cm high. These examples show that the lunar surface material is cohesive.

The sides and bottoms of footpad imprints, in many cases, are smooth. These smooth surfaces sometimes clearly reproduced, through the covering of thin aluminum foil, the aluminum honeycomb pattern of the flat bottoms and conical sides of the footpads. Laboratory tests of a Surveyor footpad (fig. 4-3) indicated that the ridges on the footpad bottom were probably 40 to 80 microns high. Thus, it is hypothesized that a large percentage of the surface material is composed of particles smaller than 60 microns in diameter (ref. 4-2).

Disturbed soil outside the imprints is darker than nearby undisturbed soil. The contrast in brightness between the disturbed and undisturbed soil was quite pronounced at all landing sites, although it was somewhat less at the Surveyor V site. Also, the smooth bottoms of footpad imprints are brighter than undisturbed soil (figs. 4-6, 4-7, and 4-9), except at small phase angles, that is, when the angle is small between the camera line of sight and the Sun direction. These photometric changes indicate that the fine structure of the soil surface is changed by physical disturbance; apparently, reflectivity decreases with increase in porosity and surface roughness.

With the exception of Surveyor VII, when some local deformation and tearing of footpad material occurred (fig. 4-12), there was no evidence of crushing of the footpad aluminum honeycomb or crushable blocks on landing. Crushing strengths were 7 N/cm^2 for footpad bottoms, 14 N/cm^2 for the footpad tops, and 28 N/cm^2 for the crushable blocks. Two small, very bright objects, which were noted on the surface near Surveyor VII, may be pieces torn from the aluminum honeycomb by a rock. One of these objects is shown in figure 4-13.

No settling of the Surveyor spacecraft after landing was detected, except in response to spacecraft commands and to compression of shock absorbers during the lunar night. No changes or movements, except those caused by spacecraft operations, were noticed in disturbed or undisturbed lunar surface material over periods of observation of up to 6 weeks.

Simulations of Landings and Hop

Computer simulations of the landings have been performed to estimate the mechanical properties of the lunar surface material at each landing site. The estimates were based on comparisons of the simulated landing data with shock-absorber force histories obtained from flight data and with footpad and crushable block surface penetration depths estimated from television pictures. Landing simulations were performed using both rigid and soft-surface models. In addition, simulations of the first two Surveyor III landings and of the Surveyor VI hop were performed to obtain estimates of

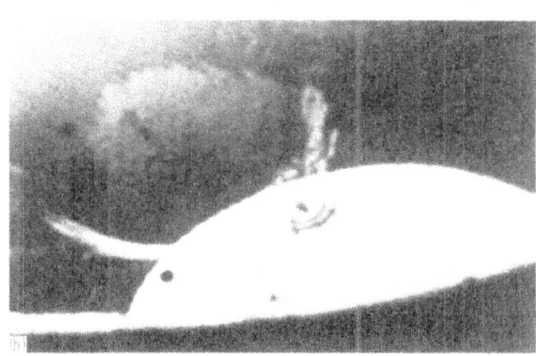

FIGURE 4-11.—Textural detail of footpad imprints. (a) Surveyor I footpad 2. Ejected soil clumps (which, for the most part, are larger than naturally occurring fragments adjacent to the footpad) demonstrate that the soil is cohesive (June 4, 1966, 06:55:26 GMT, computer processed). (b) Surveyor III footpad 2 imprint made during final landing has a waffle pattern impressed by the bottom of the footpad; low Sun angle illumination (Apr. 20, 1967, 09:05:17 GMT). (c) Surveyor III footpad 2 imprint as seen with high Sun angle illumination. Walls and bottom of imprint have higher reflectivity than natural undisturbed lunar surface. Flat surfaces and shapes of many fragments on near side of imprint indicate that the fragments have been only slightly displaced (Apr. 26, 1967, 06:05:55 GMT). (d) Surveyor V part of trench wall formed by footpad 2; cohesion permits soil to stand at angles greater than the angle of repose of loose material (Sept. 14, 1967, 04:42:06 GMT). (e) Surveyor VI footpad 2 imprint made during initial landing, showing compressed soil in imprint wall and floor (Nov. 16, 1967, 03:35:09 GMT). (f) Surveyor VI footpad 2 imprint made during hop. The smoothness of the imprint wall demonstrates the fine-grained nature of lunar soil. Particles are smaller than the

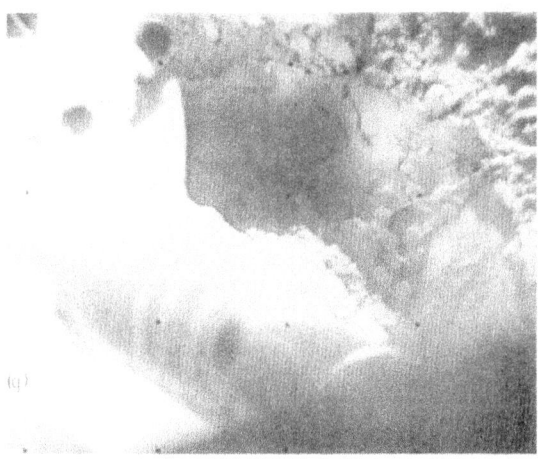

television camera can resolve (approximately 1 mm at footpad distance). Note the soil clumps that overhang the imprint rim and the lineations in the imprint wall impressed by the conical side of the footpad (Nov. 20, 1967, Catalog 6-SE-42). (g) Surveyor VII footpad 2 imprint; smoothness of wall surfaces shows that the soil is fine grained (Jan. 11, 1968, 12:28:52 GMT). (h) Surveyor VII footpad 2 imprint; the 45° wall of the imprint shows that the soil is cohesive. Note impression of the footpad honeycomb structure left in the imprint wall (Jan. 13, 1968, 10:18:45 GMT). (i) Surveyor VII footpad 2 imprint and ejecta. Soil clumps, fractured on impact after they were ejected by the footpad at landing, can be seen at lower left (Jan. 13, 1968, 10:21:44 GMT).

vernier-engine thrust histories and heights of engine nozzles above the lunar surface. These estimates have been used for surface erosion analysis.

A Surveyor footpad will not begin to crush until the pressure on it exceeds 7 N/cm^2. Therefore, the leg force histories (for up to 5 cm of crushing of the footpad bottoms) are the same for landings on surfaces with any bearing strength above 7 N/cm^2. The honeycomb blocks crush at 28 N/cm^2. Thus, any surface with a bearing strength greater than 28 N/cm^2 is essentially rigid, as far as the response of the entire spacecraft landing gear is concerned. Rigid surface landing simulations have produced shock-absorber force histories which agree reasonably well (peak forces generally higher, but deviations less than 20 percent) with

FIGURE 4-12.—Surveyor VII footpad 3 showing hole in crushable honeycomb behind magnet bracket; the hole was caused by impact on a rock during landing (Jan. 20, 1968, 20:25:20 GMT).

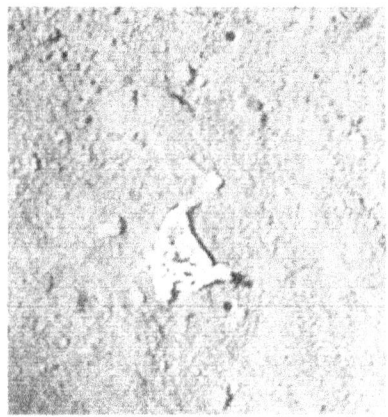

FIGURE 4-13.—Unidentified bright object on lunar surface to right of Surveyor VII footpad 2, which probably is a piece of aluminum torn from footpad or crushable block honeycomb during landing (Jan. 19, 1968, 19:42:37 GMT, computer processed).

corresponding flight data. These results indicated that force histories are relatively insensitive to surface hardness—a fact that has been confirmed by soft-surface simulations with surface hardnesses as low as 2 to 3 N/cm². Therefore, consideration of force histories alone is not sufficient; penetration data also must be considered to determine surface mechanical properties by use of soft-surface mathematical models such as described in references 4-7 and 4-3.

The soil model of reference 4-7 considers a bearing strength versus penetration depth given by $\sigma = \alpha + \beta x$, where α and β are coefficients and x is the vertical penetration. A third term ($\gamma \ddot{x}$), shown by analysis to have a negligible effect for Surveyor landings, was not included in the calculations. The surface horizontal force was assumed to be the product of a friction coefficient and the vertical resistance to penetration. After a preliminary investigation, a friction coefficient of 0.7 was adopted for the remaining analyses. The results indicated that at the Surveyor I landing site and on the scale of the Surveyor footpad the uppermost portion of the lunar soil has a bearing strength of less than 1.4 N/cm², and bearing strength increases with depth with an increasing rate and reaches a value in the range of 4.2 to 5.6 N/cm² at a depth of 5 cm. The results indicated that simulated penetrations of Surveyor I into such a surface would be approximately 4.1 cm and 3.8 cm for footpads 2 and 3, respectively. An extension of the investigation of reference 4-7, using revised penetration estimates (table 4-2), indicates that the lunar surface bearing strength reaches the range of 4.2 to 5.6 N/cm² at a depth of approximately 4 cm.

The compressible soil model defined in the Surveyor V Mission Report (ref. 4-3) and used in preparing all Mission Reports considers a bearing strength variation expressed by

$$p = p_0(1+cs) + \frac{\rho_1 \rho_2}{\rho_2 - \rho_1} \dot{s}^2$$

where

 p = pressure exerted on the penetrating object
 p_0 = static bearing pressure
 c = depth-proportionality constant
 ρ_1 = initial density of the soil
 ρ_2 = compressed density of the soil
 s = penetration

A small depth-proportionality factor ($c = 0.0328$/cm) was used; i.e., the static bearing pressure, p_0, was assumed to increase by approximately 16 percent for each 5-cm surface penetration. Consequently, this study provided values of effective average bearing strength for a depth corresponding to the maximum penetration for each landing. On this basis, a p_0 of 3.4 N/cm² gave consistently good correlation between analyses and flight data for the landings of Surveyors I, III, VI, and VII (refs. 4-1, 4-2, 4-4, 4-5, 4-8, and 4-9). (The correlation for Surveyor VI is shown in fig. 4-14.) However, for the Surveyor V landing, it

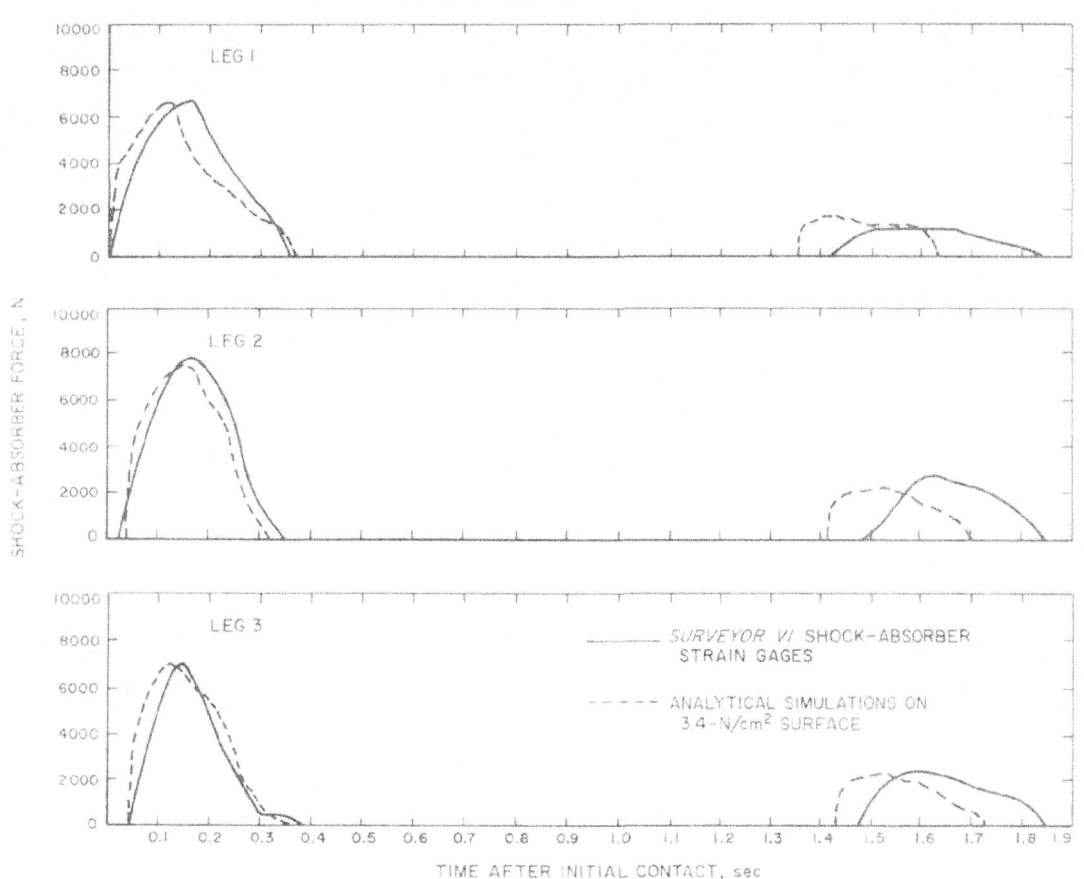

FIGURE 4-14.—Surveyor VI initial landing data and analytical shock-absorber force/time histories.

was necessary to use a p_0 of 2.8 N/cm² for acceptable correlation (ref. 4-3). This appeared to indicate a softer surface at the Surveyor V landing site. It should be pointed out, however, that the bearing capacity of a footing on a slope is smaller than that of the same soil on a horizontal surface. Consequently, the lower p_0 indicated for Surveyor V is at least partially a result of the 20° slope on which the spacecraft landed, rather than an indication of a difference in soil properties. For the posthop landing of Surveyor VI, a good force history correlation could not be achieved because of an unusual force history generated by the leg 1 shock absorber (fig. 4-4(g)). It is thought that footpad 1 struck a rock, which restricted its lateral movement during the landing, thus causing the unusual force history.

Footpad 2 penetrations resulting from landing simulations using the compressible soil model are compared in table 4-2 with the penetration values calculated from television data. From these results, it appears that penetrations obtained from landing simulations on surfaces with p_0 from 4 to 6 N/cm would bracket the actual penetrations as determined from television pictures. Thus, based on this compressible soil model, the average lunar surface bearing strength to a depth of about 6 cm is estimated to be from 4 to 6 N/cm².

During the Surveyor VI landing, a clear imprint was made by crushable block 3 (fig. 4-10(c)). A small mound of soil in the bottom of the imprint is attributed to rupture of the thin aluminum foil that covers the bottom of the cored crushable block. Tests, described in reference 4-4, were performed to establish the pressure levels (2.4 N/cm²) required to rupture

TABLE 4-2. *Footpad 2 penetrations calculated from computer compressible surface landing simulations. These values are compared with penetrations measured from television observations*

Spacecraft	Bearing strength used in simulations, N/cm² $p_0=$	Footpad 2 initial penetrations, cm	
		From landing simulations	From television observations [a]
Surveyor I	3.4	4.7	3±1
	4.9	3.7	
Surveyor III (final landing event)	3.4	5.0	2.5±1
	4.9	2.5	
Surveyor V	[b]2.8	12	12±2
Surveyor VI (initial landing)	3.4	6.9	±2
Surveyor VII	3.4	8.4	4±1
	2.0 (penetration <2.5 cm) 4.9 (penetration >2.5 cm)	3.5	

[a] Work is being performed on improving these values and will be published as soon as available.
[b] Low bearing strength value of Surveyor V site is at least partially caused by the 20° surface slope.

the aluminum foil. With the compressible soil model described above, it was calculated that the static bearing strength of the lunar soil at a depth of 1 to 2 cm is about 1.8 N/cm².

The first two Surveyor III touchdowns and the hop of Surveyor VI were simulated using a landing dynamics computer program that incorporated a mathematical model of the Surveyor flight control system. The Surveyor III landing analysis (refs. 4-10 and 4-11) produced the estimated touchdown velocities shown in table 4-1. Variations of engine thrust levels and engine height above the surface for the second landing are shown in reference 4-2. Simulations of the Surveyor VI hop were also made (ref. 4-4) to provide data on engine thrust levels and engine height above the surface. Engine thrust data are shown in figure 4-15.

Depth of Footpad Soil Penetrations

Estimating footpad penetration during first impact for each Surveyor landing is difficult since the television pictures obtained after landing show only the final position of the footpads, not their position at maximum penetration during first impact. During all the landings, the footpads experienced two or more impacts because of spacecraft rebound; imprints caused by the first impact generally were disturbed, or completely obscured, by subsequent impacts.

Two methods were used to determine the depths of footpad penetrations. The first method consisted of reconstructing footpad landing imprints in the laboratory by using a Surveyor footpad and a soil of crushed basalt. Lighting angles of a collimated light were adjusted to match the Sun's position during the lunar observations; a prototype television camera on a full-scale Surveyor model was used to view the results.

The second method was an analysis of spacecraft shadows as described in reference 4-12. In this approach, estimates relative to footpad dimensions were made of the distance and height of lunar soil features around a footpad using lengths of shadows of Surveyor television pictures taken at different Sun angles.

By combining the results of both methods, the general features of footpad imprints and ejecta rims were reconstructed and vertical dimensions of the imprints were determined.

The average footpad penetration values listed in table 4-1 are the estimated final depths below the natural lunar surface of the center of the footpads in their present position. Local lunar surface variations and differences in the way soil was ejected during impact caused significant variations in the depths of soil around each footpad. Estimates of initial

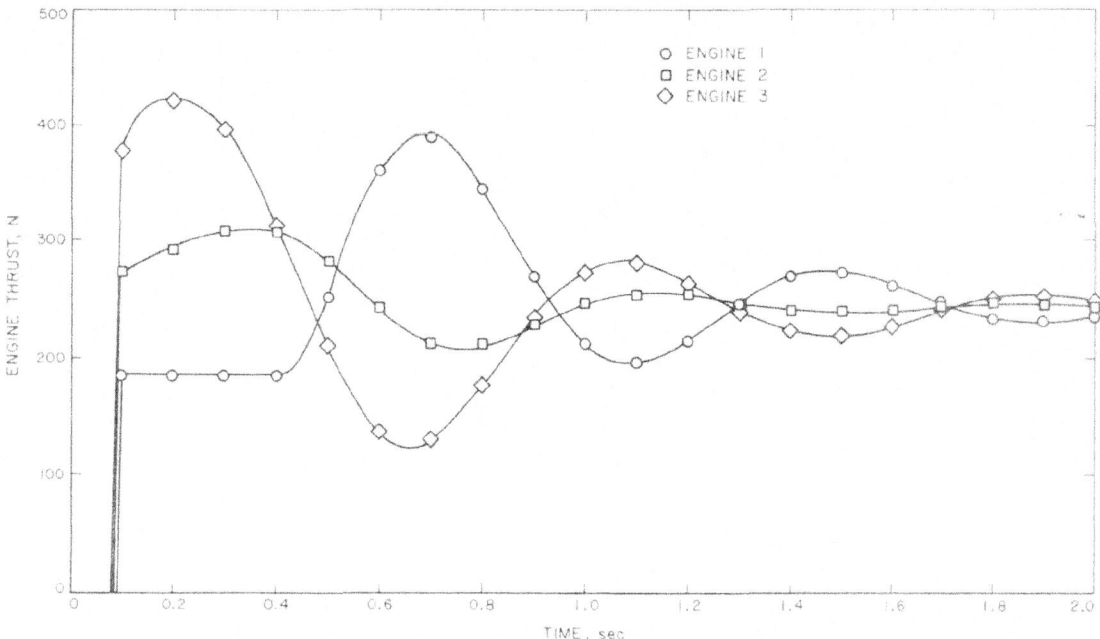

FIGURE 4-15.—Variation of Surveyor VI vernier-engine thrust with time during the hop.

footpad 2 penetrations, made during mission operations, are given in table 4-2. Improvements in these estimates are being made; results will be published as soon as available.

Elastic Properties of the Lunar Soil

Damped oscillations were observed on the strain-gage records following the final landing impacts of Surveyors I, III, and VI, and the hop made by Surveyor VI. The oscillations were most clearly recorded from the initial landing and from the hop of Surveyor VI. Clear oscillations were not observed for Surveyors V and VII.

The frequency of these oscillations can be related to the combined elastic properties of the spacecraft and the lunar soil; estimates of the elastic properties of the lunar soil can be made provided that the constants for the spacecraft are known. In addition, the rate of decay of the oscillations provides data for an estimate of the dissipative (anelastic) properties of the soil. Laboratory tests have been conducted to determine the elastic and dissipative constants of the spacecraft; analysis of these tests is presently in progress.

For a single mass-spring system with viscous (velocity) damping, the free oscillations are of the form ($f(t)$ = deflection)

$$f(t) = Ce^{-\gamma \omega_0 t} \sin \left[\omega_0^2(1-\gamma^2)\right]^{1/2} t$$

where γ is the fraction of critical damping and ω_0 is the undamped, angular resonant frequency. The quality factor Q of a resonant system is $Q = \dfrac{1}{2\gamma}$. For large Q, i.e., small γ, the damped frequency is approximately equal to ω_0. For a spacecraft on the lunar surface, ω_0 and γ depend on the stiffness and damping of the lunar surface material, as well as those of the spacecraft.

The oscillations in each leg were of the same frequency and in phase, indicating a vertical mode of vibration. During the oscillations, the maximum force developed in each leg was about 620 N, which is equivalent to about 1-mm displacement of the spacecraft center of gravity. For such a small motion, linear approximations for the spacecraft motion appear justified. However, there is some dispersion in the oscillations, with the frequency increasing slightly with decreasing amplitude. The reason for this dispersion, which is neglected in the present analysis, is not known. It may be caused by nonlinearities in the system.

For the initial landing of Surveyor VI, the average frequency of the observed oscillations is 6.3 Hz and Q is about 9; for the posthop landing, the frequency is 6.9 Hz and Q about 12. An approximate analysis indicates that a spacecraft sitting on a rigid surface would have an oscillation frequency of 8.0 Hz ± 10 percent. Provided that the constants of the spacecraft were essentially the same in both cases, this observation indicates that the lunar surface material, as loaded by the Surveyor VI footpads, had greater stiffness and contributed less damping at the second landing location than at the first landing location. For Surveyors I and III, the observed frequency was about 6.5 Hz; an estimate of Q has not yet been obtained.

In the absence of damping, the effective stiffness of the lunar surface material under one footpad is approximately

$$K_m = \frac{4\pi^2 M f_s^2 f_m^2}{f_s^2 - f_m^2}$$

where M is one-third of the spacecraft mass, f_s is the resonant frequency for the spacecraft on a rigid surface, and f_m is the observed frequency.

The rigidity modulus G for the lunar material can be estimated using the relation (ref. 4-13)

$$G = \frac{2K_m(1-\nu)}{\pi^2 r}$$

where r is the radius of the loaded area, and ν is Poisson's ratio. Since ν lies between 0 and 0.5 for all common materials, this relationship provides an estimate of G. Then, the shear wave velocity of the lunar soil, V_s, can be estimated for a given value of G and various assumed values of bulk density, ρ, since $V_s = (G/\rho)^{1/2}$. In addition, the compressional wave velocity, V_p, can be obtained from the relation

$$V_p^2 = \frac{2V_s^2(1-\nu)}{1-2\nu}$$

For Surveyors I and III, the oscillation frequency was 6.5 Hz, which results in $K_m = 4.9 \times 10^5$ dynes/cm. This implies a rigidity modulus range from 3.9×10^6 to 7.8×10^6 dynes/cm², for ν varying from 0.5 to 0, respectively, and a shear wave velocity of 16 to 28 m/sec for a bulk density of 1.0 to 1.5 g/cm³. For Poisson's ratio less than 0.45 and the same range of density, a compressional wave velocity of 33 to 70 m/sec is obtained (ref. 4-2).

By using the frequencies from the Surveyor VI landing, the value of K_m correlating with the oscillation frequency of 6.3 Hz would be reduced by about 10 percent from that given above and increased by about 60 percent for the oscillation frequency of 6.9 Hz. Corresponding changes in the velocities would be −5 percent and +30 percent, respectively.

The estimated seismic velocities are considerably lower than those expected for terrestrial soils with other mechanical properties as described in this report. Until tests on the model spacecraft have been evaluated, these results should be considered as preliminary estimates only.

Lunar Soil Erosion

During a nominal Surveyor lunar landing (engine cutoff at about 4-meter altitude), the vernier-engine exhaust gas forces acting on the lunar surface are so small that little, if any, surface erosion occurs. During the Surveyor III landing, the vernier engines continued to fire until after the spacecraft made two contacts with the lunar surface. This landing provided the first indication of the effects of rocket gases impinging on the lunar surface. To investigate the effects of the gas plume on the lunar surface, the Surveyor V vernier engines were fired for 0.55 second about 2 Earth days after landing at a thrust level less than the spacecraft lunar weight. A second erosion test was performed 6 Earth days after Surveyor VI landed. Its engines were fired for 2.5 seconds at thrust levels sufficient to cause the spacecraft to rise about 3.5 meters and land 2.4 meters from the initial landing point, thereby providing good views of the effects of the firing on the lunar surface. On Surveyors I and VI, the attitude control jets, which exert much lower forces than the vernier engines, were operated to investigate their effects on the lunar surface.

Types of Lunar Soil Erosion

Terrestrial tests have demonstrated that the vertical and horizontal shear forces exerted by rocket gases impinging on a horizontal soil surface could cause lunar surface erosion or cratering by three basic processes:

(1) Viscous erosion. Entrainment of soil particles as the exhaust gases flow over the surface (refs. 4-14 and 4-15, theoretical studies; ref. 4-16, an experimental study).

(2) Diffused gas erosion. Movement of soil caused by the outward and upward flow of gas through the pores of the soil (ref. 4-17). An eruption of the soil could occur if an engine is rapidly shut down.

(3) Bearing load cratering (also called explosive cratering). Rapid cratering caused when the exhaust gas pressure on a surface exceeds the bearing capacity of the surface (ref. 4-18). With the full expansion of Surveyor exhaust plumes in the lunar environment, this type of erosion was not likely to occur.

Vernier-Engine Firings

Surveyor III provided the first indication of the erosion effects of rocket gases on the lunar surface. The firing of the vernier engines during the Surveyor V mission was intended primarily to determine the diffused-gas eruption effects resulting from rapid engine shutdown. Surveyor VI engines were fired at a higher thrust level, and for a longer period of time, to increase the viscous erosion effects.

Observations

SURVEYOR III. The vernier engines of Surveyor III continued to fire during the first two touchdowns. The site of the second touchdown was visible to the camera from the final landed position, approximately 11 meters away. As seen in figure 4-8, not only are the imprints of the three footpads visible, but also a light streak of soil can be seen with adjacent dark soil; both light and dark soil are attributed to the firing of vernier engine 3. However, other than the indication that the vernier engines probably caused soil erosion, little additional information could be obtained.

SURVEYOR V. On September 13, 1967, 53 hours after landing, the Surveyor V vernier engines were fired at low thrust for 0.55 second. Engines 1 and 3 were fired at thrusts of 120 N; engine 2 was fired at 76 N. Study of Surveyor television pictures has shown that even though the spacecraft was resting on the inner slope of a small crater at an angle of about 20° (fig. 4-16), the firing caused no downslope motion of the spaceframe. The firing, however, did move the sensor head of the alpha-scattering instrument, which was resting on the lunar surface. During the firing, the sensor head rotated 15°, and its center of gravity moved 10 cm in a direction 45° from the direction of maximum slope. The lunar weight of the sensor head was 4.4 N. Two types of soil erosion occurred:

(1) Viscous erosion. A thin layer of soil was removed from beneath and adjacent to the vernier engines (fig. 4-17). Erosion of soil during the firing extended to distances at least up to 1.9 meters from the engines. As shown in the controlled[3] mosaics (fig. 4-18), the soil layer near vernier engine 3 and adjacent to the sensor head was substantially disturbed by the firing. Some of the soil and rock fragments moved by the firing are identified on these annotated mosaics. The largest fragment known to have been moved is 4.4 cm in diameter. Television pictures indicate that, at least in some places, soil was disturbed by viscous erosion to depths probably greater than 1 cm for distances up to 60 cm from engine 3. As shown in figure 4-18, soil at E, beside rock a, was eroded to a depth of about 1 cm. The trail (fig. 4-18(a)) left by rock H as it rolled downslope is no longer visible (fig. 4-18(b)). Figure 4-19 shows the relative distance that fragments of different sizes can be moved by gases striking the lunar surface with surface pressures equivalent to those of vernier engine 3 (ref. 4-3). This figure shows that fragments up to 4 or 5

---

[3] The controlled mosaics are composed of narrow-angle television frames mounted on a spherical surface; the center and orientation of each frame are correct relative to all other frames.

152   SURVEYOR: PROGRAM RESULTS

FIGURE 4-16.—Profile of Surveyor V and the crater in which it landed.

FIGURE 4-17.—Lunar surface beneath Surveyor VI vernier engine 3, as seen through an auxiliary mirror. (a) Prefiring picture. (b) Postfiring picture, showing the shallow crater caused by diffused gas eruption at engine shutdown.

cm in diameter were moved at distances up to approximately 20 cm; whereas, at distances of 200 cm, only fragments up to 0.4 cm in diameter were moved.

(2) Diffused gas erosion. Exhaust gases, which had diffused into the soil during the firing, caused the soil to erupt at engine shutdown and form a shallow, crescent-shaped crater (fig. 4-17(b)). The crater is 20 cm in diameter and 0.8 to 1.3 cm deep; the height of the vernier engine above the surface was 39 cm, and the maximum static pressure of the exhaust gases on the surface directly below the engine was 0.29 N/cm^2. Prefiring and postfiring pictures of the lunar surface below engine 3 are shown in figure 4-17.

SURVEYOR VI. On November 17, 1967, 177 hours after landing, the Surveyor VI vernier engines were fired for 2.5 seconds in order to lift the spacecraft from the lunar surface and to move it a short distance from the original landing site. This maneuver subjected the lunar surface to greater erosional forces from the vernier-engine exhaust gases than that exerted during the Surveyor V static firing. To achieve horizontal motion during the hop, the spacecraft's flight control system had been preset by Earth command such that the spacecraft acquired a tilt of 7° immediately following liftoff (fig. 4-20). This 7° tilt of the vernier engines caused soil eroded by the exhaust gases to be preferentially ejected to the east, away from the tilt direction (fig. 4-21).

Figure 4-22 is a mosaic of computer-enhanced pictures of the first landing site identifying the double imprints formed by the footpads and the single imprints formed by the crushable blocks during the original landing; the locations of the vernier engines before liftoff for the hop are also shown. On figure 4-22, the major areas of erosion caused by vernier-engine exhaust gases are identified with capital letters. Areas $A-E$ represent erosion principally attributed to vernier engine 2, areas $G-I$ to engine 3, and areas $K-N$ to engine 1 (ref. 4-4). Enlargements of the main erosion areas for each engine are shown in figures 4-23 through 4-25.

Some of the more pronounced erosion features, formed by the firing and visible in

MECHANICAL PROPERTIES

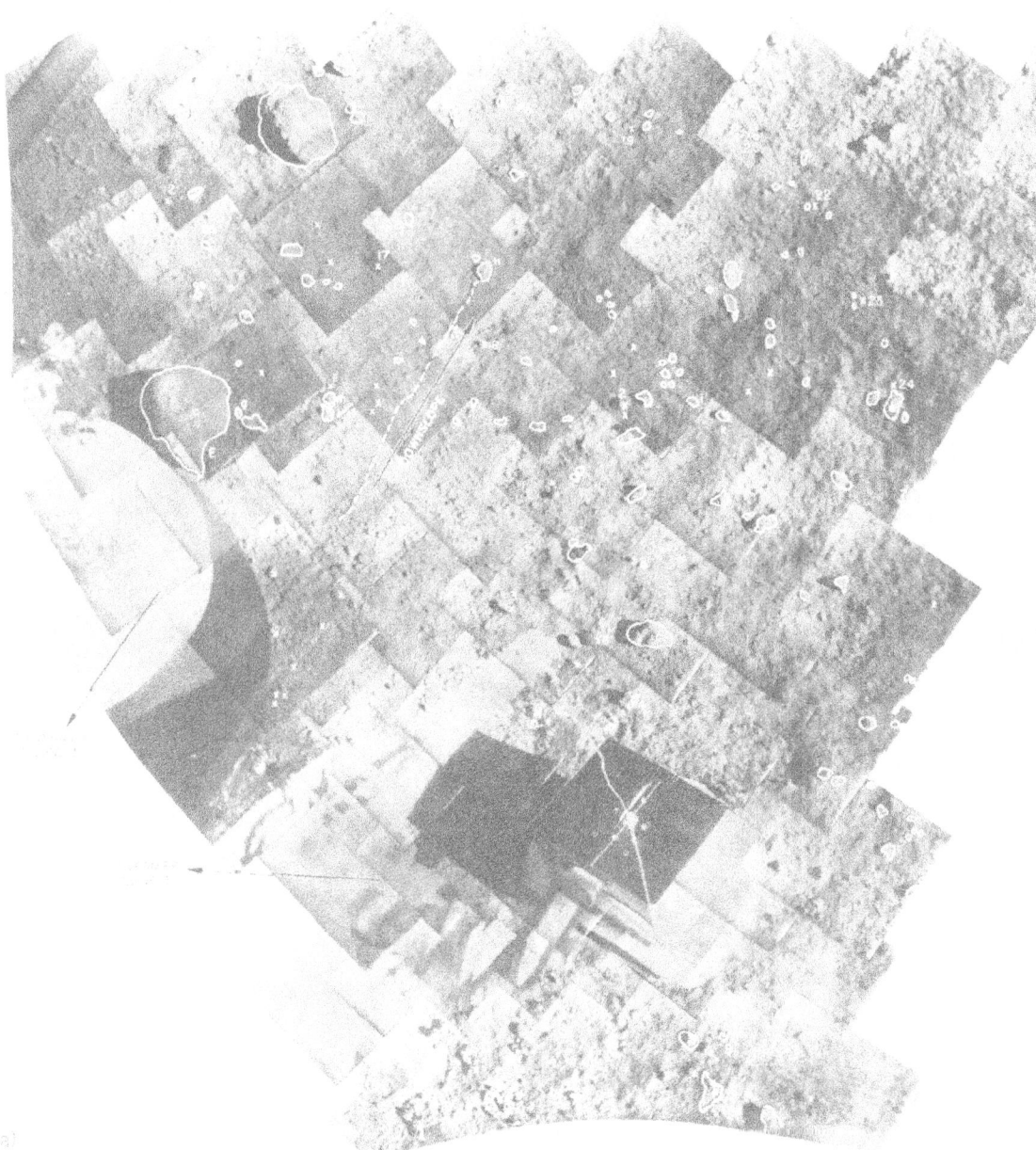

FIGURE 4-18.—(a) Prefiring, annotated mosaic of alpha-scattering-instrument area. Rock and soil fragments not moved by the firing are outlined; fragments shown by postfiring pictures to have been moved by the firing are marked with an X (Sept. 10, 1967). (b) Postfiring, annotated mosaic of alpha-scattering-instrument area. Fragments not moved by the firing are outlined; fragments that moved are marked with an X (Sept. 12, 1967)

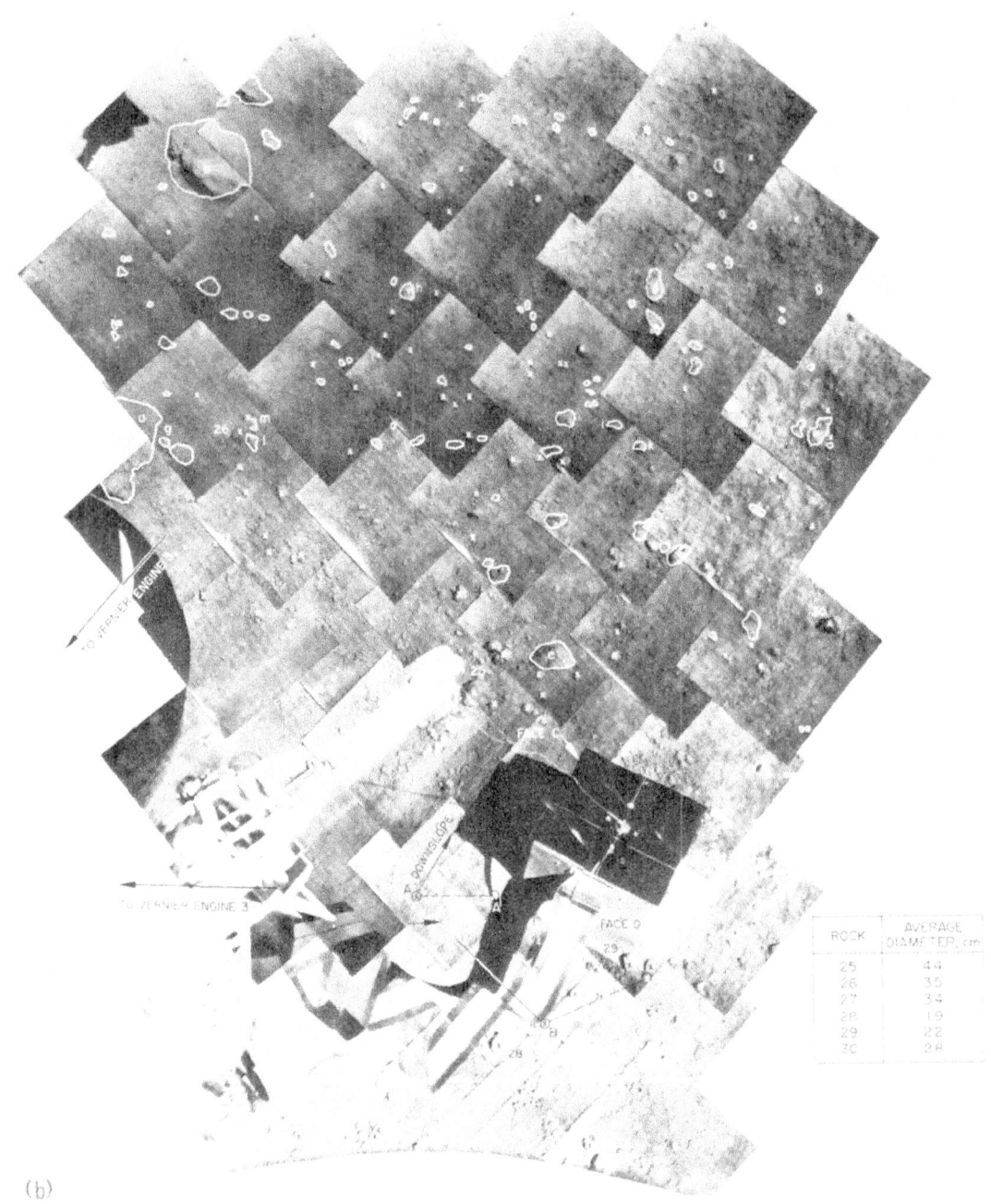

FIGURE 4-18.—Concluded.

figure 4-22, include: (1) fine, dark soil deposited in rays by engine 2 at A, B, and C; (2) the partial filling of the shallow depression at E; (3) the large number of coarse soil fragments deposited by engine 2 at D; (4) the surface with a rippled appearance at I caused by differential erosion by engine 3; and (5) the fan of fine, dark soil deposited in rays at M by engine 1. One or more of the soil clumps ejected by the firing hit the photo-

MECHANICAL PROPERTIES 155

FIGURE 4-19.—Graph of diameter versus distance for fragments moved by the Surveyor V vernier engine 3 static firing. The dashed line represents the probable maximum sizes for fragments that could be moved by the firing at distances ranging from 10 to 200 cm.

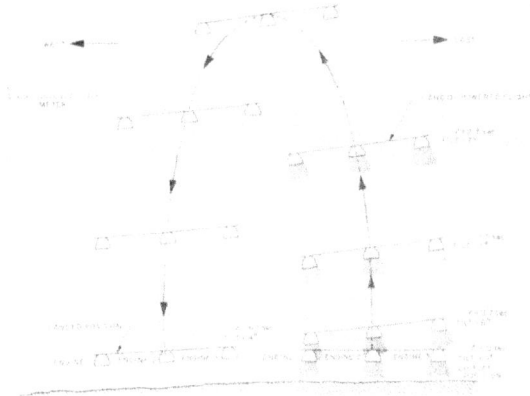

FIGURE 4-20.—Time sequence of vernier-engine positions during liftoff for the hop during the Surveyor VI mission. View is perpendicular to direction of hop.

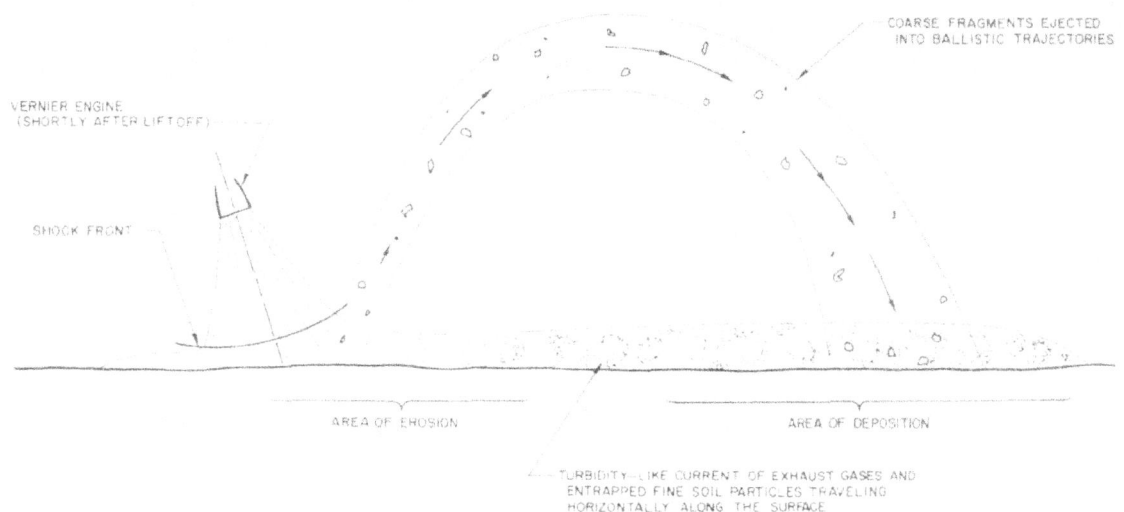

FIGURE 4-21.—Probable history of erosion in Surveyor VI vernier engine 2 area during hop.

metric target on one of the omnidirectional antennas and left a thick coating of soil adhering to the target (fig. 4-26).

Simulations and analyses.—The vernier-engine firing data and surface-pressure data for Surveyors III, V, and VI are summarized in table 4-3. The thrust levels listed were obtained from analytical simulations (Surveyors III and VI) and from strain gages on the vernier-engine support structure (Surveyor V). The minimum nozzle height listed for Surveyor III was estimated from the analytical simulations; those for Surveyors V and VI were obtained from comparisons of Surveyor pictures and photographs of laboratory simulations using a full-scale spacecraft (refs. 4-3 and 4-4). The lunar surface areas in the vicinity of Surveyor III engines 1 and 2 could not be viewed directly; therefore, their minimum nozzle heights are not included.

FIGURE 4-22.—Mosaic of posthop, computer-processed pictures showing the (1) approximate locations of the crushable blocks and vernier engines before liftoff for the hop; (2) the images of footpads above their final imprints made during the original landing; and the principal areas of erosion caused by the vernier engines during liftoff for the hop as indicated by capital letters (Nov. 15 and 16, 1967, Catalog 6-SE-22).

FIGURE 4-23.—Mosaic of computer-processed pictures, showing rays of fine, dark soil deposited by Surveyor V vernier engine 2 during the hop (Nov. 15 and 16, 1967, Catalog 6-SE-43C).

FIGURE 4-24.—Mosaic of computer-processed pictures, showing imprints of footpad 3 and crushable block 3 made during the initial landing and soil disturbance caused by Surveyor VI vernier engine 3 during the hop (Nov. 15 and 16, 1967, Catalog 6-SE-43B).

FIGURE 4-25.—Mosaic of computer-processed pictures, showing crushable block 1 impact area and soil disturbance caused by Surveyor VI vernier engine 1 during the hop (Nov. 15 and 16, 1967, Catalog 6-SE-43A).

TABLE 4-3. *Vernier-engine parameters used in computations*

Parameter	Surveyor III			Surveyor V			Surveyor VI		
	Engine 1	Engine 2	Engine 3	Engine 1	Engine 2	Engine 3	Engine 1	Engine 2	Engine 3
Maximum thrust, N	490	130	250	120	76	120	390	310	420
Nozzle exit plane height,[a] cm			25	39	39	39	32	32	32
Maximum static pressure, N/cm^2			0.50	0.29	0.18	0.29	0.69	0.92	1.35
Maximum dynamic pressure, N/cm^2			0.22	0.12	0.076	0.12	0.30	0.39	0.56

[a] Tabulated values correspond to the minimum nozzle heights used in the surface loading computations. Although some of these values were subsequently revised, the tabulated surface pressures still provide representative estimates.

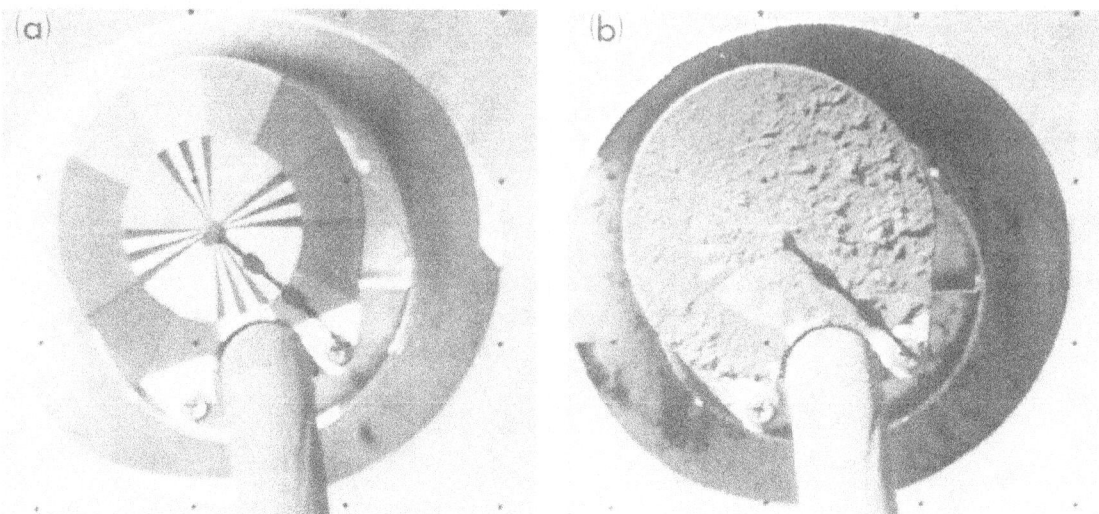

FIGURE 4–26.—Surveyor VI photometric target on omnidirectional antenna boom. (a) Prefiring picture (Nov. 15, 1967, 09:37:11 GMT). (b) Postfiring picture. Before the firing, the target was clean; after the firing, the target was coated with a layer of soil up to 0.9 mm thick. The coating probably was caused by impact of a soil clump on the target during the firing (Nov. 15, 1967, 12:30:00 GMT).

The maximum static and dynamic surface pressures listed in table 4–3 are values obtained from Roberts' theory (refs. 4–14 and 4–15). Figure 4–27 shows the relationship between these pressures and gas velocity versus radial distance. The dynamic pressure is equal to $\rho u^2/2$, where ρ is the gas mass density and u the radial velocity of the gas along the surface. The values for the Surveyor V static firing correspond to the engine thrusts and nozzle heights given in table 4–3. The pressures given for Surveyors III and VI are the maximum values encountered during the second Surveyor III landing event and Surveyor VI hop, respectively. During these maneuvers, maximum engine thrusts and minimum nozzle heights did not occur simultaneously; therefore, the listed maximum surface pressures did not correspond to both maximum thrusts and minimum nozzle heights.

The viscous erosion theory given in references 4–14 and 4–15 was used to compare the theoretical and observed crater dimensions. Theoretically, soils composed of particle sizes smaller than 500 microns would not erode as fast as observed during the Surveyor V firing. Also, for a hypothetical soil composed of 600-micron particles and with a cohesion of 0.01 N/cm^2 (selected to approximate the average erosion rate), the theoretical erosion crater diameter would be 70 cm instead of the measured 20-cm-diameter crater under the Surveyor V vernier engine 3. These calculations indicate that viscous erosion was not the major erosion mechanism for the formation of the crater. However, viscous erosion probably caused the larger soil fragments to move across the surface from positions outside the crater.

Since viscous erosion does not appear to have been the principal eroding mechanism, it is thought that diffused gas eruption occurred. This type of erosion, however, does not provide an estimate on cohesion of the surface material because the diameter of a diffused gas eruption crater is largely independent of the soil cohesion (ref. 4–17). But it can be concluded, by comparing the calculated crater diameter with the observed value, that the lunar soil must be relatively impermeable; a firing time of 0.5 second (Surveyor V) is only one-tenth of the time required to reach steady-state conditions. This is based on an assumed soil porosity between 0.3 and 0.5 and a viscosity of the exhaust gases in the soil between 1×10^{-4} and

3×10^{-4} poise, as explained in reference 4–17. From this, the permeability of the soil was calculated to be between 1×10^{-8} and 7×10^{-8} cm².

For comparison, the permeabilities of soils of different uniform grain sizes are shown in figure 4–28. This figure shows that the permeability range for the lunar surface material, probably down to a depth of about 25 cm, fits into the permeability range of silts (grain-size range from 2 to 60 microns). Lunar soil contains particles larger, and probably smaller, than this range. However, the estimated lunar permeability indicates that most of the particles are in the 2- to 60-micron size range. This estimate is in agreement with conclusions reached from light reflectance simulations of Surveyor III footpad imprints (ref. 4–2).

Estimates of soil cohesion.—Results of the Surveyor VI erosion test were used to estimate bounds for the cohesion of the lunar soil. Pictures of the Surveyor VI landing site (fig. 4–22) indicate some surface erosion, apparently of the viscous type, occurred beneath and adjacent to each engine during liftoff. There is no indication that bearing load cratering occurred. During takeoff for the Surveyor VI hop, the exhaust gas pressure on the lunar surface decreased gradually enough to prevent diffused gas eruption of the soil. Therefore, the estimates for soil cohesion are based on the conclusion that the cohesion was not large enough to prevent viscous erosion, but large enough for the soil to withstand vertical pressure loading.

According to the theory advanced in references 4–14 and 4–15, the maximum erosive shear stress occurs at the point of maximum dynamic pressure and is dependent on the effective value of the friction coefficient. Soil erosion data obtained by the Langley Research Center from soils having an initial flat surface indicated the effective friction coefficient, C_f, to be

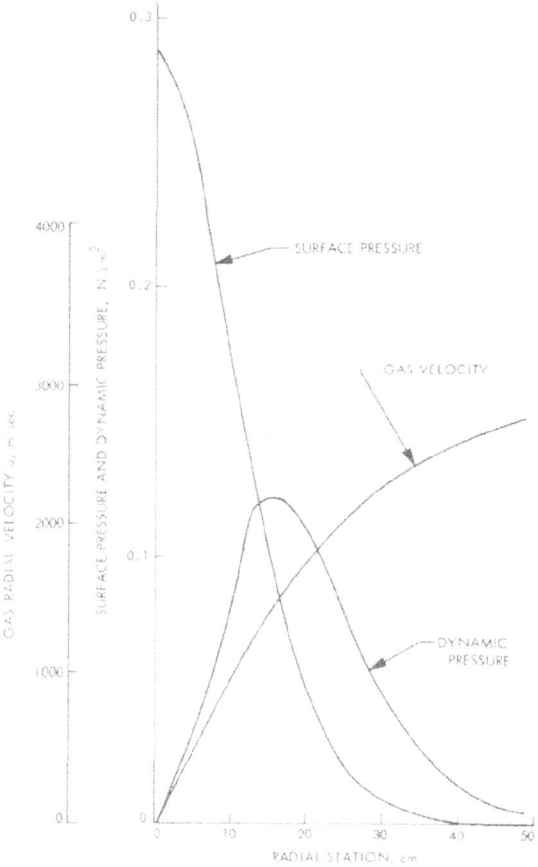

FIGURE 4–27.—Theoretical static pressure, dynamic pressure, and exhaust gas radial velocity at the surface of a plane, parallel to the engine nozzle exit plane; engine thrust = 120 N, nozzle exit plane height = 39.4 cm.

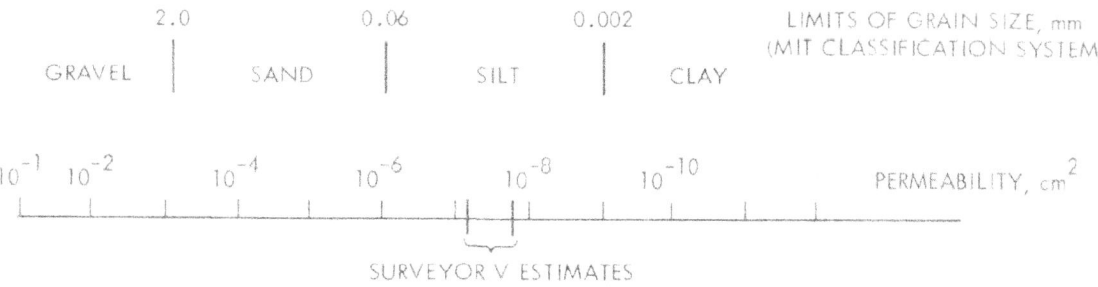

FIGURE 4–28.—Permeability of terrestrial soils versus grain size and classification related to Surveyor V results.

about 0.2, which is the value recommended in reference 4–15. For the irregular, undulating surface existing at the Surveyor VI landing site, the friction coefficient should be higher than 0.2. For the upper-bound estimate made here, C_f is taken as 0.4. According to table 4–3, the peak dynamic pressures under vernier engines 1, 2, and 3 were 0.30, 0.39, and 0.56 N/cm^2, respectively. Thus, for a friction coefficient of 0.4, the lunar surface was subjected to maximum shearing stresses of 0.12, 0.16, and 0.23 N/cm^2 by exhaust gases from vernier engines 1, 2, and 3, respectively.

Also, according to the theory in references 4–14 and 4–15, the viscous shear stresses are resisted by the frictional and cohesional forces of the soil. The resistance of the soil surface provided by the friction forces between soil grains is negligible for the small-diameter particles at the Surveyor VI landing site. As a result, erosion is essentially resisted by cohesion; therefore, upper-bound estimates of soil cohesion are equal to the maximum values of applied shearing stresses. Because each vernier engine caused some soil erosion, the minimum value for an upper-bound soil cohesion estimate is the shear stress caused by vernier engine 1. This value is 0.12 N/cm^2.

The maximum surface loading (1.35 N/cm^2) occurred under vernier engine 3. Under the assumption that Terzaghi's bearing-capacity theory (ref. 4–19) is applicable for this type of surface loading, estimates can be made for the minimum value of soil cohesion needed to prevent a bearing-capacity failure for various values of soil density and internal friction angle. An application of this theory, in conjunction with the pressure loading from Roberts' theory, indicates that a soil with an assumed weight density of 2.4×10^{-3} N/cm^3 (1.5-g/cm^3 mass density) and a soil cohesion greater than 0.0073 N/cm^2 would be sufficient to prevent a bearing-capacity failure for a soil with a 35° internal friction angle. For a friction angle of 30°, the required value of cohesion is 0.020 N/cm^2. Since a bearing capacity type of failure was not observed during the hop, this procedure indicates that the soil cohesion lower bound is 0.0073 N/cm^2.

Attitude Control Jet Operations

Observations.—Attitude control gas jets mounted on all Surveyor legs provided attitude stabilization during the flights. After landing, Surveyor I attitude control jets were operated to produce short pulses of 20-msec durations with a 30-msec pause between pulses (ref. 4–1). Pictures taken after operation of the jets revealed the presence of a small dimple crater near the attitude control jet 2 area of impingement. However, results of this test are inconclusive because no suitable prefiring pictures of the impingement area are available.

The attitude control jets on Surveyor VI were commanded to operate for a continuous burst of 4 seconds and for another burst of 60 seconds (ref. 4–4). Good television coverage of the jet impingement area on the lunar surface before, during, and after jet operation afforded clear observations of the surface erosion caused by attitude control jet 2. The nozzle of this jet was about 10.4 cm above the surface and was inclined 24° from the spacecraft vertical axis.

Comparisons of pictures taken before and after each burst (fig. 4–29) show that the disturbance of the lunar surface caused by the jet operations was minor and that no crater was formed. Some small soil fragments up to 25 cm from the impingement area were moved by the jet operation. The most conspicuous effect consisted of the movement of two lunar surface protrusions, probably soil clumps, which were 12 to 15 cm from the center of jet impingement (fragments A and B, fig. 4–29).

Simulations and analyses.—Laboratory tests were performed in which an attitude control jet was operated over soil beds in vacuum. The soil erosion caused by the jet was of the viscous type; no eruption caused by diffused gas was observed.

It was found that erosion occurred if the soil cohesion was below a limiting value. For the sandy silts used in these tests, the limiting value of the cohesion was 0.17 N/cm^2. However, these tests were conducted at a pressure of approximately 50×10^{-3} mm Hg and full expansion of the jet plume probably did not occur. Therefore, static pressure on the soil surface in the vacuum chamber probably was greater than on the lunar surface beneath the jet.

FIGURE 4-29.—(a) Mosaic of narrow-angle pictures taken shortly before the 4-second operation of the attitude control jets. Representative fragments, shown by postoperation pictures to have moved or to have been partially eroded by the firing, are circled. The dark area cutting diagonally across each picture is the camera housing. A line extending through the center of the attitude control jet is shown by the arrow. The approximate point where this line intercepts the lunar surface is shown by an X (Nov. 9, 1967, Catalog 6-MP-1).

162 SURVEYOR: PROGRAM RESULTS

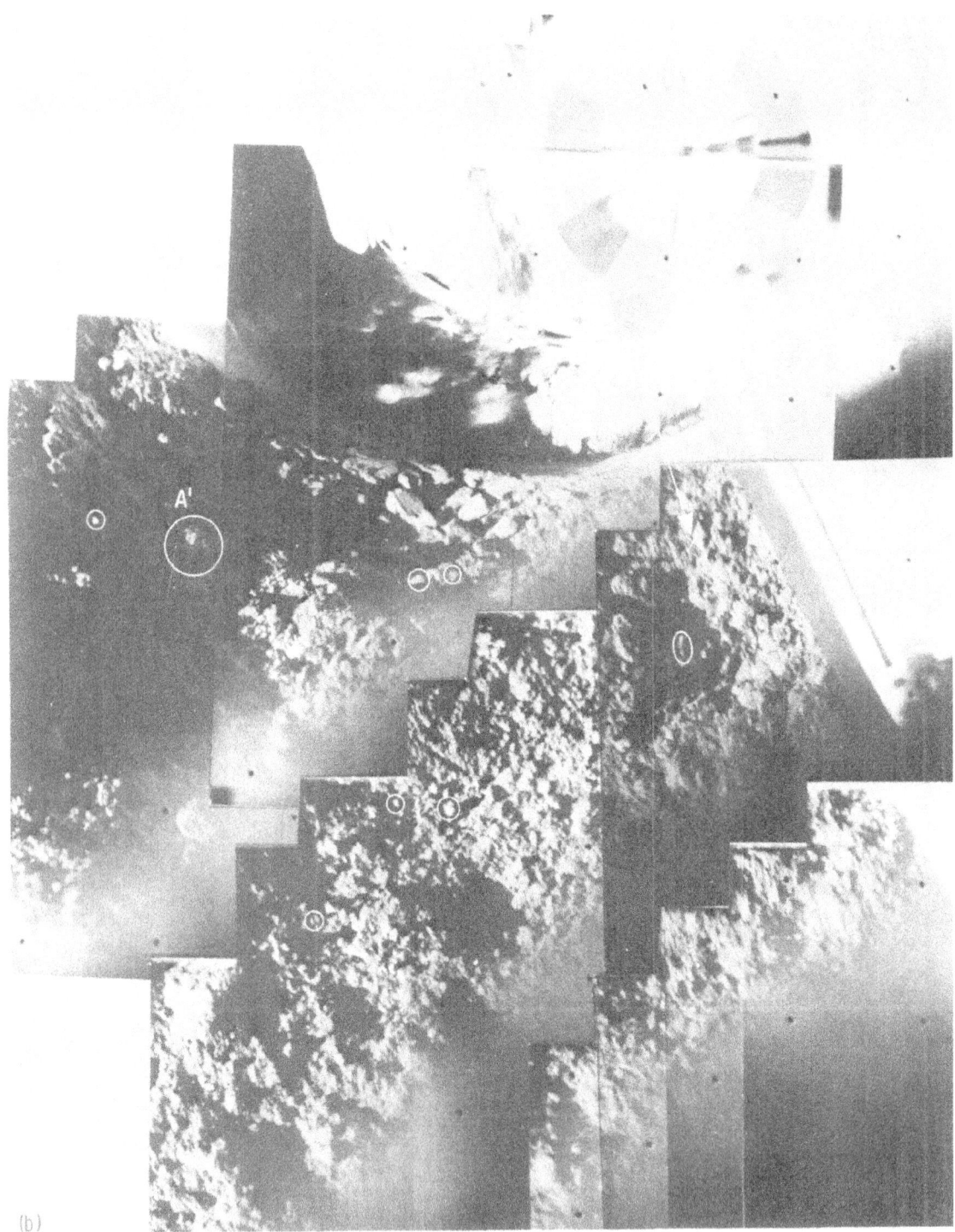

FIGURE 4-29.—Continued. (b) Mosaic of pictures taken immediately after the 4-second firing of the attitude control jets. Representative fragments that arrived at their present sites because of the firing are circled (Nov. 9, 1967, Catalog 6-MP-2).

FIGURE 4-29.—Continued. (c) Mosaic of the same pictures used in figure 4-29(b); however, the fragments circled are those fragments shown by later pictures to have been moved by the 60-second attitude control jet firing (Nov. 9, 1967, Catalog 6-MP-2).

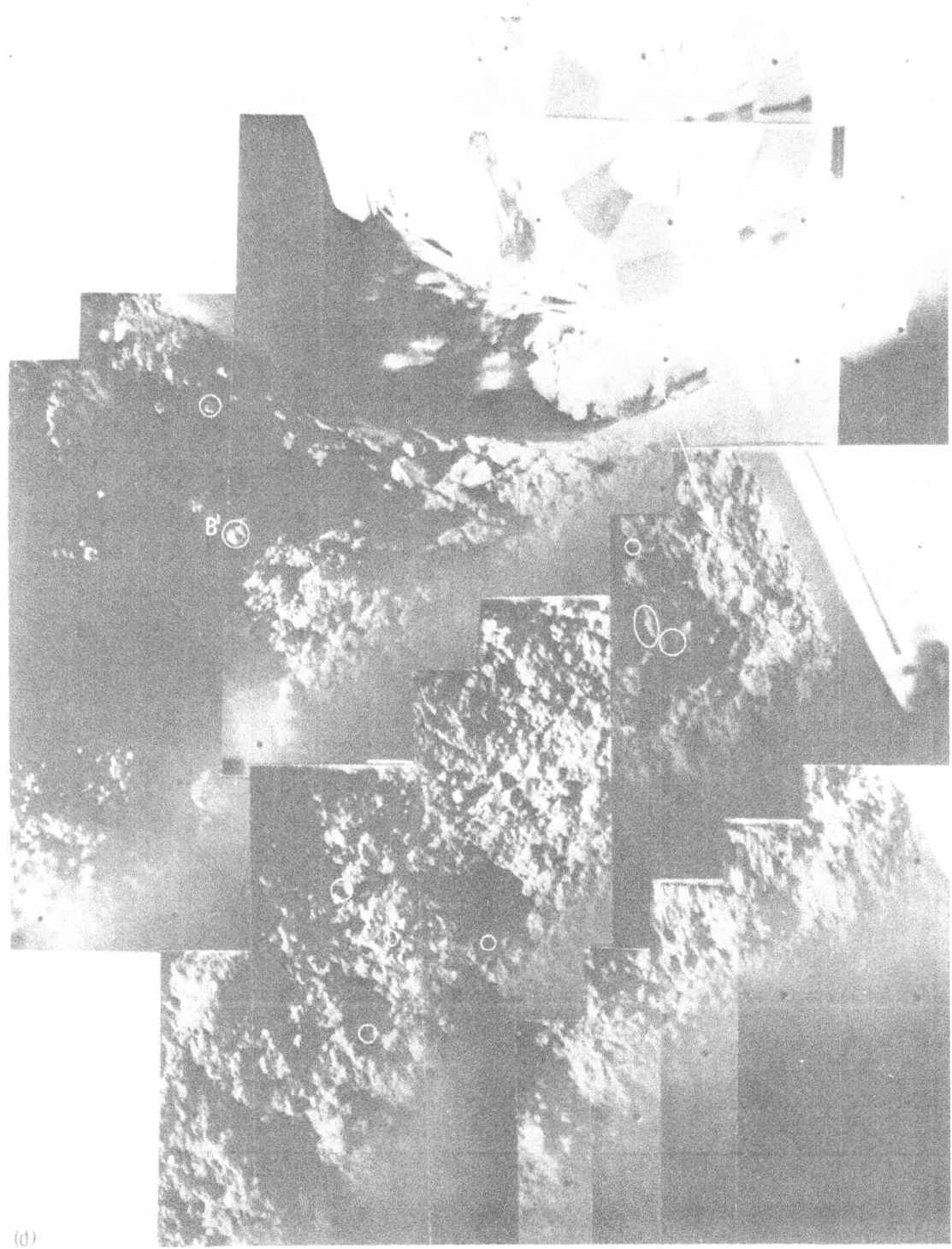

FIGURE 4–29.—Concluded. (d) Mosaic of pictures taken after the 60-second attitude control jet firing. Fragments circled arrived at their present sites after the 60-second firing (Nov. 9, 1967, Catalog 6–MP–3).

Spacecraft Contamination

During all missions, some spacecraft contamination, caused primarily by the landing impact or by vernier-engine firings, was observed (figs. 4-30 through 4-32).

Some fine material was observed on the thermal compartments of Surveyor I (fig. 4-30); however, these particles could have been deposited before the spacecraft landed.

Glare on Surveyor III pictures probably was caused by a thin layer of lunar soil deposited on the television mirror when the vernier engines were fired during the first two touchdowns.

Although Surveyor V landed on a 20° slope, no obvious contamination of the spacecraft was produced, except that the footpads plowed into the lunar surface, causing soil to be deposited on the footpad tops. Following the vernier-engine firing, a splatter of soil was

FIGURE 4-30.—Surveyor I electronic compartment top after landing. Relative lack of soil contamination is shown by the small number of soil particles.

FIGURE 4-31.—Top of Surveyor V compartment B. A clump of soil ejected onto the compartment top during the firing splattered in a direction away from vernier engine 1. (a) Prefiring picture (Sept. 10, 1967, 02:29:29 GMT). (b) Postfiring picture (Sept. 20, 1967, 05:48:58 GMT).

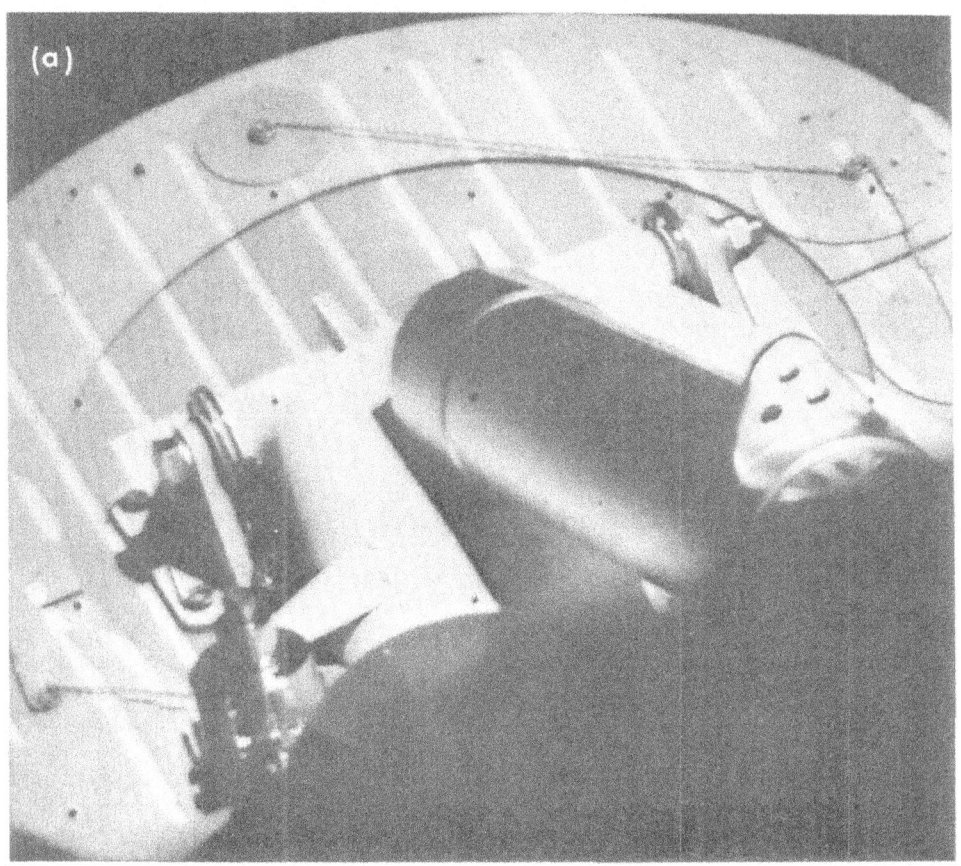
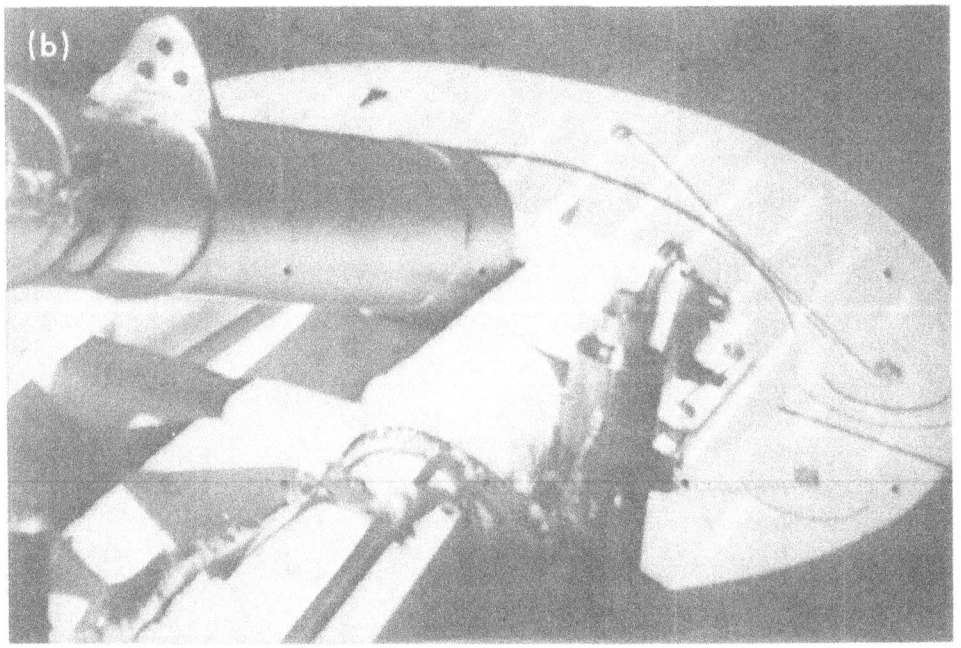

FIGURE 4-32.—Relative lack of soil contamination on Surveyor VI spacecraft after the vernier-engine firing. (a) Footpad 2; a few soil particles lie along the top edge (Nov. 17, 1967, 04:51:01 GMT). (b) Top of footpad 3 (Nov. 15, 1967, 12:23:13 GMT). (c) Top of electronic compartment A (Nov. 18, 1967, 06:04:56 GMT).

observed on top of one of the electronic compartments (fig. 4-31). This contamination was caused by impact of a soil clump, probably ejected by diffused gas eruption from beneath vernier engine 1. The soil clump must have followed a near-vertical trajectory to reach the compartment top. The firing also coated the vertical face of the sensor head of the alpha-scattering instrument with soil, destroying the reflectivity of its mirror finish.

During the hop made by Surveyor VI, some lunar soil impacted and adhered to the photometric target on one of the omnidirectional antenna booms (fig. 4-26). It is estimated that the lunar soil adhesion in this case was between 250 and 1000 dynes/cm^2. The minimum value of 250 dynes/cm^2 is based on the soil withstanding an acceleration of about 48 m/sec^2, which is the estimated minimum value of peak acceleration exerted on the photometric target during the hop landing. This value differs from that given in reference 4-4.

A layer of fine lunar soil and small soil clumps was deposited on one of the Surveyor VII auxiliary mirrors.

Summary and Conclusions

Evaluations of television and telemetry data, aided by analytical and laboratory simulations, have provided the following conclusions to date:

(1) The soil at all landing sites is predominately fine grained, granular, and slightly cohesive.

(2) At all sites, soil ejected during spacecraft landings and operations is darker than undisturbed soil; whereas, smoothed and flattened soil is brighter than undisturbed soil.

(3) Imprints of footpads and crushable blocks indicate that the soil is compressible, at least in its upper few centimeters.

(4) Static bearing strength of the lunar soil increases with depth. Bearing strengths calculated for various size penetrators are as follows: (a) In approximately the upper millimeter: less than 0.1 N/cm^2 (from imprints of small rolling fragments); (b) at a depth of 1 to 2 mm: 0.2 N/cm^2 (from imprints of the sensor head of the alpha-scattering instrument on Surveyor VII, ref. 4-5); (c) at a depth of 1 to

2 cm: 1.8 N/cm² (from Surveyor VI and VII crushable block imprints); (d) at a depth of 4 cm: between 4.2 and 5.6 N/cm² (from an extension of the Surveyor I footpad penetration analysis of ref. 4–7); (e) to a depth of 6 cm: between 4 and 6 N/cm² (average bearing strength from analyses of Surveyor footpad 2 penetrations using the compressible soil model defined in ref. 4–3).

(5) Dynamic bearing stress developed on a crushable block exceeded 2.4 N/cm² during penetration to a depth of 3 cm.

(6) Estimates of the soil shear wave velocity are between 15 and 35 m/sec and of the compressional wave velocity between 30 and 90 m/sec. These estimates, based on oscillations in the spacecraft landing leg forces, are lower than those expected for terrestrial soils with other mechanical properties as reported in this section.

(7) Viscous soil erosion, i.e., erosion by the entrainment of soil particles as gas flows over the surface, occurred during the vernier-engine firings and attitude control jet operations. Fine soil eroded during the Surveyor VI vernier-engine firing was picked up by exhaust gases moving in a horizontal sheet along the lunar surface and was redeposited at distances up to several meters. Some soil clumps and fragments ejected from the exhaust gas impingement areas rose at least 1 meter above the lunar surface and traveled at least 4 meters.

(8) During the Surveyor V vernier-engine static firing, diffused gas eruption produced a crater 20 cm wide and 0.8 to 1.3 cm deep. The Surveyor VI dynamic firing did not cause diffused gas eruption of the soil.

(9) Based on the Surveyor V vernier-engine firing, the permeability of the soil, to a depth of about 25 cm, is estimated to be in the range from 1×10^{-8} to 7×10^{-8} cm². This corresponds to the permeability of Earth silts and indicates that most of the lunar soil is in the 2- to 60-micron particle-size range.

(10) Soil cohesion is estimated to be between 0.007 and 0.12 N/cm² (from vernier-engine firings).

(11) Lunar soil, eroded by impingement of vernier-engine exhaust gases, adhered to the spacecraft. The most conspicuous examples are: (a) Surveyor III: Dust on the mirrors; (b) Surveyor V: Soil on the polished vertical surface of the alpha-scattering-instrument sensor head; (c) Surveyor VI: A thick layer of soil on the photometric target. Adhesive strength of the soil impacting, and adhering to, the Surveyor VI photometric target is estimated to be between 250 and 1000 dynes/cm². (d) Surveyor VII: A layer of fine soil and small soil clumps on one of the auxiliary mirrors.

(12) Soil at the Surveyor VII highland site is similar in mechanical properties to the soil at the mare landing sites. The higher rock population at this site did not cause an increase in bearing strength of the soil.

(13) Within an 18-meter radius of Surveyor VII, 0.6 percent of the area is covered by rocks larger than 20 cm in diameter, 1.2 percent by rocks larger than 10 cm, and 2.8 percent by rocks larger than 5 cm (ref. 4–5).

References

4–1. CHRISTENSEN, E. M.; BATTERSON, S. A.; BENSON, H. E.; CHANDLER, C. E.; JONES, R. H.; SCOTT, R. F.; SHIPLEY, E. N.; SPERLING, F. B.; and SUTTON, G. H.: Lunar Surface Mechanical Properties. Surveyor I Mission Report. Part II: Scientific Data and Results, Tech. Rept. 32-1023, Jet Propulsion Laboratory, Pasadena, Calif., Sept. 10, 1966, pp. 69–85.

4–2. CHRISTENSEN, E. M.; BATTERSON, S. A.; BENSON, H. E.; CHOATE, R.; JAFFE, L. D.; JONES, R. H.; KO, H. Y.; SPENCER, R. L.; SPERLING, F. B.; and SUTTON, G. H.: Lunar Surface Mechanical Properties. Surveyor III Mission Report. Part II: Scientific Results, Tech. Rept. 32-1177, Jet Propulsion Laboratory, Pasadena, Calif., June 1, 1967, pp. 111–153.

4–3. CHRISTENSEN, E. M.; BATTERSON, S. A.; BENSON, H. E.; CHOATE, R.; HUTTON, R. E.; JAFFE, L. D.; JONES, R. H.; KO, H. Y.; SCHMIDT, F. N.; SCOTT, R. F.; SPENCER, R. L.; and SUTTON, G. H.: Lunar Surface Mechanical Properties. Surveyor V Mission Report. Part II: Science Results, Tech. Rept. 32-1246, Jet Propulsion Laboratory, Pasadena, Calif., Nov. 1, 1967, pp. 43–88.

4–4. CHRISTENSEN, E. M.; BATTERSON, S. A.; BENSON, H. E.; CHOATE, R.; HUTTON, R. E.; JAFFE, L. D.; JONES, R. H.; KO, H. Y.; SCHMIDT, F. N.; SCOTT, R. F.; SPENCER, R. L.; SPERLING, F. B.; and SUTTON, G. H.: Lunar Surface Mechanical Properties. Surveyor VI Mission Report. Part II: Science Results, Tech. Rept. 32-1262, Jet Propulsion Laboratory, Pasadena, Calif., Jan. 10, 1968, pp. 47–108.

4-5. CHOATE, R.; BATTERSON, S. A.; CHRISTENSEN, E. M.; HUTTON, R. E.; JAFFE, L. D.; JONES, R. H.; KO, H. Y.; SPENCER, R. L.; and SPERLING, F. B.: Lunar Surface Mechanical Properties. Surveyor VII Mission Report. Part II: Science Results, Tech. Rept. 32-1264, Pasadena, Calif., June 15, 1968, pp. 77-134.

4-6. RENNILSON, J. J.; DRAGG, J. L.; MORRIS, E. C.; SHOEMAKER, E. M.; and TURKEVICH, A.: Lunar Surface Features. Surveyor I Mission Report. Part II: Scientific Data and Results, Tech. Rept. 32-1023, Jet Propulsion Laboratory, Pasadena, Calif., Sept. 10, 1966, pp. 32-1023.

4-7. SPERLING, F.; and GARBA, J.: A Description of the Surveyor Lunar Landing Dynamics and an Evaluation of Pertinent Telemetry Data Returned by Surveyor I, Tech. Rept. 32-1035, Jet Propulsion Laboratory, Pasadena, Calif., 1967.

4-8. Surveyor Spacecraft System Bimonthly Progress Summary, Rept. SSD 68179R, Hughes Aircraft Co., Culver City, Calif., Aug. 23, 1966.

4-9. Surveyor Spacecraft System Bimonthly Progress Summary, Rept. SSD 68218R, Hughes Aircraft Co., Culver City, Calif., Oct. 24, 1966.

4-10. Surveyor Spacecraft System Bimonthly Progress Summary, Rept. SSD 68218-6, Hughes Aircraft Co., Culver City, Calif., Aug. 23, 1967.

4-11. Surveyor III Flight Performance Final Report, Rept. SSD 68189-3, Hughes Aircraft Co., Culver City, Calif., 1967.

4-12. SPENCER, R. L.: Determination of Footpad Penetration Depth from Surveyor Spacecraft Shadows, Tech. Rept. 32-1180, Jet Propulsion Laboratory, Pasadena, Calif., 1967.

4-13. TIMOSHENKO, S.; and GOODIER, D.: *Theory of Elasticity*, second edition, McGraw-Hill, New York, p. 367, 1951.

4-14. ROBERTS, L.: The Action of a Hypersonic Jet on a Dust Layer. Institute of the Aerospace Sciences, paper no. 63-50.

4-15. ROBERTS, L.: The Interaction of a Rocket Exhaust with the Lunar Surface. The Fluid Dynamic Aspects of Space Flight, vol. 2, pp. 269-290, Proceedings of the AGARD-NATO Specialists Meeting, Marseilles, France, Apr. 20-24, 1964. Gordon and Breach, Science Publishers, New York, London, Paris. [Sponsored by the Fluid Dynamics Panel of the Advisory Group for Aeronautical Research and Development (AGARD).]

4-16. LAND, N. S.; and CLARK, L. V.: Experimental Investigation of Jet Impingement on Surfaces of Fine Particles in a Vacuum Environment, NASA TND-2633, Feb. 1965.

4-17. SCOTT, R. F.; and KO, H. Y.: Transient Rocket-Engine Gas Flow in Soil. AIAA J., vol. 6, no. 2, pp. 258-264, Feb. 1968.

4-18. ALEXANDER, J. D.; ROBERDS, W. M.; and SCOTT, R. F.: Soil Erosion by Landing Rockets. Final Report 1301, Hayes International Corporation, July 15, 1966.

4-19. TERZAGHI, K.: Theoretical Soil Mechanics. John Wiley & Sons, Inc., New York, ninth printing, Article 46, June 1959.

ACKNOWLEDGMENTS

Appreciation is extended to the numerous individuals at many organizations (especially JPL and HAC) who designed, built, tested, and operated the spacecraft and the associated launch vehicle and ground stations; to HAROLD BENSON, CLAUDE CHANDLER, FRED SCHMIDT, EDWARD SHIPLEY, and JOHN STALLKAMP who participated in some of the missions as members of the Lunar Surface Mechanical Properties Working Group; to CHARLES GOLDSMITH, WILLIAM PEER, ALEX IRVING, ALBERT PLESCIA, and LLOYD STARKS, JPL, for assistance during mission operations and assembly of mosaics; to DAVE CONAWAY, MARGARET DOVE, and JOHN HINCHEY, HAC, and JOHN GARBA and DANIEL BOOKSTEIN, JPL, for assisting in the landing dynamic simulations; to ROBERT BRESHEARS, JOHN STOCKY, and CHARLES DODGE, JPL, for their vernier-engine performance analyses; and to LOUIS SIDWELL, JPL, for his attitude control jet performance analyses.

5. Soil Mechanics Surface Sampler

R. F. Scott (Principal Investigator) and F. I. Roberson

The soil mechanics surface-sampler experiment was conducted on Surveyors III and VII; tests were performed successfully on both missions. The flawless performance of Surveyor VII allowed more discrimination in the performance of lunar tests, and the more quantitative data obtained also provided a means for further interpretation of Surveyor III data. A discussion of the physical differences between the missions is presented here, and some conclusions are drawn regarding the difference in the characteristics of the two landing sites.

The Subsystem

Development of the surface sampler as a soil mechanics instrument (fig. 5-1) involved the study and incorporation of sensors for direct measurements of position, force, and deceleration (ref. 5-1). The evolution of the Surveyor program and schedule led to a change in subsystem design (ref. 5-2); a modified surface sampler was flown on Surveyor III and, following further changes in the surface sampler/spacecraft interface, on Surveyor VII.

Mechanism

The Surveyor VII surface-sampler mechanism was identical to that flown on Surveyor III (ref. 5-3), with the exception of two magnets, which were placed in the bearing plate of the scoop door (ref. 5-4).

Electronics

The electronics auxiliary for the Surveyor VII surface sampler incorporated a 7.5-watt heater, larger than the 5-watt unit for the Surveyor III electronics. The design of the electronics was the same for both missions.

Because Surveyor VII was a later-generation spacecraft, more telemetry channels were available to the surface sampler. The addition of two temperature sensors, attached directly to the retraction and elevation motors, was one reflection of this increased capability. The motor-current readout on Surveyor III was assigned to a single commutator frame, which yielded a maximum of eight motor-current samples for a 2-second motor actuation. More commutator frames were available on Surveyor VII, and the motor-current signal was fed to five symmetrically spaced commutator frames on the mode 4 engineering commutator. This provided, at the highest spacecraft bit rate of 4400 bits/second, a potential (depending on the random start and stop time of a pulse) maximum of 40 motor-current samples for a 2-second motion. This 50-msec sampling rate provided an excellent envelope of the motor-current pulse (fig. 5-2) and led to a more

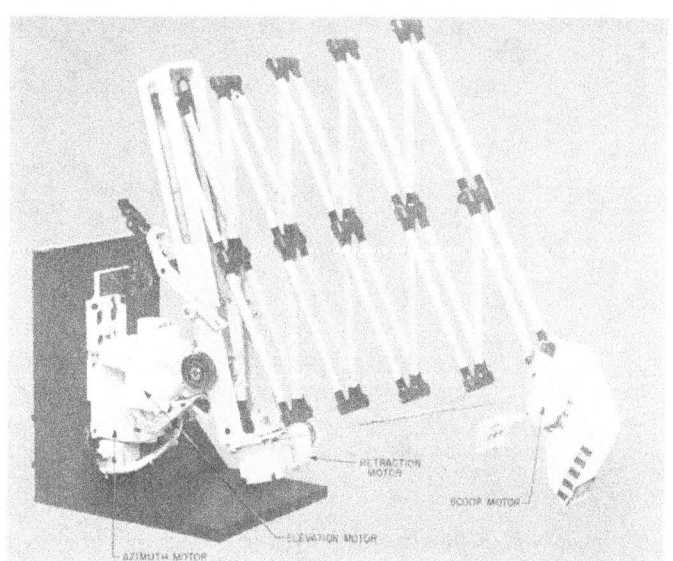

FIGURE 5-1.—Surface sampler on stand, partly extended.

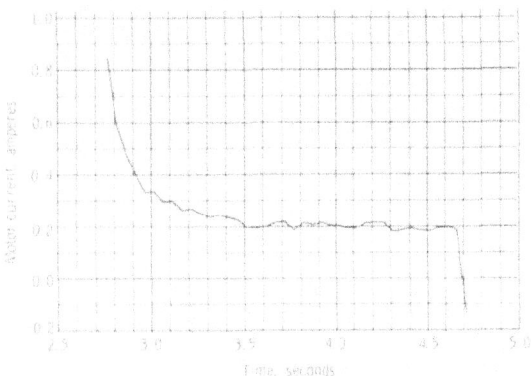

FIGURE 5-2.—Typical plot of motor current.

accurate indication of forces applied to the lunar surface.

The Missions

Surveyor III landed in the southeast part of Oceanus Procellarum, a site typical of the maria (refs. 5-5 through 5-8). Surveyor VII landed, 8½ months later, near the crater Tycho, on an ejecta blanket of material considered to be much younger than the mare material. During the interval between the missions, adjustments were made in operations plans, and some new techniques were devised.

Surveyor III

During the Surveyor III operations for the first lunar day, the surface sampler performed 7 bearing tests, 4 trench tests, and 13 impact tests (fig. 5-3). During the performance and analysis of these tests, additional ideas for improved operations techniques were evolved, and ultimately led to generation of new command tapes for use on the Surveyor VII mission. The total operating times and the number of commands issued during the mission are listed in table 5-1.

Surveyor VII

At the time of Surveyor VII touchdown, the operations plans called for the deployment of the alpha-scattering instrument before any surface-sampler operations to prevent any possible disturbance of the lunar surface before a chemical analysis could be obtained. Because the alpha-scattering instrument was not deployed normally to the lunar surface, the surface sampler was used to assist in the deployment (ref. 5-9). Initial sampler contact with the lunar surface occurred at bearing point 1 (fig. 5-4) and, as shown in figure 5-5, was remarkably similar to the initial bearing test on Surveyor III (bearing test 1, fig. 5-3). Using command tapes especially designed to operate the surface sampler in a bearing test mode, as well as tapes to provide pictures between each command during a trenching test in order to construct a motion picture, successful soil tests were continued throughout the first lunar day. The total operating times and number of commands issued during the mission are shown in table 5-1; 16 bearing tests, 7 trenching tests, and 2 impact tests were performed. After initial deployment to the undisturbed surface, the alpha-scattering instrument was redeployed to a second sample position (a rock) and then to a third sample position (disturbed surface). Figure 5-6 shows the surface sampler in the process of moving the alpha-scattering instrument from the second to the third lunar sample. The locations of the soil tests and of these sample positions are shown in figures 5-4 and 5-7. The results of data analysis from bearing test 2 on Surveyor VII are shown in figure 5-8. This diagram is a revision of figure 5-5 of reference 5-9, following postmission studies.

Temperature data and motion increment measurements for the surface-sampler motors on Surveyor VII showed that the motors did not change their operating characteristics significantly (particularly as far as distance per command is concerned), even though they operated under a wide range of temperatures.

TABLE 5-1. *Summary of surface-sampler operations*

Operation	Surveyor III	Surveyor VII
Total power on time	18 hr 22 min	36 hr 21 min
Total spacecraft commands	5879	12 639
Total functions	1898	6956
Total bearing tests	7	16
Total trenching tests	4	7
Total impact tests	13	2

The similarity of the elevation motors on the surface samplers on both missions can be seen from the curves of force versus motor current shown in figure 5-9. From these curves, it is clearly evident that the bearing forces applied by the surface samplers were nearly equal under stall conditions. Thus, direct comparisons could be made of the soil at both landing sites.

The surface-sampler subsystem was operated during the second lunar day, thus demonstrat-

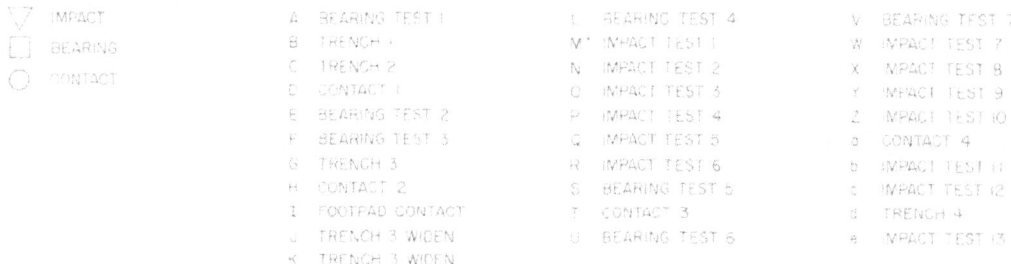

▽	IMPACT	A	BEARING TEST 1	L	BEARING TEST 4	V BEARING TEST 7
☐	BEARING	B	TRENCH 1	M IMPACT TEST 1	W IMPACT TEST 7
○	CONTACT	C	TRENCH 2	N IMPACT TEST 2	X IMPACT TEST 8
		D	CONTACT 1	O IMPACT TEST 3	Y IMPACT TEST 9
		E	BEARING TEST 2	P IMPACT TEST 4	Z IMPACT TEST 10
		F	BEARING TEST 3	Q IMPACT TEST 5	a CONTACT 4
		G	TRENCH 3	R IMPACT TEST 6	b IMPACT TEST 11
		H	CONTACT 2	S BEARING TEST 5	c IMPACT TEST 12
		I	FOOTPAD CONTACT	T CONTACT 3	d TRENCH 4
		J	TRENCH 3 WIDEN	U BEARING TEST 6	e IMPACT TEST 13
		K	TRENCH 3 WIDEN		

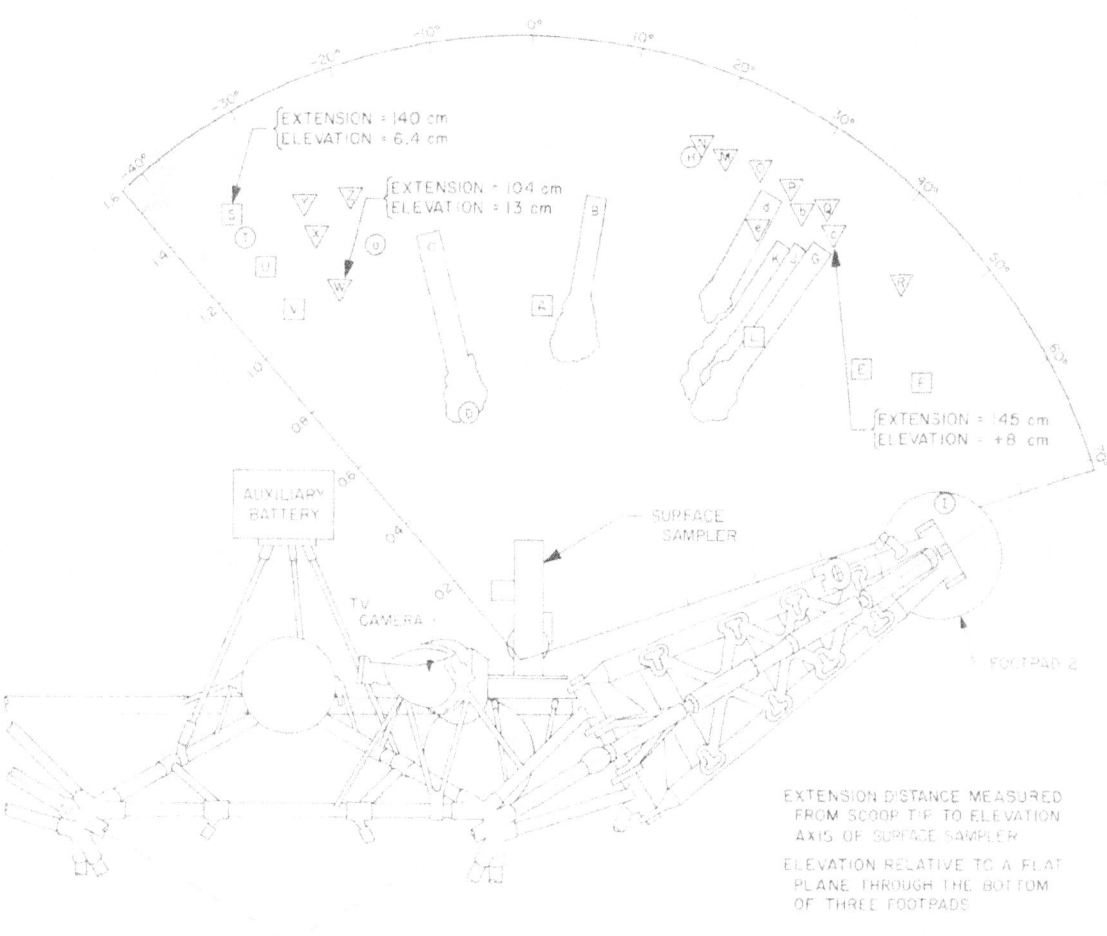

FIGURE 5-3.—Plan view of surface-sampler operations area, showing locations of all tests performed during the first lunar day of the Surveyor III mission.

ing that the unit and auxiliary had survived the lunar night.

The Operations: Comparison of Tests

Surveyor III

The landing of Surveyor I (ref. 5-3) demonstrated that the lunar surface was composed of a granular, soillike material and gave some indication of the properties of the soil. The Surveyor III surface sampler had been arranged and calibrated so that measurements of the current supplied to its motors would give an indication of the force applied to the lunar surface in testing operations. The state of the spacecraft telemetry following landing on the lunar surface, however, precluded any measurements of

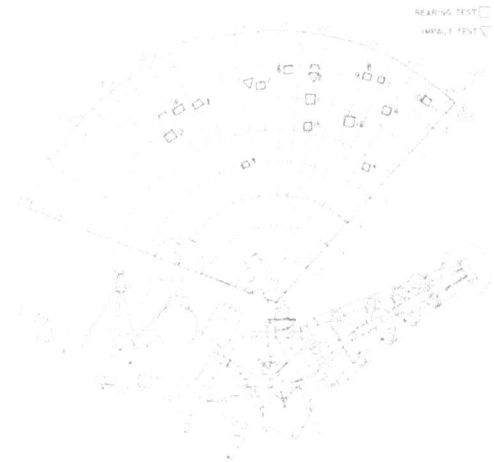

FIGURE 5-4.—Locations of bearing and impact tests performed on Surveyor VII.

FIGURE 5-5.—Similarity of initial bearing tests performed. (a) Surveyor III (day 112, 05:17:27 GMT). (b) Surveyor VII (day 011, 03:55:42 GMT).

the motor currents and, therefore, the forces, during subsequent operations. The maximum forces exerted by the elevation and retraction motors under stall conditions had been measured in terrestrial tests and related to temperatures. Since the motor temperatures could be estimated during Surveyor III operations, the surface tests were conducted until deliberate stalling of the motor was achieved. In this manner, it was possible to estimate the forces applied to the surface. From this information and from various tests conducted with Surveyor III, the properties of the lunar surface soil were estimated.

The lunar surface material appears to have the properties and behavior of a fine-grained, terrestrial soil possessing a small amount of cohesion and an angle of internal friction corresponding to a medium-dense soil. The density is apparently within the range of ordinary terrestrial soil (ref. 5-2). Most of the lunar tests, with the exception of the bearing tests, indicated a relatively homogeneous soil material in lateral extent, but one that increased somewhat in strength as a function of depth. This was observed by the greater difficulty of deepening trenches as the depth exceeded a few centimeters.

Relatively few rocks were accessible to the surface sampler on Surveyor III, and only one of these rocks was picked up for a closer exami-

FIGURE 5-6.—Surface sampler holding the alpha-scattering instrument during deployment to third lunar sample on the Surveyor VII mission (day 022, 11:21:12 GMT).

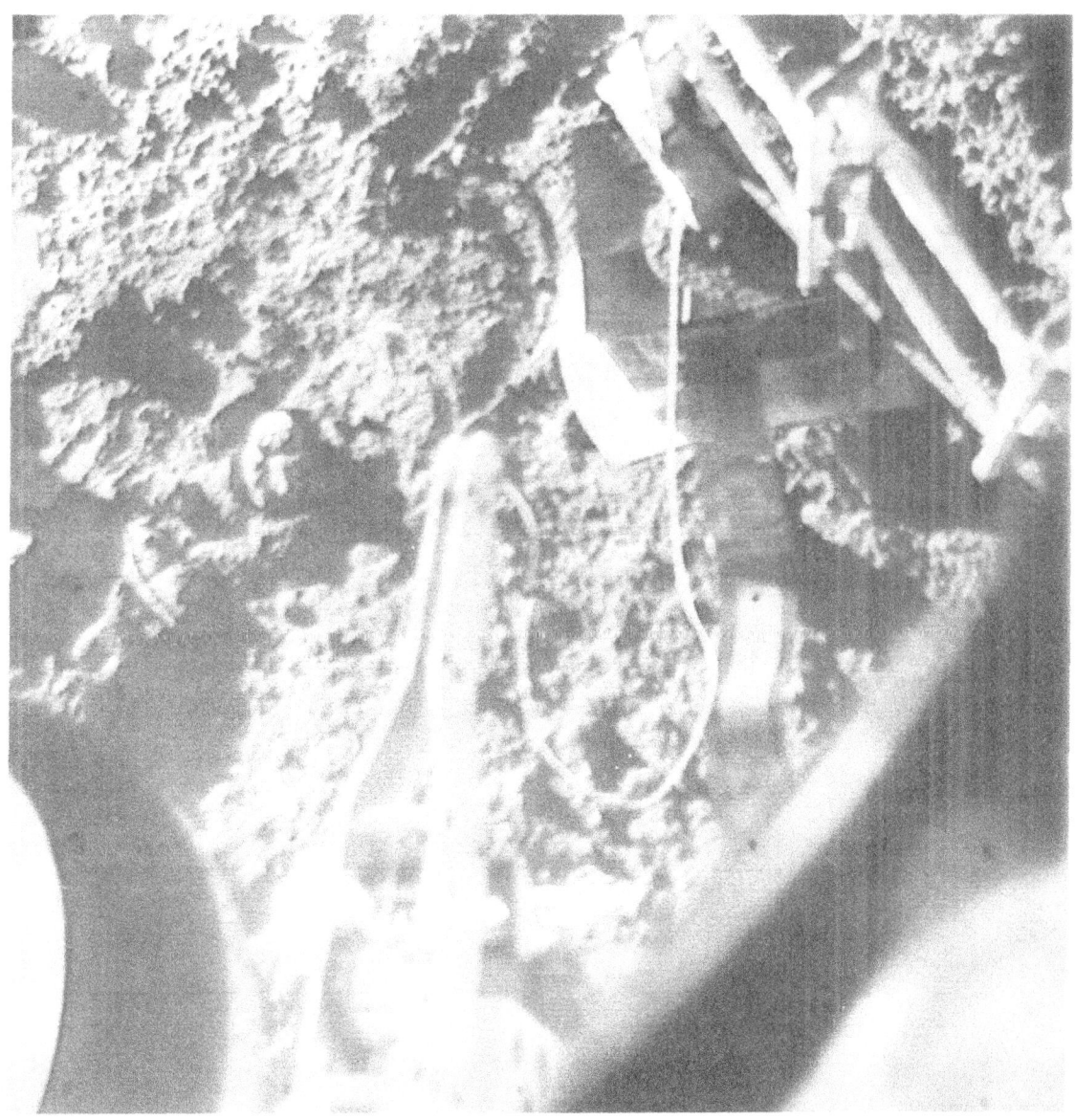

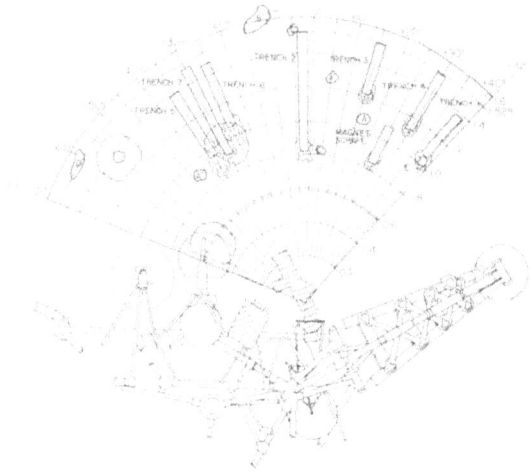

FIGURE 5-7.—Plan view of Surveyor VII surface-sampler operations showing locations of trenches, rocks, and positions of alpha-scattering instrument.

nation. Its dimensions, however, were too small to permit the weight of the rock to be determined.

Impact tests were performed on Surveyor III; however, the relatively small spring constant of the torque spring precluded the determination of density from these tests.

Surveyor VII

Surveyor VII landed on an ejecta blanket north of the rim of Tycho. Telemetry was used extensively for the measurement of motor currents during subsequent surface operations. As a consequence, it was possible to determine the force versus depth of penetrations during bearing tests (ref. 5-9) and trenching operations. The soil behaved qualitatively and quantitatively in a fashion similar to that observed at the Surveyor III site; relatively minor differences were observed. Results from bearing tests show that the mechanical properties cited for the mare material can be applied to the Tycho blanket material.

During Surveyor VII operations, obstructions were encountered in the trenches. A large fragment of rock, which could not be moved with the force available to the surface sampler in retraction, was encountered at a depth of about 3 cm. Therefore, it appeared that the soil cover at the Surveyor VII site was generally thinner than that encountered by Surveyor III. This soil cover probably constitutes only the upper few centimeters of the fragmental debris layer that constitutes the Tycho ejecta blanket.

In one of the trenches formed by Surveyor VII, a depth of 20 to 24 cm was attained in successive passes; subsequent analysis of the motor-current data indicated relatively little increase of strength with depth in this excavation. In fact, from the trenching information, the soil seems to be relatively uniform to the maximum depth. This difference in behavior from Surveyor III may be because of the different chemical nature of the soil at the two sites, a different age of the material since its deposition, or to other factors (ref. 5-10).

Comparison of Mare and Highland Sites

Qualitatively, the soil at the two sites exhibited certain differences in deformational behavior. In the mare area, a penetration test of the lunar surface caused cracking and splitting of the surface material as it was pushed up by the soil displaced by the surface sampler. Although this cracking was manifested to some small extent at the Surveyor VII site, the process of deformation was much more one of plastic working of the soil. It can be concluded, therefore, that the soil at the Surveyor III site exhibits a brittleness lacking in the soil in the vicinity of Tycho. That is to say, the soil at the Surveyor III site possesses some cohesion among the individual surface grains, which can be

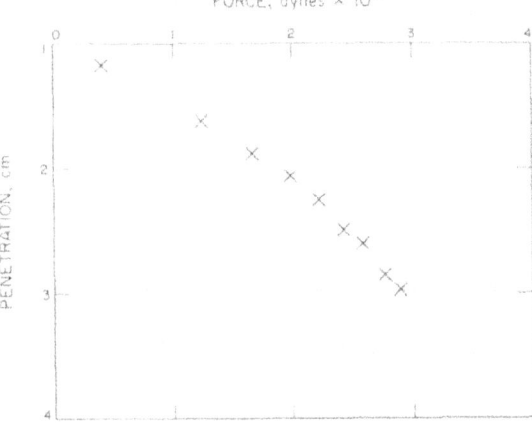

FIGURE 5-8.—Force versus penetration, bearing test 2.

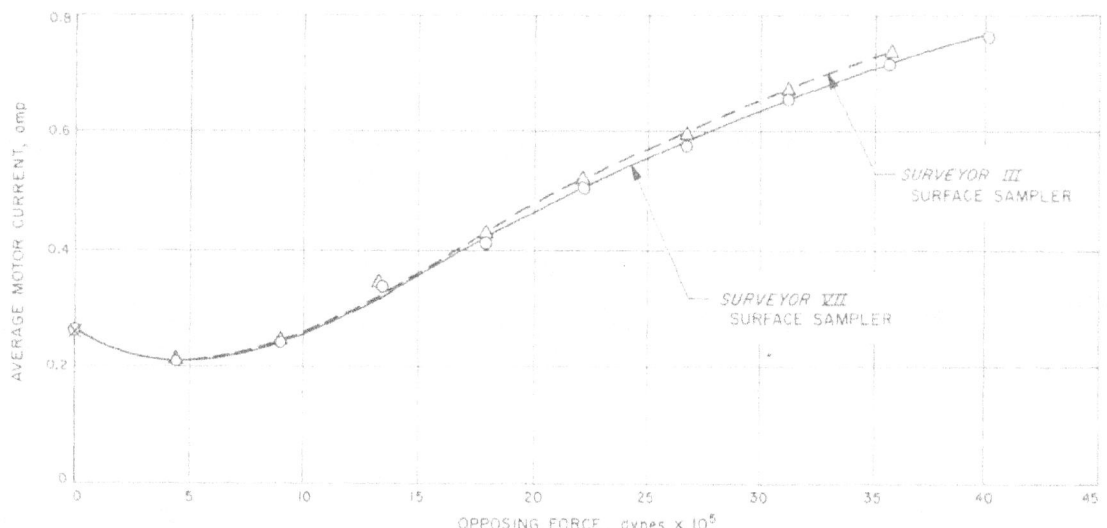

FIGURE 5-9.—Force versus motor current for surface-sampler bearing modes on Surveyors III and VII.

broken and may not be reestablished to the same degree. This can be confirmed by observations of the behavior of the material during trenching operations at both landing sites. At the Surveyor III site, a trenching operation produced relatively large chunks of lunar surface material which could be easily crushed by the surface sampler. At the Surveyor VII site, the trenching operation produced relatively small clods of soil, and the material appeared to deform without cracking or forming large lumps. Since both soils are cohesive, it indicates a somewhat different nature of the cohesiveness of the two sites.

It must be emphasized that the soil at the Surveyor III site also indicated a tendency to be cohesive even after it was disturbed, so that some adhesion among the individual fragments must be of a reversible nature. To some extent, it can be concluded that, at the Surveyor III site, a crust of somewhat brittle material existed to a depth of 2½ to 5 cm below the surface. Below this depth, the soil at both sites exhibited no substantial difference, with the exception of the increase in strength with depth observed at the Surveyor III site. With possible modifications, because of the depth of the underlying rock fragments, the soil at the Surveyor VII site also appeared to be quite uniform laterally over the entire region tested by the surface sampler.

One rock fragment was squeezed in the jaws of the Surveyor III surface sampler; rock fragments were picked up by the Surveyor VII surface sampler, and one rock fragment was broken by an impact. Since the rock picked up by the surface sampler on Surveyor III remained intact when subjected to a force of several hundred newtons per square centimeter by the surface-sampler jaws, it was concluded to be a rock fragment rather than an aggregate of soil particles. The range of densities obtained by picking up and weighing a fragment adjacent to the Surveyor VII spacecraft confirms this conclusion. The density was in the range 2.4 to 3.1 g/cm³, with the most probable value about 2.8 to 2.9 g/cm³. Another rock fragment of apparently similar appearance was fractured by a moderately strong blow by the Surveyor VII surface sampler. The magnitude of the impact indicates that either the sample was weakened by fracture planes or discontinuities within it, or that it was not a particularly strong rock initially (ref. 5-9).

On both missions, some lunar soil adhered to the surface sampler; this adhesion seemed to occur more frequently toward the end of the lunar day. On Surveyor VII, soil fragments were observed to adhere to the outside of the surface-sampler scoop by the end of lunar operations. The adhesion was not strong apparently, as it was observed that soil deposited on the mirror surface of the alpha-scattering sensor head slid fairly readily across the surface after 24-hour contact with the sensor head.

On Surveyor VII, the surface sampler carried two magnets embedded in the scoop door; some of the operations conducted were performed for the purpose of studying the interaction of the magnets with the lunar material. Magnetic material in granular form and one apparently magnetic fragment were picked up during these tests.

No alpha-scattering instrument was carried on Surveyor III; therefore, all of the surface-sampler time was devoted to lunar-surface tests. On the Surveyor VII mission, a large amount of time was devoted to interaction with the alpha-scattering instrument to insure that it successfully sampled the lunar surface.

Summary and Conclusions

The surface sampler, a versatile and extremely useful apparatus for performing a variety of mechanical functions, also proved its value as a remote manipulation tool.

Although relatively complex operations plans had been established before the launch and touchdown of Surveyors III and VII, it was found that successful surface-sampler operations were conducted by evaluating in real time the obstacles and tasks to be performed, and by operating the surface sampler in conjunction with the television camera to accomplish the required tasks. The use of motor current proved a useful addition to the process of manipulating the surface sampler, which responded correctly to all commands transmitted during operations. The design of the mechanism and its electronic auxiliary was more than adequate for all lunar-surface operations.

Data acquired from the soil mechanics surface-sampler experiment on the Surveyor III and VII missions provided significant information on the mechanical properties of the lunar soil. The range of properties has been reduced greatly from the ranges postulated before Surveyor. A summary of soil parameters, which are presented and discussed in reference 5-9, is given below:

(1) The lunar surface at the Surveyor VII landing site is covered with a fine-grained soil whose depth over rock, or rock fragments, varies from 1 or 2 cm to at least 15 cm. Many rock fragments ranging in size to 10 cm lie on the surface within the surface-sampler operations area. This differs from the Surveyor III site principally in the rock distribution, as only small fragments, partially buried, were found there.

(2) The surface soil at the two sites exhibits similar properties. The behavior of the soil at a depth of several centimeters is, therefore, consistent with a material possessing a cohesion on the order of 0.35 to 0.7×10^4 dynes/cm^2, an angle of friction of 35° to 37°, and a density of about 1.5 g/cm^3.

(3) The resistance of the soil to penetration and, therefore, its strength, increases with depth in the top 1 or 2 cm.

(4) To a depth of several millimeters at the lunar surface, the soil appears less dense, softer, and more compressible than the underlying material.

(5) The bearing capacity of the lunar soil to the 2.54-cm-wide area of the closed scoop of the surface sampler was about 2.1×10^5 dynes/cm^2, at a penetration of about 3 cm.

(6) Qualitatively, the soil at the Surveyor VII site was less brittle than at the Surveyor III site; there was less general cracking, and tests and trenching operations provided smaller lumps or aggregates of lunar soil.

(7) The density of a single rock, which was picked up and weighed during the Surveyor VII mission, was in the range 2.4 to 3.1 g/cm^3.

(8) The excavation of one partially buried rock during Surveyor VII operations revealed that the subsurface part was angular in contrast to the rounded visible part.

(9) One apparently intact rock was fractured by a blow from the Surveyor VII surface sampler.

(10) The adhesion of lunar soil to the surface-sampler scoop appeared to increase with time on the lunar surface.

References

5-1. Scott, R. F.: Soil Mechanics Surface Sampler Experiment for Surveyor. J. Geophys. Res., vol. 72, no. 2, Jan. 15, 1967, pp. 827-830.

5-2. Scott, R. F.; and Roberson, F. I.: Soil Mechanics Surface Sampler: Lunar Tests, Results, and Analyses. Surveyor III Mission Report. Part II: Scientific Data and Results, Tech. Rept. 32-1177, Jet Propulsion Laboratory, Pasadena, Calif., June 1, 1967, pp. 69-110.

5-3. Rouze, E. R.; Clary, M. C.; Le Croissette, D. H.; Porter, C. D.; and Fortenberry, J. W.: Surveyor Surface Sampler Instrument. Tech. Rept. 32-1223, Jet Propulsion Laboratory, Pasadena, Calif., Feb. 1, 1968.

5-4. de Wys, J. Negus: Electromagnetic Properties: Magnet Test. Surveyor VII Mission Report. Part II: Science Results, Tech. Rept. 32-1264, Jet Propulsion Laboratory, Pasadena, Calif., Mar. 15, 1968.

5-5. Surveyor I Mission Report. Part II: Scientific Data and Results, Tech. Rept. 32-1023, Jet Propulsion Laboratory, Pasadena, Calif., Sept. 10, 1966.

5-6. Surveyor III Mission Report. Part II: Scientific Data and Results, Tech. Rept. 32-1177, Jet Propulsion Laboratory, Pasadena, Calif., June 1, 1967.

5-7. Surveyor V Mission Report. Part II: Science Results, Jet Propulsion Laboratory, Pasadena, Calif., Nov. 1, 1967.

5-8. Surveyor VI Mission Report. Part II: Science Results, Tech. Rept. 32-1262, Jet Propulsion Laboratory, Pasadena, Calif., Jan. 10, 1968.

5-9. Scott, R. F.; and Roberson, F. I.: Soil Mechanics Surface Sampler. Tech. Rept. 32-1264, Jet Propulsion Laboratory, Pasadena, Calif., Mar. 15, 1968.

5-10. Surveyor VII Mission Report. Part II: Science Results, Tech. Rept. 32-1264, Jet Propulsion Laboratory, Pasadena, Calif., Mar. 15, 1968.

6. Lunar Surface Temperatures and Thermal Characteristics

J. W. Lucas (Chairman), W. A. Hagemeyer, J. M. Saari, L. D. Stimpson, and J. M. F. Vickers [1]

Each of the five landed Surveyor spacecraft transmitted data back to Earth for at least 2 weeks. The Surveyor landing sites are shown in figure 6-1; table 6-1 lists the selenographic location, time of landing, local slope, and Sun elevation above the eastern horizon at landing.

In addition to the selenographic location differences, the local terrain on which each of the spacecraft landed was different. Surveyor I landed on a relatively smooth, nearly level surface, encircled by hills and low mountains. Surveyor III landed about halfway down the slope of a crater about 200 meters in diameter and 15 meters deep. The local slope was about 12½° from the lunar horizontal. Surveyor V landed with one leg on the rimless edge of a 9- × 12-meter crater, which was 1.5 meters deep, and the other two legs within the crater. The local slope was about 20° from the lunar horizontal. Surveyor VI landed on a relatively smooth, nearly level, flat surface. The local slope was less than 1° from the lunar horizontal, but after the hop made by the spacecraft, the local slope on the new site was about 4°. Surveyor VII landed in extremely rough terrain, but with a local slope of only about 3°.

The behavior of the various spacecraft on the lunar surface varied. Surveyor I gave excellent data for two successive lunar days, and partial data were obtained as late as the fifth and sixth lunar days. The spacecraft operated for 48 hours into the first lunar night.

Surveyor III landed with the vernier propulsion system still at a thrust level almost equal to the lunar weight. It lifted off after initial touchdown and remained aloft for about 24 seconds. Liftoff also occurred after the second touchdown, the spacecraft remaining aloft for 12 seconds before third touchdown. At the time of second touchdown, all analog telemetry signals (which included all but two of the temperature telemetry channels, both associated with the television camera) became erroneous. The anomaly was localized in the signal-processing analog-digital converters, and it was found that most of the analog data obtained in the lowest rate mode (17.2 bit/sec) was fairly reliable and could be corrected with simple calibration factors. However, the overall accuracy of telemetered temperatures from Surveyor III was estimated at ±6° K compared with that for the other spacecraft of ±4° K. Surveyor III experienced a solar eclipse (by the Earth) during its first lunar day on April 24, offering the first opportunity to observe such an event from the

[1] With the exception of J. M. Saari of the Boeing Scientific Research Laboratories, the authors are affiliated with the Jet Propulsion Laboratory.

FIGURE 6-1.—Surveyor landing sites.

TABLE 6-1. *Positional characteristics of Surveyor spacecraft*

Spacecraft	Selenographic coordinates Atlas/ACIC system		Selenographic location	Touchdown time		Sun elevation above eastern horizon at touchdown, deg	Approximate local slope, deg
	Longitude	Latitude		GMT date	GMT hr:min:sec		
Surveyor I	2.46° S	43.23° W	Southwest part of Oceanus Procellarum (Ocean of Storms).	June 2, 1966	06:17:36	28.5	<1
Surveyor III	2.99° S	23.34° W	Southeast part of Oceanus Procellarum (Ocean of Storms).	Apr. 20, 1967	a00:04:17	11.8	12.4
Surveyor V	1.4° N b	23.2° E b	Mare Tranquillitatis (Sea of Tranquillity).	Sept. 11, 1967	00:46:42	16.4	20
Surveyor VI	0.51° N	1.39° W	Sinus Medii (Central Bay).	Nov. 10, 1967	01:01:04	2.8	c<1
Surveyor VII	40.88° S	11.45° W	Ejecta blanket of Crater Tycho.	Jan. 10, 1968	01:05:36	12.5	3

a Initial touchdown; second touchdown was at 00:04:41 GMT; final touchdown was at 00:04:53 GMT.
b Approximate; not precisely located.
c Before the hop. After the hop, the slope was about 4°.

Moon. Surveyor III did shut down almost immediately after sunset (2 hr) on the first lunar day.

Surveyor V, which operated for about 115 hours into the first lunar night, also experienced a solar eclipse (by the Earth) on the second lunar day on October 18 and operated for about 215 hours into the second lunar night. It operated for a short period of time during the fourth lunar day, transmitting 200-line television pictures.

The vernier rocket engines on Surveyor VI were fired on the lunar surface during the first lunar day, causing the spacecraft to lift off from the lunar surface and to hop 2.4 meters. Surveyor VI operated for about 40 hours into the lunar night; it was revived on the second lunar day, but gave thermal data for only a short time.

Surveyor VII, which operated for about 80 hours into the first lunar night, was successfully revived on the second lunar day, giving good thermal data during the day; however, contact with the spacecraft was lost before sunset on the second lunar day.

Surveyor I presented the first opportunity to obtain in situ estimates of the lunar surface temperature and thermophysical characteristics, in addition to engineering data on the thermal behavior of the spacecraft during operation on the lunar surface. It should be emphasized that none of the Surveyor spacecraft carried any instruments, as such, to measure lunar surface temperatures or surface thermal characteristics. For operational reasons, the spacecraft were thermally isolated from the lunar surface to the greatest extent possible. Fortunately, there were temperature sensors on the outer surfaces of two electronic compartments, on the solar panel, and on the planar array, which were highly dependent on the local thermal radiation environment and only partially dependent on other spacecraft equipment. These spacecraft temperatures have been used to estimate the average brightness temperature of those portions of the surface viewed by each sensor. In this report, brightness temperature is used in the usual sense, that is, the experimentally observed temperature a surface with unit emissivity must have to produce the measured response.

The Earth-based (telescopic) eclipse measurements used for comparison were performed to a resolution of either 8 or 10 seconds of arc

(14 or 18 km at the disk center). The error of matching Earth-based infrared maps with lunar photographs is 4 to 8 km based on a least-squares approximation using 30 to 50 identifiable hot spots. The location of the Surveyor spacecraft, relative to nearby features on Lunar Orbiter photographs, is known to 1 meter. The Surveyor spacecraft provided estimates of surface temperature out to about 18 meters from the compartments. Thus, compared with the best previous infrared telescopic observations, this is an improvement in ground resolution by a factor of 1000. The derived temperatures after sunset, and during the two eclipses, were used to estimate the thermal characteristics of the lunar surface at each site.

Earth-Based Thermophysical Observations

The surface temperatures calculated from the spacecraft thermal data were compared with Earth-based measurements and theoretical thermophysical models. In the following paragraphs, a summary is given of the pertinent Earth-based measurements of the various landing site regions.

Albedo

With regard to the theoretical models, it is necessary to know the bolometric albedo so that during illumination the amount of solar radiation absorbed by the surface can be calculated. If the small amount of energy conducted in or out of the surface during illumination is ignored, then the Lambertian temperature (with unit surface emissivity assumed) T_L is defined by the expression

$$\sigma T_L^4 = (1-A) S \sin \psi \qquad (1)$$

where
- σ = Stefan-Boltzmann constant, $W/m^2 \, °K^4$
- A = bolometric or total solar albedo, dimensionless
- S = solar irradiation, W/m^2
- ψ = elevation angle of the Sun to the surface

By this definition, the Lambertian temperature is that which a perfectly diffuse blackbody surface would have to radiate the same energy as is absorbed. Actually, as discussed later in this chapter, the lunar surface exhibits directional effects in its emission; however, it has been found that the Lambertian temperature provides a useful comparison to the spacecraft data.

In order to calculate T_L, the bolometric albedo of each site must be known. For this purpose, the simultaneous infrared and photometric scan data of reference 6-1 were used. Of particular interest was the scan at full Moon ($-2°$ phase angle) just prior to the December 19, 1964, eclipse. The data show the brightness[2] temperature changes with the photometric brightness on adjacent regions. This allows the calculation of the relationship between the photometric brightness on this scan and the bolometric albedo.

Now, because of the directional emission of the lunar surface, the observed brightness temperature T_b differs from the Lambertian temperature because of the angle of view so that

$$T_b(\psi) = D(\psi) T_L \qquad (2)$$

which defines the directional factor $D(\psi)$. For the full-Moon scan, if the bolometric albedo A is assumed proportional to the measured photometric brightness B, then

$$A = KB \qquad (3)$$

where K is a constant, which if known allows the determination of A for any point. To determine K, measurements were made on two areas, 1 and 2, of differing brightness at the same ψ, so from equations (1) and (2)

$$\left. \begin{array}{l} \sigma \left[\dfrac{T_{b,1}}{D(\psi)} \right]^4 = (1-KB_1) S \sin \psi \\ \\ \text{and} \\ \\ \sigma \left[\dfrac{T_{b,2}}{D(\psi)} \right]^4 = (1-KB_2) S \sin \psi \end{array} \right\} \qquad (4)$$

Eliminating $D(\psi)$ between these two equations and solving for K, we find

$$K = \frac{T_{b,2}^4 - T_{b,1}^4}{B_1 T_{b,2}^4 - B_2 T_{b,1}^4} \qquad (5)$$

---

[2] With unit surface emissivity assumed.

It was thought that K could possibly be a function of ψ, so many pairs of points of different brightness were measured over the disk. The results showed that K was essentially independent of ψ.

For each landing site region, B was measured from the scan data and the bolometric albedo calculated with the value of K determined above (see table 6-2). Because the measurements were made with a resolution of 10 seconds of arc (18 km at the center of the disk) and with a location accuracy of 4 to 8 km, the albedo of the region in the immediate vicinity of a spacecraft could depart considerably from the quoted values.

Thermophysical Properties of the Surveyor Landing Sites

The thermophysical properties can be determined only from postsunset eclipse or lunation cooling curves. The most extensive eclipse measurements are those of references 6-2 and 6-3 made during the December 19, 1964, eclipse. Data on isotherms during totality for the equatorial region have been published (ref. 6-4); the measurements revealed anomalous cooling of features of a wide range of sizes, varying from kilometer-size craters to entire maria. It would not, therefore, be surprising if thermal heterogeneity were found to dimensions much smaller than possible to measure by the Earth-based eclipse measurements; for example, any local areas strewn with sizable boulders should cool more slowly than unstrewn areas.

Isotherms in the region of the Surveyor I landing site have been transferred to the lunar aeronautical chart (fig. 6-2). In that region,

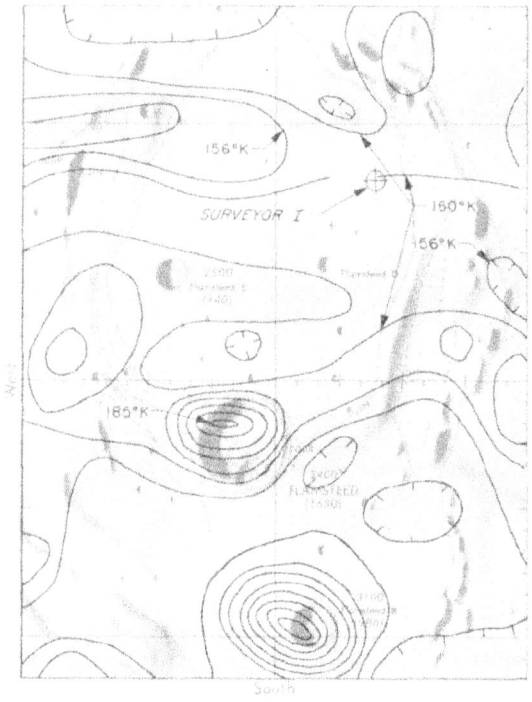

FIGURE 6-2.—Isothermal contours for landing site region of Surveyor I obtained during totality of lunar eclipse.

the craters Flamsteed and Flamsteed B are prominent hot spots. It can be seen that the area in which Surveyor I landed is one with small horizontal thermal gradients; thus, it contains the highly insulating properties that typify the general lunar surface.

Isothermal contours in the region of the Surveyor III site during totality of the December 19, 1964, eclipse (ref. 6-2) are shown in figure 6-3. The region is relatively bland. Iso-

TABLE 6-2. *Mission characteristics*

Characteristic	Surveyor I	Surveyor III	Surveyor V	Surveyor VI	Surveyor VII
Landing site	Oceanus Procellarum.	Oceanus Procellarum.	Mare Tranquillitatis.	Sinus Medii.	Ejecta blanket of Tycho.
Solar constant, W/m²	1352	1386	1375	1423	1442
Bolometric albedo $(A) \equiv$ lunar reflectivity (ρ_2).	0.052	0.076	0.077	0.084	0.17

TEMPERATURES AND THERMAL CHARACTERISTICS

FIGURE 6-3.—Isothermal contours for landing site region of Surveyor III obtained during totality of lunar eclipse.

thermal contours for the Surveyor V and VI landing site regions from the same eclipse data are shown in figures 6-4 and 6-5, respectively. Again both regions appear to be relatively bland at the limit of resolution of the Earth-based measurements.

The crater Tycho is an outstanding thermal anomaly on the lunar surface from the standpoint of the temperature difference over its environs and the size of the area affected. Isotherms of the region from reference 6-3 are shown in figure 6-6 and indicate that there are three maxima in the temperature distribution within the crater and that the anomaly extends about one crater diameter beyond the rim. The Surveyor VII landing site is within the anomalous area surrounding the crater. During eclipse totality, the central peak is about 62° K warmer than the environs, whereas the Surveyor VII landing site area is only 14° K warmer.

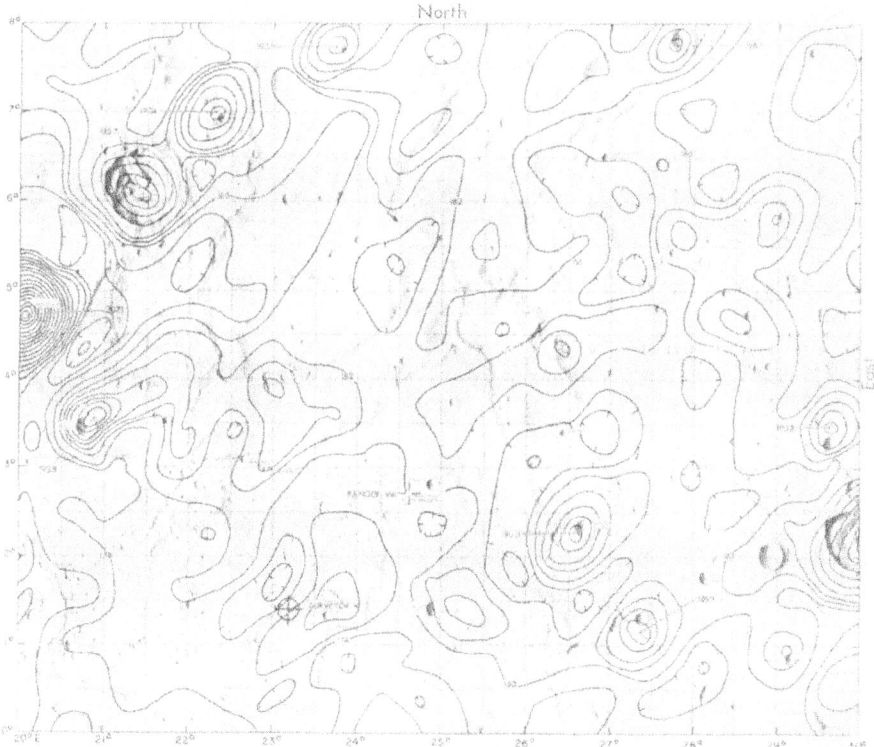

FIGURE 6-4.—Isothermal contours for landing site region of Surveyor V obtained during totality of lunar eclipse.

FIGURE 6-5.—Isothermal contours for landing site region of Surveyor VI obtained during totality of lunar eclipse.

Earth-Based Predictions of Lunar Surface Temperatures of the Surveyor Landing Sites

The spacecraft data were compared with Earth-based measurements of the illuminated lunar surface made during the December 19, 1964, eclipse. The latter measurements, it is to be noted, are influenced by the directional effects of infrared emission determined by the direction from which the site regions were observed on Earth. Lunation calculations (ref. 6-5) of the homogeneous model were used assuming constant thermophysical properties. These properties are characterized by the thermal parameter $\gamma = (k\rho c)^{-1/2}$, where k is thermal conductivity, ρ is density, and c is specific heat. This constant γ model, however, can not adequately represent the Earth-based measurements during both eclipse and postsunset, since the former requires a much larger constant γ than the latter. During illumination, the model predicts temperatures essentially in agreement with equation (1) if γ is greater than 500. Recently a particulate model of the lunar soil has been proposed (ref. 6-6) which agrees with both the eclipse and postsunset cooling.

The bolometric albedos used in the following calculations are those given in table 6-2 for each landing site. The temperatures were corrected for the appropriate Moon/Sun distance. Also, the normal to each surface element was assumed coincident with the local vertical.

A γ value of 800 (ref. 6-5) is typical for the

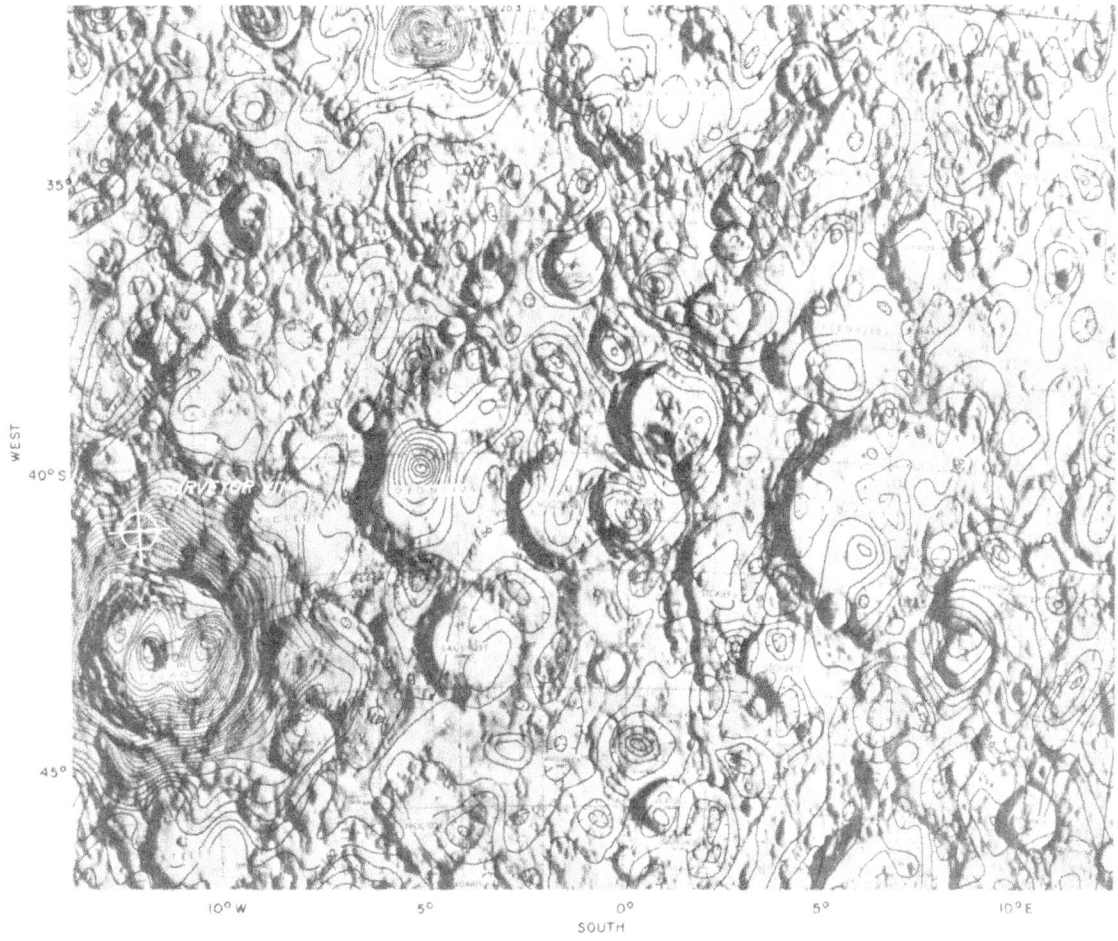

FIGURE 6-6.—Isothermal contours for landing site region of Surveyor VII obtained during totality of lunar eclipse.

lunation of the equatorial Surveyor sites and was derived from Earth-based postsunset measurements of maria in the eastern sector (ref. 6-7). The larger γ values given in the following paragraphs resulted from Earth-based eclipse measurements. The difference in γ is thought to be a consequence of heat exchange from only the uppermost millimeters of soil during an eclipse, whereas a different type of soil at a lower depth is involved during the lunation warming and cooling phases.

The calculated lunar surface Lambertian temperatures (ref. 6-5)[3] for the homogeneous model at the Surveyor I landing site are shown in figure 6-7. The specific values for solar constant and lunar reflectivity (albedo) used for each mission are given in table 6-2. The time scale was fixed assuming a flat Moon surface at sunset. The $\gamma=800$ intermediate curve in figure 6-7 is considered most representative of the site. It is of some interest to note that a γ value of 1350 had been inferred for the Surveyor I site from Earth-based eclipse measurements. Temperatures calculated for the Surveyor III site are shown in figure 6-8.

Thermal measurements were made of the Surveyor III site during the April 24, 1967, eclipse. Figure 6-9 shows a predicted cooling

---

[3] B. P. Jones calculated the Lambertian curves for the different Surveyor sites including postsunset where the differentiation due to γ is significant.

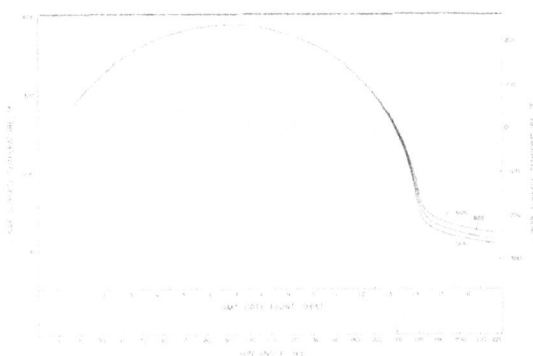

FIGURE 6-7.—Calculated Lambertian temperature for Surveyor I landing site.

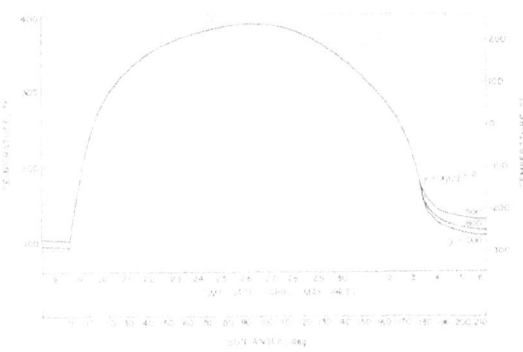

FIGURE 6-8.—Calculated Lambertian temperature for Surveyor III landing site.

FIGURE 6-9.—Predicted eclipse Lambertian temperature for Surveyor III landing site.

curve for the site from Earth-based measurements obtained during the December 19, 1964, eclipse (ref. 6-2). When this curve was compared with the theoretical eclipse cooling curves for a homogeneous model (ref. 6-8), it was possible to infer a value for γ of 1400. Values of γ in this range, as determined from eclipse calculations, are representative of the insulating material that characterizes much of the lunar surface. The warming curve in figure 6-9 represents calculated equilibrium surface temperatures corresponding to the insolation at each time.

The calculated Lambertian temperature for the Surveyor V landing site is shown in figure 6-10. Figure 6-11 is a predicted eclipse cooling curve for the site, from Earth-based measurements. By using the theoretical eclipse cooling curves for a homogeneous model (ref. 6-8), a γ of 1350 was obtained for the lunar surface material.

The calculated Lambertian temperatures and Earth-based temperatures (ref. 6-1) at the Surveyor VI landing site are shown in figure 6-12. Each value has been plotted at that time in November 1967 when the elevation angle of the Sun was the same as when the measurement was made. These Earth-based measurements show the directionality of lunar infrared emission; near local noon, when the surface was observed from the same general direction as the Sun (i.e., when the phase angle was small), the measured temperatures were higher than the calculated Lambertian temperatures. Earth-based eclipse observations show cooling during

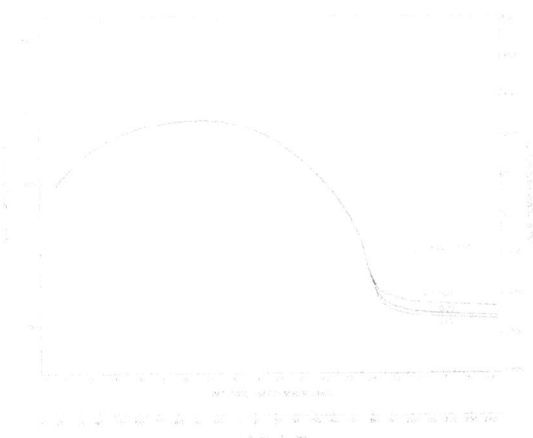

FIGURE 6-10.—Calculated Lambertian temperature for Surveyor V landing site.

made measurements of Tycho to a 9-seconds-of-arc resolution up to a few minutes before the end of totality. These eclipse observational data fit the cooling curve for a homogeneous model with $\gamma=450$ inside the crater and with $\gamma=1100$ outside the crater by 30 seconds of arc. It should be noted that a $\gamma=1091$ was erroneously used in reference 6–10 for postsunset temperatures outside the crater instead of 800 from actual postsunset measurements (ref. 6–7).

Although no Earth-based measurements of the Surveyor VII landing site region were made during the lunar night, it was possible to obtain a postsunset cooling curve by interpolation in the following manner:

(1) Earth-based eclipse cooling curves were obtained from the data of reference 6–2 for the crater itself, the landing site region, and the environs outside the anomalous region surrounding the crater. These curves showed the landing site region had a temperature difference over the environs only 0.27 as large as that for the crater itself.

(2) Postsunset cooling curves were available for the crater (ref. 6–1); for the environs, a theoretical curve for the homogeneous model with $\gamma=800$ was assumed.

(3) A postsunset curve for the landing site region was determined by interpolating 0.27 of the way from the environs curve to the crater curve, resulting in the predicted X curve shown in figure 6–13. This postsunset curve corresponds to a γ of 550 for the landing site region.

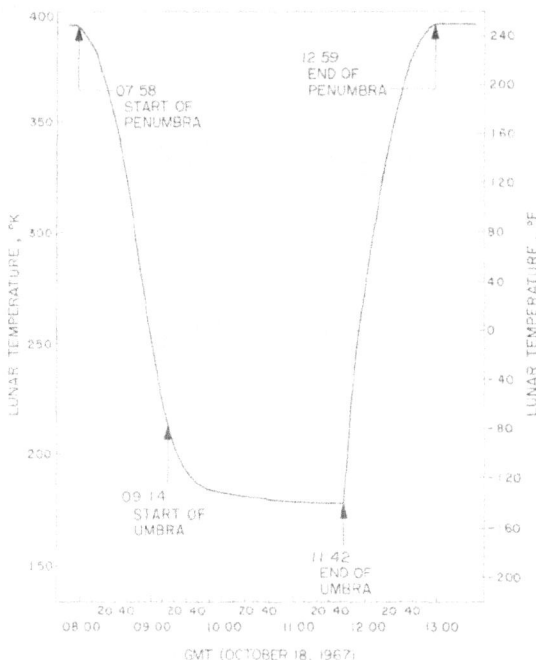

FIGURE 6–11.—Predicted eclipse temperature for Surveyor V landing site.

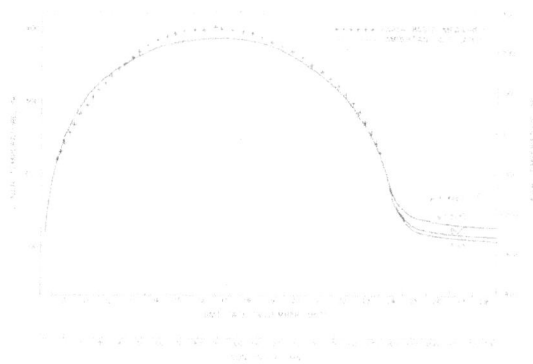

FIGURE 6–12.—Earth-based and calculated temperatures for Surveyor VI landing site.

totality comparable to that for a homogeneous model with a γ of 1100.

The calculated Lambertian temperatures for the Surveyor VII landing site are shown in figure 6–13. Also shown are the Earth-based measured temperatures, which again show a directional effect distributed over a larger portion of the lunar day. During the December 19, 1964, eclipse, Ingrao et al. (ref. 6–9)

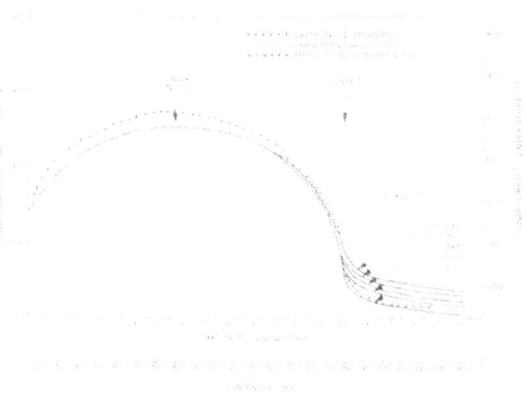

FIGURE 6–13.—Earth-based, calculated, and predicted temperatures for Surveyor VII landing site.

Directional Effects

It has been determined that when the lunar surface is illuminated by the Sun the observed brightness temperature is not constant for different angles of observation, i.e., the surface does not behave like a Lambertian surface (ref. 6-11). This effect, ascribed to surface roughness, causes the brightness temperature to be higher when the phase angle is small (i.e., when the Sun/surface/observer angle is small) than when it is large. (Qualitatively, the emission is greater when viewing the lunar surface with the Sun over one's shoulder.) Such directionality will have an effect on the radiation received by the compartments on the Surveyor spacecraft to a degree depending on the scale of the local surface roughness.

To correct the calculations for directional effects, Earth-based measurements over the entire lunar disk for three Sun angles were used. For a Sun elevation angle of 90°, the measurements of Sinton (ref. 6-12) were taken which show the variation in radiance from the subsolar point as a function of the angle of observation. For two other Sun angles of 30° and 60°, the infrared scan data for different phases made by Shorthill and Saari were used. Albedo corrections for each point were made from the full-Moon photometric data. The directional factor was determined from equation (2) by using a calculated Lambertian temperature at each point.

Directional factors obtained in this manner were referenced to a lunar surface element by a coordinate system with azimuth and elevation angles for the direction of observation defined as follows. Azimuth angles were measured from the normal projection of the Sun direction onto the surface. Elevation angles were measured from the surface in the plane of observation. Directional factors obtained over the globe were referenced to this azimuth-elevation angle system. A least-squares spherical harmonic fit, symmetrical with respect to plus and minus azimuth angles, was then computed for the data. A contour plot of the directional factor for a Sun elevation angle of 60° is shown in figure 6-14. Directional factors were, of necessity, obtained from global measurements made on a variety of features. It is possible, therefore,

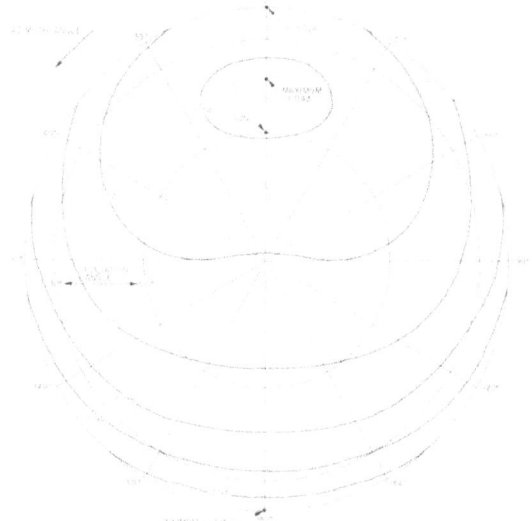

FIGURE 6-14.—Contours of directionality for Sun elevation angle of 60°.

for a small area such as a Surveyor landing site to have different directional effects than the average surface if the local roughness or surface configuration differed significantly from the average.

Spacecraft Thermal Measurements

Spacecraft Description

The Surveyor spacecraft (fig. 6-15) had a basic structural frame of tubular aluminum which served as a tetrahedral mounting structure for the electronic gear and propulsion system. The three spacecraft legs were attached at the three corners of the base. The planar array antenna and solar panel, mounted on a mast about 1 meter above the apex of the structure, cast varying shadow patterns on the spacecraft and on the lunar surface throughout the lunar day. Changes in shadow patterns occurred as a result of the commanded repositionings of the planar array antenna and solar panel and from the apparent movement of the Sun (about 0.5°/hr).

Generally, the spacecraft components in the Sun-illuminated areas had white painted surfaces that provided a low-solar-absorptance and high-infrared-emittance thermal finish. The

TEMPERATURES AND THERMAL CHARACTERISTICS

outboard face, i.e., the surface facing the blanket of each compartment (see figs. 6-16 and 6-17). The blanket isolated the panels from the inside of the compartments. Because the outboard faces of the compartments had a strong radiative coupling to the lunar surface, but were virtually shielded from view of other spacecraft components, an analysis of lunar surface brightness temperatures was possible.

Some differences existed in the thermal blanket design (75 layers of aluminized Mylar)

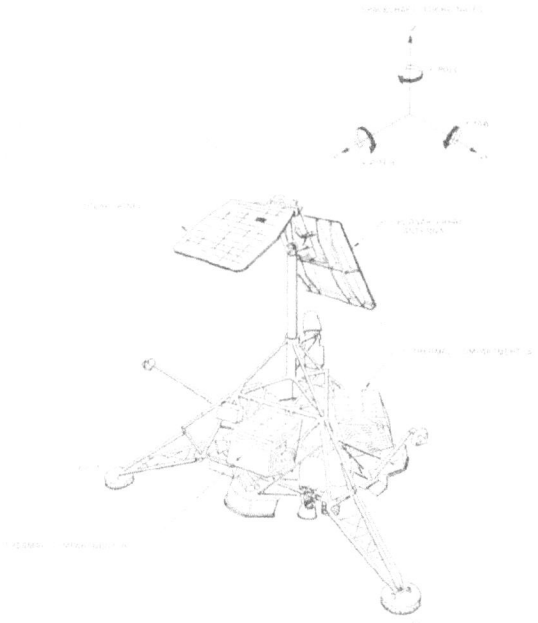

FIGURE 6-15.—Surveyor spacecraft configuration.

polished-aluminum underside thermally isolated the spacecraft from the lunar surface.

The temperature data of various points in the spacecraft were provided by platinum resistance temperature sensors. Each sensor was calibrated individually to $\pm 2°$ K; other nominal system inaccuracies degraded the overall accuracy to $\pm 4°$ K.[4] Most of the 75 sensors measured internal spacecraft temperatures. Some, however, were externally located and were responsive to the lunar surface radiation; four were located on the outside panels of the two main electronic components, on the solar panel, and on the planar array antenna.

Compartment canisters. Compartments A and B housed the spacecraft electronics and battery. A thermal blanket of multilayer insulation surrounded the components in each compartment and, in turn, was covered with an aluminum panel. A temperature sensor was bonded to the polished-aluminum inner surface of the

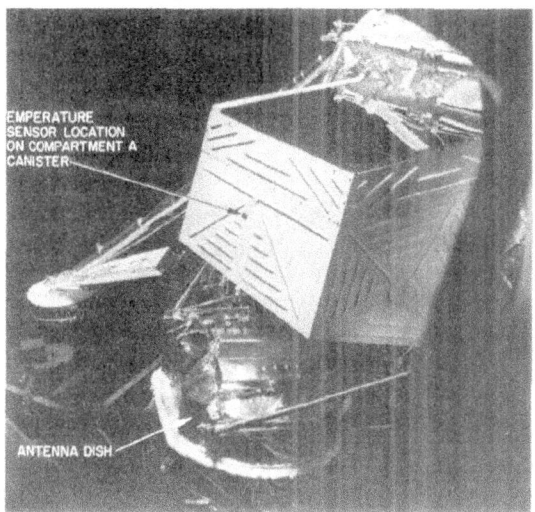

FIGURE 6-16.—Surveyor model, compartment A.

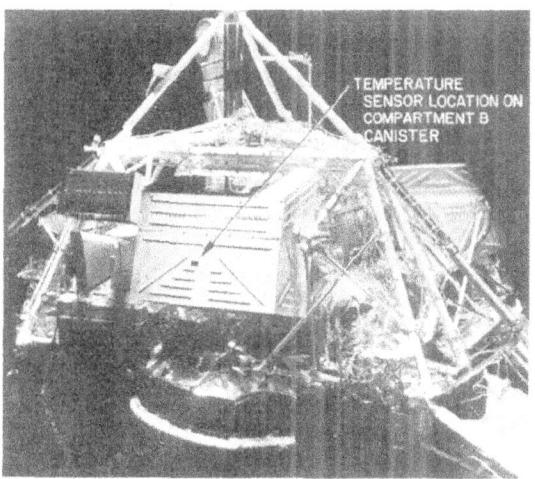

FIGURE 6-17.—Surveyor model, compartment B.

---

[4] These temperature sensors were low resolution; other sensors, critical for spacecraft performance assessment, were calibrated to $\pm 1°$ C with an overall accuracy of $\pm 3°$ C over a narrow temperature range.

on the different missions. Surveyors I and III had a blanket that was integral (taped and interleaved) to the compartment wall. Surveyors V, VI, and VII had a new blanket design installed as a separate item. Both designs provided excellent insulation. The specific parameters needed to obtain lunar surface temperatures from the compartment temperature data are given in references 6-10 and 6-13 through 6-16.

Solar panel and planar array antenna. The solar panel and planar array antenna were relatively low-heat-capacity planar surfaces. Temperature data measured by these two surfaces also may be used to derive lunar surface brightness temperatures. Additional complexity is introduced in the analysis, since these subsystems thermally interact with some spacecraft equipment.

Two different solar panel designs were used; one for Surveyors I and III and another for Surveyors V, VI, and VII. The specific parameters needed to obtain lunar surface temperatures from the solar panel and planar array antenna temperatures are given in references 6-10 and 6-13 through 6-16. The planar array antenna as a source of temperature data is not now considered very practical because of the additional support structure attached to it.

Surface Views

The landed orientations for the different Surveyor spacecraft are compared in figure 6-18. The surface views from the compartments are depicted along with local downslope tilt. It should be noted that preliminary values were used for calculation purposes as shown and were based on solar panel and planar array antenna positional data.

The lunar surface temperatures were found to be dependent primarily on the Sun elevation angle to the local lunar surface slope. Thus, for Surveyors III and V which landed on sloping surfaces, simple time translations of the lunar noon to a local zenith resulted in improved temperature distributions. The lunar surface temperature measured by each compartment sensor also was influenced by terrain features and shadowing of the lunar surface by the spacecraft.

Surveyor I. The assumed landed orientation of Surveyor I is shown in figure 6-19. The azimuth of leg 1, the Y axis, is given as 1° south of west; the vertical spacecraft $-Z$ axis is taken to be tilted 0.5° toward the west. To insure early-morning coverage, the science bay (television camera) was directed eastward, which resulted in compartment A viewing southwest and compartment B viewing northwest. The normal to the outer canister face of

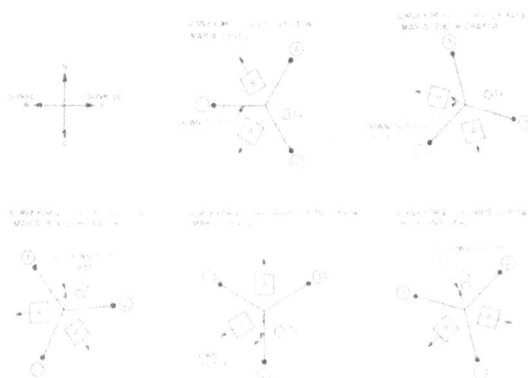

FIGURE 6-18.—Spacecraft landed orientations.

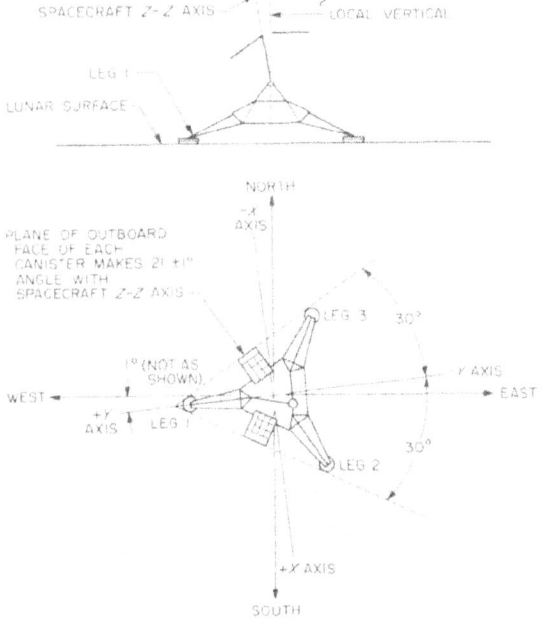

FIGURE 6-19.—Surveyor I landed orientation.

each compartment made an angle of $69\pm1°$ with the spacecraft $-Z$ (vertical) axis. The view factors from compartments A and B to the lunar surface (defined later in eq. 6 as F_{12} plus F_{13}) were approximately 0.28 and 0.29, respectively. The temperature data measured by thermal sensors on the outboard faces of the compartments were given in reference 6–13.

The solar panel was stepped throughout the lunar day so that it would be nearly normal to the Sun vector. The solar panel temperature data are presented in figure 6–20, since they are planned for future use and were not presented earlier in reference 6–13.

Surveyor III. The assumed orientation of Surveyor III with respect to lunar coordinates is given in figure 6–21. The normal to the compartment A outer canister face was lying in a vertical plane of azimuth 9° east of south and was inclined at an angle of 65° to the local vertical. The normal to the compartment B outer canister face had an azimuth 16° north of west and was inclined at an angle of 81° to the local vertical. The spacecraft $-Z$ axis approximated the direction of the local surface normal; both compartment normals were inclined at 69° to this direction.

The Surveyor III landing site (about 45 meters southeast of the crater center) is shown in figure 6–22. The surface area viewed by each compartment was limited by the canister face orientation and crater rim. On this basis, compartment A viewed a maximum projected surface area of 1.3×10^4 m², and compartment B an area of 2.6×10^4 m². The resulting view

FIGURE 6–21.—Surveyor III landed orientation.

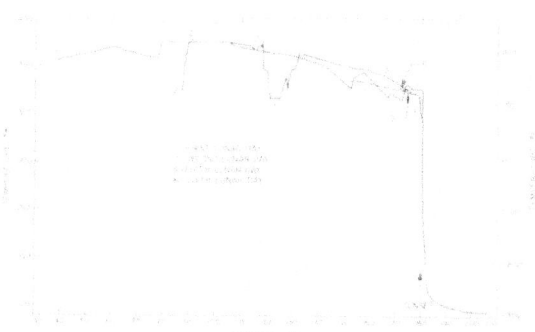

FIGURE 6–20.—Temperatures of Surveyor I solar panel on subsequent lunar days.

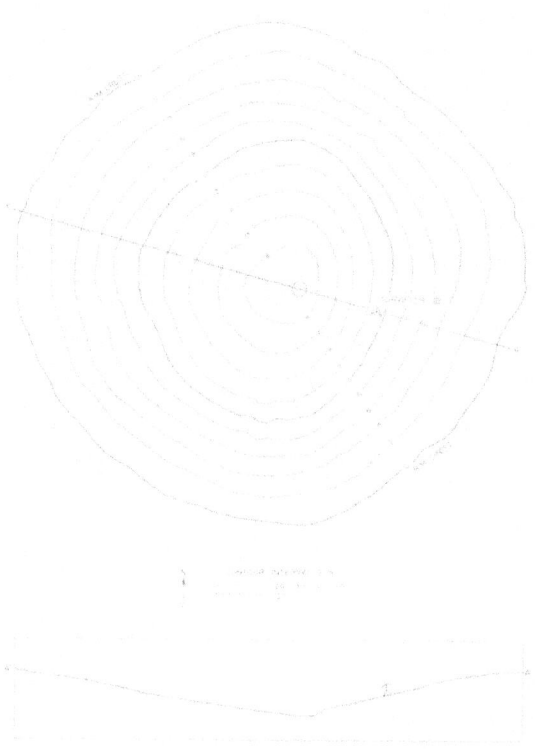

FIGURE 6–22.—Contour map and profile of crater in which Surveyor III landed.

factors from compartments A and B to the lunar surface were 0.31 and 0.41, respectively. The compartment, solar panel, and planar array temperature data are given in reference 6–14.

Surveyor V. Surveyor V landed in a small (9 x 12 m) crater, with leg 1 positioned near the crater rim and legs 2 and 3 downslope on the southwest wall of the crater. Figure 6–23 shows the assumed orientation of Surveyor V with respect to the lunar coordinates after landing and after sunset of the first lunar day. At approximately sunset of the first lunar day, the shock absorbers on legs 2 and 3 compressed, placing the spacecraft even more downslope. During the second lunar day, the spacecraft assumed the orientation it had during the first lunar day.

The location of the spacecraft within the crater profile is shown in figure 6–24. Compartment A primarily viewed the east side of the crater, the surface beyond the crater, and space, with an overall view factor of 0.247 to the lunar surface. Compartment B viewed the west side of the crater, the surface beyond the crater rim, and space, with an overall view factor of 0.255 to the lunar surface. The compartment, solar panel, and planar array temperature data are given in reference 6–15.

Surveyor VI. Surveyor VI, after the initial landing and later after the hop, was situated

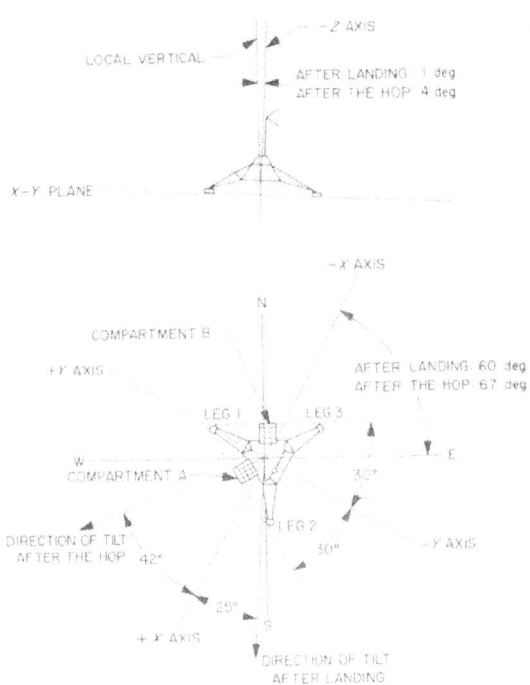

FIGURE 6–23.—Surveyor V landed orientation.

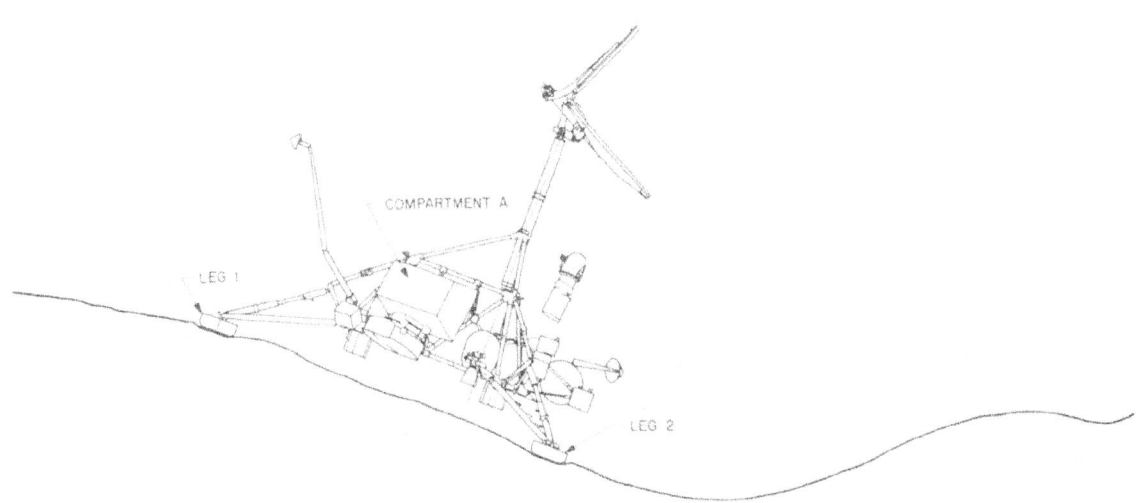

FIGURE 6–24.—Profile of crater in which Surveyor V landed.

TEMPERATURES AND THERMAL CHARACTERISTICS 195

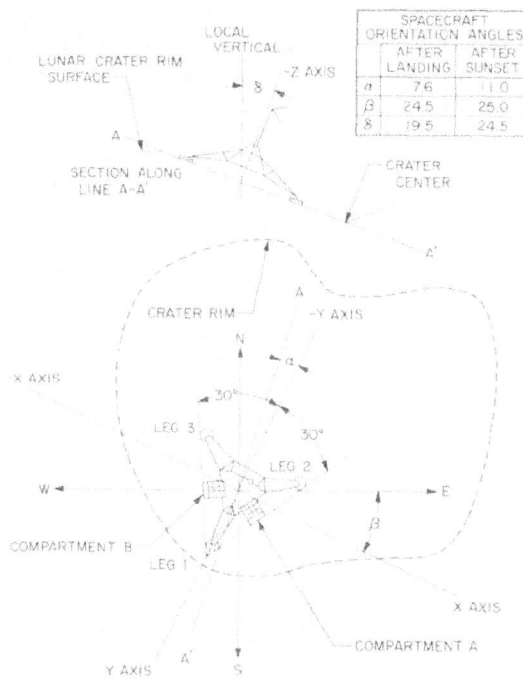

FIGURE 6-25.—Surveyor VI landed orientations.

on a generally level, flat surface. The hop occurred on November 17 when the vernier engines were fired for 2.5 seconds, causing the spacecraft to rise and to land 2.4 meters from the original landing point. The assumed orientations of the spacecraft with respect to the lunar coordinates are shown in figure 6-25. Compartment A viewed the area to the southwest with a view factor of 0.321 to the lunar surface after the initial landing and 0.350 after the hop. Compartment B viewed the area to the north, with a view factor of 0.318 to the lunar surface after the initial landing and 0.316 after the hop. The compartment temperature data are given in reference 6-16. The solar panel temperature data are presented in figure 6-26, since they were not included in reference 6-16.

Surveyor VII. Surveyor VII landed on a generally level surface in a highland area. The assumed orientation of the spacecraft with respect to lunar coordinates is shown in figure 6-27. Compartment A viewed the east,

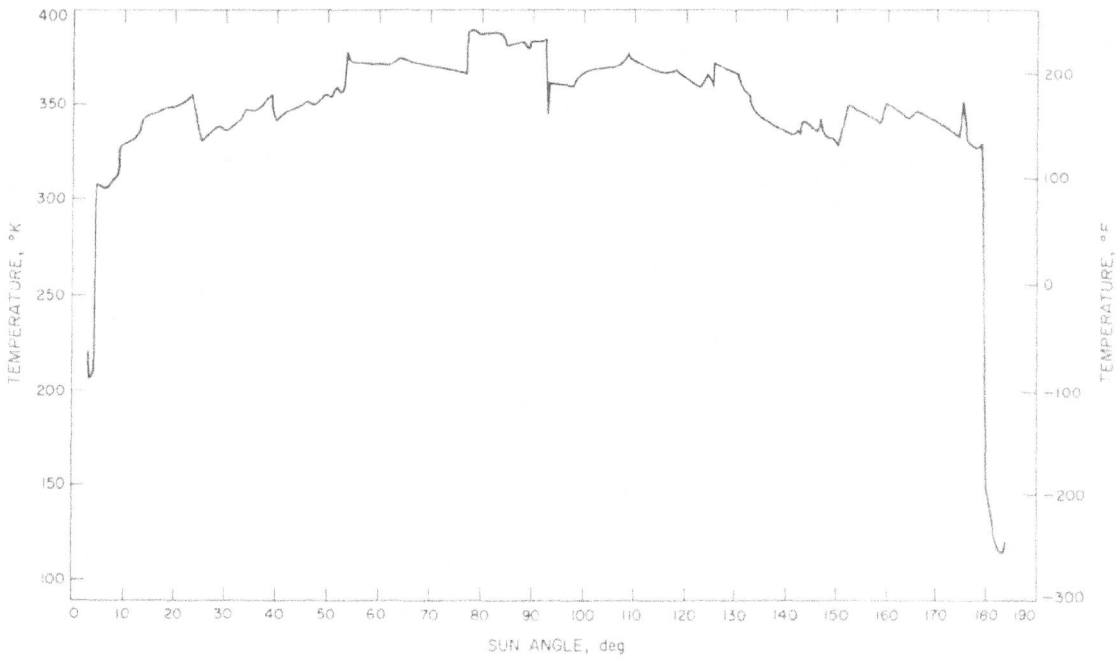

FIGURE 6-26.—Temperature of Surveyor VI solar panel.

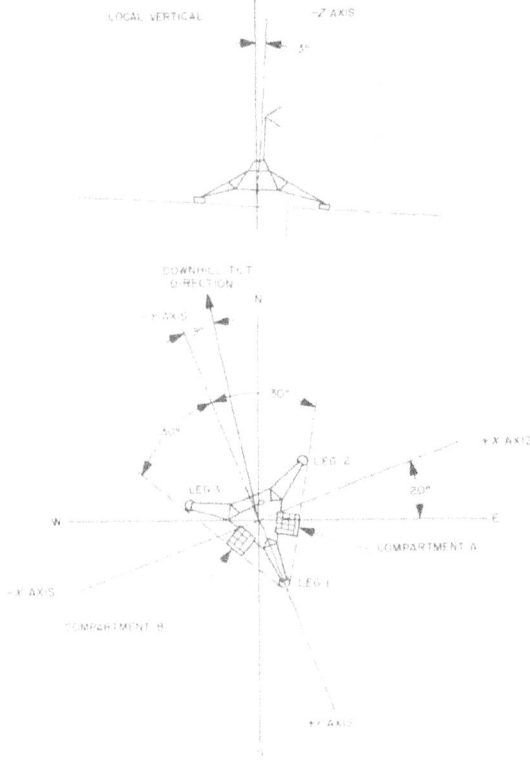

FIGURE 6-27.—Surveyor VII landed orientation.

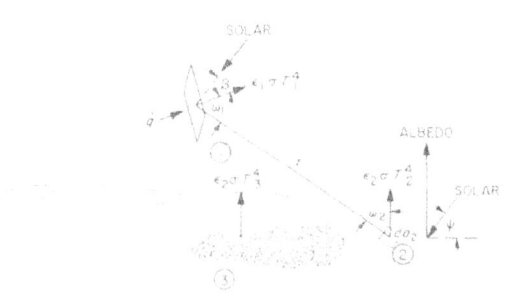

FIGURE 6-28.—Heat exchange for a compartment outboard face.

with a view factor of 0.337 to the lunar surface. Compartment B viewed the area to the southwest, with a view factor of 0.333 to the lunar surface. During the lunar night, shock absorber 2 compressed, resulting in a slope of the spacecraft vertical axis about 6° with respect to the local vertical during the second lunar day. The compartment, solar panel, and planar array temperature data are given in reference 6-10.

Surface Temperature Calculations

The lunar surface brightness temperatures were determined from the spacecraft compartment outboard face temperatures. This was accomplished by evaluating the radiative heat balance required between the compartment face and the lunar surface viewed using the following equation (depicted in fig. 6-28):

$$\sigma T_2^4 = \frac{\sigma T_1^4}{\epsilon_2 F_{12}} - \frac{F_{13}}{F_{12}} \sigma T_3^4$$

$$- \frac{\alpha_{1S}}{\epsilon_1 \epsilon_2 F_{12}} \Big[F_{12} \rho_2 \sin \psi + \cos \beta \Big] \quad (6)$$

$$- \frac{\dot{q}}{\epsilon_1 \epsilon_2 F_{12}}$$

where

T_1 = compartment surface temperature, °K
T_2 = lunar surface brightness temperature in sunlight, °K
T_3 = lunar surface brightness temperature in shadow
  = 200° K was used in the calculations
S = solar irradiation constant, W/m²
F_{12} = geometric view factors from compartments to sunlit portion of lunar surface, dimensionless
F_{13} = geometric view factors from compartments to shadowed portion of lunar surface, dimensionless
$\dot{q}$ = heat flux from inside to outside of compartment wall, W/m²
σ = Stefan-Boltzmann constant
  = 5.675×10^{-8} W/m² °K⁴
ϵ_1 = compartment surface emittance, dimensionless
  = 0.87 ± 0.02
ϵ_2 = lunar surface emittance, dimensionless

= 1.0 (brightness temperature assumption)

α_{LS} = compartment surface solar absorptance, dimensionless
  = 0.20 ± 0.02

β = angle between direction of Sun and normal to compartment surface, deg

ψ = elevation angle of the Sun to the lunar surface, deg

$\rho_2 \equiv A$ = lunar reflectivity to solar irradiation, dimensionless

Table 6–2 shows some of the parameter values used in the lunar-surface temperature calculations for each mission. Other parameters used, such as view factors and Sun angles, and the detailed results of these calculations, are given in references 6–10 and 6–13 to 6–16.

Shading Test

A differential shading test was performed on Surveyor III compartment B to try to establish the degree of thermal isolation of the compartment outboard face from the Vycor mirrors and the radar and doppler velocity sensor (RADVS) antenna. The top of the compartment on all Surveyor spacecraft was covered with Vycor mirrors which had a low α/ϵ (=0.12/0.79) permitting heat flow out of the compartment during the high lunar day temperatures. (Thermal switches internal to the compartment isolated the interior from the mirrors during the lunar night.) The RADVS antenna was located under the outer compartment face.

At the beginning of the shading test, all components were unshaded, including the compartment outboard face, the Vycor mirrors, and the RADVS antenna. Next, with all the components shaded from the direct Sun by the solar panel, the outboard face temperature dropped 6.7° K (12° F). Subsequently, each component was unshaded in succession: the first was the RADVS antenna, which resulted in a $\Delta T = 0.6°$ K (1° F); the outboard face, which resulted in a $\Delta T = 6.1°$ K (11° F); and the Vycor mirrors, which resulted in a $\Delta T = 0.8°$ K (1.5° F). (The sum is greater than the original 6.7° K (12° F) drop because of the increasing Sun angle during the test.) By far the greatest temperature change occurred when the outer face was unshaded, with small changes resulting when either the RADVS antenna or the compartment top was exposed. The results, while encouraging, were not conclusive because of the high Sun angle at the time of the test.

Data Analysis

Preliminary Results

Temperature data measured by thermal sensors located on the outboard panels of the electronic compartments, on the solar panel, and on the planar array antenna, for the most part, are presented in references 6–10 and 6–13 to 6–16. (Figs. 6–20 and 6–26 contain solar panel temperature data not previously reported.) Included are data taken during the first lunar day, during eclipses, after sunset, and on some subsequent lunar days; auxiliary Sun/panel/spacecraft relative positional information is also included.

Lunar surface brightness temperatures were calculated using the compartment outboard panel temperatures and equation (6), depicted essentially as a radiation balance relationship in figure 6–28. These preliminary results were presented in references 6–10 and 6–13 through 6–16 shortly after each Surveyor mission. In some cases, lunar surface brightness temperatures were calculated using the solar panel temperatures; however, they were not presented because of the incomplete effort.

The lunar surface temperatures derived from the Surveyor I compartment data using equation (6) were found to differ from Earth-based predictions. Higher morning temperatures were evident and lower values of γ resulted from Surveyor. Similar discrepancies occurred on the subsequent Surveyor missions, and especially large differences were noted during the eclipses on the Surveyor III and V missions. These Surveyor-based eclipse lunar surface temperatures were found to be higher than Earth-based predictions by 50° and 80° K, respectively.

Another discrepancy also appeared when lunar surface temperatures derived from the solar panel temperature measurements were found to differ from the results obtained from compartment data. In fact, the solar panel results were closer to Earth-based predictions. This comparison is shown in figure 6-29 during the Surveyor III eclipse.

Under the early conviction that compartment data were more reliable, efforts were concentrated on reducing these data, as well as attempting to explain the above discrepancies. During some of the missions, the compartment-based results were as much as 25° K higher during the lunar morning or afternoon than Earth-based predictions. At night, the compartment-based lunar surface temperatures also were higher, resulting in γ values near 500, compared with Earth-based eclipse predictions for γ of about 1350. Recent Earth-based measurements taken during the lunar night by Wildey, Murray, and Westphal (ref. 6-7) have resulted in γ values less than 1000, compared with the earlier Earth-based measurements taken during eclipses by Saari and Shorthill (refs. 6-2 through 6-4), which resulted in γ values averaging about 1350.

Several effects have been considered in modeling the lunar surface characteristics in order to explain the difference between eclipse and postsunset measurements. Effects given serious consideration are directional thermal emission, variation of thermal conductivity with soil depth, and variation of density with depth. Winter and Saari (ref. 6-6) have developed a "cube" model that varies thermal conductivity with depth, and they have been able to match the Wildey, Murray, and Westphal lunar postsunset data as well as eclipse measurements with this model. Jones (ref.

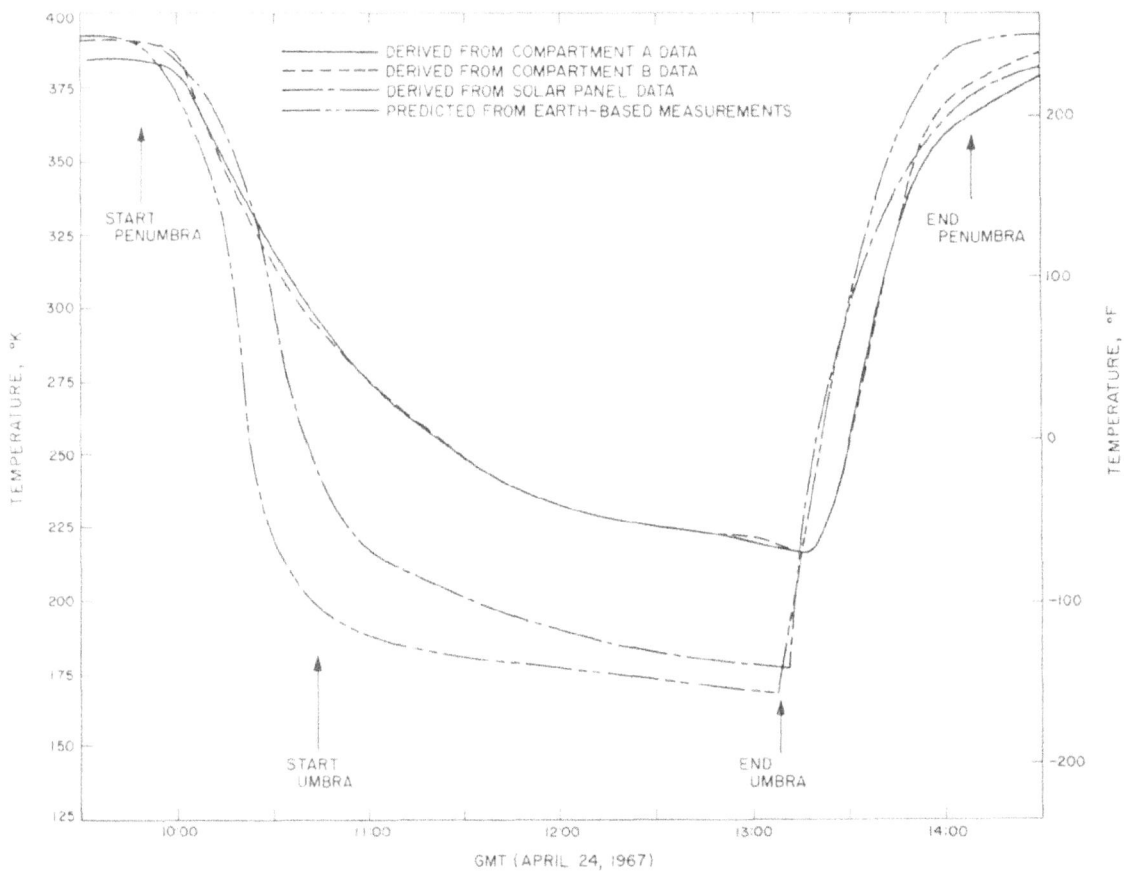

FIGURE 6-29.—Lunar surface eclipse temperatures from different sources for Surveyor III landing site.

6-17) recently suggested a model that includes density variation with depth and conductivity as a function of depth and temperature.

Error Analyses

Early efforts were made during the mission operations time period to explain the discrepancies in lunar surface temperatures derived from compartment, solar panel, and Earth-based data. These included the possible effects that could result from dust, paint degradation, crater cavities, heat capacity of rocks, and directionality. Of these, only the last two effects were found to be of interest; rocks may have had the effect of maintaining a higher postsunset temperature for compartment B on Surveyor VII, and directionality partially explained some of the higher morning or afternoon temperatures (ref. 6-14). Additional efforts to discretize the lunar surface into many nodes to obtain more accuracy were of limited value. Further discretization efforts should be restricted to the immediate foreground, particularly where craters or rocks exist.

Efforts subsequent to the mission operations time period were devoted to searching for other possible errors and determining their influence on the calculated lunar surface temperatures. The first errors investigated were those within equation (6), where it was confirmed that the temperature sensor inaccuracy and the view factor uncertainty were the significant error contributors during the lunar day; also, the uncertainty in compartment internal heat loss, $\dot{q}$, was significant after sunset. The inaccuracy in the sensor measurement was established as being the most significant error source with a possible error of $\pm 2°$ K by itself and $\pm 4°$ K including other telemetry system inaccuracies.

Relative errors, which were due to the various uncertainties, were separately calculated at many different times during both the lunar day and the lunar night. When these results were applied as error bands to the compartment and solar panel data, the resulting lunar surface temperatures from these two sources still could not be made to overlap without assuming some unreasonably large initial error sources. However, the slope of the postsunset surface temperature derived from compartment data on Surveyor V matched the slope of Earth-based surface temperatures by using a value of $\dot{q}$ 80 percent of its nominal.

Attention was then given to the solar panel to see whether, by varying the initial errors, one could possibly make it more nearly match the compartment results. Heat conduction from the mast, where there were thermal sensors on two of the drive motors, increased the difference between the two spacecraft predictions of the lunar surface temperatures. Allowing for change of emittance of the back surface paint at cryogenic temperatures produced a slightly favorable reduction in this difference. Most of the other possible contributors increased the energy received by the solar panel, which further separated the two results.

Later, it was realized that since the compartment side faces were nearly vertical they would view approximately one-half of the lunar surface and one-half of the cold space, whereas an outboard face containing the temperature sensor was tilted back about 20° and would view more cold space than warm lunar surface. The importance of this effect was estimated for Surveyor V during the lunar night, where data were available for 200 hours after sunset and where transient effects were at a minimum. Simplified calculations were made, assuming that the sides and outboard face were isolated from each other but were conductively coupled to the interior of the compartment via the same value for $\dot{q}$ as that previously used for the outboard face. These calculations showed that the side faces could be 10° K warmer than the outboard face and thus could significantly affect the outboard face temperature.

Then, using nominal lunar surface temperatures for $\gamma=500$ and $\gamma=750$, the temperature of the outboard face was computed including conduction from the side faces and nominal $\dot{q}$ from the interior. These computed temperatures were compared with the actual Surveyor V canister temperatures, as shown in figure 6-30. Agreement was good between the outer face temperatures computed assuming a $\gamma=500$ and the actual flight data; without the side face conduction, a γ of about 400 previously had been obtained.

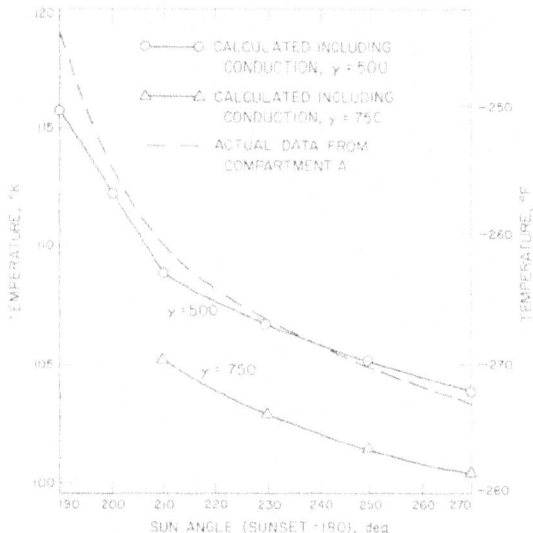

FIGURE 6-30.—Surveyor V outboard face postsunset temperature with conduction from sides.

Based on this agreement between actual and calculated outer face temperatures, the comparison of results from different data sources shown in figure 6-31 was made. The error band derived previously for the compartments was placed around a lunar surface $\gamma=500$ temperature curve, recognizing that the error band may be larger than that shown when the conductive effect is included. Note that lunar surface temperatures calculated from the solar panel fall just below this error band. If an error band were placed on the solar panel results, these bands would overlap, particularly well late into the lunar night. Also shown are the results of the new cube model (ref. 6-6) for the lunar surface during the night, which has an average γ value of about 850. The cube model had been adjusted to fit the Wildey, Murray, and Westphal data (ref. 6-7) also shown in the figure. This lies within the probable error band for the solar panel lunar surface temperature results, and lies between the solar panel and the compartment results until well

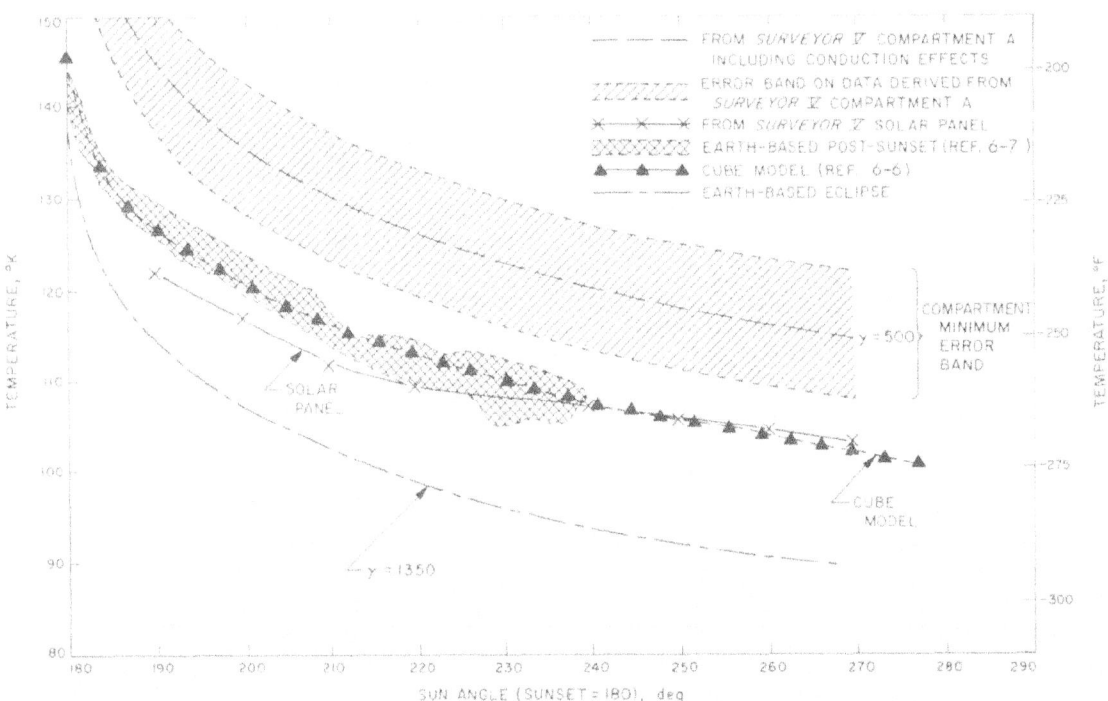

FIGURE 6-31.—Lunar surface postsunset temperatures for Surveyor V.

into the night. Figure 6-31 also presents the curve for $\gamma = 1350$, originally predicted from Earth-based eclipse measurements, which indicates that this value of γ is much too large for the night.

During the lunar day, one of the two compartment side faces was nearly always illuminated by the Sun. In certain cases, the Sun can illuminate the polished-aluminum inboard face at low Sun elevation angles. This face then would become quite hot and conduct heat to the outboard face via the side and bottom faces. Approximate calculations indicate that the outboard face temperature might be elevated as much as 30° K in certain cases and that the lunar surface temperature could be correspondingly reduced.

Conclusions

The errors associated with compartment-based lunar surface temperature calculations appear to be significant, particularly during the lunar day. Heat conduction from the other compartment faces is the most important factor to be included in the revised compartment model. The solar panel errors appear to be less and the equations simpler than those for the revised compartment model.

The postsunset lunar surface temperatures derived from Surveyor V data and Earth-based measurements appear to be in agreement within the error bands. This is the only case investigated in some depth. Similar detailed calculations are required for the other spacecraft, including the lunar days and eclipses.

The lunar soil is highly insulating, as indicated by the values of the thermal parameter obtained. It is estimated the Surveyor-based γ values given in references 6-10 and 6-13 through 6-16 should be increased at least 100 (cm^2 sec$^{1/2}$ °K/g-cal).

Directional emission of the lunar surface may be a partial cause of temperature differences between those derived from Surveyor compartment data and those predicted assuming Lambertian emission.

No dust layer that had an appreciable thermal effect was apparent on any of the spacecraft, since it would have been detected in the thermal sensors.

References

6-1. SAARI, J. M.; and SHORTHILL, R. W.: Isothermal and Isophotic Atlas of the Moon. NASA CR-855, Sept. 1967.

6-2. SAARI, J. M.; and SHORTHILL, R. W.: Thermal Anomalies of the Totally Eclipsed Moon of December 19, 1964. Nature, vol. 205, 1965, pp. 964-965.

6-3. SAARI, J. M.; and SHORTHILL, R. W.: Physics of the Moon. American Astronautical Society, Tarzana, Calif., Science and Technology Series, vol. 13, 1967, pp. 57-99.

6-4. SAARI, J. M.; and SHORTHILL, R. W.: Isotherms in the Equatorial Region of the Totally Eclipsed Moon. Boeing Scientific Research Laboratories, 1966.

6-5. JONES, B. P.: Diurnal Lunar Temperatures. Paper 67-289 presented at AIAA Thermophysics Specialist Conference, New Orleans, La., Apr. 17-19, 1967.

6-6. WINTER, D. F.; and SAARI, J. M.: A New Thermophysical Model of the Lunar Soil. Document D1-82-0725, Boeing Scientific Research Laboratories, 1968.

6-7. WILDEY, R. L.; MURRAY, B. C.; and WESTPHAL, J. A.: Reconnaissance of Infrared Emission from the Lunar Nighttime Surface. J. Geophys. Res., vol. 72, no. 14, 1967, pp. 3743-3749.

6-8. JAEGER, J. L.: Surface Temperature of the Moon. Australian J. Phys., vol. 6, no. 10, 1953.

6-9. INGRAO, H. C.; YOUNG, A. T.; and LINSKY, J. L.: A Critical Analysis of Lunar Temperature Measurements in the Infrared. The Nature of the Lunar Surface. Proceedings of the 1965 IAU-NASA Symposium, Johns Hopkins Press, Baltimore, 1966, pp. 185-211.

6-10. VITKUS, G.; GARIPAY, R. R.; HAGEMEYER, W. A.; LUCAS, J. W.; JONES, B. P.; and SAARI, J. M.: Lunar Surface Temperatures and Thermal Characteristics. Surveyor VII Mission Report. Part II: Science Results, Tech. Rept. 32-1264, Jet Propulsion Laboratory, Pasadena, Calif., Mar. 15, 1968.

6-11. SIX, F.; MONTGOMERY, C.; SAARI, J. M.; and SHORTHILL, R. W.: Directional Characteristics of the Lunar Thermal Emission. Paper presented at AIAA Thermophysics Specialist Conference, New Orleans, La., Apr. 17-19, 1967.

6-12. SINTON, W. M.: Temperatures of the Lunar Surface. Physics and Astronomy of the Moon (S. Kopal, ed.), Academic Press, New York, ch. 11, 1962.

6-13. LUCAS, J. W.; CONEL, J. E.; GARIPAY, R. R.; HAGEMEYER, W. A.; and SAARI, J. M.: Lunar Surface Thermal Characteristics. Surveyor I Mission Report. Part II: Scientific Data and

Results, Tech. Rept. 32-1023, Jet Propulsion Laboratory, Pasadena, Calif., Sept. 10, 1966.

6-14. LUCAS, J. W.; CONEL, J. E.; HAGEMEYER, W. A.; JONES, C. B.; SAARI, J. M.; and WANG, J. T.: Lunar Temperatures and Thermal Characteristics. Surveyor III Mission Report. Part II: Scientific Results, Tech. Rept. 32-1177, Jet Propulsion Laboratory, Pasadena, Calif., June 1, 1967.

6-15. LUCAS, J. W.; GARIPAY, R. R.; HAGEMEYER, W. A.; SAARI, J. M.; SMITH, J.; and VITKUS, G.: Lunar Surface Temperatures and Thermal Characteristics. Surveyor V Mission Report. Part II: Science Results, Tech. Rept. 32-1246, Jet Propulsion Laboratory, Pasadena, Calif., Nov. 1, 1967.

6-16. VITKUS, G.; GARIPAY, R. R.; HAGEMEYER, W. A.; LUCAS, J. W.; and SAARI, J. M.: Lunar Surface Temperatures and Thermal Characteristics. Surveyor VI Mission Report. Part II: Science Results, Tech. Rept. 32-1262, Jet Propulsion Laboratory, Pasadena, Calif., Jan. 10, 1968.

6-17. JONES, B. P.: Density-Depth Model for the Lunar Outermost Layer. J. Geophys. Res. (in press).

ACKNOWLEDGMENTS

We wish to acknowledge the authors of the Mission Reports (refs. 6-10 and 6-13 through 6-16) from which much of this material was drawn: J. A. CONEL, JPL, and R. R. GARIPAY, Hughes Aircraft Co.; B. P. JONES, MSFC; C. B. JONES, JPL, and J. W. SMITH, JPL; G. VITKUS, Northrop Corp.; and J. T. WANG, JPL.

We also wish to acknowledge the important early contributions of H. KNUDSEN, Hughes Aircraft Co., for the original suggestion to use compartment outer face temperature data, and W. JAWORSKI, Northrop Corp., for the initial data analysis.

7. Lunar Surface Electromagnetic Properties

D. O. Muhleman, W. E. Brown, Jr., L. Davids, J. Negus de Wys, and W. H. Peake

Radar Data and Data Reduction

D. O. MUHLEMAN (CHAIRMAN), W. E. BROWN, JR., L. DAVIDS, AND W. H. PEAKE

Analysis and interpretation of the Surveyor data on electromagnetic properties of the lunar surface material are discussed here. The text is in three parts:

(1) Methods of reduction and presentation of the radar data obtained for the various Surveyor spacecraft during the landing phase.

(2) Interpretation of the radar data in terms of lunar properties.

(3) Analysis and interpretation of the magnet data.

Radar System

Radar signal-strength data were taken with the radar altimeter and doppler velocity sensor (RADVS) for approximately the final 3 minutes of Surveyor flights I, III, V, VI, and VII. Identical systems were used on each flight; these systems consisted of four narrow beams, three of which were oriented at angles of 25° with respect to the roll axis (at a wavelength of 2.3 cm) and the fourth along the roll axis. Beam 4 was frequency modulated to obtain slant-range information. The sole design criterion for the RADVS system was determined by guidance and control requirements. Nevertheless, telemetered values of the various signal strengths have yielded unique and significant information concerning the radar reflection properties of the lunar surface in the various landing regions.

Data from each flight were obtained during the final 2 to 3 minutes before touchdown. During this time, the range varied from about 0 to 20 km, the beam angles of incidence from a maximum of 60° off normal, the beam surface intercept radius from about 1 km; the beam was swept laterally along the lunar surface to distances of up to 20 km. Since the approach geometry was slightly different for each flight, the initial orientations of the radar beams were correspondingly different; in each case, the final minute of descent was performed with the spacecraft roll axis essentially perpendicular to the mean lunar surface and with essentially zero lateral velocity. Consequently, during this time period, three of the beams remained at an incidence angle of 25° to the mean surface, and beam 4 was approximately at normal incidence.

A very brief description of the radars is given in this report. The velocity beams divide the output of a 7-watt, two-cavity klystron, radiated by two parabolic antennas in the three beams. The antennas are divided by septums into receiving and transmitting sections. The return signals are detected in quadrature by pairs of microwave mixers that are excited by the transmitter signal, converting to the doppler frequencies in the range of about 100 to 80 000 Hz. The doppler signals are amplified in matched pairs of preamplifiers that employ automatic gain control, each in three discrete gain states differing by about 25 dB.

The preamplifier outputs are processed by frequency trackers, which provide estimates of the center frequency of each doppler spectrum. The three doppler frequencies are combined in frequency converters to give three orthogonal velocity components in spacecraft coordinates as analog voltages.

The altimeter transmits about 250 MW from a reflex klystron, which is frequency modulated at the repeller with a sawtooth waveform. The return signal is processed in the same way as in the velocity beams by microwave mixers, preamplifiers, and a frequency tracker. The signal frequency consists of the sum of a component proportional to range (arising from the frequency modulation) and a doppler component. The doppler component is subtracted in the converter, using data derived from the velocity beams. An analog voltage proportional to slant range is provided at the output.

The instrumentation for signal strength consists of rectifiers and filter amplifiers in each frequency tracker. The preamplifier outputs are processed by single-sideband modulators, which are driven by the frequency f_c+f_0, where f_c is a reference frequency (600 kHz) and f_0 is the tracker estimate of the input signal frequency. The modulators are designed to produce the lower sideband, causing the signal spectrum to occur at about f_c. The translated spectrum is amplified in an IF amplifier with a center frequency of 600 kHz and a bandwidth of 10 kHz. The signal-strength instrumentation is at the output of this amplifier.

The rectifier is a half-wave diode rectifier operating at a sufficiently high level to provide reasonably linear performance. The output is filtered by a single-section RC filter with an 0.05-second time constant. The output is scaled to provide a full-scale voltage of 5-V dc.

Unambiguous interpretation of the signal-strength analogs requires knowledge of the preamplifier gain state, indicated by discrete outputs provided for each beam, and signal frequency, which is necessary because the preamplifiers employ low-frequency rolloff. The frequencies are computed by the inverse of the transformation in the converter of the radar, using telemetered radar velocity and range output data. In the case of the altimeter, the range-frequency scale factor undergoes a discrete change at 304.8 meters (1000 ft), this event being indicated by a 304.8-meter (1000 ft) mark.

This particular design, which was entirely adequate for its guidance and control function, caused certain difficulties in the interpretation of the signal-strength data for scientific purposes, particularly the use of discrete gain states. During periods in which a beam was normal to the lunar surface (usually beam 4), the radar echo apparently contained two components intermittently: a (strong) coherent return and a (nominal) incoherent component. The receiver systems attempted to follow these signal variations, thus causing rapid gain-state switching. This effect was particularly common during the last 30 seconds of flight when the nominal signal strength also rapidly increases with time.

The radar data reduction consists of converting the telemetered signal-strength voltages to the actual signal strengths in watts (accomplished by using the telemetry calibrations and gain-state information) and by computing a radar cross section with a "radar equation" that removes the range effect and includes the measured numerical values of the radar parameters, i.e., the antenna gains and transmitter powers. The accuracy of this process is difficult to estimate. Errors will arise from the uncertainties in the telemetry calibrations, the numerical values of gains and transmitter powers, incorrect gain-state decisions, and any possible difficulty in the theoretical radar equation. In all, we are concerned with 20 nearly independent radar systems, all of which can only be preflight calibrated under a tight time schedule. This situation should be contrasted to that for an Earth-based radar system where many months are available for antenna pattern and gain measurements and where the transmitter power can be continuously monitored during the experiments. Nevertheless, the Surveyor signal-strength data are highly consistent among the various beams on each flight, strongly suggesting that a corresponding flight-to-flight consistency exists. In the worst case, one beam was found to differ by 2 dB from the other three beams; in all other cases, the various beams appear to be consistent to better than 1 dB.

Trajectory Reconstruction

A necessary requirement for accurate evaluation of lunar radar data is a complete knowledge of the terminal-descent trajectory. With this knowledge, a time history of the slant range to the surface, incidence angle, and relative velocity of the individual radar beam(s) may be determined. In addition, where the precise touchdown point can be determined by using Lunar Orbiter photographs or by other means, then the trace of beam incidence points on the lunar surface can be reconstructed as a function of time. By this means, both geometric and terrain effects could be removed from the telemetered radar signal-strength measurements.

While the reconstructed trajectory is necessary for analysis of lunar surface electrical properties, the primary purpose of the Surveyor postflight, terminal-descent trajectory-analysis effort was to recognize anomalous spacecraft behavior, discover the source(s), and suggest appropriate system modifications for succeeding spacecraft. The accuracies of trajectory reconstruction, therefore, were based on these engineering requirements rather than on the scientific requirements of lunar surface analysis. However, the reconstruction accuracies proved adequate for both purposes.

In preparation for terminal-descent, postflight analysis, the development of a set of computer programs was initiated based on the concept of obtaining a least-squares best fit of telemetry data. The program was also designed to solve for both system and telemetry errors. The technique required as a base, a three-dimensional, six-degree-of-freedom simulation of the terminal descent. An existing six-degree-of-freedom computer program (6DOF), which contained radar and flight-control subsystem models, including weight and moment of inertia changes, was used for this purpose. Figure 7-1 presents a functional block diagram of the 6DOF program. The integration of rigid-body rotational and translational dynamic equations are performed using simple trapezoidal numerical integration. The resulting computational accuracies obtained have, in general, been several times better than the most severe requirements.

The control system model consists of the altitude, acceleration, and range/velocity command descent control loops. The radar subsystem serves as the navigational sensor in the latter control loop and must provide continuous and accurate data on range and velocity so that the spacecraft may successfully negotiate a soft landing. Also included in this block is a model of the propulsion system. Although it is not required for an accurate trajectory reconstruction, the subsystem dynamics included time constants of at least 0.1 second in magnitude.

By the time Surveyor I was launched, the least-squares program had been developed to the point that all telemetry-error sources had been modeled and initial work had begun on modeling the spacecraft system errors. During program checkout, it was found that large system errors (out of specification errors) would cause the program to diverge from the solution, while small errors were resolved with high accuracy. To achieve convergence in the first case, it was necessary to modify the model so that the large system error became a small relative error, which, of course, negated the reason for such a program. Lack of convergence resulted primarily from a highly nonlinear relation between spacecraft-system parameters and trajectory characteristics and high correlation between available telemetry data. At this same time, it was found from Surveyor I postflight work that the 6DOF program could be perturbed by hand with very quick convergence and matching of the telemetry data.

Since, in either case, the basic system model required modification for large system errors, the least-squares best-fit approach was dropped in favor of the much simpler technique of manual perturbation. To support this approach, effort was expended on increasing the flexibility of the 6DOF program and developing analysis techniques utilizing the comparison of telemetered descent time events (burnout, segment

FIGURE 7-1.—Functional block diagram of the 6DOF computer program.

intercept, segment acquisition, etc.) to determine system errors. This avenue of postflight analysis was used with success and, in the case of Surveyor V with its highly nonstandard descent, it probably was the only reasonable method.

In practice, the 6DOF trajectory reconstruction was required to match closely (to at least the expected accuracy of the radar system) the best postmission estimates of range, velocity (from telemetry data), and incidence angle for the entire terminal-descent phase. The times spent in any discrete phase such as the descent segments were required to be within 0.5 second of the actual time; the total descent time from ignition to touchdown was required to be within 1.0 second of the known total time. An indication of the goodness of fit to the actual trajectory was the subject of a study which showed that independent errors of approximately 1 percent in range and velocity would cause from 0.5- to 1.0-second errors in the time interval of a descent segment and even greater variance in total descent time. Based on this analysis, the errors between the 6DOF simulation trajectory and the true trajectory are estimated to be 2 percent in beam range, 1 percent in beam velocity, and 1° in beam incidence angle (1 σ). The 1-σ error in spacecraft roll attitude is estimated to be 0.3°. These errors in the trajectory reconstruction cause errors on the order of 0.5 dB or less in the radar signal-strength calculations. It is important to realize that the incidence angles obtained in this way are relative to the mean lunar surface. Errors during the short steering phase following retro burnout (about 2 min before touchdown) may be somewhat larger since trajectory parameters change rapidly during this period. These errors were larger during the final 10 seconds of the Surveyor V descent.

The details of the data-reduction procedures and programs are given in reference 7–1. The radar equation used in all of the reductions is

$$P_r = \frac{P_t G^2 \lambda^2}{(4\pi)^3 R^4} \left[\pi (R\phi)^2 \right] \gamma(\theta) \qquad (1)$$

where
P_r = power received
P_t = power transmitted
G = antenna gain
R = range along beam center
θ = angle of incidence
λ = wavelength (23 cm)
ϕ = beam half-angle

Clearly, $\lambda(\theta)$ is the angularly dependent radar cross section projected perpendicularly to the beam axis. For a gaussian approximation for the antenna pattern, it can be shown that

$$\phi^2 = \frac{6750}{G} \left(\frac{\pi}{180} \right)^2 \qquad (\text{rad})^2 \qquad (2)$$

All the data in this section of this report have been reduced from the measured signal strengths P_r, using equations (1) and (2), and the parameters given in table 7–1. The half-power beam angle is about 3°. In principle, the actual power received is the integral of the beam pattern over the illuminated surface. Strictly speaking, equation (1) is correct only if the actual lunar angular scattering law is flat. This error is negligible for angles of incidence greater than, say, 5°. The actual scattering law apparently is sufficiently peaked near normal incidence to cause an approximate 25 percent error at normal incidence by using equation (1).

TABLE 7–1. *Surveyor radar parameters*

Spacecraft	Gain, dB				Transmitter power, dBW			
	Beam 1	Beam 2	Beam 3	Beam 4	Beam 1	Beam 2	Beam 3	Beam 4
Surveyor I	28.00	28.05	28.55	29.20	33.55	31.85	33.44	24.97
Surveyor III	28.90	28.10	28.40	29.10	33.96	33.29	33.82	25.07
Surveyor IV	28.55	28.45	28.50	28.05	33.79	34.35	33.87	24.64
Surveyor VI	28.20	28.10	27.70	28.50	34.70	33.60	33.05	25.05
Surveyor VII	28.27	27.87	27.66	28.44	34.02	34.92	34.07	24.32

Variations in Radar Cross Section

All signal-strength data have been reduced to the cross section $\gamma(\theta)$, expressed variously as a function of time to touchdown, height above the mean surface, and distance along the lunar surface to the touchdown point. Complete sets of such curves for Surveyor I may be found in reference 7–1, Surveyor III in reference 7–2, and Surveyor VII in reference 7–3. The quantity of data is far too great to present in this report; therefore, typical examples only are presented here.

The gamma radar cross sections for Surveyor I are shown as a function of time to touchdown in figure 7–2. The approach geometry for this flight was such that beam 4 was almost always normal and beams 1, 2, and 3 were at an incidence angle of 25°. Thus, we would expect the gamma values for these beams to be nearly constant except for real variations in the lunar surface. The beam 4 data in figure 7–2 clearly show the normal-incidence, rapid signal-strength variations over an approximate 10-dB range. It is impossible to estimate a meaningful average value of the beam 4 data, but a straight numerical mean is about 0.4 (−4 dB).

The same data representation for Surveyor III is shown in figure 7–3. In this case, the beam incidence angles do not reach the final touch-

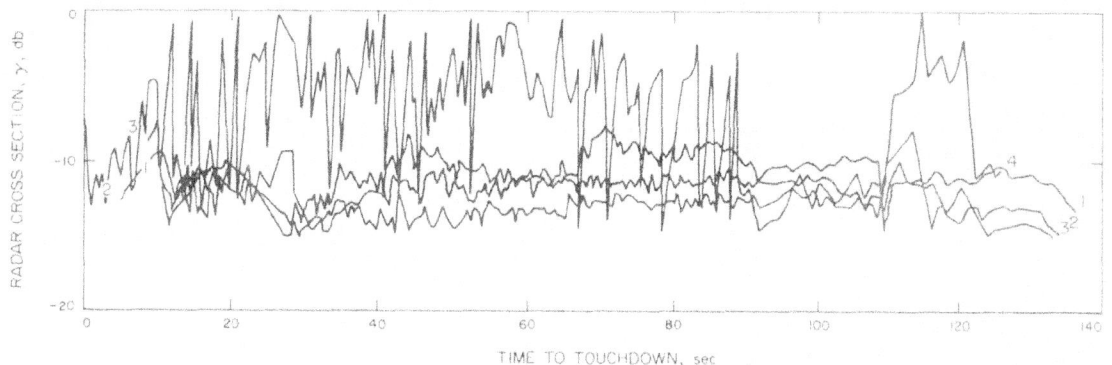

FIGURE 7–2.—Radar cross section for Surveyor I as a function of time to touchdown.

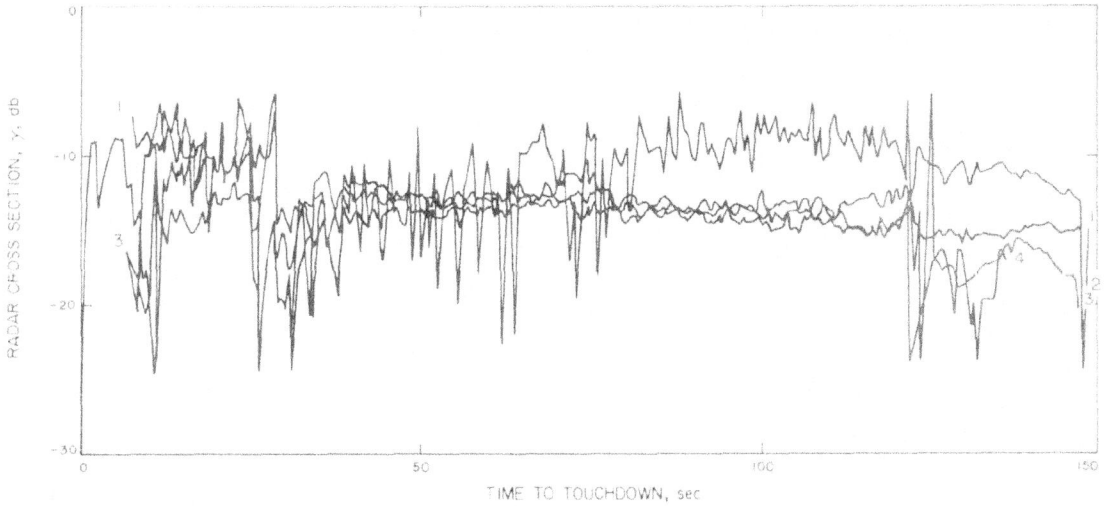

FIGURE 7–3.—Radar cross section for Surveyor III as a function of time to touchdown.

down configuration until about 100 seconds before touchdown; some evidence of the angular variation in gamma can be seen before this time in figure 7-3. After 100 seconds, beams 1, 2, and 3 are in close agreement and with nearly constant γ. Beam 4 actually approached normal incidence quite slowly, reaching $\theta=2°$ at 60 seconds where, again, fading becomes noticeable.

Figure 7-4 shows the time variations for Surveyor V beams which had continuously vary-

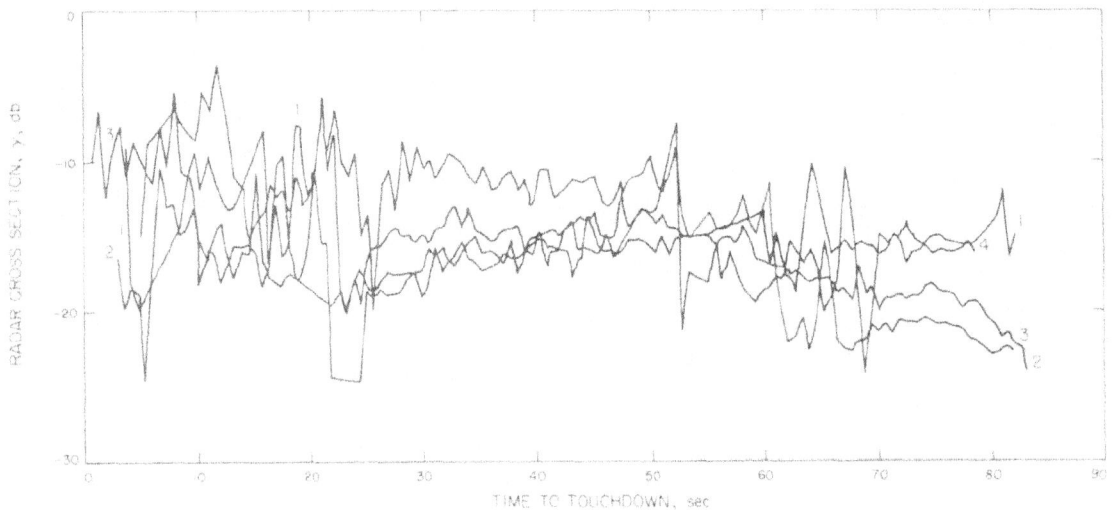

FIGURE 7-4.—Radar cross section for Surveyor V as a function of time to touchdown.

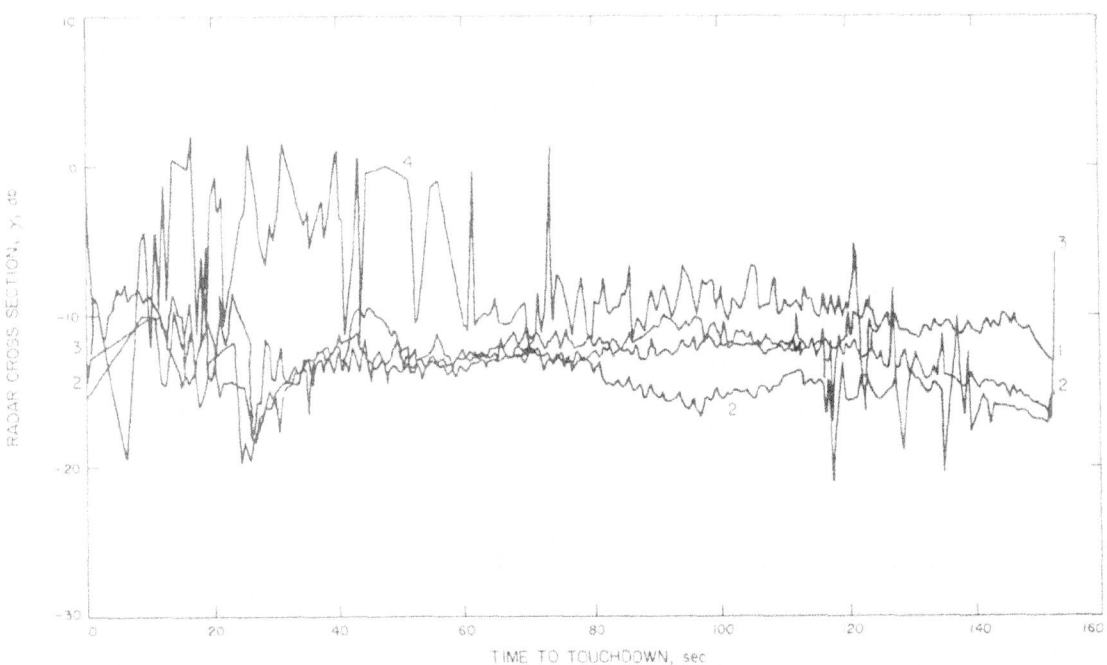

FIGURE 7-5.—Radar cross section for Surveyor VI as a function of time to touchdown.

ing incidence angles that did not reach the touchdown configuration until a few seconds before the touchdown. Beam 4 did not reach $\theta = 2°$ until about 10 seconds; very little evidence of fading can be seen in figure 7-4. This flight exhibited the weakest echos of all Surveyor flights (compare fig. 7-4 with figs. 7-2 and 7-3). The cross-section data for Surveyor VI are

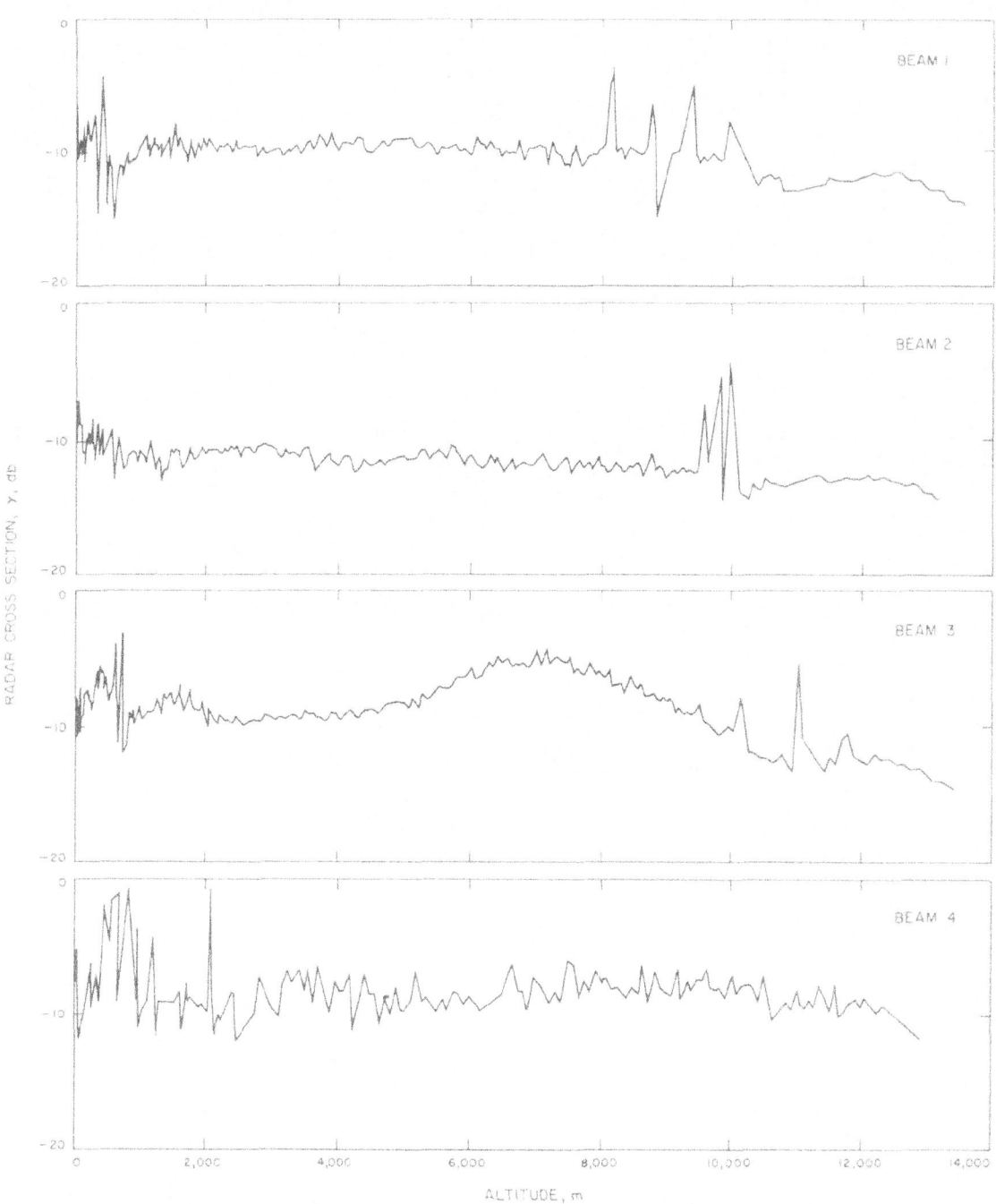

FIGURE 7-6.—Radar cross section for Surveyor as a function of altitude.

shown in figure 7-5. The landing configuration was reached rapidly for this flight (by 120 sec). A significant dip, which probably can be explained by a terrain variation, can be seen around 100 seconds for beam 2. Strong fading occurs on beam 4 after 80 seconds. The significant decrease in γ for beams 1, 2, and 3 between approximately 30 and 20 seconds occurred on all flights and appears to be associated with gain-state switching; i.e., the dip probably is a characteristic of the radar system.

An example of the altitude dependence of γ is shown in figure 7-5 with the Surveyor VII data. Except for the last 1000 meters of the flight (time, 30 sec to 0) where the system is strongly affected by gain switching and a rapid increase in signal strength, no altitude effects are in evidence. The variations in γ above 10 km are explained easily in terms of the known changes in the incidence angles. Considerable real structure is evident with beam 3 data, which show strong returns when that beam crossed a group of mountains and a crater shortly before touchdown (see ref. 7-3). Figure 7-6 shows a strong beam 3 echo between the altitudes of 10 to 5 km; this echo is caused by the local terrain, which is oriented more perpendicularly to the beam than the mean surface. The altitudes have nothing to do with this phenomenon; altitude was selected as the independent variable in figure 7-6. The rise in gamma at 2.0 km is the typical response when a beam crossed a crater that appears "fresh" in the associated Lunar Orbiter photograph. This will be discussed more fully below.

Radar Cross-Section Dependence on Incidence Angle

The quantity of particular interest that may be estimated from the data is the variation in the radar cross section, either γ or σ, as a function of the angle of incidence θ. The values of γ plotted in figures 7-2 through 7-6 are functions of the incidence angle relative to local surface illuminated by a given beam. The incidence angles relative to the mean lunar surface as a function of time have been obtained with the 6DOF computer program. Only if the local-plane incidence angle data were available to us could we cross-plot the γ versus θ data to obtain a determination of true angular dependence of γ. Clearly, the use of the mean-plane incidence angle data assumes that the surface is flat over the entire region covered by the beams. If a given landing region has a general slope of, say, 1° or 2°, it will significantly affect interpretation of $\gamma(\theta)$, particularly for data taken near normal incidence where the backscattering law is peaked. However, in most cases, the Surveyor landing sites are free of slopes and "rolling hills" and the interpretation of $\gamma(\theta)$ is relatively straightforward. A notable exception is the Surveyor VII beam 3, which did pass over "hills" that were large compared to the beam size.

A possible explanation of the flight-to-flight variation of the degree of fading on beam 4 (near normal incidence) may be formulated from these ideas. If the intermittent strong echoes observed on beam 4 data are coherent returns (truly specular), they would be absent if the local surface covered by the beam were sloped 1° or 2° from the horizontal. In this case, the specular scattering direction would not be back toward the radar and only noncoherent echoes would be detected.

The angular cross-section data $\gamma(\theta)$ are shown in figures 7-7 through 7-12 where the values of the independent variable θ were taken from the 6DOF program. Most of the data occur at θ of about 25° and, to a lesser extent, near 0°. (All Surveyor I radar data are at these two angles and will not be shown here; however, it can be inferred from fig. 7-2.) The Surveyor III data are shown in figure 7-7 with all of the beams

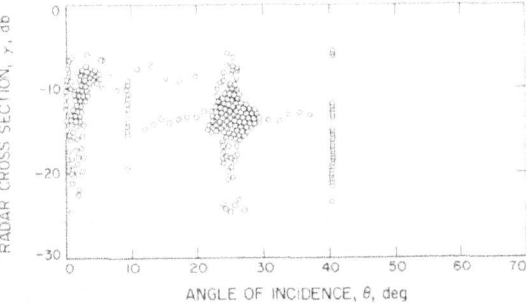

FIGURE 7-7.—Radar cross section for Surveyor III as a function of incidence angle.

superimposed. It can be seen that at any given angle, e.g., 40°, the values of γ are widely spread. This variation may be caused by any of the following factors:

(1) Beam is illuminating different regions of the lunar surface, which may have varying local slopes and true differences in electric reflectivity.

(2) Signal may be strongly fading at any given time.

(3) Receiver may be instantaneously out of lock.

(4) For any given data point, we may have selected the wrong gain state in the data reduction.

The cluster of data around θ=25° has the additional effect of superimposing the signal strengths of three radars for which our numerical values of the gains and transmitter powers may be slightly in error (±1 dB). Since all of the data have been included in figure 7-7, the effects of the strongly nonlinear behavior of the radars during the final 30 seconds of flight are quite evident. The clustering near θ=0° is more difficult to explain. Strong "up-fades" were absent for Surveyor III and, in this sense, figure 7-7 is not typical. However, there is a considerable presence of very weak echoes (also nontypical from θ=0° to 3°). It is not known whether these echoes are caused by the receiver dropping out of lock or by many incorrect gain-state decisions in the data reduction. It is certain, however, that these data do not represent the true lunar cross section near normal incidence. (See section on radar data, p. 219).

The data for each beam are shown separately for Surveyor V in figure 7-8. Beam 4, in particular, appears to yield a good estimate of the backscatter function for this region, partly because the geometry of the trajectory swept the beam through a wide range of θ in a near-optimum way. Beams 1, 2, and 3 exhibit the usual wide spread in values near θ~25° for apparently the same reasons as discussed above. A superposition of all the data is shown in figure 7-9. (See section on radar data for interpretation.)

The cross sections for the various beams for Surveyor VI are shown in figure 7-10 and the

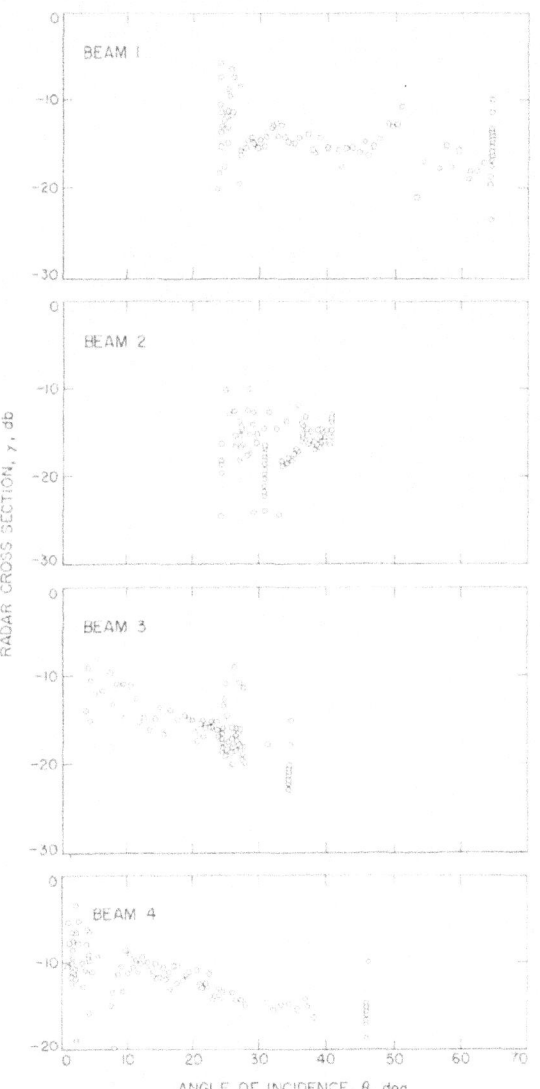

FIGURE 7-8.—Radar cross section for Surveyor V as a function of incidence angle.

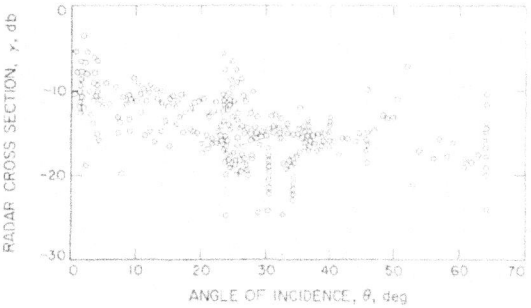

FIGURE 7-9.—Radar cross section for Surveyor V as a function of incidence angle. All beams are superimposed.

composite in figure 7-11. A double clustering of data near normal incidence is evident, but not understood. It does not appear possible to interpret the two clusters as coherent and incoherent, since they should differ by about 18 dB. Furthermore, it is not at all clear which group, if any, represents the "normal incidence cross section."

The Surveyor VII composite data are shown in figure 7-12 with evidence of two clusters near normal incidence. The most notable characteristic of these data is that the cross sections are significantly higher at all angles of incidence (by nearly +2 dB). This, apparently, represents a true difference in the radar cross section in the Tycho area (Surveyor VII) relative to the mare regions.

Correlations With Lunar Orbiter Photographs

Because the various Surveyor landing sites were selected in regions that appeared "smooth" in lunar photographs, only a small variation in lunar topography was explored by the radar systems.

Consequently, few topographical features are suggested by the Surveyor signal-strength data, with the exception of beam 3 on Surveyor VII. However, in several cases, anomalously strong echoes have been correlated with certain craters visible on the Lunar Orbiter photograph of the appropriate region. The first such detection occurred with Surveyor I, which is discussed in

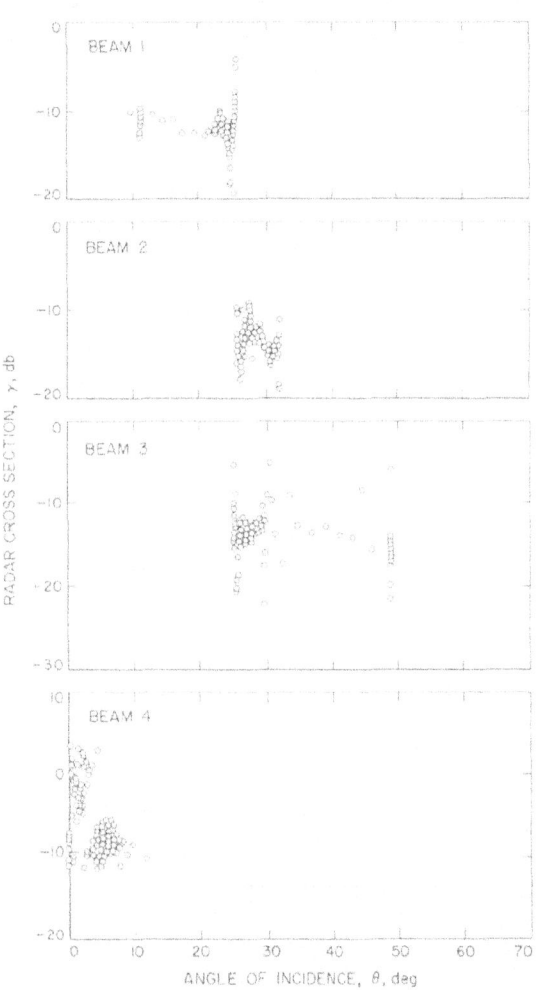

FIGURE 7-10.—Radar cross section for Surveyor VI as a function of incidence angle.

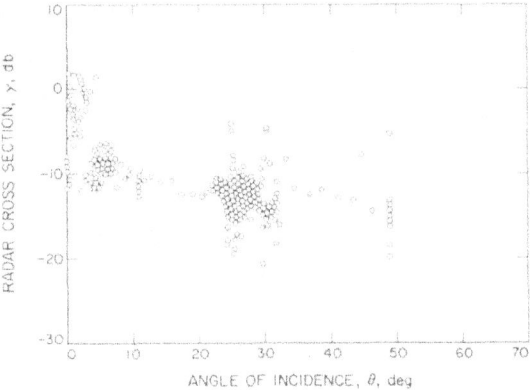

FIGURE 7-11.—Radar cross section for Surveyor VI as a function of incidence angle. All beams are superimposed.

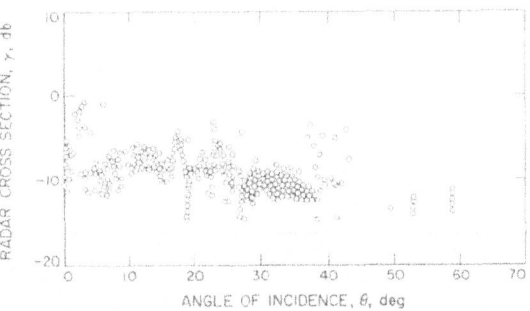

FIGURE 7-12.—Radar cross section for Surveyor VII as a function of incidence angle. All beams are superimposed.

reference 7-1. In most cases, it is intuitively clear from inspection of the photographs why a certain crater was detectable in the radar response and why others were not.

The correlations of the beam traces on the lunar surface with the corresponding Lunar Orbiter photographs are shown in figures 7-13 through 7-18. The cartesian coordinates of the interception of the beam center with the mean lunar surface were computed with the 6DOF program. The beam patterns for Surveyor I are shown in figure 7-13 (recall that beam 4 is always normal for this flight). A close inspection of this Lunar Orbiter photograph indicates that beam 1 (to the left) crossed two anomalous craters, whereas the other beams (2, 3, and 4) crossed nothing so prominent. Figure 7-14 shows an enlargement of the beam 1 trace with a scaled graph of the γ cross section. The correlations between the craters of interest and the cross sections are clear, and no significant correlations with other craters can be defended. These effects cannot be explained in a quantitative way, but we can say that they probably arise from a combination of focusing toward the radar, more compacted material or less granulated material, and higher intrinsic dielectric constant.

The Surveyor III region is shown in figure 7-15 along with a cross-section graph. The only detectable event occurs on beam 1, which rises vertically along the center. Clearly, the crater that lies about 1200 meters from the spacecraft is responsible for the radar feature. The very complex beam traces for Surveyor V are shown in figure 7-16. Even though many craters are traversed, no events can be seen on the cross-section data.

The Surveyor VII photograph is shown in figure 7-17 where it can be seen that beam 3 (from left to right) crossed a significant crater about 700 meters from the touchdown point. The effect on the radar cross section is easily seen in figure 7-6 (beam 3) at the altitude coordinate of 2000 meters. Figure 7-18 shows a larger view of the Surveyor VII landing site (again with beam 3 to the left). Beam 3 traversed a "hill" tilted in the direction toward the Sun (to the right) which created the broad and very strong feature seen in figure 7-6 during the altitude range from 9 to 5 km. This event occurred about 5 km from the touchdown point.

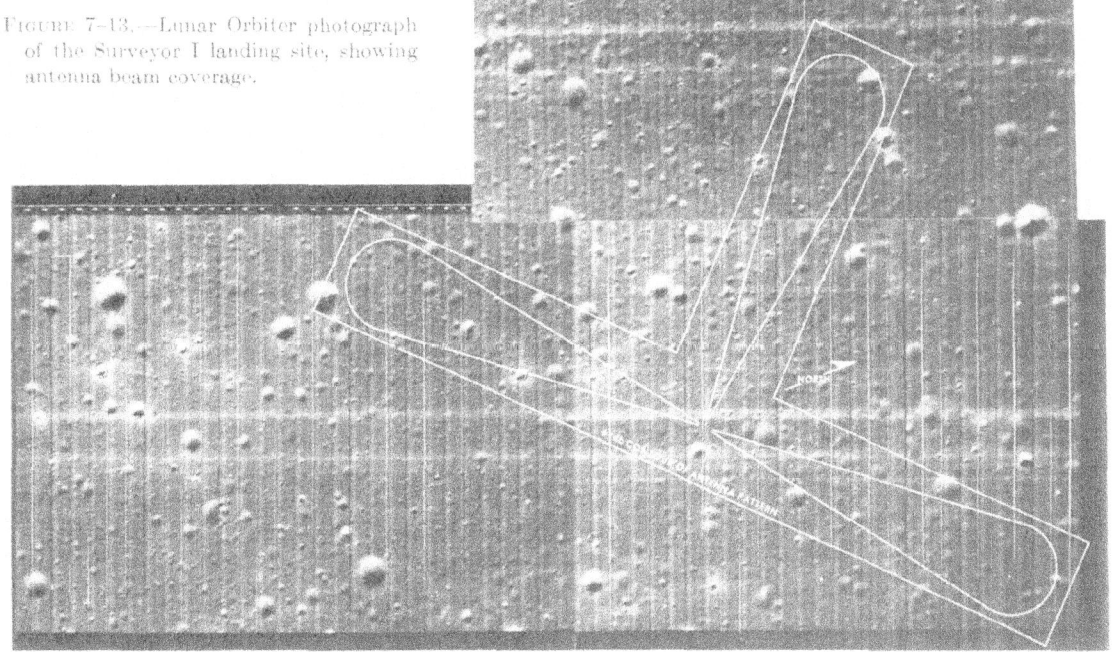

FIGURE 7-13.—Lunar Orbiter photograph of the Surveyor I landing site, showing antenna beam coverage.

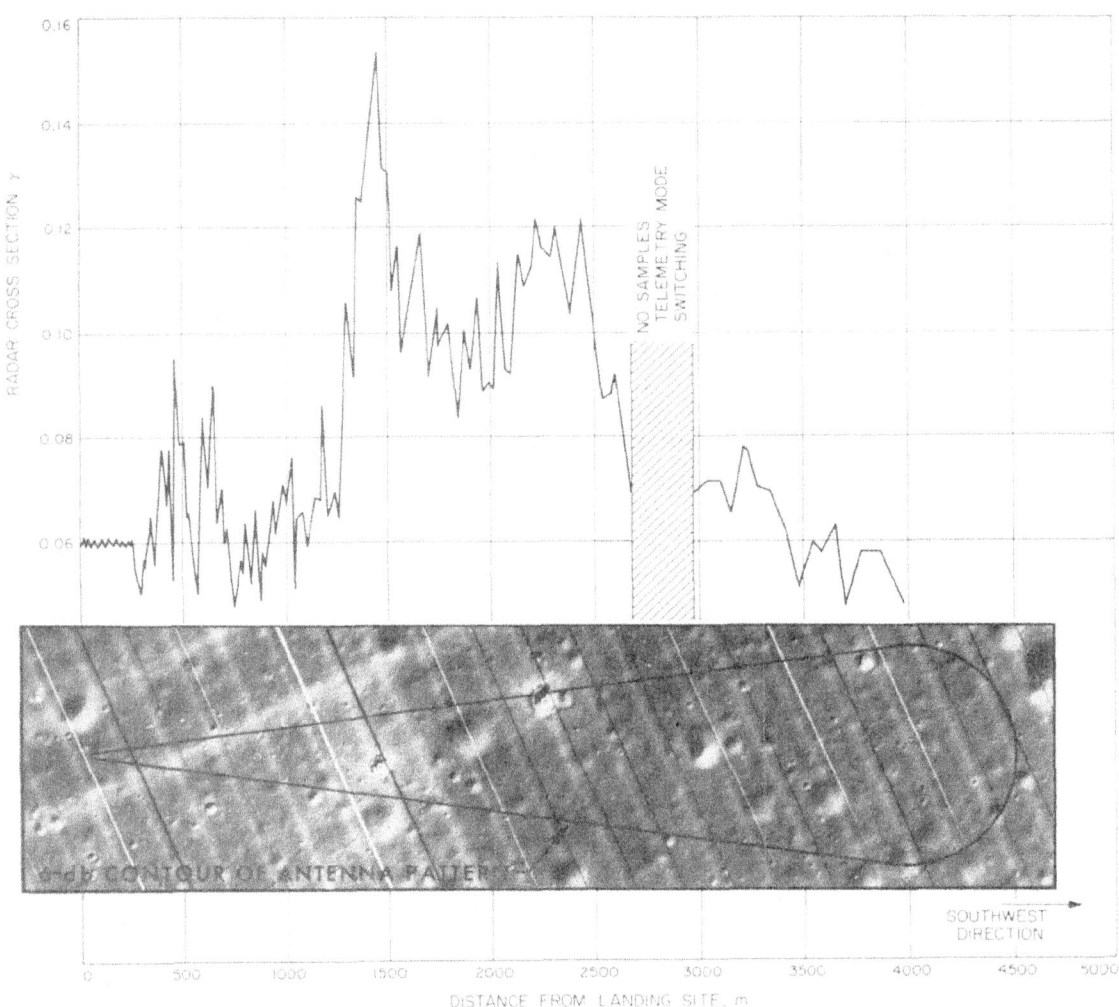

FIGURE 7-14.—Correlation between Surveyor I, beam 1, radar coverage and the radar cross section.

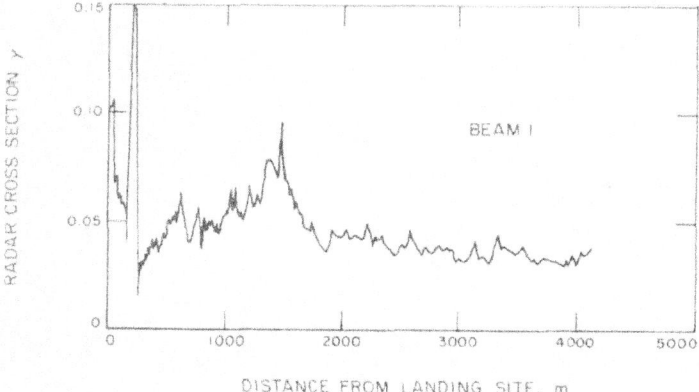

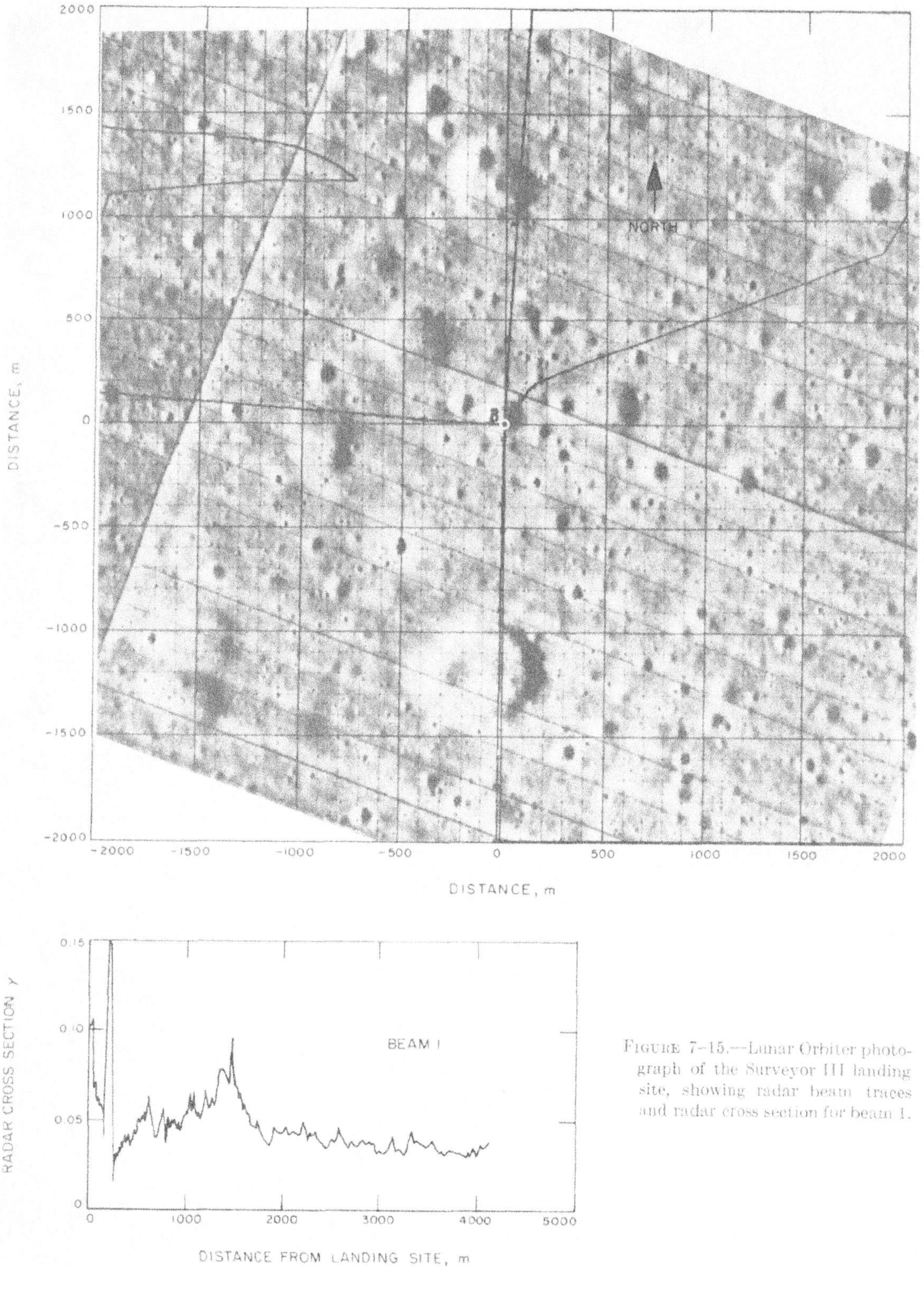

FIGURE 7-15.—Lunar Orbiter photograph of the Surveyor III landing site, showing radar beam traces and radar cross section for beam 1.

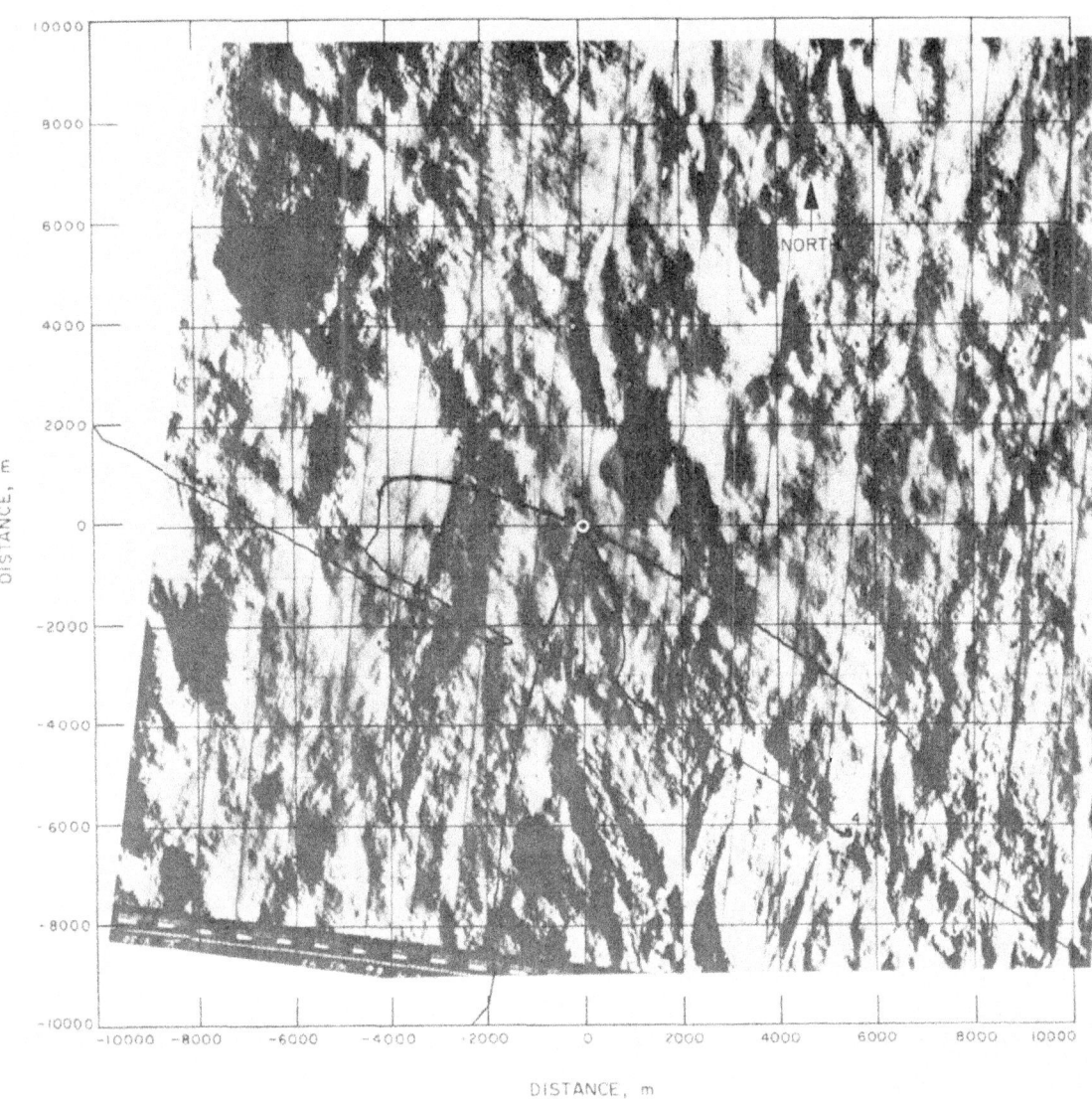

FIGURE 7-16.—Lunar Orbiter photograph of the Surveyor V landing site, showing radar beam traces.

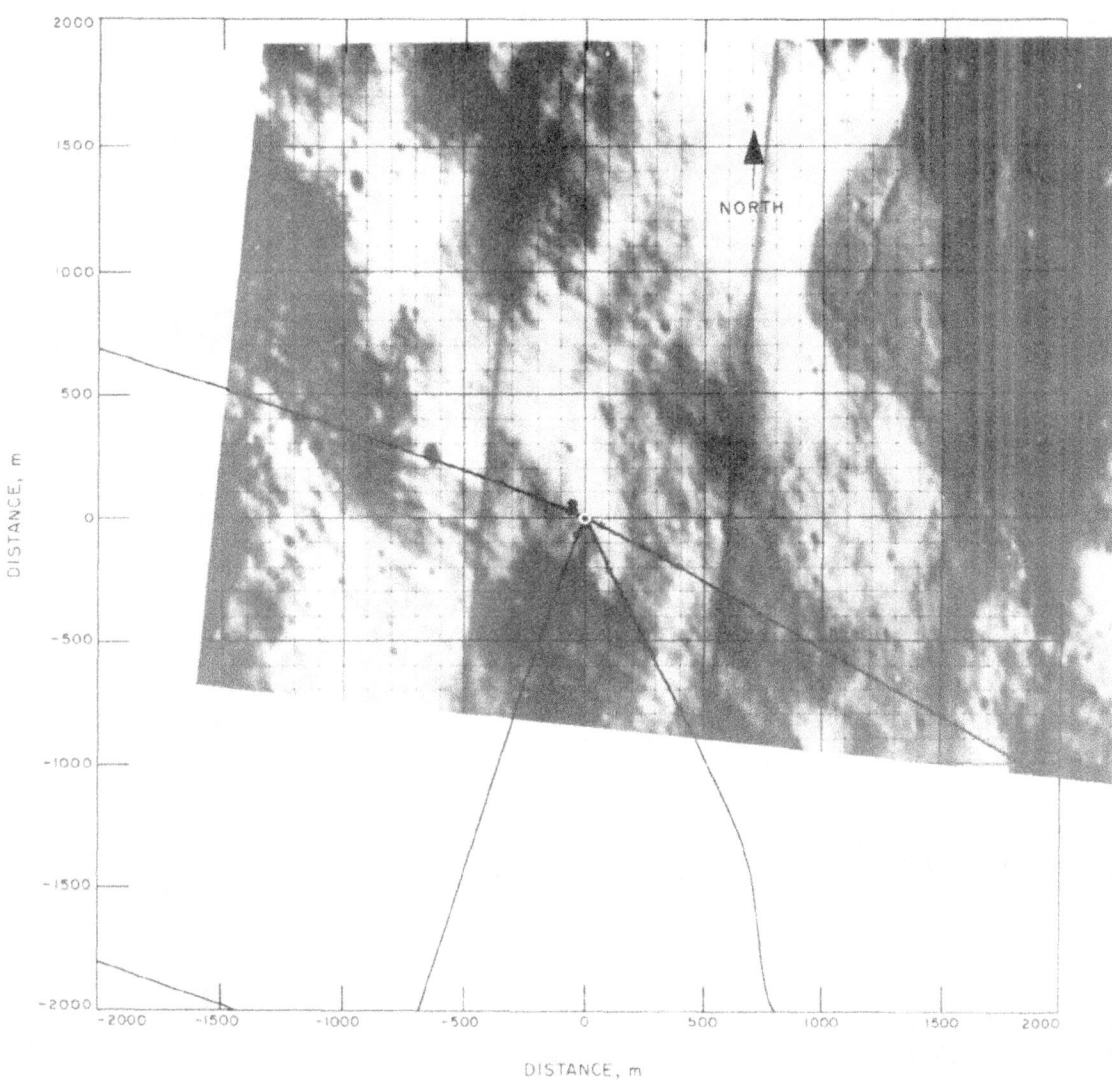

FIGURE 7-17. Lunar Orbiter photograph of the Surveyor VII landing site, showing radar beam traces, with beam 3 crossing the radar-anomalous crater.

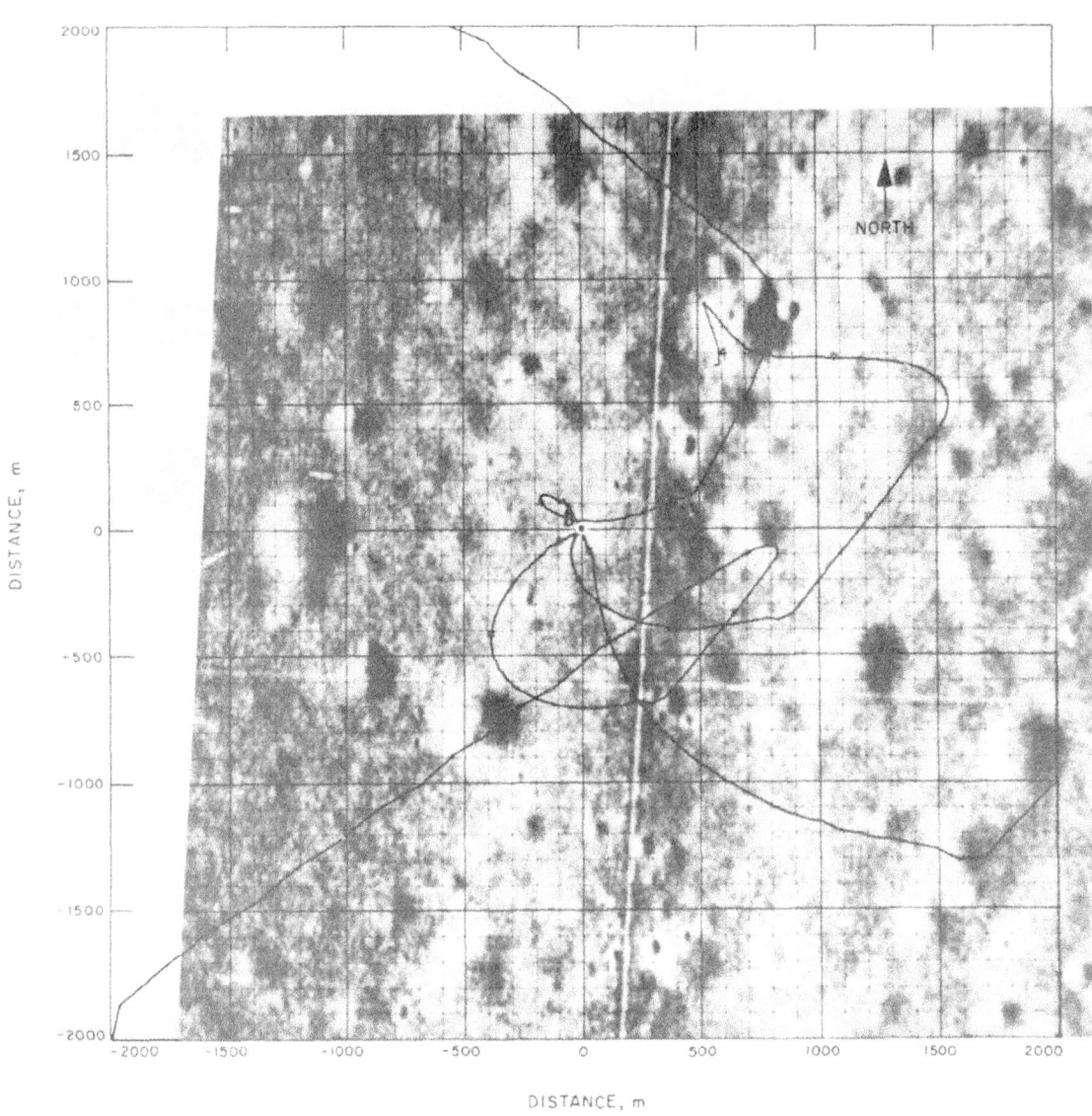

FIGURE 7-18.—Total beam coverage for Surveyor VII.

Interpretation of Surveyor Radar Data

D. O. Muhleman

In principle, the measurement of the radar cross section for the complete range of incidence angles can be interpreted in terms of the electrical parameters of the surface materials (assuming that they are homogeneous to several skin depths) and surface topographic statistics. In practice, these two effects are very difficult to separate even with "perfect" measurements. In measurements of terrestrial surfaces, it is possible to measure the electrical parameters from laboratory samples and to carry out topographic surveys. Although many such tests have been conducted, relatively little theoretical insight concerning the radar-scattering behavior for such test regions has been obtained.

In lunar or planetary radar tests, all the information must be extracted from the reflectivity data; consequently, one must adopt some theoretical scattering theory to remove the effects of surface roughness in order to estimate the surface electrical parameters; in particular, the dielectric constant. The general structure of the lunar backscatter characteristics is fairly well known from radar measurements from the Earth. We will assume, a priori, that the backscatter law for a local region on the lunar surface is the same as that for the average lunar disk as seen from the Earth.

We defined the function $F(\theta)$ and the parameter η such that the power backscattered per unit solid angle (per watt of power per square square meter illuminated) is

$$\eta \frac{F(\theta)}{4\pi}$$

Then the power backscattered from a surface element at angle θ (measured from the sub-Earth point for the case of illumination from the Earth) is

$$P_r(\theta) = \eta \frac{P_t G A_r}{4\pi r^4} \frac{R^2}{2} F(\theta) \cos\theta \qquad (3)$$

where G is the antenna gain, A_r the effective antenna area, r the distance from the radar to the lunar surface, and R the radius of the Moon. If the Moon were a perfectly conducting smooth sphere, the power received would be

$$P_{r_s} = \frac{P_t G A_r}{(4\pi)^2 r^4} \pi R^2 \qquad (4)$$

which, when used to normalize equation (3), yields the angular radar cross section $\sigma(\theta)$:

$$\frac{P_r(\theta)}{P_{r_s}} = \sigma(\theta) = 2\eta F(\theta) \cos\theta \qquad (5)$$

We notice that the gamma cross section of chapter 7 (pp. 203–218) is related to $\sigma(\theta)$ by the relationship

$$\sigma(\theta) = \gamma(\theta) \cos\theta \qquad (6)$$

If we call the total observed cross section of the Moon ρ, then it must be that

$$2\eta \int_0^{\pi/2} F(\theta) \cos\theta \sin\theta \, d\theta = \rho \qquad (7)$$

which supplies the normalization condition for the backscattering functions $F(\theta)$ or $F(\theta) \cos\theta$.

We must now consider the case where a local (flat) region of the Moon is illuminated by a radar beam of known pattern $G(\phi)$, where ϕ is the angle measured from the electrical axis and the beam is assumed to be symmetrical in azimuth about this axis. Assuming that the beam is gaussian with small half-power angle ($\sim 3°$), the power received from a surface element is

$$P_r(\theta) = \frac{P_t}{(4\pi)^2} \left(\frac{G}{2}\right) \lambda^2 \eta F(\theta) \qquad (8)$$

Thus, if we know that angular behavior of the scattering law, $F(\theta)$, and use equation (7), we may estimate η from measured values of $P_r(\theta)$; i.e., the Surveyor radar data.

There are two scattering theories available that represent the Earth-based lunar radar data well. The first theory, which we will call $F_1(\theta)$, derived by Hagfors, Beckmann, and others (see ref. 7–1), is based on an approximate application of exact Huygens' theory and assumes that the surface autocorrelation function is exponential (really linear) and that surface-height variations are gaussianly distributed about some mean level. In this case

$$F_1(\theta) \cos \theta = \frac{K_1}{(\cos^4 \theta + C \sin^2 \theta)^{3/2}} \quad (9)$$

where C is a wavelength-dependent parameter related to the surface statistics. The normalization constant K_1 must be found with the use of equation (7). The second theoretical expression, derived by Muhleman (see ref. 7-2), assumes that the echo power is incoherent and that surface-height variations from some mean level are distributed exponentially. The scattering function is then

$$F_2(\theta) \cos \theta = \frac{K_2 \cos \theta}{(\sin \theta + \alpha \cos \theta)^3} \quad (10)$$

where α is also a wavelength-dependent statistical surface parameter. Although equation (10) represents lunar echoes all the way to the lunar limb, equation (9) overestimates the limb power, and a correction derived by Beckmann is usually applied to fit lunar data. However, this correction will not seriously affect our normalization (eq. (7)) and will not be applied.

If we use either $F_1(\theta)$ or $F_2(\theta)$ in equation (7), K_1 and K_2 can be computed only if we know the relation between η and ρ. We will assume that they are equal, which cannot be far from correct, and treat η as the Fresnel reflection coefficient for normal incidence. This should be recognized as a serious assumption.

The functions $F_1(\theta)$ and $F_2(\theta)$ are quite similar and, apparently, the only significant difference is that $F_1(\theta)$ is based on gaussian and $F_2(\theta)$ on exponential surface-height statistics. The practical difference in the two functions is that $F_2(\theta)$ is more peaked at normal incidence. One could hope to resolve this question with the Surveyor data, but the lack of normal-incidence cross-section data prevents it, as will be seen below.

The cross-section data for Surveyor V versus incidence angle are shown in figure 7-19 along with equation (9) with $C=8$. The theory fits the data well from 0° to 30°. The agreement for larger angles can be improved by using a smaller value of C, but then the curve will fall well below the data for small angles. The use of several values of C seems theoretically unreasonable. The same data are shown in figure 7-20 along with $F_2(\theta)$ for $\alpha=0.38$ and 0.42. In this case, most of the (poor) near-normal incidence data fall below the curves. The situation is even more complex for Surveyor VII, as shown in figure 7-21 along with $F_1(\theta)$ for $C=6$ and $F_2(\theta)$ for $\alpha=0.46$. These

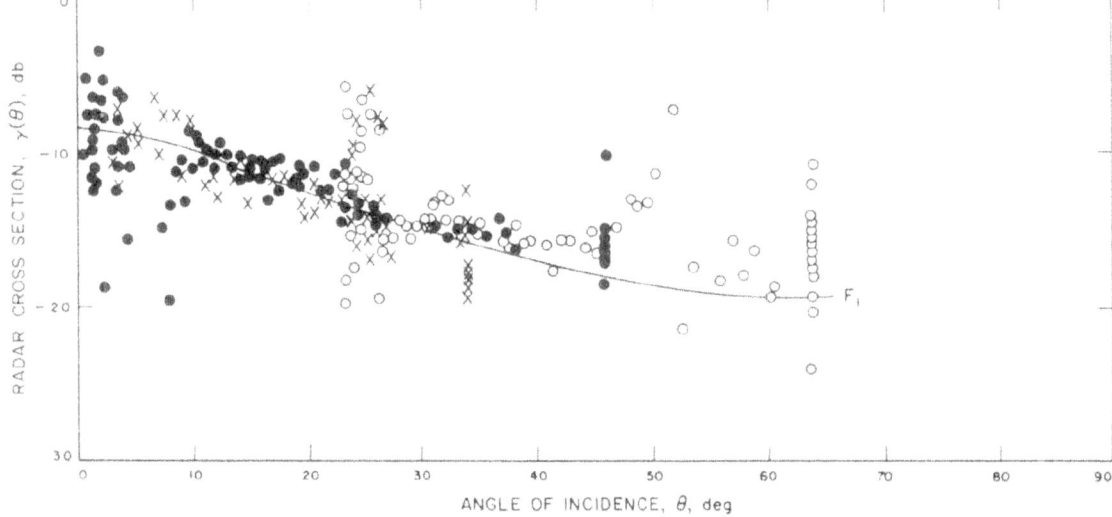

FIGURE 7-19.—Radar cross section versus angle of incidence for Surveyor V. $F_1(\theta)$ scattering model, with $C=8$.

parameter estimates imply that the Surveyor VII region (Tycho rim) is somewhat "rougher" than the mare landing sites. A larger value of α could be chosen such that $F_2(\theta)$ would pass through the lower cluster of near-normal incidence data; i.e., the surface assumed even rougher. Apparently, it is impossible to determine which cluster truly represents the Moon in this region.

All the $\gamma(\theta)$ data were treated in this way to obtain estimates of C and α and then to estimate η with the use of equations (7) to (10). The data for Surveyors I, III, and VI are even more difficult to treat than those for V and VII. See table 7-1 for formal results.

The values listed in table 7-2 represent our estimates of the surface reflectivities where we have attempted to remove the effects of surface roughness. The results for the two scattering theories agree to within the estimated errors. Consistent with the above assumptions, particularly the interpretation of η as the Fresnel reflection coefficient, the estimates of the dielectric constants can be computed from the Fresnel equation for normal incidence

$$\epsilon = \left(\frac{1+\eta^{1/2}}{1-\eta^{1/2}}\right)^2 \qquad (11)$$

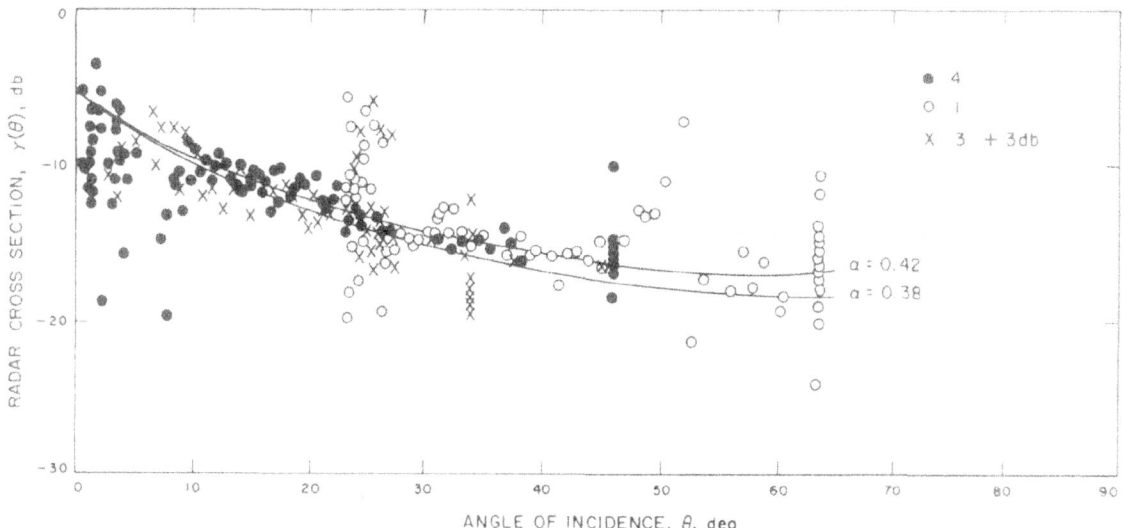

FIGURE 7-20.—Radar cross section versus angle of incidence for Surveyor VI. $F_2(\theta)$ scattering model, with $\alpha=0.38$ and 0.42.

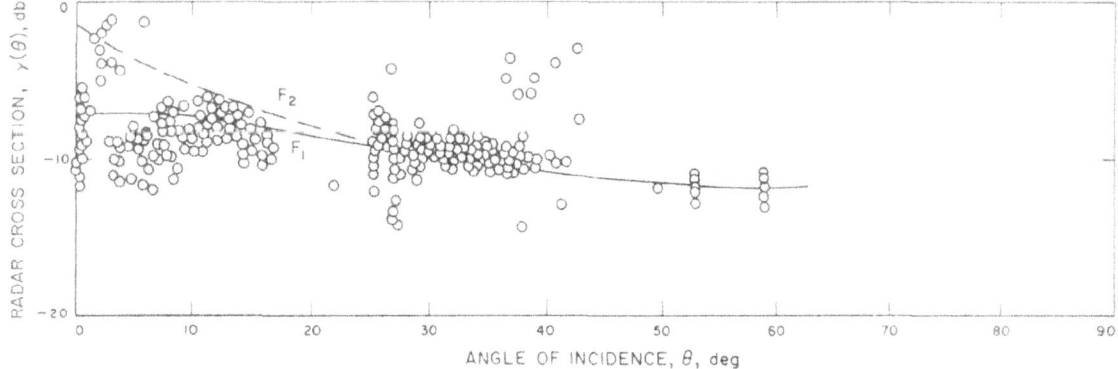

FIGURE 7-21.—Radar cross section versus angle of incidence for Surveyor V. $F_1(\theta)$ and $F_2(\theta)$ scattering models are indicated.

We have taken the unweighted average of η from the two theories (and for their uncertainties) for each case. Application of equation (11) then yields the results of table 7–3. These results are quite consistent with an estimate for the average Moon of about $\epsilon = 2.7$ from radar (Earth based) and 2.0 to 2.5 from Earth-based, radio-emission polarization effects.

Finally, in a far more speculative calculation, we can estimate a range of material densities using the well-known relation between the dielectric constant of a material in a solid form and that for a granulated form; i.e., variation in ϵ with packing factor. Furthermore, we will assume that the density of the material in the "solid" form is 2.90 g/cm^3, and that the lunar basalt has a dielectric constant of $\epsilon_s = 4, 5, 6,$ or 7. The latter factor is actually completely unknown! These results are shown in table 7–4.

TABLE 7–2. *Reflectivities*

Spacecraft	η, with $F_1(\theta)$	η, with $F_2(\theta)$
Surveyor I	0.040 ± 0.020	0.053 ± 0.020
Surveyor III	.030 ± .005	.033 ± .006
Surveyor V	.027 ± .005	.031 ± .005
Surveyor VI	.033 ± .008	.047 ± .010
Surveyor VII	.072 ± .015	.094 ± .020

TABLE 7–3. *Estimates of dielectric constants*

Spacecraft	Dielectric constant
Surveyor I	2.40 ± 0.50
Surveyor III	2.07 ± .11
Surveyor V	2.00 ± .16
Surveyor VI	2.27 ± .20
Surveyor VII	3.28 ± .40

TABLE 7–4. *Speculative estimates of the lunar surface densities at the landing sites*

Spacecraft	Measured ϵ	Assumed ϵ (solid)	$\left(\dfrac{density}{2.90}\right)$	Density (with $\rho_s = 2.90$ g/cm^3), g/cm^3
Surveyor I	2.40 ± 0.50	4	0.57	1.70
		5	.48	1.40
		6	.42	1.20
		7	.33	.95
Surveyor III	2.07 ± .11	4	.47	1.40
		5	.40	1.15
		6	.35	1.00
		7	.32	.95
Surveyor V	2.00 ± .16	4	.45	1.30
		5	.38	1.10
		6	.33	.95
		7	.31	.90
Surveyor VII	3.28 ± .40	4	.81	2.35
		5	.67	1.95
		6	.58	1.70
		7	.52	1.50

Magnet Data

J. NEGUS DE WYS

Before the Surveyor program, no direct data were available on the composition of lunar surface materials. Surface models consisting of materials quite different from terrestrial surface rocks and soils were defended theoretically in the literature, as were models based on terrestrial analogs.

For several centuries, lunar albedo differences in the mare and highland areas, which were observed from the Earth, had been interpreted as indicative of possible differences in composition and roughness. Color-reflectance photographs, taken with red and green filters, showed differences within mare areas—some parts appeared redder, some greener (ref. 7-4). Highland areas appeared much lighter, with many craters and rays resulting in near-white images. It is interesting to note that the area around the crater Tycho appeared reddish, similar to many mare areas; the Surveyor VII landing site was within this reddish halo. Later reflectance studies using black-and-white photographs also showed distinct areas of different reflectance within the maria. Figure 7-22 shows the Surveyor landing sites on a slightly gibbous Moon.[1]

Although the Lunar Orbiter photographs showed detailed surface morphology and indicated abundant craters as well as structures suggestive of flow, nothing definite could be determined concerning the composition of the lunar surface. However, from some of the surface features, an internal source for some surface material seemed a possibility.

Radar returns showed greater reflectivity from cratered areas. This could be related to possible compositional differences in terms of densities and dielectric properties, but could also be related to slope in terms of reflectance planes and rock distribution. Thus, these data also could not result in definite conclusions regarding composition. However, it should be noted that the loss tangent-to-density ratio derived from the landing radar (RADVS echo

[1] E. A. Whitaker, private communication, 1967.

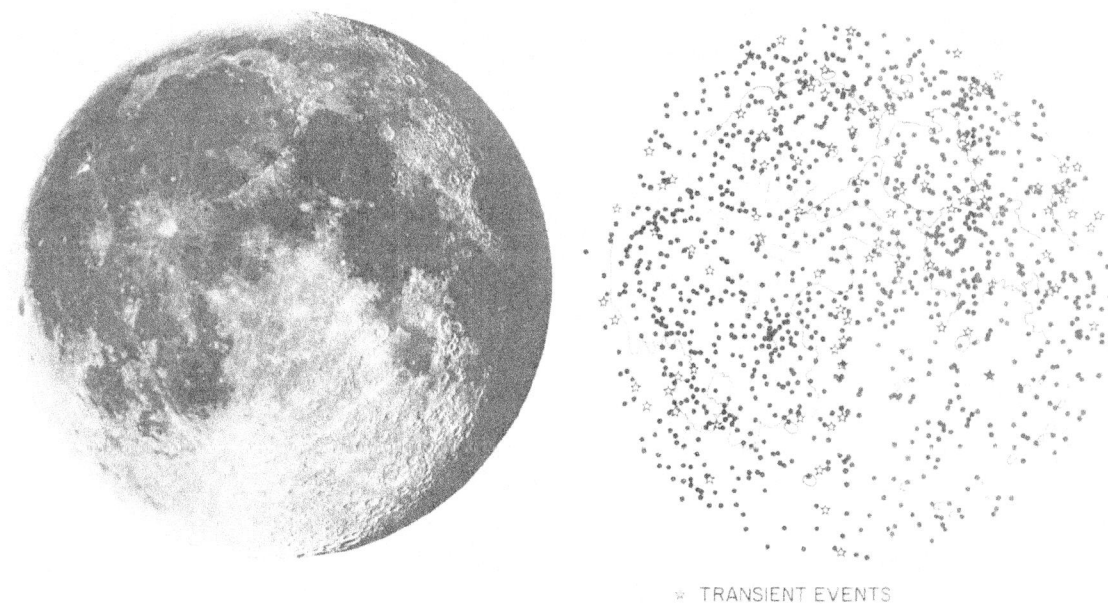

☆ TRANSIENT EVENTS
• INFRARED ANOMALIES (REF. 7-5)

FIGURE 7-22.— Slightly gibbous Moon showing reflectance differences within the mare areas. A comparison of Surveyor landing sites shows a similarity in reflectance properties.

data) on Surveyor III (ref. 7-2) was suggestive of basalt.

Infrared scanning of the Moon at a wavelength of 10 to 12 microns resulted in the identification of over 1000 "hot spots," or thermally anomalous areas. Of 300 anomalies analyzed, 80 percent were found to coincide with craters.[2] Figure 7-22 shows the distribution of thermal anomalies observed in 10- to 12-micron scanning. Thermal parameter (γ) values of 20 to about 1400 have been suggested. The Tycho area is the brightest anomaly observed (ref. 7-5). This range of values is comparable with values for solid terrestrial rock at one extreme and fine powder on the other. Analysis has not shown evidence for an internal heat source, but has not excluded such a possibility.[3] Compositional deductions cannot be made from these data.

Nearly 600 lunar transient events[4] ranging from obscure hazes to lightninglike streaks, which have been reported during the last 300 years, have been collected and evaluated (ref. 7-6); over 200 of these events occurred in one area, the Herodotus-Shroeter Valley–Cobra Head complex. Such repetitious activity is suggestive of volcanism or outgassing; Kosyrev (ref. 7-7) has reported obtaining a spectrographic plate of the crater Alphonsus during a gaseous emission from the central peak. From preliminary comparisons, about 20 of the lunar event sites appear to coincide with areas showing thermal anomalies. A volcanic interpretation of these events would suggest basaltic composition. (See fig. 7-22 for a comparison of areas showing thermal anomalies at 10 to 12 microns and areas from which visual events have been reported.)

Besides the problem of composition, the origin of lunar craters has long been a subject of debate. On the one extreme, meteoritic impact and accretion of the lunar surface are suggested; on the other extreme, volcanic activity and associated phenomena are considered dominant in the formation of lunar surface material and surface morphology.

This controversy over causative agents of morphology also involves composition. If the surface material were primarily the result of volcanism and the craters predominantly the result of endogenic processes, then, by analogy with the Earth, the expected composition over most of the lunar surface might be basaltic, since the volume of terrestrial basalt equals about five times the volume of all other extrusives combined on Earth (ref. 7-8). Such a lunar model assumes a similar basic gabbroic parental magma and a similar degree, or less, of differentiation in resultant melts.

However, if the surface material and cratering were results of impact and accretion, the surface composition would reflect some alteration toward the composition of the meteoritic flux; some addition of meteoritic nickel-iron would be expected (93 percent of observed falls are chondrites, about 6 percent irons). Furthermore, pulverization of surface material by impact should produce a size distribution of particles and fragments in the particulate lunar surface covering to some depth rather than a purely fine-grained homogeneous material (ref. 7-9),[5] overlain by material with a rock and particle distribution of considerable heterogeneity.

Magnetite, the magnetic material found in basalt, usually ranges from about 5 to 12 percent in terrestrial basalts and is usually finely disseminated throughout the rock. Meteoritic nickel-iron in the form of kamacite (about 5 percent Ni) or taenite (13 to 65 percent Ni) might be expected to occur on the lunar surface as separate small particles or fragments resulting from impact. Churning of the upper lunar surface layer might be expected to mix thoroughly these fragments with surface material to some depth. These magnetic particles would be free particles, rather than embedded in silicates as is the magnetite in basalt. Laboratory studies show a considerable difference between the results of magnet contact with powdered basalt and results of contact with powders to which separate pure powdered iron has been added. Thus, a magnet test could—

---

[2] J. Saari, private communication, 1968.
[3] D. F. Winter, private communication, 1968.
[4] B. M. Middlehurst, private communication, 1968.

[5] J. A. O'Keefe, private communication, 1968.

(1) Suggest a rock type.
(2) Indicate addition of free magnetic fragments.
(3) Give some indication of whether endogenic or exogenic agents may have been dominant on the lunar surface.

It should be noted that, in considering the observed falls as a model meteoritic flux, the assumption is made that the flux was relatively constant during a large period of geologic time (about 4.5 billion years). Such an assumption may or may not be valid.

Purpose of Magnet Tests

The purpose of the magnet test was to determine—
(1) Presence and amount of magnetic material on the lunar surface.
(2) Differences in amount of magnetic material in the maria and in the highlands.
(3) Differences in amount of magnetic material between the surface tests and the subsurface tests (to a depth of about 20 cm).
(4) Presence of larger magnetic fragments on the lunar surface, if any (on Surveyor VII).

Data derived from operations of the alpha-scattering instrument indicate the elemental percent in the Fe-Ni-Co group; the determina-

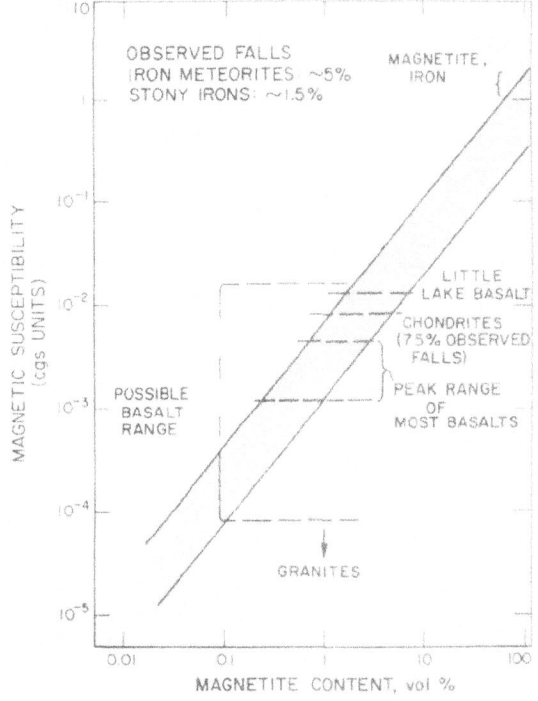

FIGURE 7-23.—Magnetic susceptibility of common terrestrial rocks is largely a function of magnetite content. Sedimentary rocks and gneisses also fall in the range of granites. Pure iron may go considerably higher than unity.

FIGURE 7-24.—Magnet assembly on footpad 2 of Surveyor V before launch at Cape Kennedy. Similar assemblies were attached to footpad 2 on Surveyor VI and to footpads 2 and 3 on Surveyor VII.

tion of whether the iron content occurs in pyroxenes, hornblendes, or in a magnetic form (Fe, Fe_3O_4, or Fe-Ni combinations) can be derived only from the magnet data.

Magnet-Assembly Design

Magnetic susceptibility of most terrestrial rocks is proportional to the magnetite content (ref. 7-10); see figure 7-23, where the abscissa scale is based largely on information found in reference 7-11. Thus, magnetic data may suggest a rock type on the basis of magnetite content and may indicate an anomalous addition of magnetic material to a normal terrestrial rock type.

Footpad magnet assemblies.—The footpad magnet assemblies, attached to footpad 2 on Surveyors V, VI, and VII (also on footpad 3 of Surveyor VII), consisted of a magnetized Alnico V bar, 5 x 1.27 x 0.32 cm, and a similar-size, unmagnetized bar of Inconel X-750. Figure 7-24 shows the footpad 2 magnet assembly on Surveyor V before launch at Cape Kennedy.

Magnetic flux readings along pole edges ranged from about 300 to 800 gauss. Plots of 120 readings on each bar were made to assist in the interpretation of the results. Magnets with this strength strongly attract the common ferrimagnetic mineral, magnetite (Fe_3O_4), and the ferromagnetic metals, nickel, iron, and cobalt.

Surface-sampler magnets.—In addition to the magnets on footpads 2 and 3, two rectangular horseshoe magnets, 1.9 x 0.96 x 0.32 cm, were embedded side by side in the back of the surface-sampler door in Surveyor VII. Magnetic flux readings at the pole faces were about 700 gauss; the south poles were adjacent (see fig. 7-25). On the Moon, this configuration is capable of attracting and picking up a fragment of nickel-iron meteorite, or magnetite about 12 cm in diameter. Thus, a test of magnetic material at trench depths was possible on Surveyor VII, as well as the capability of choosing the test area and testing larger fragments.

Data

Surveyors V and VI landed in mare areas, Mare Tranquillitatis and Sinus Medii, respectively. In Mare Tranquillitatis, the landing involved a meter-long slide through the lunar surface material, at a depth of about 7 to 12 cm on the inner slope of a 9- x 12-meter crater. The magnet assembly on Surveyor V contacted material through most, if not all, of this slide (refs. 7-12 and 7-13). In sunset lighting, the amount and distribution of magnetic material were clearly visible. A relatively small amount of magnetic material, with a particle size below camera resolution (1 mm), was observed to be present on the lunar surface in Mare Tranquillitatis (see fig. 7-26).

The second magnet landing in a mare area (Sinus Medii) achieved a successful test following the hop performed by Surveyor VI (ref. 7-14). During the landing, the lower seven-eighths of the bar magnet made momentary contact with the lunar surface material. A small amount of magnetic material, less than that on the Surveyor V magnet, was observed to be present. The lesser amount was probably the result of the difference in landing modes. Again, the particles were smaller than camera resolution (see fig. 7-27).

Surveyor VII made the first and only landing in a highland area. It was anticipated that this area might show a difference in the extent of differentiation in the surface material. A more acidic rock would contain considerably less magnetite; this would be quite obvious in a magnet test. On the other hand, a higher contribution of magnetic material from meteoritic impact would also be obvious. Neither footpad (2 or 3), to which magnets were attached, penetrated the surface sufficiently (more than 6 cm) to provide a magnet test (see fig. 7-28). However, the magnets on the surface sampler provided an opportunity for surface tests as well as for depth tests (ref. 7-15). Figures 7-29 through 7-31 show the results after the first two bearing strength tests with the surface sampler. A small amount of magnetic material is observed to outline the horseshoe magnets after the first bearing strength test, with a slight increase following the second test. After trenching at a depth of up to 20 cm (fig. 7-32), an obvious increase in the amount of magnetic material was observed. Figure 7-33 shows the area before and after trenching efforts.

ELECTROMAGNETIC PROPERTIES

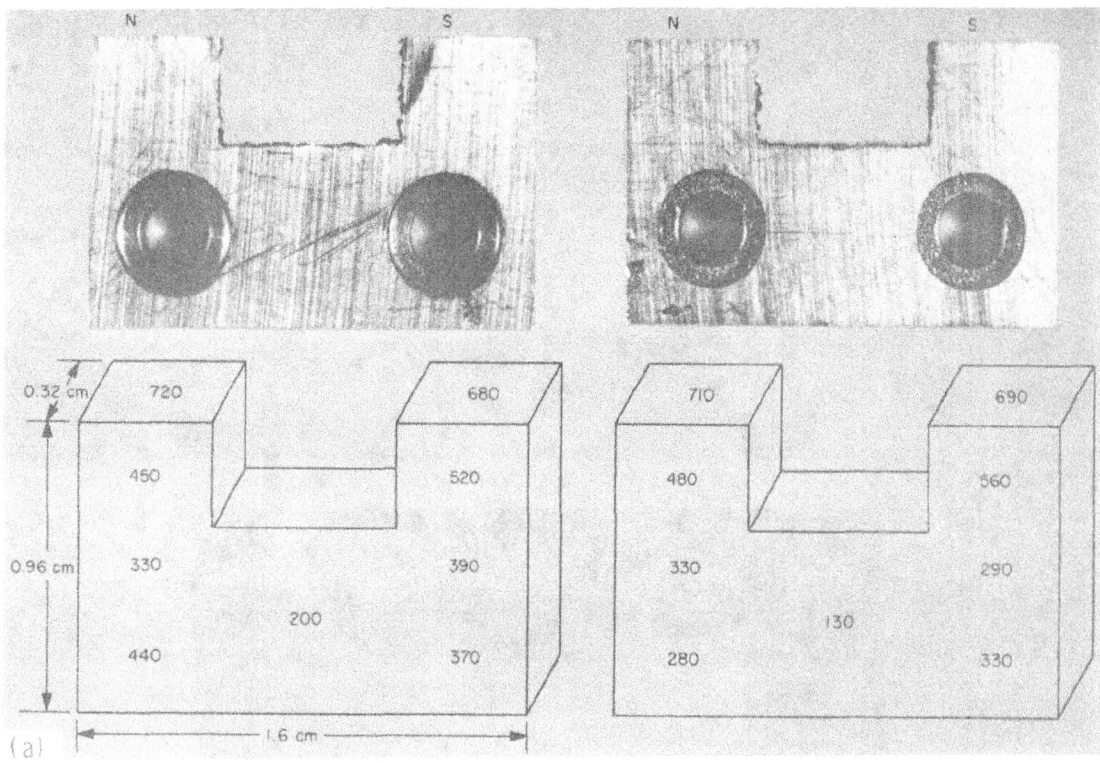

FIGURE 7-25.—(a) Magnets embedded in the door of the surface sampler on Surveyor VII. Below are the magnetic flux readings. (b) Closeup of surface-sampler magnet assembly; after mounting, the entire assembly is painted blue.

FIGURE 7-26.—(a) Meter-long slide in surface material was performed by footpad 2 in the Surveyor V landing (Sept. 12, 1967, 00:18:35 GMT; Sept. 11, 1967, 23:38:23 GMT). (b) Surveyor V magnet assembly in sunset lighting shows a relatively small amount of magnetic material adhering to the magnet (Sept. 23, 1967, 09:54:20 GMT).

ELECTROMAGNETIC PROPERTIES

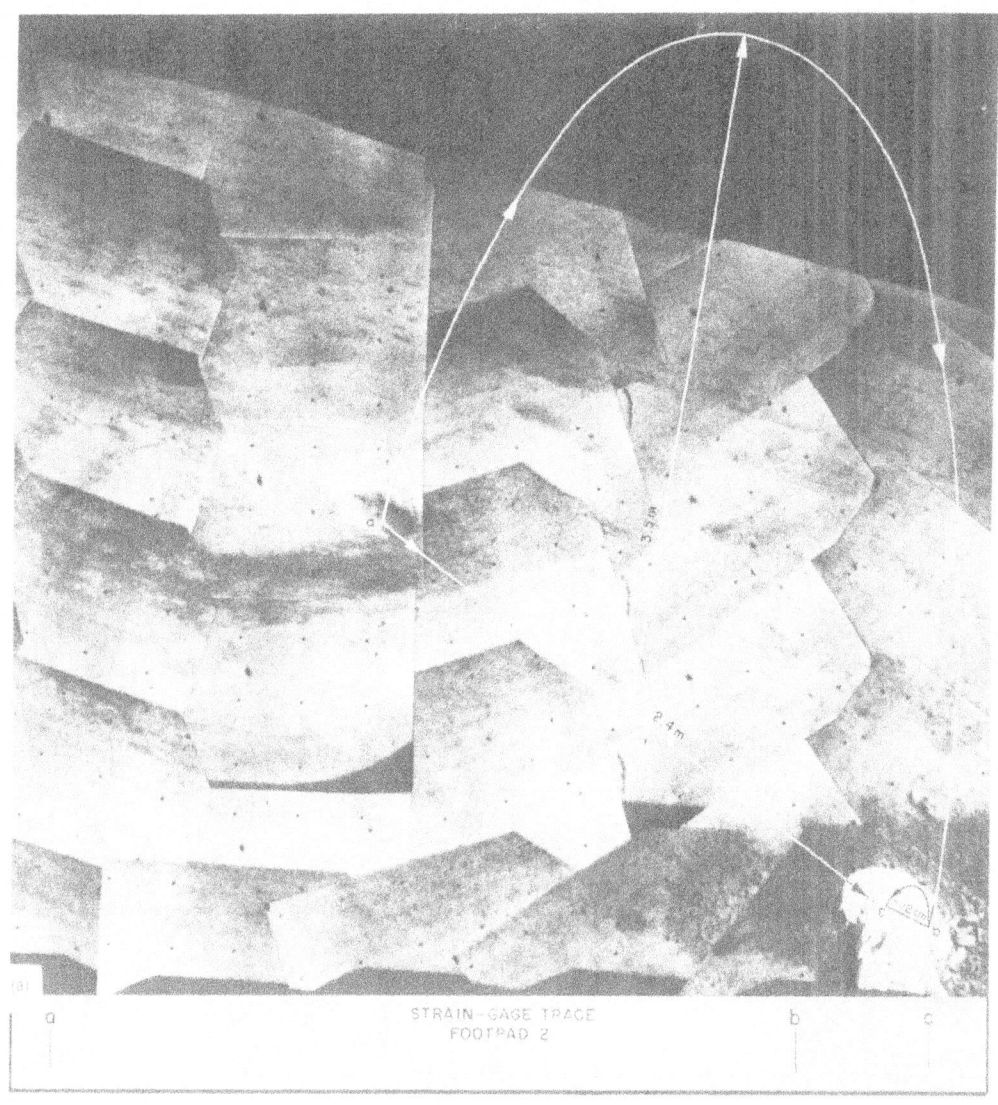

FIGURE 7-27.—(a) Hop performed by Surveyor VI caused footpad 2 to follow this trajectory. In landing, a short bounce occurred, and momentary contact with the lunar surface material was made by the magnet. The strain-gage data shown at the bottom indicate that the movement from b to c was a bounce rather than a slide. (b) After the hop, the Surveyor VI magnet assembly shows a small amount of fine material collected on the magnet (Nov. 19, 1967, 04:59:57 GMT, computer processed).

ELECTROMAGNETIC PROPERTIES 231

FIGURE 7-28.—Surveyor VII footpads after landing north of Tycho. (a) Footpad 2 (Catalog 7-SE-7). (b) Footpad 3. Because the penetration depths of the footpads were less than 6 cm, neither magnet contacted lunar surface material.

FIGURE 7-29.—Views of the surface-sampler magnets, before, during, and after the first bearing strength test. The depth of the test was about 5 cm. Some magnetic material can be seen around the magnetic pole ends. (See fig. 7-33 for test area.)

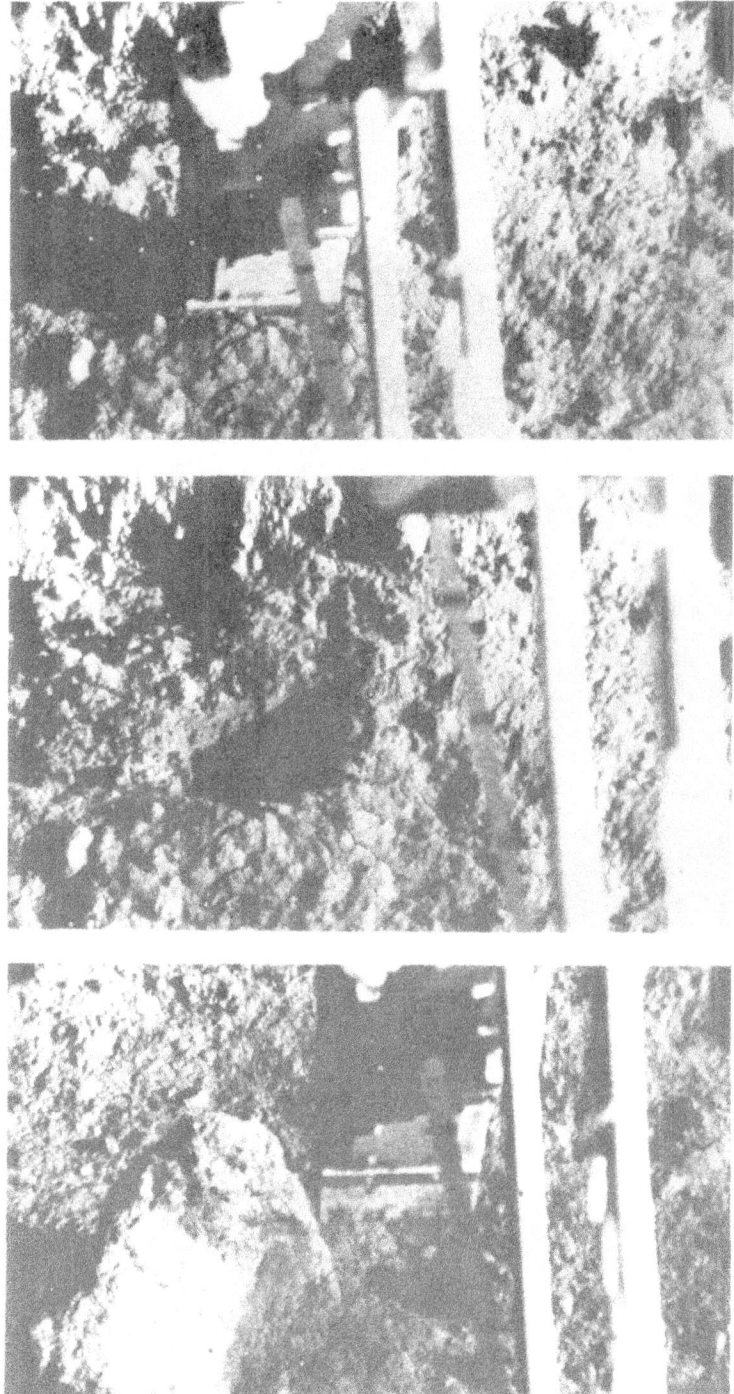

FIGURE 7-30.—Second bearing strength test showing surface-sampler magnets before, during, and after test. The depth of the test was about 5 cm. A small increase is seen in the amount of magnetic material that outlines the magnetic pole ends.

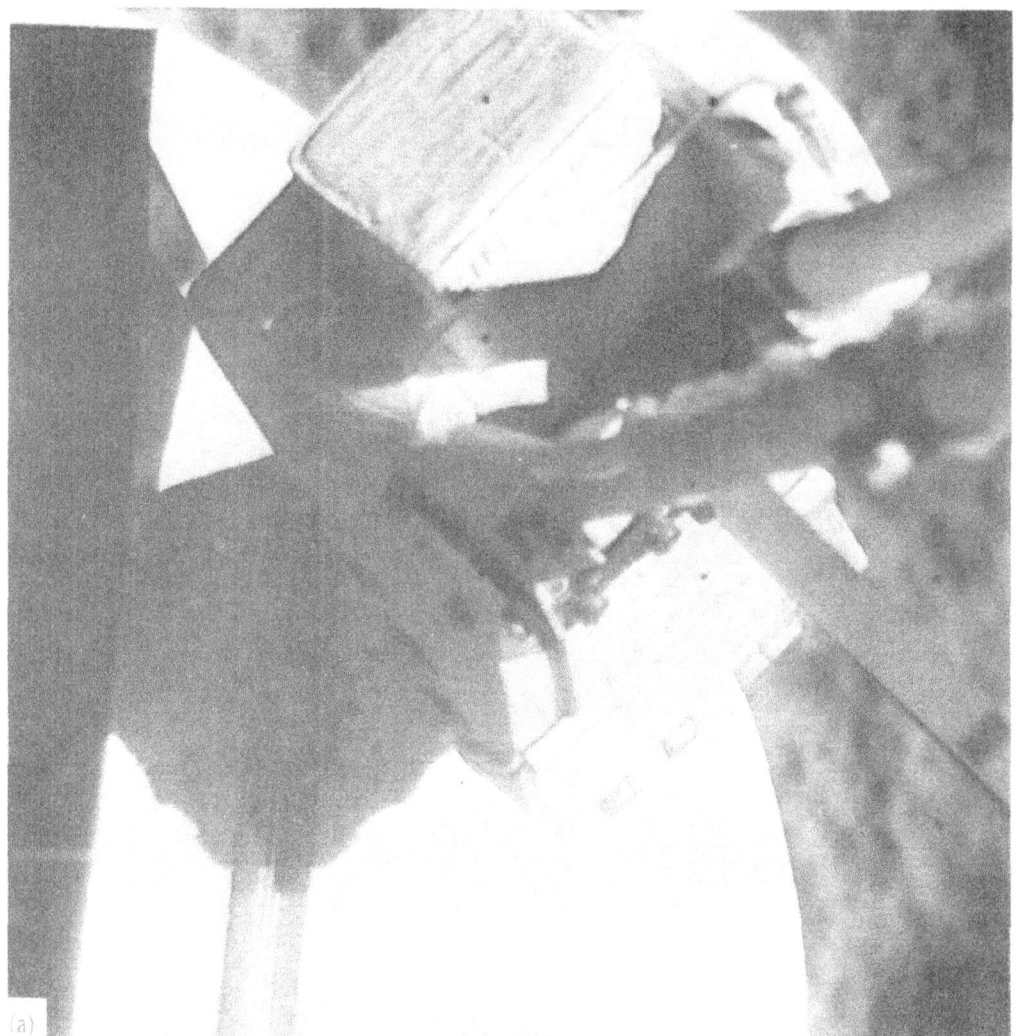

FIGURE 7-31.—Closeups of surface-sampler magnets following two bearing strength tests. (a) Pole edges (Jan. 20, 1968, 08:36:18 GMT). (b) Side of magnet outlined by magnetic material.

FIGURE 7-32.—After trenching at depths up to 20 cm, an obvious increase in magnetic material was observed on the surface-sampler magnets (Jan. 20, 1968, 15:08:28 GMT).

ELECTROMAGNETIC PROPERTIES 237

FIGURE 7-33.—(a) Early view of surface-sampler operations area showing bearing strength test sites, a, and object, b, which was selected for magnetic-attraction test on the basis of its luster, shape, and gray tone. (b) Surface-sampler operations area after trenching efforts. (Catalog 7-SE-16).

An object, selected to be tested for magnetic attraction, was chosen on the basis of shape, gray tone, and luster in comparison with other rocks in the area (fig. 7-34). The object is about 1.2 cm in diameter, appears to be dense, fairly smooth, and shows a slight luster. The surface sampler was dragged on the surface toward the object, which was attracted and picked up by the horseshoe magnets (fig. 7-35). From the motions of the surface sampler and the relative positions of the object, it was concluded that the object was adhering to the scoop door because of magnetic attraction.

The morphology of the object is not in disagreement with that of a nickel-iron meteorite,[6] or of a nodular magnetite fragment. Whatever the actual composition of the fragment and its source, it is evidence of possible availability of iron on the Moon, since only ore-grade material would be attracted by the magnet assembly used.

Laboratory Studies

Various laboratory studies have been conducted for comparison with the results of the magnet tests on the lunar surface.

The distance from which a magnet with a magnetic flux of about 500 gauss will attract powdered iron or magnetite on the Earth is about 1.9 cm (refs. 7-12 and 7-13).

The bar magnets and horseshoe magnets on the surface sampler were recorded photographically following contact with rock powders with a wide range of compositions—powdered CaOH with additions of iron and powdered basalt with additions of iron. Tests conducted in a 10^{-6}-torr vacuum included impact in rock-powder samples as well as jet-exhaust tests.

Landing modes were simulated in the laboratory using 37- to 50-micron basalt powder. These tests assisted in determining the depth of footpad penetration and dynamics of contact with the lunar surface material (refs. 7-12 and 7-13).

Following the successful pickup of a magnetic fragment north of the crater Tycho, the

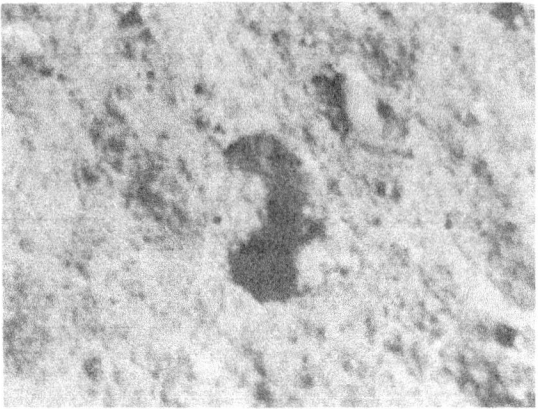

FIGURE 7-34.—Object selected for magnetic-attraction test. Note luster, dark appearance, smooth shape, and apparent dense quality (Jan. 13, 1968, 02:06:28 GMT).

motions of the surface sampler were reproduced in the laboratory. Various meteorite and basaltic fragments were used. The only objects that could be attracted and picked up by the surface-sampler magnets were magnetite and meteoritic fragments of nickel-iron. Figure 7-36 shows some of the laboratory tests in rock powders. Two landing-mode simulations, with their lunar counterparts and simulations of surface-sampler motions during fragment pickup, are also shown.

Discussion

Two different categories of magnetic particles are points of concern in the lunar magnet results and the laboratory studies:

(1) The magnetite content and mode of occurrence in a rock type.

(2) The amount of free kamacite or taenite particles added to a rock powder by meteoritic impact.

As shown in figure 7-23, magnetic susceptibility of a rock can be plotted versus magnetite content, with a resultant grouping of rock types. Granites, gneisses, and sedimentary rocks are usually low in magnetic susceptibility because of the low magnetite content (<0.1 percent). Although basalts have a wide range of magnetic susceptibility above 10^{-4}-cgs units, the range in which most basalts fall is more

---

[6] H. Brown, private communication, 1968.

FIGURE 7-35.—Object attracted and picked up by the adjacent magnetic south pole ends of the surface-sampler magnets was about 1.2 cm in diameter, fairly smooth, with some irregularities that appear suggestive of exfoliation (Jan. 20, 1968, 14:58:39 GMT).

restricted. Magnetic susceptibility may vary somewhat with a change only in the mode of occurrence of the magnetite, i.e., particle size and grouping of particles, and with the amount of pure Fe_3O_4 relative to titanium replacement of iron (see fig. 7-37). Both chondritic meteorites and the Little Lake basalt fall in the upper part of the magnetic susceptibility range for basalts. Pure magnetite falls at unity, or higher, as do taenite, kamacite, and pure iron which may fall much higher. Particle size and grouping affects the magnetic susceptibility of a rock. In Little Lake basalt, the magnetite occurs as particles, 10 to 25 microns in diameter, which are disseminated throughout the rock (fig. 7-38).

SCORIACEOUS BASALT

BASALT + 2% Fe

BASALT

BASALT + 3% Fe

BASALT + 1% Fe

BASALT + 4% Fe

FIGURE 7-36.— Laboratory studies. (a) Magnet-assembly contact with rock powders in atmosphere.

ELECTROMAGNETIC PROPERTIES

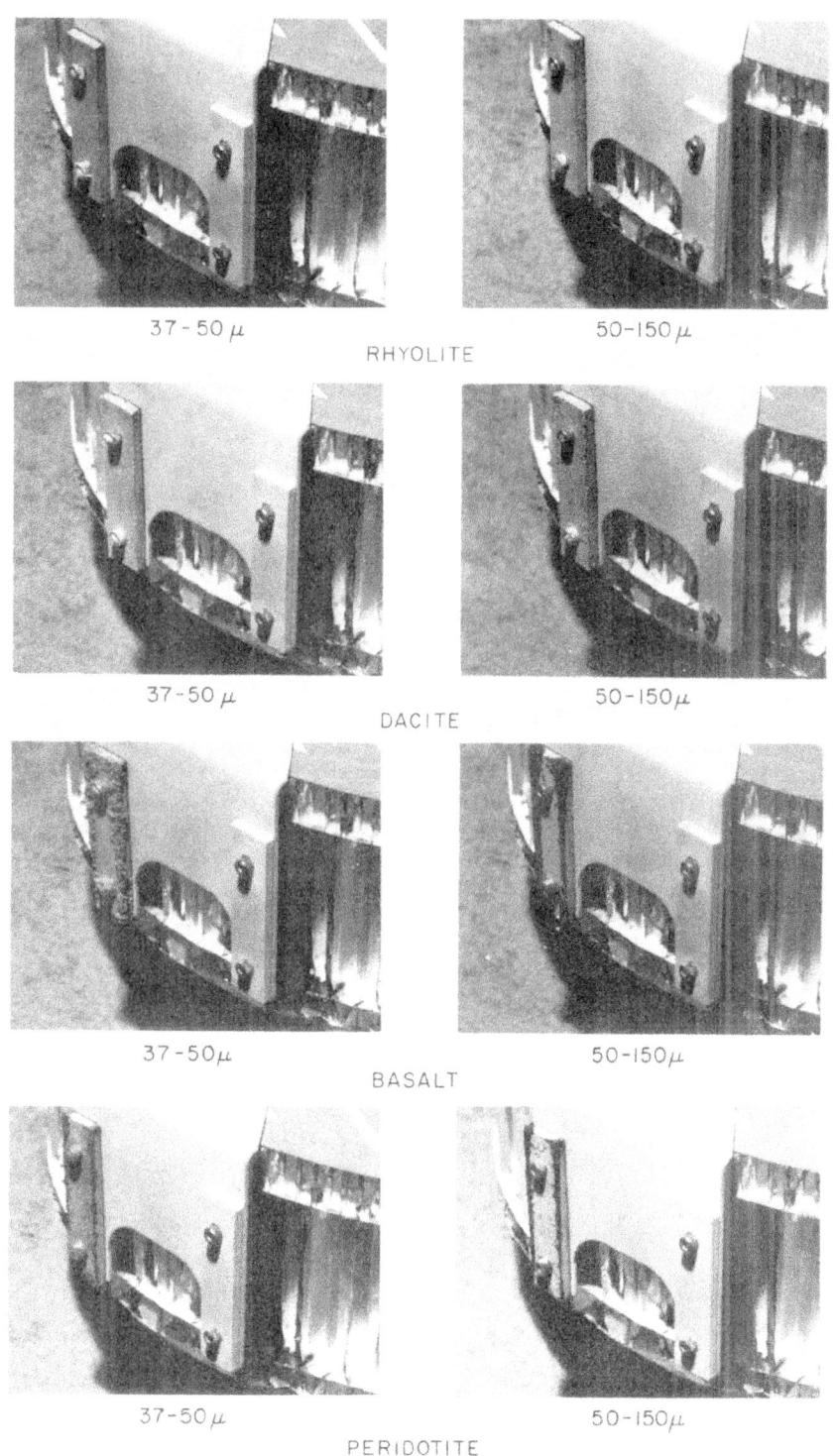

37-50 μ 50-150 μ
RHYOLITE

37-50 μ 50-150 μ
DACITE

37-50 μ 50-150 μ
BASALT

37-50 μ 50-150 μ
PERIDOTITE

FIGURE 7-36.—Continued. (b) Magnet-assembly contact with powdered basalt with added powdered iron in atmosphere.

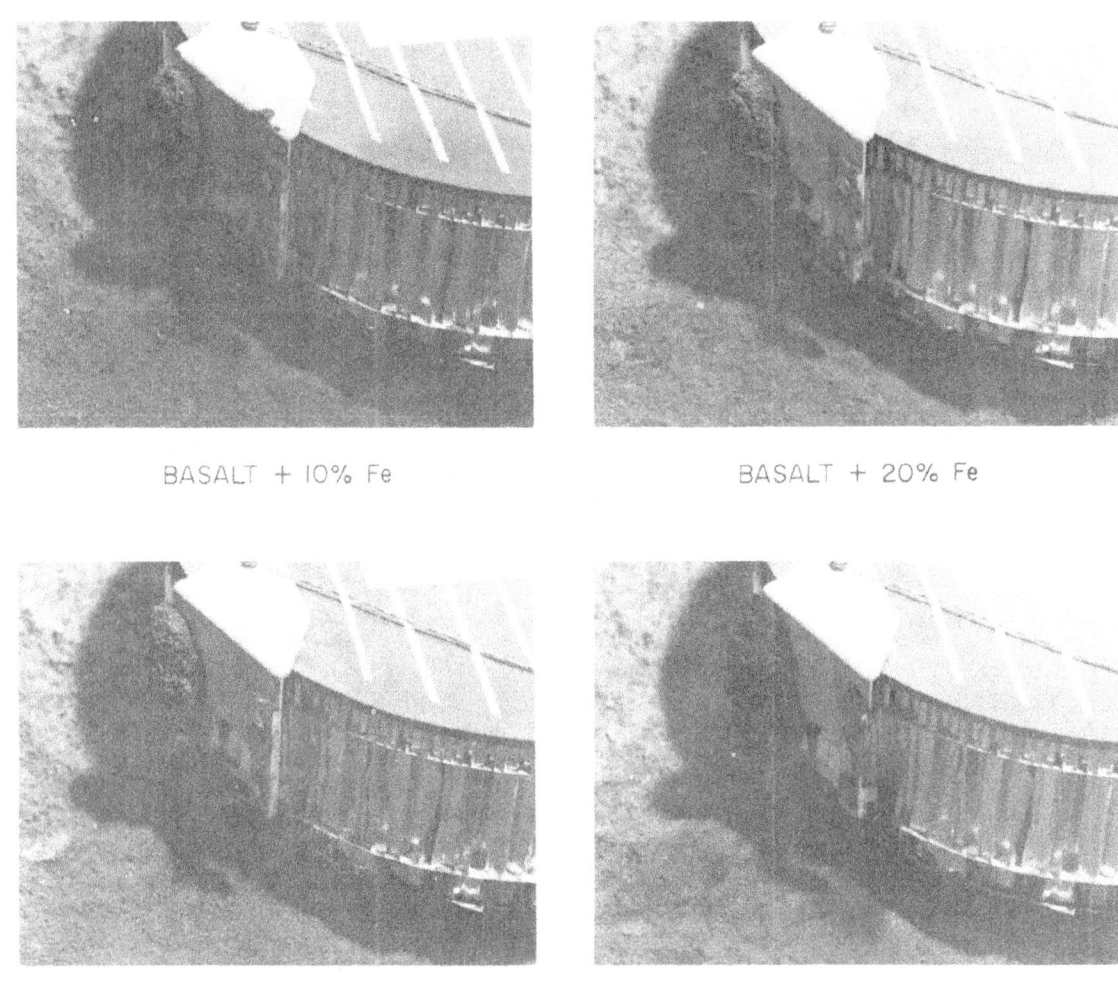

FIGURE 7-36.—Continued. (c) Magnet-assembly contact with powdered basalt with added powdered iron in atmosphere.

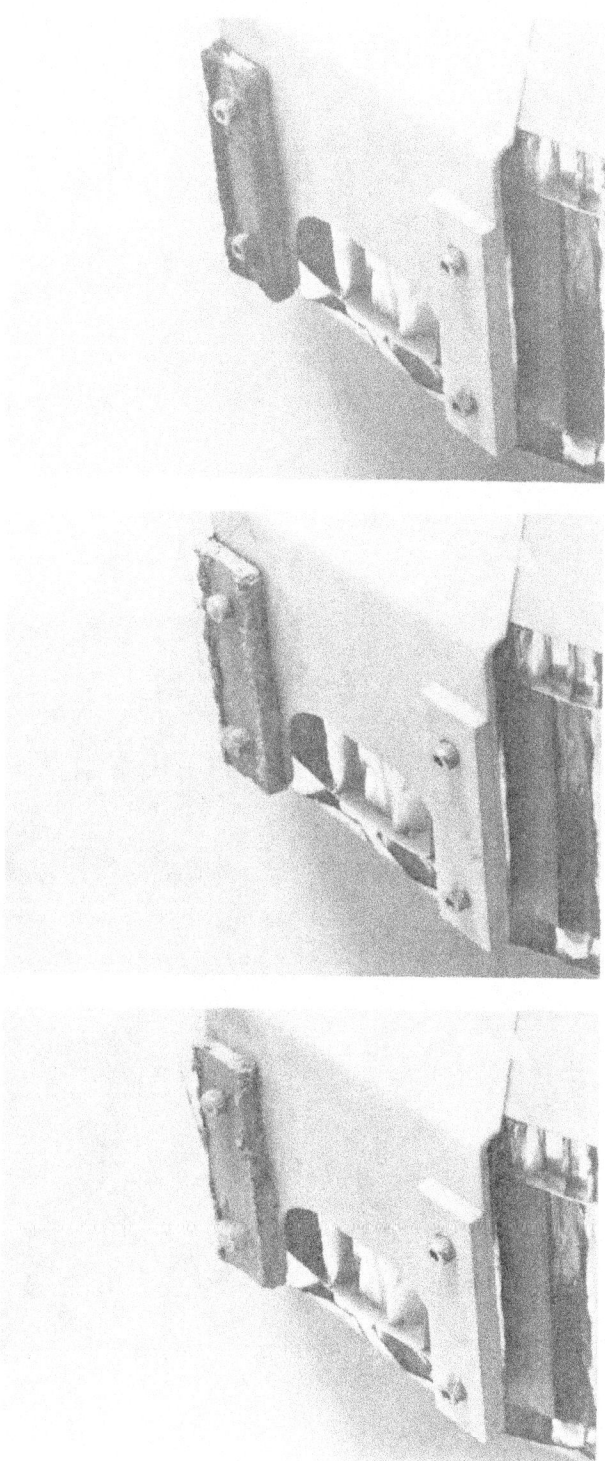

FIGURE 7-36.—Continued. (d) Magnet-assembly contact with powdered basalt in atmosphere. Top: 37 microns; middle, 37 to 50 microns; bottom: 50 to 150 microns.

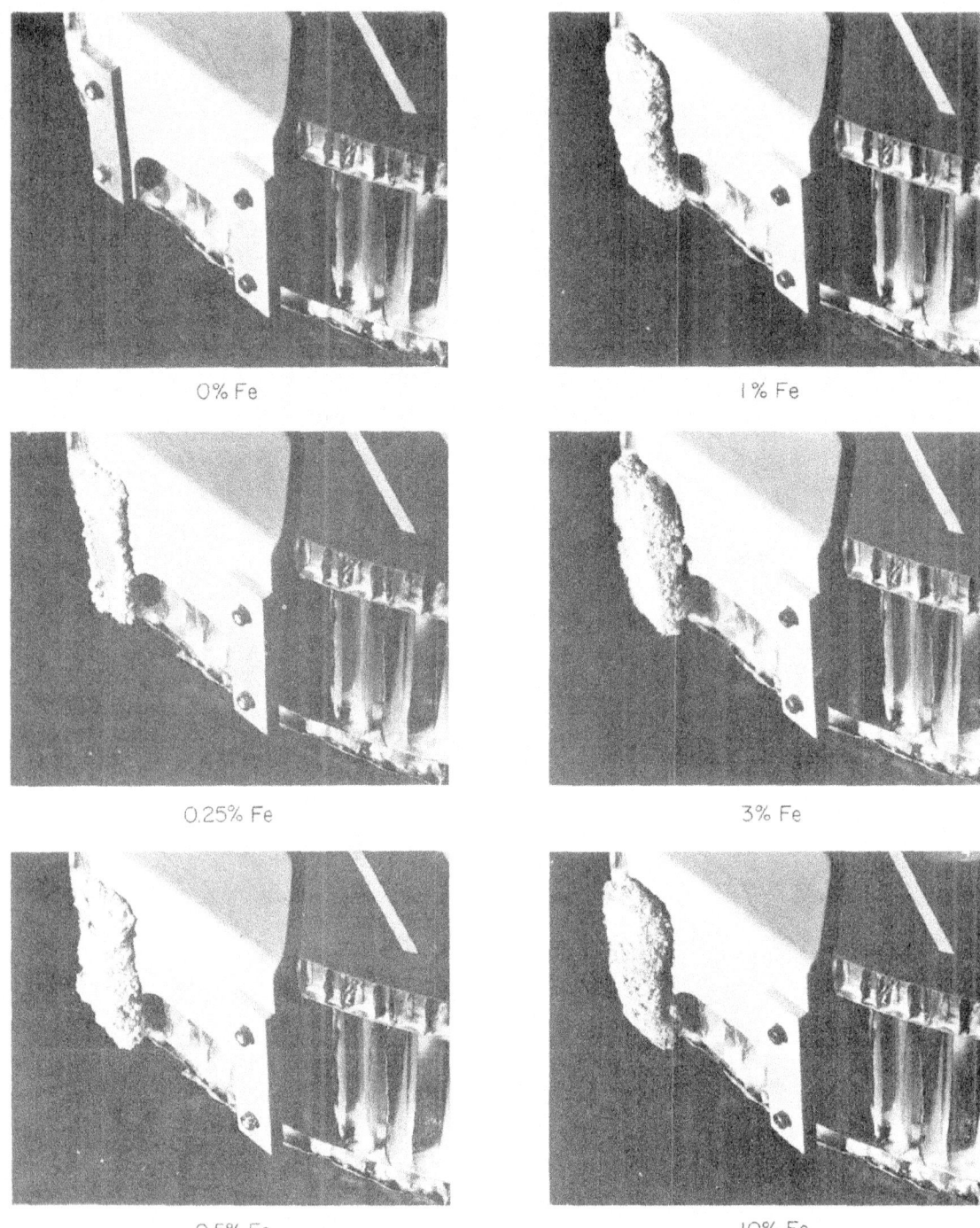

FIGURE 7-36.—Continued. (e) Magnet-assembly contact with powdered CaOH with additions of powdered iron in atmosphere.

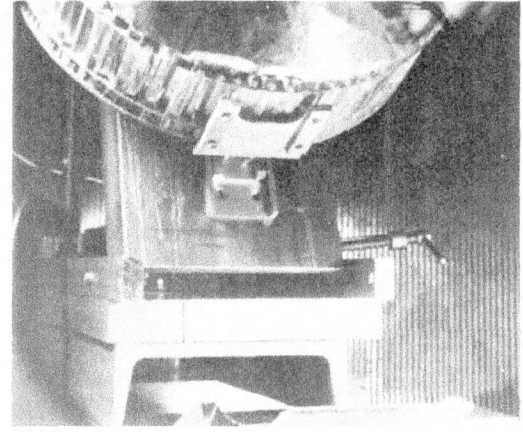
100% BASALT

BASALT + 2% Fe

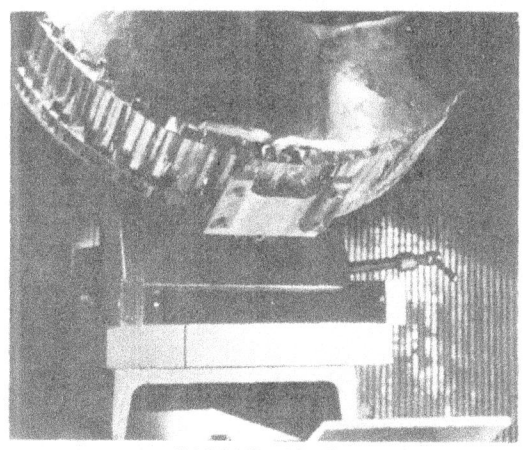
BASALT + 1% Fe

BASALT + 5% Fe

FIGURE 7-36.—Continued. (f) Magnet-assembly contact with powdered basalt (37 to 50 microns) with additions of powdered iron in 10^{-6}-torr vacuum.

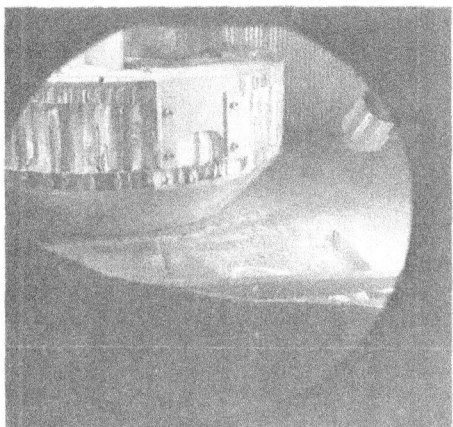
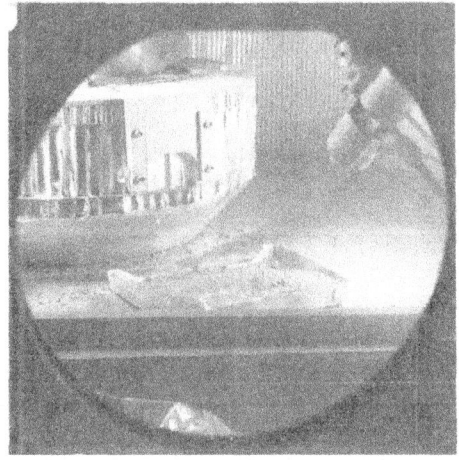

FIGURE 7-36.—Continued. (g) Results of firing the attitude-control jet into 37 to 50 microns basalt with varying percentages of iron in 10^{-6}-torr vacuum. Top, before firing; middle, after firing into 37 to 50 microns basalt with 1 percent added iron; bottom, after firing into 37 to 50 microns basalt with 4 percent added iron.

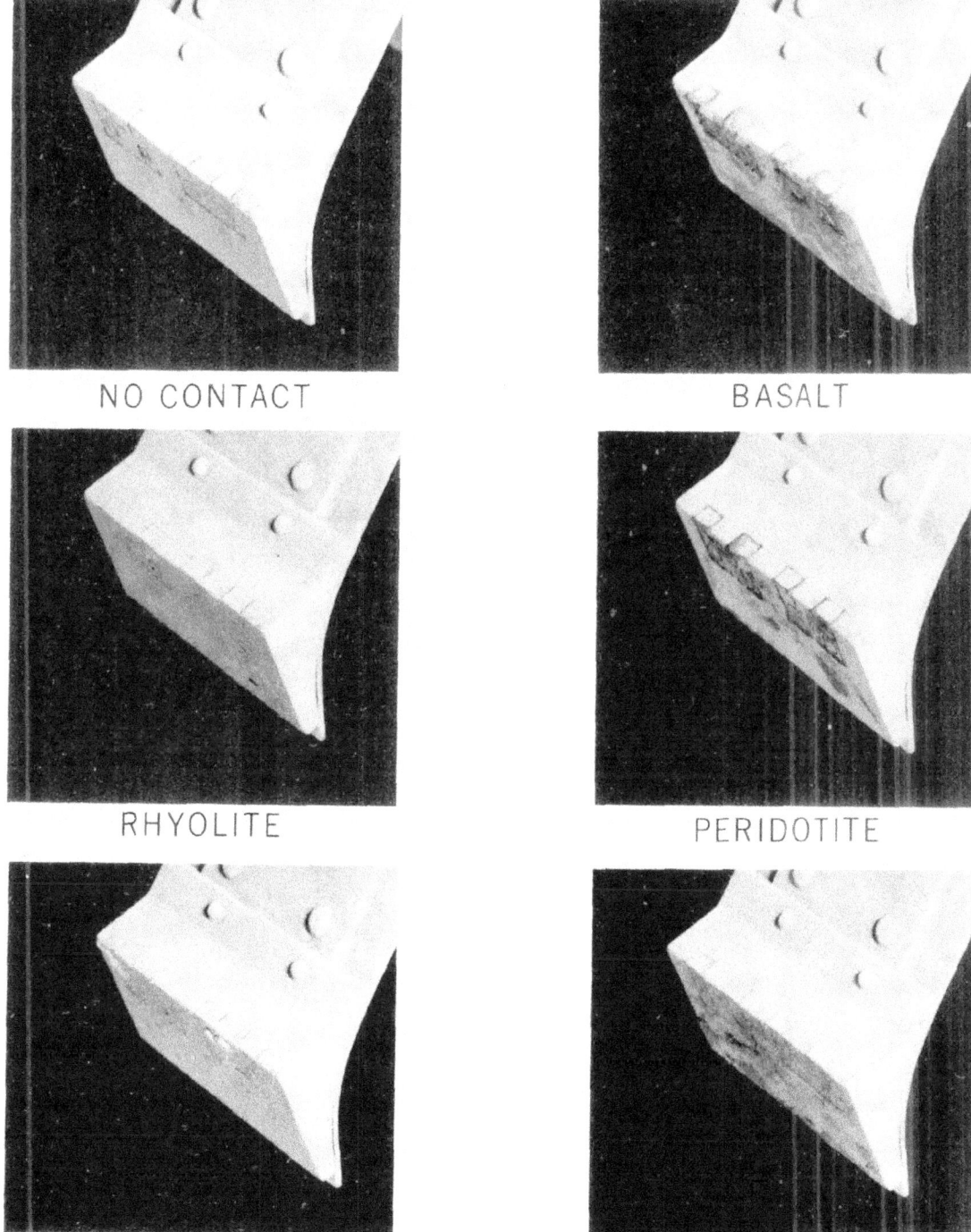

FIGURE 7-36.—Continued. (h) Surface-sampler magnets after contact with various powdered rocks in atmosphere.

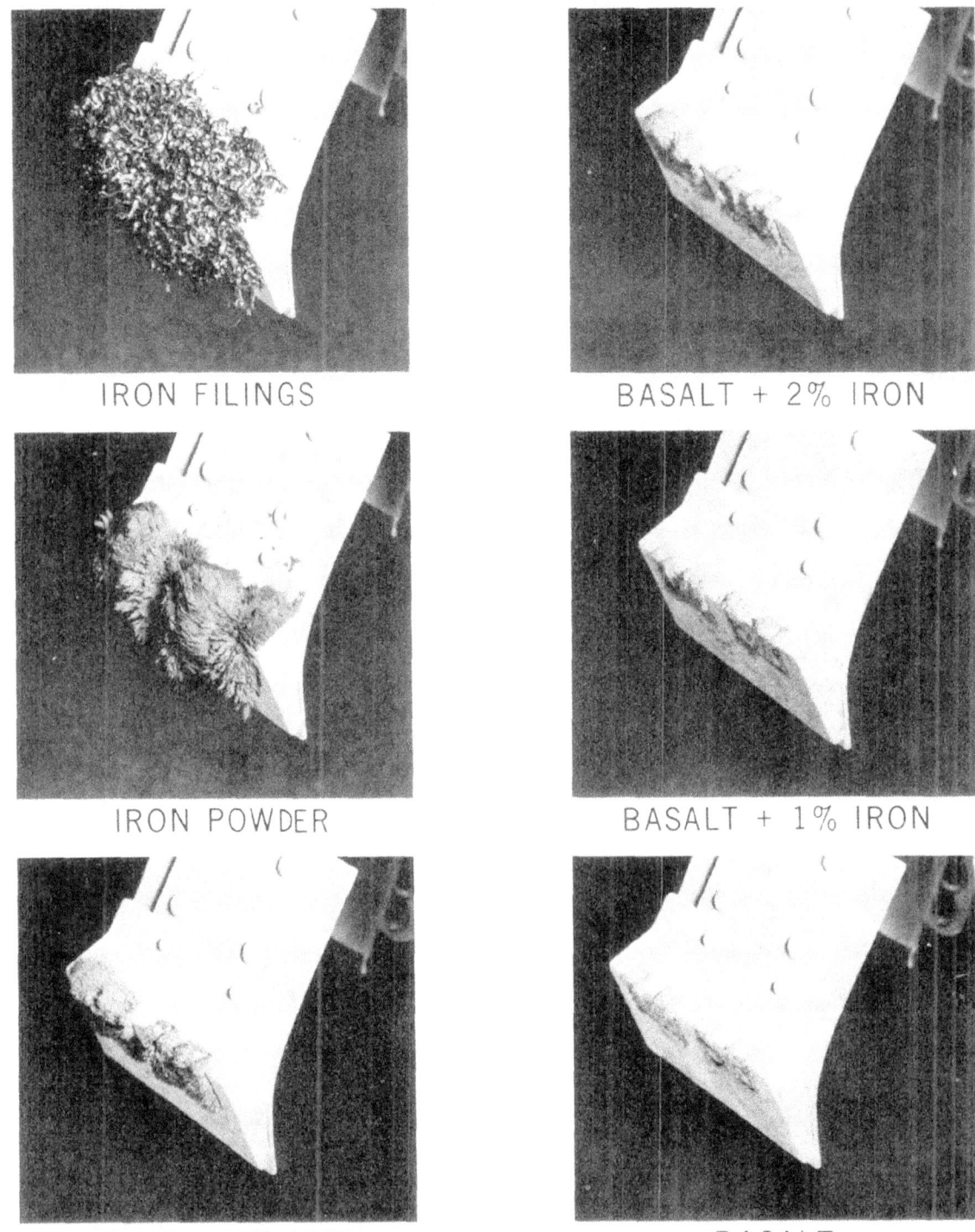

FIGURE 7-36.—Continued. (i) Surface-sampler magnets after contact with various sizes of iron particles and basalt with and without iron additions.

ELECTROMAGNETIC PROPERTIES

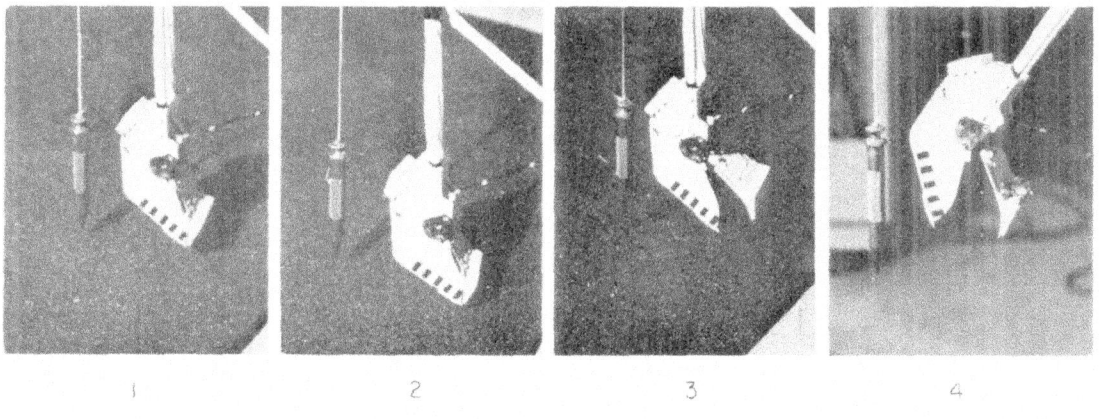

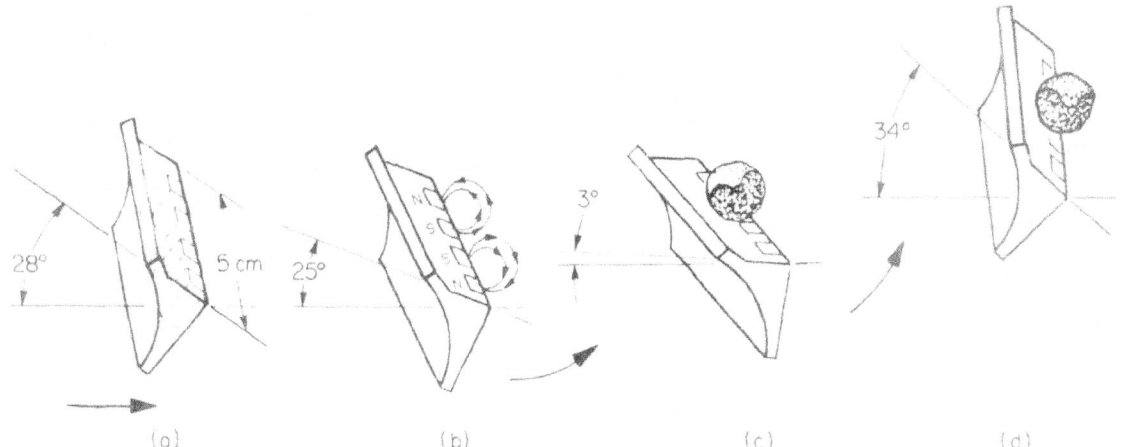

FIGURE 7-36.—Concluded. (j) Laboratory simulation of angular relationships of surface-sampler magnets to the lunar surface during pickup of magnetic fragment. Note plumb bob.

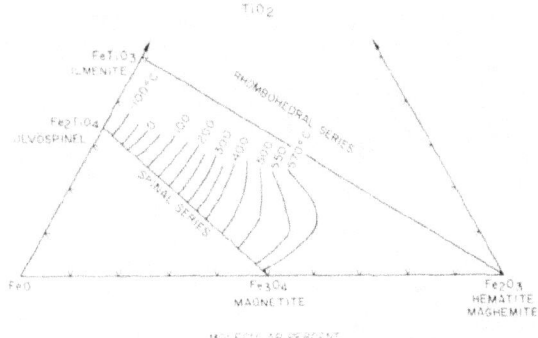

FIGURE 7-37.—Phase diagram showing FeO-Fe_2O_3-TiO_2 relationships. Only material on the lower Fe_2TiO_4-Fe_3O_4 composition line would be attracted by the Surveyor magnets. Curie temperatures are indicated for various compositions.

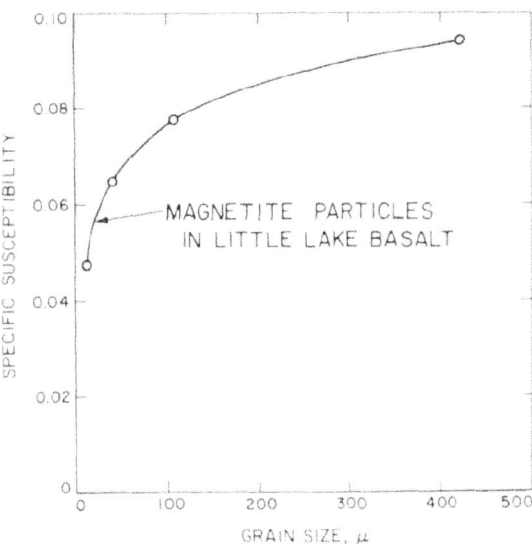

FIGURE 7-38.—Magnetite particle size versus magnetic susceptibility.

Laboratory studies in which a magnet with a magnetic flux strength of 600 gauss was placed in contact with various powdered-rock types show that the amount of material adhering to the magnet has a direct correlation with the magnetite content, as would be expected from the magnetic susceptibility plot.

The material that adhered to the magnet in these tests is not all magnetite, but a mixture of about the same proportions as the powder being tested. Thus, the material on the magnet has a similar albedo, texture, grain-size distribution, and overall appearance to the parent sample. Practically no material is attracted by the magnet in contact with rock-powder samples with less magnetite than that found in basaltic rocks (rhyolite and dacite powders). When present, magnetite usually occurs as very finely disseminated particles throughout the rock.

The other extreme to be considered is that all magnetic material in a sample occurs as free particles not embedded in the rock grains. This situation would be more analogous to meteoritic infall, causing addition of magnetic kamacite and taenite particles. Tests in which powdered iron was added to CaOH powder closely resemble this situation. From comparison with lunar results, a volumetric addition of 0.25 percent iron powder is an extreme upper limit for the amount of free magnetic material that could be present at the lunar sites tested, if no inherent magnetite were present. Thus, it becomes obvious that the meteoritic nickel-iron addition to the lunar sites must be very small and, in fact, is not in evidence in the magnet tests if the similarity of the lunar tests to a contact with basalt powder with no iron addition is accepted.

The presence of magnetic material is thus established in the mare areas tested and in the landing site north of Tycho. From comparison with the various laboratory studies, a basaltic composition is suggested, and thus a basaltic range for magnetite content and magnetic susceptibility. A magnetite content of 8 ± 2 percent is suggested, which indicates a magnetic susceptibility of about 2×10^{-2}-cgs units (see fig. 7-24).

Without considering the differences in landing modes and testing methods, the Surveyor V landing site (Mare Tranquillitatis) would appear to have more magnetic material to a depth of about 10 cm than the Surveyor VI site (Sinus Medii) or the Surveyor VII site (Tycho). On the same basis, the Surveyor VI site would appear to have the least amount of magnetic material to a depth of about 8 cm.

Still not considering the testing-mode differences, in the tests of the surface-sampler magnets on Surveyor VII, the magnetic material appears to increase considerably with depth from surface tests to trench bottom (about 20 cm; see figs. 7-32 and 7-33).

However, if test modes are considered, the increase in amount of the material on the Surveyor V magnet over that displayed by Surveyor VI is explained adequately by dragging a magnet 1 meter through the material rather than exposing the magnet to a momentary contact.

In the case of the surface-sampler magnets, both the mode of testing and possibly the packing of material may be involved. In the bearing strength tests, the surface sampler was pushed into the surface to a depth of about 5 cm. Trenching reached a depth of about 20 cm, which could conceivably represent a sample with different particle packing. However, no indication of this was seen in the power needed for trench passes at the bottom

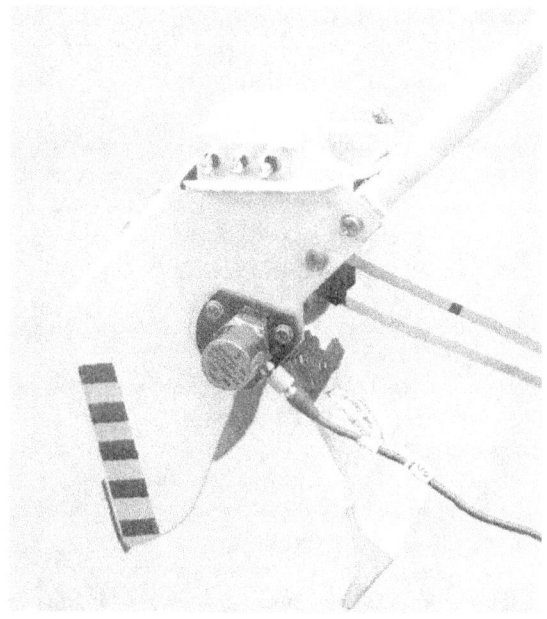

FIGURE 7-39.—Accelerometer attached to surface sampler to determine forces exerted on rock.

FIGURE 7-40.—Movement of surface sampler. (a) During rock pickup operation. (b) Subsequent movement that caused the object to be dislodged. The numbers indicate the sequence of command. Movements are quite jerky.

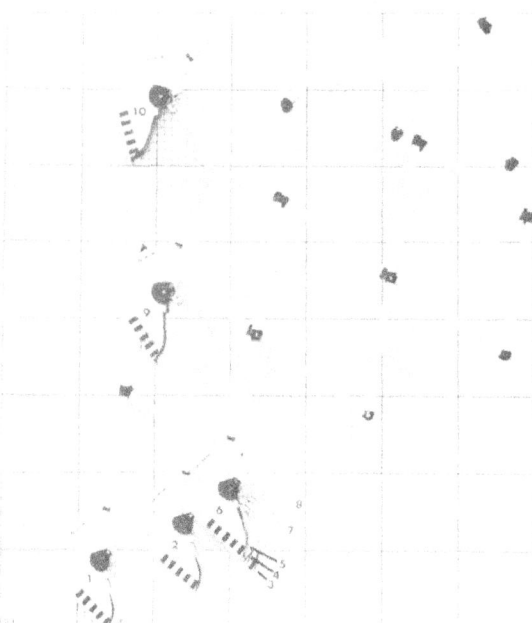
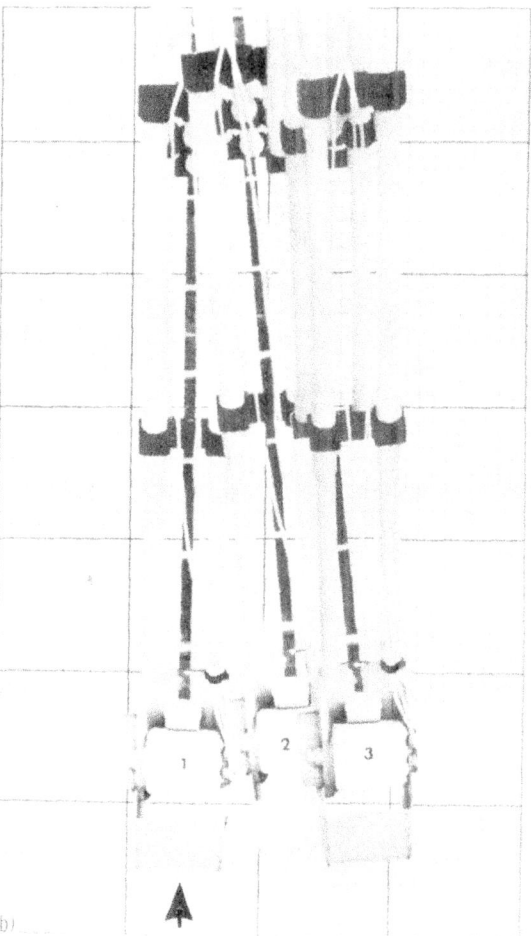

in comparison to surface passes. A 10-percent difference in packing probably is necessary to cause a difference in power requirements.[7]

The effects of particle-packing differences are under laboratory study using basaltic powders with magnetite contents spanning the range of magnetic susceptibility for basalts. Finer fractions of iron-powder additions and powder additions of assumed meteoritic flux compositions also are being studied.

Magnetic attraction of vapor deposits of iron on silicate particles has been investigated with iron thicknesses of a few angstroms on silicate grains 25 microns in diameter. No attraction was observed. Thus, the presence of such a deposition presumably could not be tested by the Surveyor magnets.

In rock form, rather than powder, the magnets will attract material with magnetic susceptibilities approaching unity or higher; i.e., magnetite, nickel-iron meteorites, and pure iron. On the Moon, objects up to approximately 12 cm could be attracted and suspended by the two small horseshoe magnets on the surface sampler. The object picked up by the surface sampler at the Surveyor VII landing site was about 1.2 cm in diameter; it was suspended by the two adjacent south poles at an angle of about 35° to the horizon, and was thrown off by a side movement of the surface sampler. From tests conducted with an accelerometer attached to the surface sampler, the force that dislodged the fragment was concluded to be about 5×10^2 dynes (see figs. 7–39 and 7–40). The force needed to suspend it was about 1.8×10 dynes. Thus, the magnetic susceptibility lies between 3×10^{-1} and 8×10^{-1} cgs units. Studies conducted with meteoritic fragments, rocks, and minerals also indicate that the susceptibility must be about 1, suggesting that the object could be composed of magnetite or meteoritic nickel-iron.

A comparison of the lunar-magnet test results with other data from the landing sites is of geologic interest in the interpretation of the magnet data. As observed from photographs taken with red and green filters (ref. 7–4), all mare landing sites are in areas that appear reddish. The Surveyor VI landing site is more mottled than that of Surveyor V, but distinctly shows higher reflectance in the red. The Surveyor VII site, north of Tycho, is in a reddish halo around that crater, suggesting that the surface material there may not be typical of highlands. Most lunar highlands are very light to white in the composite photograph referenced. Thus, from this reflectance work alone, a similarity in material at all Surveyor landing sites is suggested.

Panoramas of the mare landing sites and the Tycho rim (fig. 7–41) show an apparent increase in rock distribution in the Tycho area, and a perceptible increase in rolling hills and steeper slopes. However, the gross morphology appears to be somewhat similar.

Rock shapes in the maria as well as on the Tycho rim range from rounded and porous to blocky, dense, and angular. Light and dark rocks are seen in both types of areas, with lighter rocks predominant. Rock shapes typical of volcanic activity are seen in all locations (see fig. 7–42). Such shapes range from those suggestive of ash and lappilli to densely crystalline igneous rocks, which would be expected as throwout material at some depth with volcanic activity. Experiments with magma in a high vacuum suggest that any magma extruded on the lunar surface would be extremely porous (ref. 7–16).

A consideration of surface-rock distribution and subsurface cross sections is pertinent. All footpad imprints and trenches reveal a very fine-grained, homogeneous material with no particles larger than 1 mm; i.e., camera resolution. No perceptible packing difference is observed in the bottoms of trenches (see footnote 7). Rock fragments are distributed over the surface in all states of burial, from balanced perch positions to near burial. It might be expected that this areal distribution would be similar at near-surface depths, or that its density would diminish with depth. However, no rocks appear to be present below the surface to the depths viewed (fig. 7–42). This is more suggestive of depositions of volcanic ash and later of rocks, rather than of churning and pulverization by bombardment.

---

[7] F. I. Roberson, private communication, 1968.

ELECTROMAGNETIC PROPERTIES

FIGURE 7-41.—The panoramas of the mare areas were quite similar. Although more rocks, hills, and steeper slopes occur north of Tycho, the general morphology is similar to that of the maria. (a) Surveyor I.

FIGURE 7-41.—Continued. (b) Surveyor III.

ELECTROMAGNETIC PROPERTIES

FIGURE 7-41.—Continued. (c) Surveyor V.

FIGURE 7-41.—Continued. (d) Surveyor VI.

FIGURE 7-41.—Continued. (c) Surveyor VII.

258 SURVEYOR: PROGRAM RESULTS

FIGURE 7-41.—Concluded. (e)—Concluded.

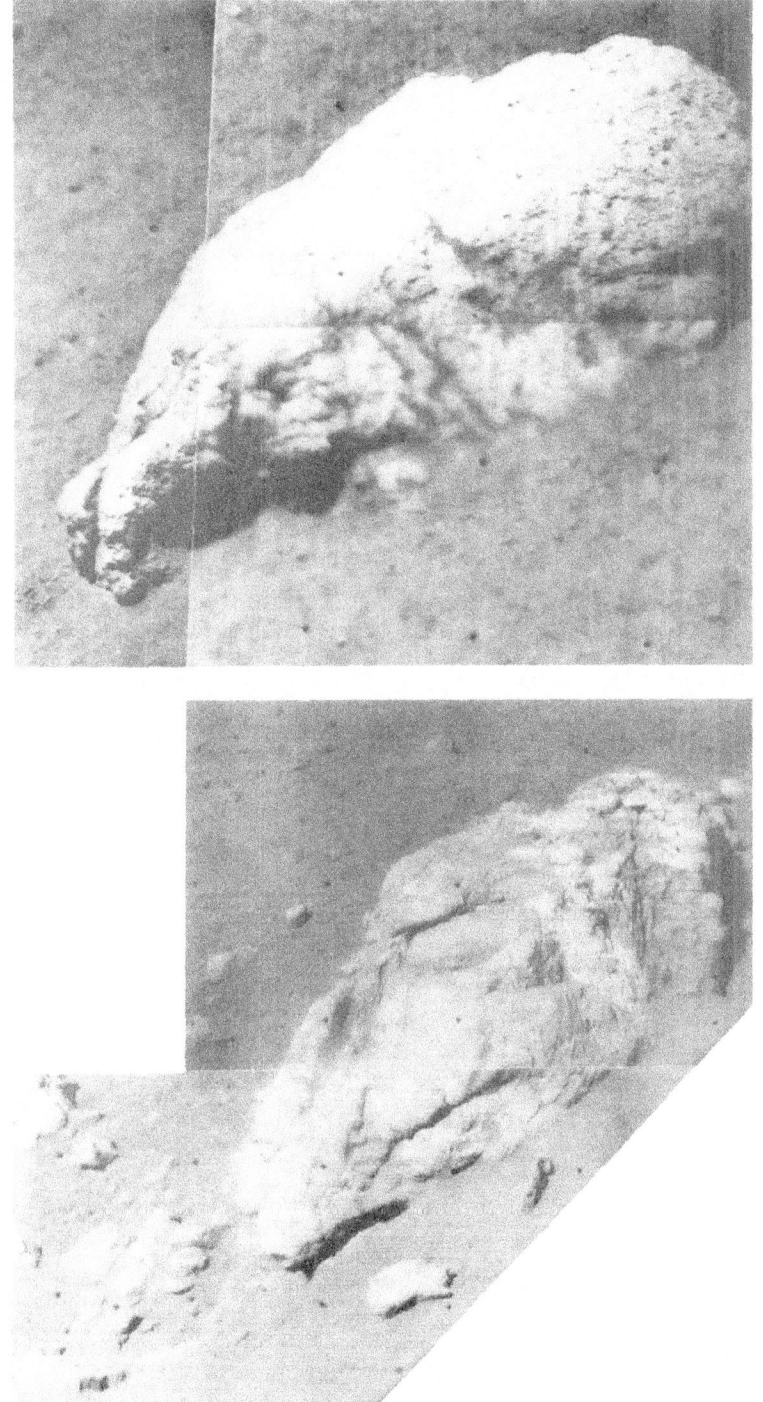

FIGURE 7-42.—Rocks with morphology were present at all Surveyor landing sites. (a) Surveyor I. Top: note the vesicular structure; bottom, note the lavalike structure, basaltlike pattern of breakage, tubelike fragments.

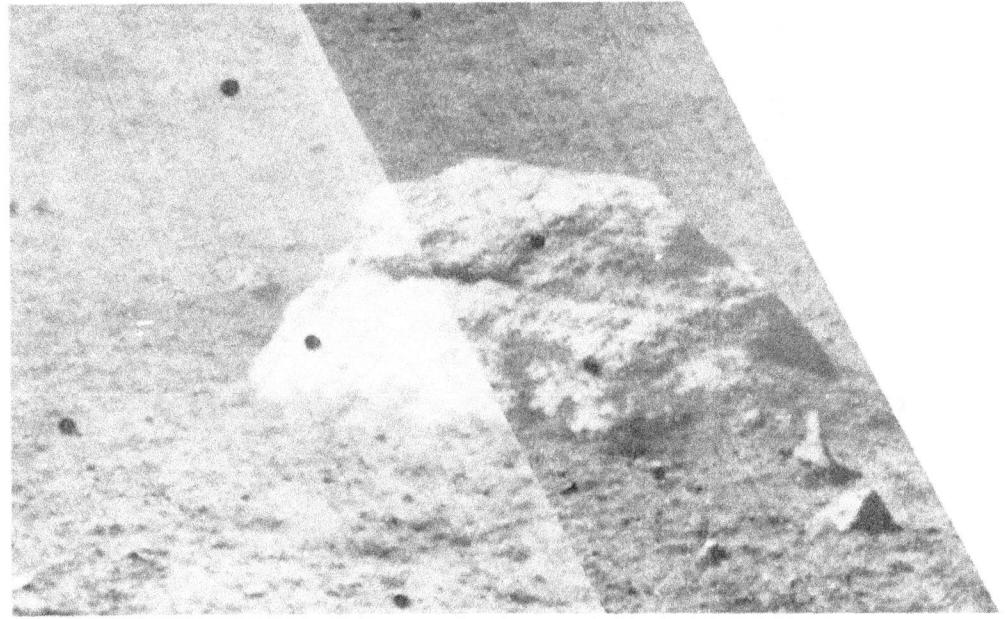

FIGURE 7-42.—Continued. (b) Surveyor III. Note the vesicular structure in both rocks.

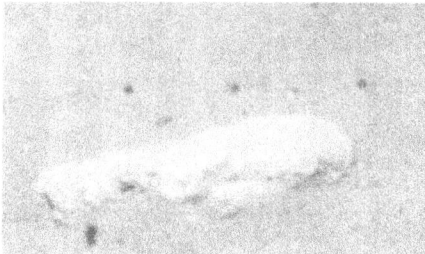

FIGURE 7-42.—Continued. (c) Surveyor V. Both of these rocks may be vesicular.

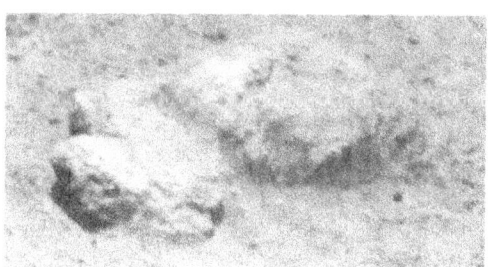

FIGURE 7-42.—Continued. (d) Surveyor VI. The picture at the right is computer processed. Both of these rocks may be vesicular.

FIGURE 7-42.—Concluded. (e) Surveyor VII. Top left: note the light flecks that may be phenocrysts in a darker matrix; bottom left: note the fracture typical of igneous rocks; above: rock rubble showing morphology typical of terrestrial basalt.

From the mare sites, the alpha-scattering instruments produced data that were compatible with basaltic composition (refs. 7-17 and 7-18). A decrease by about a factor of 2 in heavy elements and a slight increase in aluminum were observed at the Surveyor VII site (ref. 7-19). Such a decrease in heavy elements, if because of iron content, is not observable in the magnetic iron data. Alpha-scattering data may be reflecting contamination of one rock type (basalt) with another more acidic-rock type. The numerous rocks scattered on the surface are predominantly lighter than the surface layer on which they rest. Dust from these rocks is undoubtedly present and would probably show much lower iron content. Rocks turned over by the surface sampler evidence rounding and weathering only on exposed rock tops. The depth to which the surface sampler penetrated in bearing strength tests was about 5 cm. All the material subsurface appears to be homogeneous, fine-grained, darker material (see fig. 7-43). The Surveyor VII magnet contacts represent tests of this dark subsurface material, and are very similar to the tests from the maria and to laboratory tests in basalt.

Hapke (ref. 7-20) has compared the optical characteristics of the lunar surface with those of rocks and meteorite powders. Intensity, polarization, spectrum, and albedo of the Moon were matched only by those of basic rocks containing lattice iron, but little or no free iron. These studies are consistent with extrapolation of conclusions from the Surveyor mare landing sites to the mare areas in general.

Conclusions

From comparison of the lunar results with laboratory studies, the following conclusions are possible:

(1) A small amount of magnetic material is present at all Surveyor landing sites tested (Mare Tranquillitatis, Sinus Medii, and north of the crater Tycho) (fig. 7-38). The amount of magnetic material, if existing as free particles, is less than 0.25 percent by volume.

(2) All lunar results are consistent with laboratory studies using finely powdered basalts. The Little Lake basalt, which was used in laboratory studies, contains about 10 to 12 percent by volume, magnetite disseminated in grains about 25 microns in diameter. Studies in various powder sizes of basalt show that slight differences in particle sizes at the landing sites and in landing modes are sufficient to explain the relatively slight differences in the amount of magnetic material observed on the lunar magnets.

(3) The object attracted by the surface-sampler magnets is probably magnetite or meteoritic nickel-iron.

It is the author's opinion that a volcanic source is suggested for the lunar surface material. Because of the lack of an indication of any powdered meteoritic nickel-iron and because of the uniform, fine-grained surface material in all areas, both in footpad imprints and trench walls, the surface material probably does not represent a meteoritic pulverization product. This suggests that the material was either fine grained at the time of deposition, i.e., ash, or was pulverized by a process not yet identified. The thermal regime and solar wind are possible contributors to this process.

Meteoritic impact undoubtedly occurs on the Moon. However, any addition of meteoritic nickel-iron that may be present at the Surveyor landing sites must be less than amounts perceivable by the methods employed.

FIGURE 7-43.—Comparison of material in imprints, trenches, and slide walls show that the subsurface material is fine grained and homogeneous, with no large fragments. (a) Surveyor 1. This computer-processed picture shows the imprint made by footpad 2.

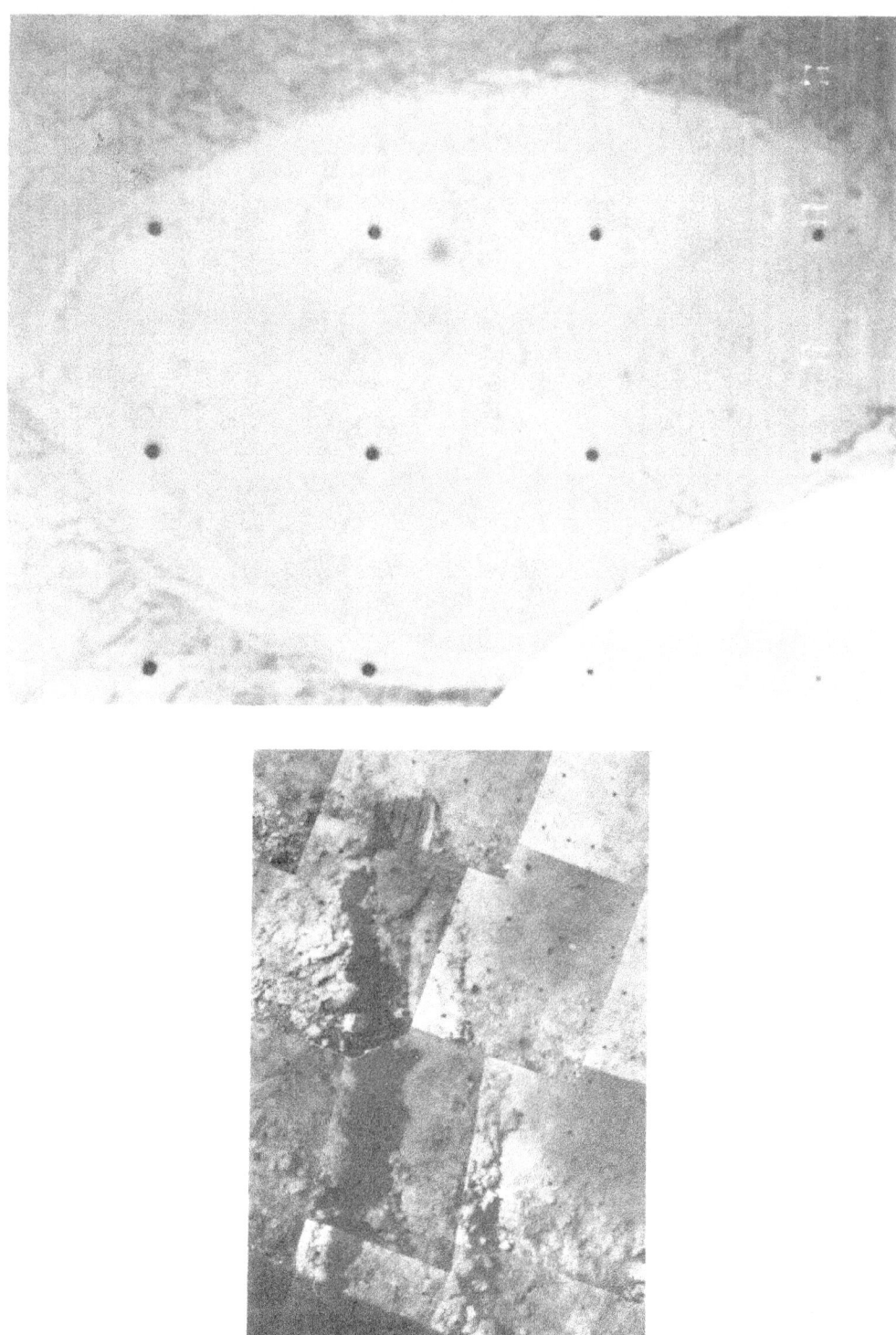

FIGURE 7-43.—Continued. (b) Surveyor III. Top: footpad 2 imprint; bottom: note the smooth trench wall.

FIGURE 7-43.—Continued. (c) Surveyor V. Note the smooth imprint and slide wall. Also see figure 7-26(a).

FIGURE 7-43.—Continued. (d) Surveyor VI. Top: imprint of footpad 2 made during initial landing; bottom: imprint of footpad 2 made during the hop.

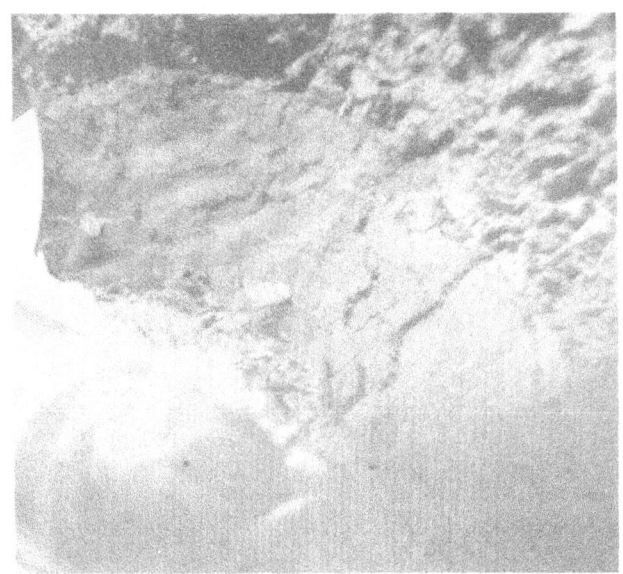

FIGURE 7-43.—Concluded. (e) Surveyor VII. Top: imprint of footpad 2; bottom: closeup of a trench wall. All trench walls showed a similar smooth texture (see fig. 7-33).

References

7-1. BROWN, W. E., JR.: Lunar Surface Surveyor Radar Response. J. Geophys. Res., vol. 72, no. 1, Jan. 15, 1967.

7-2. BROWN, W. E., JR.; DIBOS, R. A.; GIBSON, G. B.; MUHLEMAN, D. O.; PEAKE, W. H.; and POEHLS, V. J.: Lunar Surface Electrical Properties. Surveyor III Mission Report. Part II: Scientific Results, Tech. Rept. 32-1177, Jet Propulsion Laboratory, Pasadena, Calif., June 1, 1967, pp. 189-194.

7-3. MUHLEMAN, D. O.; BROWN, W. E., JR.; and DAVIDS, L.: Radar Reflectivity Analysis. Surveyor VII Mission Report. Part II: Science Results, Tech. Rept. 32-1264, Jet Propulsion Laboratory, Pasadena, Calif., Mar. 15, 1968, pp. 32-1264.

7-4. MIETHE, A.; and SEEGERT, T. B.: Schmidt Reflector Photograph Using Two Color Filters, 1911.

7-5. SHORTHILL, R. W.; and SAARI, G. M.: Recent Results of Lunar Eclipse Measurements Showing Hot Spots. Advances in Astro-Science, vol. 20, 1967, pp. 545-556.

7-6. MIDDLEHURST, B. M.; and BURLEY, J. M.: TR-X 641-66-178, Goddard Space Flight Center, Greenbelt, Md., 1966.

7-7. KOSYREV, N. A.: The Moon, 1962, pp. 263-274.

7-8. DALEY, R. A.: Igneous Rocks and the Depth of the Earth. McGraw-Hill Book Co., Inc., New York, 1933.

7-9. MELOY, T.; and O'KEEFE, J. A.: Size Distribution of Lunar Surface Materials. J. Geophys. Res., vol. 73, no. 6, Mar. 15, 1968, pp. 2229-2301.

7-10. CLARK, S. P., JR., ed.: Handbook of Physical Constants. Memo. 97, Yale University, New Haven, Conn., 1966.

7-11. BALSLEY, J. R.; and BUDDINGTON, A. F.: Iron-Titanium Oxide Minerals, Rocks, and Aeromagnetic Anomalies of the Adirondack Area, New York. Economic Geol., vol. 53, no. 7, 1958, pp. 777-805.

7-12. DE WYS, J. NEGUS: Surveyor V Magnet Experiment. Science, vol. 158, no. 3801, Nov. 3, 1967, pp. 632-633.

7-13. DE WYS, J. NEGUS: Lunar Surface Electromagnetic Properties: Magnet Experiment. Surveyor V Mission Report. Part II: Science Results, Tech. Rept. 32-1246, Jet Propulsion Laboratory, Pasadena, Calif., Nov. 1, 1967, pp. 151-175.

7-14. DE WYS, J. NEGUS: Electromagnetic Properties: Magnet Test. Surveyor VI Mission Report. Part II: Science Results, Tech. Rept. 32-1262, Jet Propulsion Laboratory, Pasadena, Calif., Jan. 10, 1968, pp. 155-169.

7-15. DE WYS, J. NEGUS: Lunar Surface Electromagnetic Properties. Surveyor VII Mission Report. Part II: Science Results, Tech. Rept. 32-1264, Jet Propulsion Laboratory, Pasadena, Calif., Mar. 15, 1968, pp. 224-240.

7-16. DOBAR, W. I.: Simulated Basalt and Granite Magma in Vacuum. Icarus, vol. 5, 1966, pp. 399-405.

7-17. TURKEVICH, A. L.; FRANZGROTE, E. J.; and PATTERSON, J. H.: Chemical Analysis of the Moon at Surveyor V Landing Site: Preliminary Results. Surveyor V Mission Report. Part II: Science Results, Tech. Rept. 32-1246, Jet Propulsion Laboratory, Pasadena, Calif., Nov. 1, 1967, pp. 119-149.

7-18. TURKEVICH, A. L.; FRANZGROTE, E. J.; and PATTERSON, J. H.: Chemical Analysis of the Moon at Surveyor VI Landing Site: Preliminary Results. Surveyor VI Mission Report. Part II: Science Results, Tech. Rept. 32-1262, Jet Propulsion Laboratory, Pasadena, Calif., Jan. 10, 1968, pp. 127-153.

7-19. FRANZGROTE, E. J.; PATTERSON, J. H.; and TURKEVICH, A. L.: Chemical Analysis of the Moon at the Surveyor VII Landing Site: Preliminary Results. Surveyor VII Mission Report. Part II: Science Results, Tech. Rept. 32-1264, Jet Propulsion Laboratory, Pasadena, Calif., Mar. 15, 1968, pp. 241-266.

7-20. HAPKE, B.: Lunar Surface Composition Inferred From Optical Properties. Science, vol. 159, Jan. 5, 1968.

ACKNOWLEDGMENT

Appreciation is extended to G. MASTERS for his successful programing of the radar data processing for the Science Computer Facility; to V. POEHLS and R. HARRINGTON for material concerning the radar system presented in this section of this report; and to L. DAVIDS, who utilized the six-degree-of-freedom program.

8. The Alpha-Scattering Chemical Analysis Experiment on the Surveyor Lunar Missions

A. L. Turkevich (Principal Investigator), W. A. Anderson, T. E. Economou, E. J. Franzgrote, H. E. Griffin, S. L. Grotch, J. H. Patterson, and K. P. Sowinski[1]

Surveyors V, VI, and VII carried alpha-scattering instruments to obtain chemical analyses of lunar surface material. The lunar locations of these analyses are indicated in figure 8-1. At the Surveyor V landing site (Mare Tranquillitatis), two samples were analyzed; one sample was analyzed at the Surveyor VI site (Sinus Medii), another mare region. With the cooperation of the soil mechanics surface sampler experiment, three samples were analyzed at the Surveyor VII terra landing site (near the crater Tycho). Values derived from analyses of the first sample on each of these missions have been reported in references 8-1 through 8-3. The preliminary results given in these references were obtained from portions of the data received by Teletype. These data were analyzed by computers programed for real-time mission support using a library of only eight elements. These preliminary analyses permitted only partial temperature and gain-shift corrections, and made no provision for nonstandard sample geometry. The results, though assigned rather large errors at present, represent the first direct chemical analyses of an extraterrestrial body, permitting significant conclusions to be drawn concerning the nature of the lunar surface material and its history.

History

The alpha-scattering method of analysis makes use of the characteristic energy spectra of alpha particles rebounding at nearly 180° after collision with atomic nuclei. The energy spectra of protons, produced by nuclear reactions of alpha particles with some light elements, are also used.

The phenomenon of large-angle scattering of alpha particles by matter was first reported by Geiger and Marsden in 1909 (ref. 8-4). Rutherford (ref. 8-5) showed that this behavior could not be explained by the atomic theories of that time, and used it as a basis for his nuclear model of the atom. The relation of the fraction of the kinetic energy remaining with the alpha particle after impact, T_m/T_0, to the angle of scattering, θ, and the mass number of the nucleus, A, was derived by Darwin (ref. 8-6) on the assumptions that kinetic energy and momentum are conserved and that the scattering nucleus is initially at rest. This relation is:

$$\frac{T_m}{T_0} = \left[\frac{4 \cos \theta + (A^2 - 16 \sin^2 \theta)^{1/2}}{A+4} \right]^2 \quad (1)$$

From equation (1), the theoretical possibility of applying this phenomenon to chemical analysis is obvious. However, the lack of high-intensity sources of monoenergetic alpha particles and of convenient methods for alpha-particle-energy measurement made application impractical for several decades after these discoveries. Starting in 1950, several investigators suggested the use of charged particles from an accelerator for chemical analysis (ref. 8-7). Rubin and coworkers (ref. 8-8) developed a method of analyzing thin samples using accelerated alpha particles and protons, with a

---

[1] The affiliations of A. Turkevich, E. Franzgrote, and J. Patterson are listed in app. C of this report. With the exception of H. Griffin of the Argonne National Laboratory and S. Grotch of the Jet Propulsion Laboratory, the remaining authors are from the University of Chicago.

FIGURE 8-1.—Earth-based telescopic photograph of the Moon showing locations of the three Surveyor chemical-analysis sites.

magnetic spectrometer for energy analysis. In 1960 one of the present authors, at the suggestion of the late Prof. Samuel K. Allison of the University of Chicago, found, in a preliminary investigation, that an alpha-scattering method of analysis using isotopic alpha-particle sources was feasible (ref. 8-9). On the basis of these results, the development of an analytical instrument for the Surveyor program was started. A research apparatus containing one curium-242 alpha source and one silicon semiconductor detector was designed and used to obtain much fundamental data on the alpha-scattering response of elements and compounds (ref. 8-10).

These data showed that the minimum sensitivity of the method was for elements in the region of sodium. The measurement of the energy spectra of protons, which are obtained from nuclear interactions of alpha particles with elements in this region, was therefore introduced to improve the sensitivity for these geologically important elements. These (α, p) reactions are also associated with the name of Rutherford, who reported the first artificial transmutation of an element by such a reaction with nitrogen (ref. 8-11).

A breadboard model of an instrument for a lunar mission was built for further study of the characteristics of the alpha and proton spectra (refs. 8-10 and 8-12). Then a prototype instrument was designed to be compatible with a sample-preparation system on a lunar softlanding spacecraft. This prototype was not tested extensively because of a change in the Surveyor program, which required direct deployment of an instrument to the Moon's surface instead of the analysis of prepared samples.

A series of four flight prototypes of a deployable instrument, followed by four flight models, was originally planned. These (actually only seven instruments were built) were designed and built at the University of Chicago's Laboratory of Applied Science and later by that university's Laboratory of Astrophysics and Space Research (ref. 8-13). Prototypes 1 and 2 (P-1 and P-2) were used for developmental work on the instrument and on the method of analysis. On the basis of results from the program at this stage, postponements in the launch schedule and a clearer identification of the hazards to the experiment under lunar conditions, significant improvements in the experiment were made. These improvements included the introduction of a commandable electronic-pulse generator that could be used to calibrate the energy scale of the instrument, the introduction of a crude ratemeter to monitor the solar and cosmic-ray proton background during a mission, and a redesign of the outer envelope of the deployable part of the instrument to provide better temperature control. Prototype 3 (P-3), which incorporated these changes, was used as a type-approval model to insure the capability of the instrument to function properly under mission conditions. Shortly before the Surveyor V mission, P-3 was used for a practice loading of sources and testing at Cape Kennedy. Prototype 4 (P-4) eventually became the instrument used on Surveyor V, while flight instruments 1 and 2 (F-1 and F-2) were used on Surveyors VI and VII, respectively. Flight instrument 3 (F-3) was used as the backup or spare instrument for all three flights. More details on the instrument development and history are given in reference 8-14.

Methods

In an elemental sample only a few atoms thick, the spectrum of alpha particles scattered backward in a vacuum would consist of a narrow peak, whose energy, T_m, is determined by equation (1). For a thicker sample, alpha particles scattered at various depths below the surface suffer a loss of energy in the material before and after the scattering event. The relationship of thin and thick target spectra is shown in figure 8-2. The energy distribution of alpha particles scattered from thick samples, then, is a continuous spectrum, terminating sharply at the energy, T_m, determined by the mass number of the element (eq. 1).

Figure 8-3 shows a series of plots of T_m/T_0 as a function of scattering angle for various mass

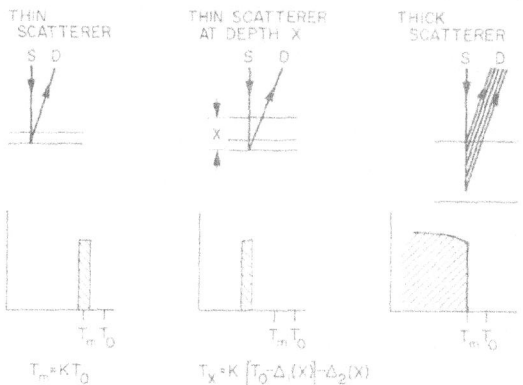

FIGURE 8-2.—Comparison of spectra of alpha particles scattered from thin and thick samples. The upper diagrams show the geometry of scattering; the lower indicate the corresponding shapes of associated spectra.

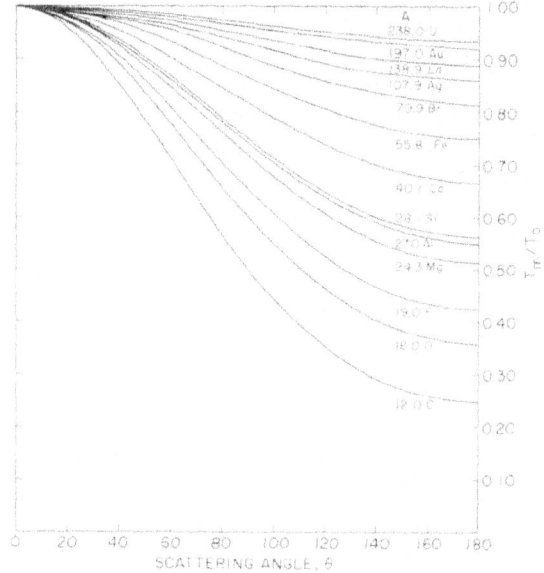

FIGURE 8-3.—Dependence of the ratio of the maximum energy of scattered alpha particles to the initial energy, T_m/T_o, on the laboratory angle of scattering, θ, and on the mass number, A.

numbers, according to equation (1). It can be seen that—

(1) The greatest resolution between mass numbers is at a scattering angle of 180° (although there is very little change above 160°).

(2) The resolution decreases rapidly with increasing mass number.

The practical resolution limit for distinguishing individual elements for the instrument used in the Surveyor program is approximately at mass number 40 (Ca).

The proton spectra from (α, p) reactions in thick samples are also characteristic of the elements making up the sample. Such reactions are energetically impossible in the major isotopes of elements such as carbon and oxygen; they have negligible yields, because of the high coulomb barrier for 6.1-MeV alpha particles, in elements heavier than about calcium. The yields of protons are relatively high for the geochemically important elements sodium, magnesium, and aluminum, which have a low scattering probability for alpha particles.

Figure 8-4 is a set of alpha and proton spectra (called a library) of most of the elements that may be expected as major constituents of common rocks. In this figure, the intensities, after background subtraction, of the scattered alpha particles and protons from various elements are plotted on a logarithmic scale as a function of the pulse-height-analyzer channel number. The intensities have been normalized to a measurement time of 1000 minutes and a source strength of 10^{11} disintegrations/min. In figure 8-4, the channel number is related to the energy deposited by the particle in the detector by the relations:

$$N_\alpha = 18.94 E_\alpha - 10.9$$
$$N_p = 18.6 E_p - 10.7$$

where N is the channel number; E is the energy, in million electron volts, deposited in the detector. These spectra were obtained in pre-

FIGURE 8-4.—A library of elemental spectra: (a) C, O; (b) B, N, F; (c) Na, Mg, Al; (d) Si, S, K; (e) Ca, Ti, Fe, Ni, Ag, Au.

CHEMICAL ANALYSIS

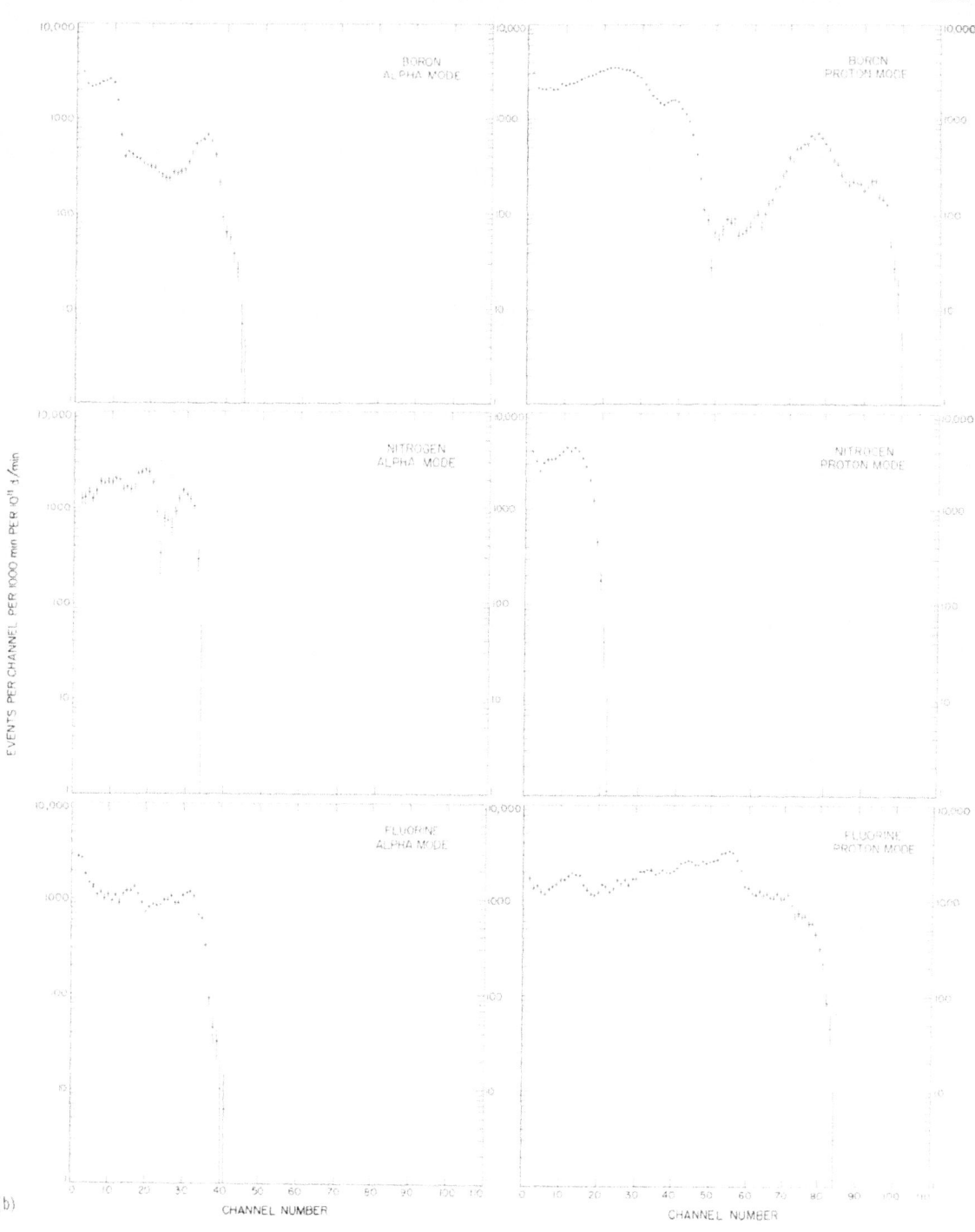

FIGURE 8-4.—Continued.

FIGURE 8-4.—Continued.

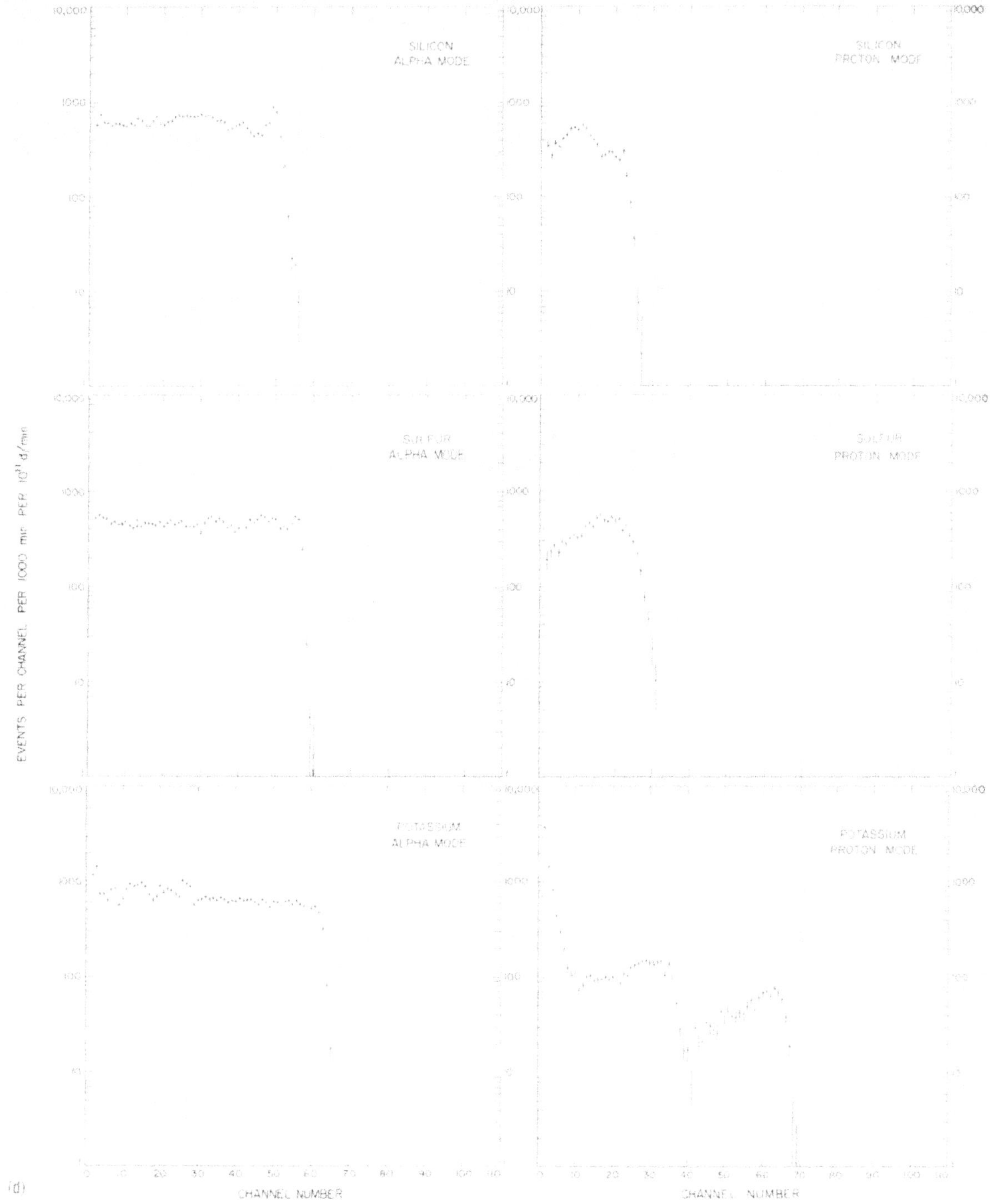

CHEMICAL ANALYSIS

FIGURE 8-4.—Continued.

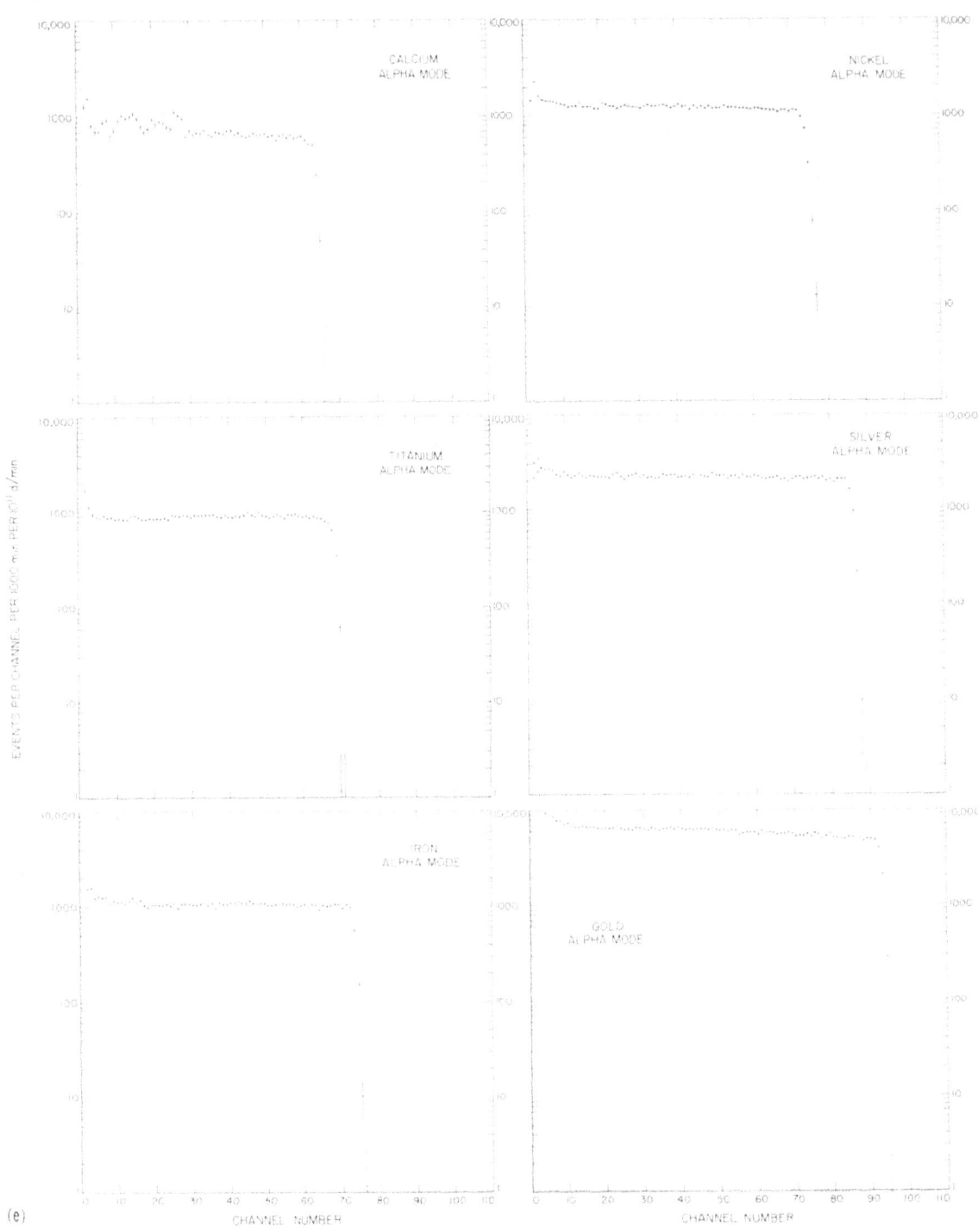

FIGURE 8-4.—Concluded.

mission calibrations with the Surveyor flight instruments.

The typical shape of the scattered-alpha spectrum of an element heavier than calcium (fig. 8-4(e)) is a relatively flat plateau terminating in a sharp dropoff to zero intensity at T_m, as determined by equation (1). The intensity at a given channel number in the spectrum is determined by the scattering cross section for alpha particles of the nuclei of the material and the energy loss of the alpha particles caused by ionization in the sample (ref. 8-10). The fact that the plateau for a heavy element has very nearly a constant intensity indicates that the effects of the variation of these properties with energy are very nearly balanced over a range of several million electron volts. Because the spectra of the heavier elements are nearly identical in shape, only a few representative spectra are included in this library.

Figure 8-5 shows the variation of intensity of alpha scattering (near T_m) with atomic number, both plotted on logarithmic scales. The units of the ordinate are the same as shown in figure 8-4. The solid points are values obtained directly from spectra of elemental samples, while the open points are derived from spectra of simple compounds of the element. The line drawn with a slope of 3/2 through the points for elements of higher atomic number shows the approximate power dependence of intensity on atomic number, predicted by simple theory (ref. 8-10).

In figure 8-5, it is seen that the intensity of scattering is greatly enhanced above the 3/2 power line for elements lighter than sodium. This is associated with direct interaction of alpha particles with nuclei of low charge. This interaction has very high scattering cross sections at characteristic resonance-energy regions. For this reason, the spectra of these light elements are not only greatly increased in intensity but also show more structure than in the cases where Rutherford scattering predominates. The intensities of scattering are often very strongly dependent on scattering angle, with the effect especially large at angles close to 180°. This resonance-scattering phenomenon gives the method a very high sensitivity for carbon and oxygen (fig. 8-4(a)), whereas the sensi-

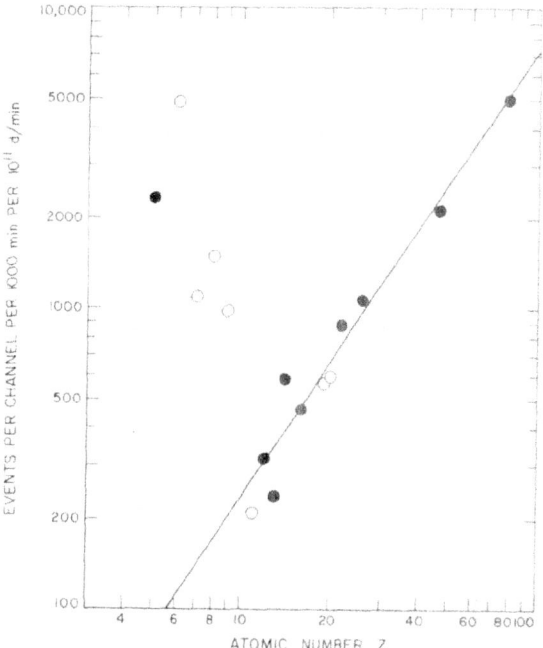

FIGURE 8-5.—Dependence of intensity of alpha scattering on atomic number in a Surveyor-type instrument. The ordinate is the intensity of scattering near the high-energy endpoint by the various elements included in the library of figure 8-4. Solid circles indicate intensities of scattering from elemental samples. Intensities derived using compounds of certain elements (for example, O, Na) are indicated by open circles.

tivity would be very poor for these elements if the intensity had followed the $Z^{3/2}$ law.

In elements that undergo (α, p) reactions, some of the structure in the spectra from the alpha detectors (for example, fig. 8-4 (b), (c), and (d)) is caused by the protons. Because the range of the protons is often greater than the sensitive depth of the silicon semiconductor detectors used for alpha particles, the peaks that appear in the alpha spectra do not always represent the total energy of the protons, but merely the amount deposited in the alpha detector. For this reason, variation of the sensitive depth of the detectors can cause shifts in the positions of the peaks that are caused by the protons.

The proton spectra of figure 8-4 have been obtained with semiconductor detectors covered with gold foil of a thickness just sufficient to absorb the scattered alpha particles. This also reduces the energy of the protons, completely absorbing those with energy less than 1.5 MeV. The proton detectors had sensitive depths great enough to measure the full energy of the entering proton. Because the cross sections for the (α, p) reactions are usually much lower than for alpha scattering, the total active area of the proton detectors had to be greater to obtain adequate intensities for the proton spectra.

Because the protons are produced by nuclear interactions, the proton spectra consist of a series of peaks corresponding to resonance interactions of alpha particles with the nuclei of elements emitting protons. The peaks are usually greatly broadened, however, by the large variation in the geometrical relationships of the proton detectors in the instrument to the various parts of the sample where the protons are produced. The elements producing protons include boron, nitrogen, fluorine, sodium, magnesium, aluminum, silicon, phosphorus, sulfur, chlorine, potassium, and titanium (see fig. 8-4 (b), (c), and (d)). The spectrum for aluminum (fig. 8-4(c)) extends to the highest energy of all proton-producing elements; its presence in significant amounts in a sample can be readily detected by a cursory examination of the proton spectrum. Other elements that produce high-energy protons include sodium, fluorine, boron, phosphorus, and potassium. Elements such as nitrogen, magnesium, silicon, and sulfur produce protons whose highest energy is less than 3.5 MeV. These properties, in addition to the differences in the characteristic shapes of the spectra, allow resolution of the various elements that produce protons.

In the analysis of spectra from samples that contain more than one element, use is made of the fact that for both protons and alpha particles, to quite a good approximation, the composite spectra have the shapes of the individual elemental spectra additively combined. Figure 8-6 illustrates this for the system Si, C, and SiC. In such a complex system, the contribution of an individual element has been shown (ref. 8-10) to be proportional to (1) its contribution as a pure element (the library spectrum), (2) the atomic fraction of the element in the sample, and (3) the energy-loss cross section of the element. It was also shown, experimentally, that assuming the energy-loss cross sections to be proportional to the square roots of the atomic mass numbers is adequate for preliminary analysis. (A functional dependence of the energy-loss cross section on the atomic number, rather than on the mass number, seems to be more justified theoretically.) Thus, for each energy channel j of the alpha and proton spectra, an equation may be set up of the form

$$_jI = (k/\langle A^{1/2}\rangle) \sum_i x_i {}_jI_i A_i^{1/2} \qquad (2)$$

where

$_jI$ = total intensity of the sample spectrum in channel j

x_i = atom fraction of an element i in the sample

$_jI_i$ = intensity in channel j of the library spectrum of element i

$A_i^{1/2}$ = square root of the mass number of element i

$\langle A^{1/2}\rangle = \sum_i x_i A_i^{1/2}$, average square root of the mass numbers of the elements in the sample

k = factor to take into account differences in the conditions (for example, the effective source strengths and sample distances) under which the library spectra and the sample measurements were obtained. The recognized, slightly different effect of distance on the response in the alpha and proton modes and the effect of distance on the shapes of the proton spectra are ignored at this stage of analysis.

Equation (2) may be rearranged into the form:

$$_jI = \sum_i f_i {}_jI_i \qquad (3)$$

where

$$f_i = \frac{k x_i A_i^{1/2}}{\langle A^{1/2}\rangle} \qquad (4)$$

The f_i values are the factors by which the library spectra are multiplied to obtain a com-

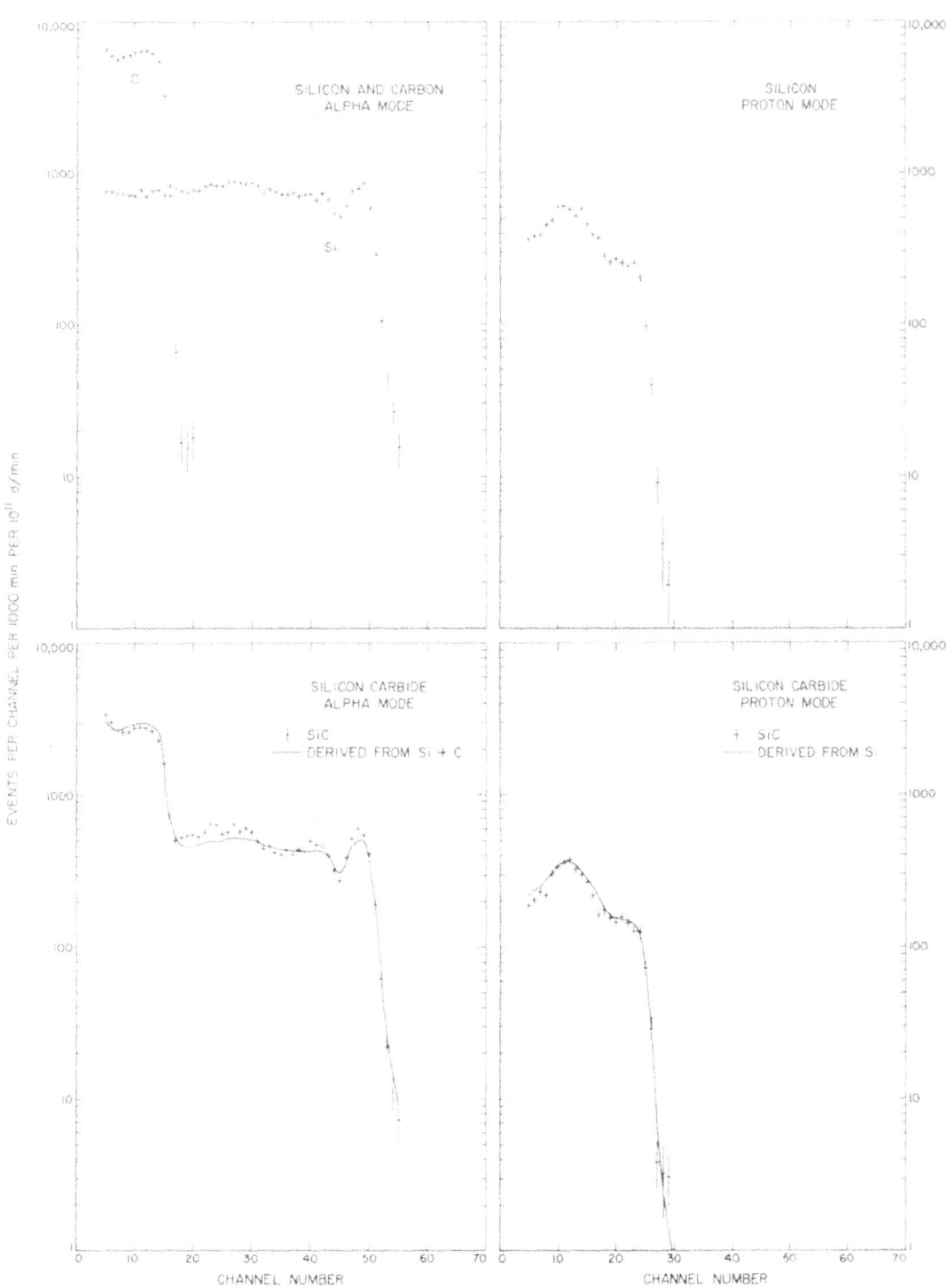

FIGURE 8-6.—Alpha and proton spectra for the system Si, C, and SiC (both actual and derived from Si and C).

posite spectrum that best corresponds to the observed sample spectrum. They are obtained by a least-squares treatment of the data by means of equation (3).

From equation (4), it is seen that the values of $f_i/A_i^{1/2}$ for a given sample are proportional to the atomic fractions x_i in the sample.

Thus, considering only chemical elements represented in the library, the fraction of the atoms that are of element i is given by

$$x_i' = \frac{f_i/A_i^{1/2}}{\sum_i f_i/A_i^{1/2}} \qquad (5)$$

Because hydrogen, helium, and lithium are not determined by this method, the analytical results are in terms of fractions of atoms heavier than lithium.

Thus, it is seen that a library of elemental spectra, such as that in figure 8-4, is essential to the analysis of the alpha-scattering data. For elements that are not readily available in their pure solid state, the library spectra may be derived from spectra of their compounds. This is best accomplished using the spectrum of a simple compound containing only one element for which the library spectrum is not known. Equation (2) may then be solved for $_j^0I_i$ of this element for each energy channel, as this is the only quantity in the equation that is unknown. In figure 8-4, the spectra of carbon, nitrogen, oxygen, fluorine, sodium, potassium, and calcium were derived using this technique. Some systematic errors are apparent below channel 30 in the alpha spectra of sodium, potassium, and calcium. They were introduced in these examples by the method of derivation using spectra from carbonates.

This method of chemical analysis by use of alpha particles was tested on a variety of rocks. Table 8-1 shows the results obtained using the P-1 instrument on eight rocks of a reasonable range in composition (ref. 8-15). The data are presented in terms of atomic percent, together with a measure of the uncertainty. The uncertainty is derived from the statistical errors of the measurement and the least-squares analysis. The table also presents results of analysis by conventional techniques.

Table 8-2 lists, in atomic percent, the standard deviations between the alpha-scattering method and the conventional analyses for the most abundant elements (ref. 8-15). The less than 1-atomic-percent standard deviations for sodium and aluminum depend strongly on the data from the proton mode, which indicates its importance in the analysis of these elements.

In interpreting the results from this analytical technique, certain inherent characteristics must be kept in mind. The inability to determine directly elements lighter than beryllium and, in particular, hydrogen, has already been mentioned. Another characteristic is that the penetration of alpha particles into the material is so small (at least with 6.1-MeV alpha particles) that the results apply strictly to the surface of the sample. Moreover, the depth to which the particles penetrate before interacting depends on the nucleus and processes involved in the interaction. The lack of resolution for the heavier elements has also been mentioned, as well as the only moderate sensitivity and accuracy of the results in practical situations. In addition, the interpretation depends on whether the average atomic stopping power of the material investigated is essentially constant over the sample examined. However, the technique, because of its relative comprehensiveness and the lightness and ruggedness of the equipment, together with the low-power requirements, is particularly suitable for early chemical-analysis missions to extraterrestrial bodies.

Instrument

The instrument for the Surveyor alpha-scattering experiment consists of a sensor head, which is deployed directly to the lunar surface, and an electronics package, which is housed in a thermally controlled compartment on the spacecraft. Figure 8-7 shows the alpha-scattering sensor head, electronics, and the mechanism for attaching the sensor head to the spacecraft and deploying it to the lunar surface. The alpha-scattering instrument delivers to the spacecraft, for transmission to the Earth, information on the number and energy of alpha particles scattered back from a sample and on the number and energy of protons produced in the sample by (α, p) reactions.

Figure 8-8 shows the Surveyor VII spacecraft

TABLE 8-. *Comparison between alpha and conventional analyses of some rocks (in atomic percent)*

Rock element	W 1		G 1		Dunite		Hamlet	
	Alpha	Conventional[a]	Alpha	Conventional[a]	Alpha	Conventional[a]	Alpha[b]	Conventional[a]
Hydrogen	0.4 ±0.4	1.1	1.0 ±0.6	0.9		4.3	0.6 ±0.3	0.1
Carbon	58.8 ±0.9	.03		.04	(c)	.05	54.1 ±0.7	(d)
Oxygen		60.4	61.8 ±1.3	61.7	58.2 ±0.5	56.0		36.6
Fluorine	0.0 ±0.05	.02	(c)	.08	0	.01		(d)
Sodium	2.1 ±0.3	1.44	2.2 ±0.5	2.2	0.02 ±0.06	.01	0.5 ±0.14	.78
Magnesium	1.8 ±0.8	3.54	1.1 ±1.2	.2	19.4 ±0.5	23.1	12.8 ±0.6	14.8
Aluminum	6.6 ±0.4	6.28	5.9 ±0.6	5.7	9.12 ±0.07	.19	1.3 ±0.2	1.23
Silicon	20.3 ±0.9	18.9	23.5 ±1.2	25.6	15.0 ±0.6	13.9	15.9 ±0.7	16.2
Phosphorus	(c)	.04	(c)	.03	0.02 ±0.06	0	(c)	.14
Sulfur	2.9 ±0.9	.01	1.0 ±1.2	.01	2.3 ±0.5	.01	3.8 ±0.7	1.45
Chlorine	(c)	(d)	1.6 ±0.8	.01	(c)	(d)	0.1 ±0.4	(d)
"Calcium"[e]	3.4 ±0.6	4.5	1.8 ±0.6	2.9	(c)	.1	(c)	.92
"Iron"[f]	3.1 ±0.2	3.78	0.3 ±0.1	.59	2.38 ±0.8	2.32	8.4 ±0.2	7.9
"Barium"[g]	(c)	.01	0.03 ±0.01	.06	0.000 ±0.006	(d)	0.007 ±0.04	(d)

Rock element	Tektite		Limestone[a]		Sulfide ore[a]		Syenite[a]	
	Alpha	Conventional[a]	Alpha	Conventional	Alpha	Conventional	Alpha	Conventional
Hydrogen		0.8	16.8 ±0.4	(d)	0.7 ±0.3	8.52	0.3 ±0.4	0.15
Carbon	59.7 ±0.5	.006		16.4	15.8 ±0.8	(d)	60.6 ±0.9	.43
Oxygen	(c)	63.9	59.3 ±0.8	59.9	0.01 ±0.04	16.1		60.0
Fluorine	(c)	.01	(c)	.255	0.76 ±0.22	(d)	(c)	(d)
Sodium	1.0 ±0.2	.73	0.39 ±0.15	1.11	3.1 ±0.7	.745	2.4 ±0.4	2.05
Magnesium	1.14 ±0.65	1.44	0.2 ±0.5	1.64	3.9 ±0.3	2.59	(c)	2.12
Aluminum	4.2 ±0.3	4.8	1.6 ±0.2	4.73	13.2 ±0.9	4.50	3.5 ±0.5	3.88
Silicon	26.1 ±0.7	25.2	5.7 ±0.6	.05	0.1 ±0.2	14.7	22.5 ±1.2	21.0
Phosphorus	0.1 ±0.2	.01	0.1 ±0.1	.17	5.6 ±0.9	.033	(c)	.06
Sulfur	0.6 ±0.7	(d)	1.3 ±0.6	(d)		8.63	(c)	.047
Chlorine	0.4 ±0.5	.006	0.4 ±0.4	15.2	4.5 ±0.7	2.12	(c)	(d)
"Calcium"[e]	1.7 ±0.4	2.1	15.4 ±0.5	.47	10.3 ±0.3	12.0	4.6 ±1.4	5.12
"Iron"[f]	1.6 ±0.1	1.63	0.75 ±0.09	(d)	0.000 ±0.006	.03	3.1 ±0.6	4.95
"Barium"[g]	0.003 ±0.006	.001	(c)				0.16 ±0.07	.17

[a] Results of conventional analyses are from the U.S. Geological Survey. [e] These samples (with results of conventional analyses) were obtained from the National Bureau of Standards.
[b] This analysis of the meteorite Hamlet was performed by H. B. Wiik.
[c] The least-squares calculation gave slightly negative answers. The value was set to zero, and the calculation repeated.
[f] "Calcium" represents potassium and calcium; "iron" denotes the elements titanium through zinc, inclusive; "barium" means all the elements heavier than zinc.
[d] No results of conventional analyses are available.

TABLE 8-2. *Statistical summary of comparison between alpha and conventional analyses, showing standard deviations between the results of the two types of analyses*

Rock element	Standard deviation, atomic percent
Oxygen	2.1
Sodium	.3
Magnesium	2.0
Aluminum	.4
Silicon	1.3
"Calcium"	1.2
"Iron"	.7

in its landed configuration. The sensor head (identified in fig. 8-8) is mounted on the outer edge of the spaceframe, midway between legs 2 and 3. The electronics equipment associated with the instrument is in the compartment directly above the sensor head (identified in fig. 8-8). The position of the sensor head, relative to the television camera and surface sampler (on Surveyor VII), is also noted in figure 8-8. The television camera was used to obtain pictures of the instrument and of the local terrain to which it would be deployed; the surface sampler could be extended to the alpha-scattering instrument area.

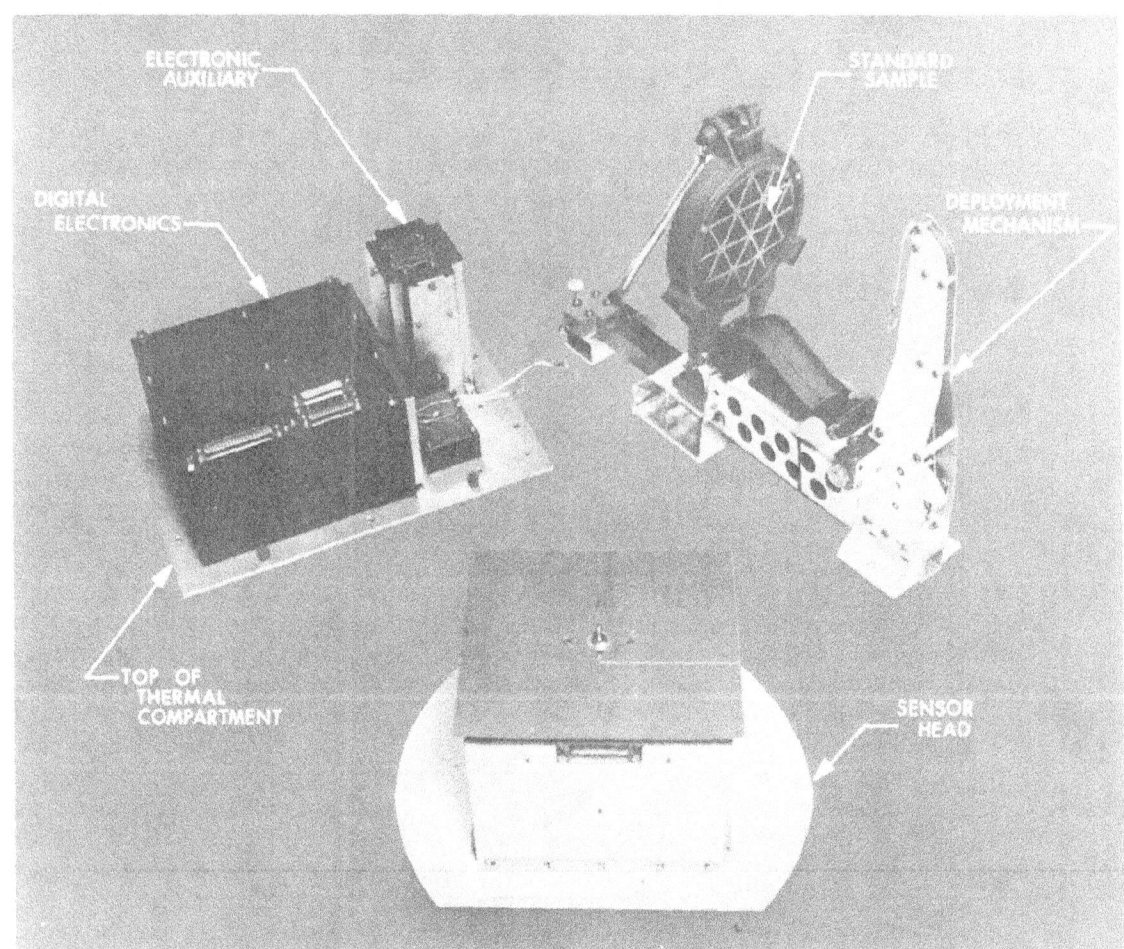

FIGURE 8-7.—Alpha-scattering instrument and auxiliary hardware. The sensor head is the part of the instrument that was lowered to the lunar surface. In its stowed position on the spacecraft, the sensor head was held on the deployment mechanism on top of the standard sample. The digital electronic and electronic auxiliary were contained in electronics compartment C for thermal control.

CHEMICAL ANALYSIS 285

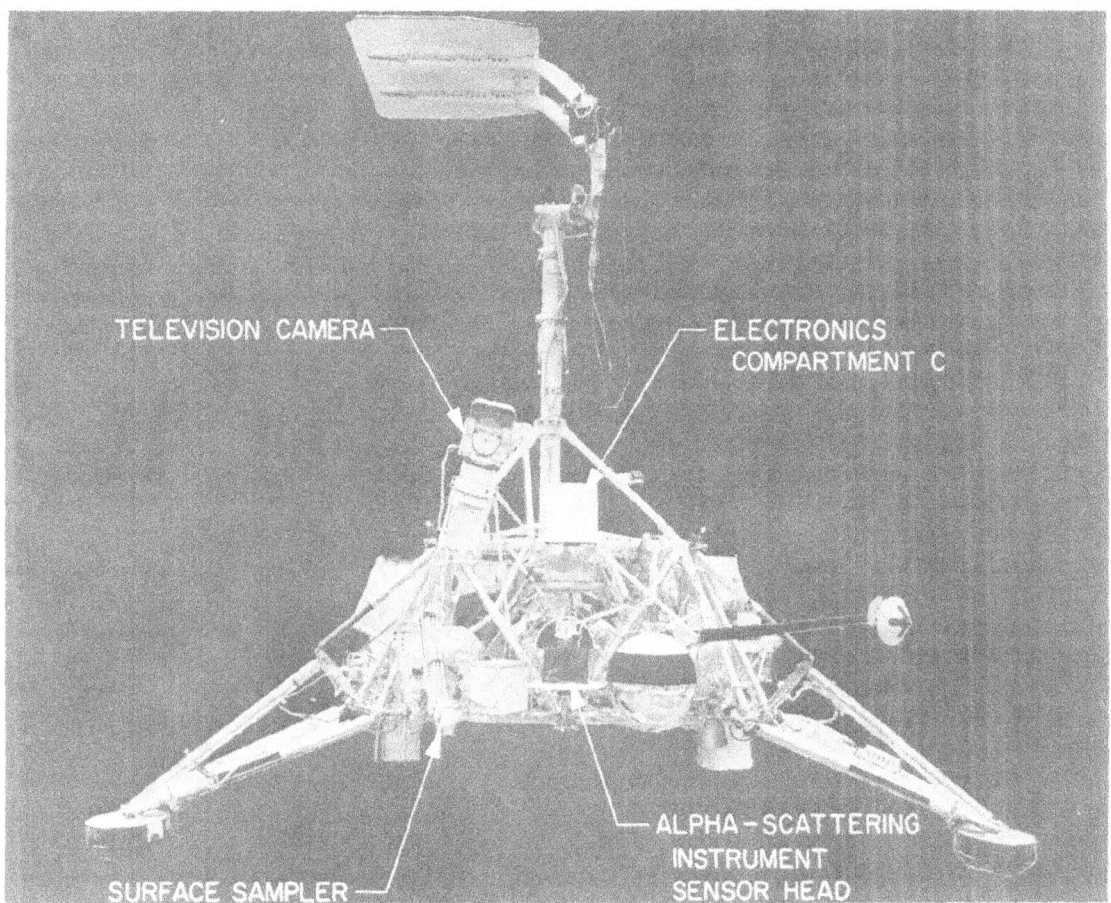

FIGURE 8-8.—Photograph of the Surveyor VII spacecraft showing the location of the alpha-scattering instrument relative to other parts of the spacecraft.

During the flight and lunar touchdown, the sensor head was attached to the spacecraft by means of the deployment mechanism. This mechanism consisted of a platform that held the sensor head and a device that could lower the sensor head to the lunar surface on command from Earth.

The total weight of the instrument and associated equipment (deployment mechanism, supporting substructure, thermally insulating compartment, spacecraft interfacing, and cabling) on the spacecraft is 13 kilograms. The operating power is less than 2 watts. An additional 15 watts are consumed by heaters in the sensor head and in the electronics compartment when both are operating. The power for the electronics is supplied at 29 volts by the spacecraft; power for the heaters is supplied at 22 volts.

Sensor Head

The sensor head is shaped like a box, 17.1 by 15.8 by 13.3 cm high, with a 30.5-cm-diameter plate on the bottom. (See fig. 8-7 for an overall view of the sensor head.) It contains six source capsules that provide the alpha particles, two detectors to register the number and energy of the scattered alpha particles, four detector systems for the protons, and part of the instrument electronics. Figure 8-9 shows a view of the bottom of the sensor head. The

FIGURE 8-9—Bottom view of the alpha-scattering sensor head showing the relative position of sources and detectors and the port through which the lunar surface was examined.

10.8-cm circular opening is the port through which the sample is analyzed; the sources and detectors can be seen through the sample port. Figure 8-10 is a cutaway view of the sensor head showing the operational relationships of these components.

The geometrical arrangement of sources and detectors, as seen in figures 8-9 and 8-10, was chosen to satisfy several requirements. The sources and alpha detectors were mounted as close together as practical so that particles scattered from the sample could be measured at the desirable large angles. (See "Method" and fig. 8-3.) A minimum of two independently operable alpha detectors was thought to be necessary. To maintain instrument delivery schedules and to avoid developmental problems, the same types of source capsules and alpha detectors used in earlier versions of the instrument were retained. Thus, the choice of the number and geometry of these components was reduced to a compromise between their most compact arrangement and the highest total intensity of sources.

The number, size, and location of the proton detectors were chosen to maximize reliability and the measured event rate. Because the range of protons of the same energy is larger than

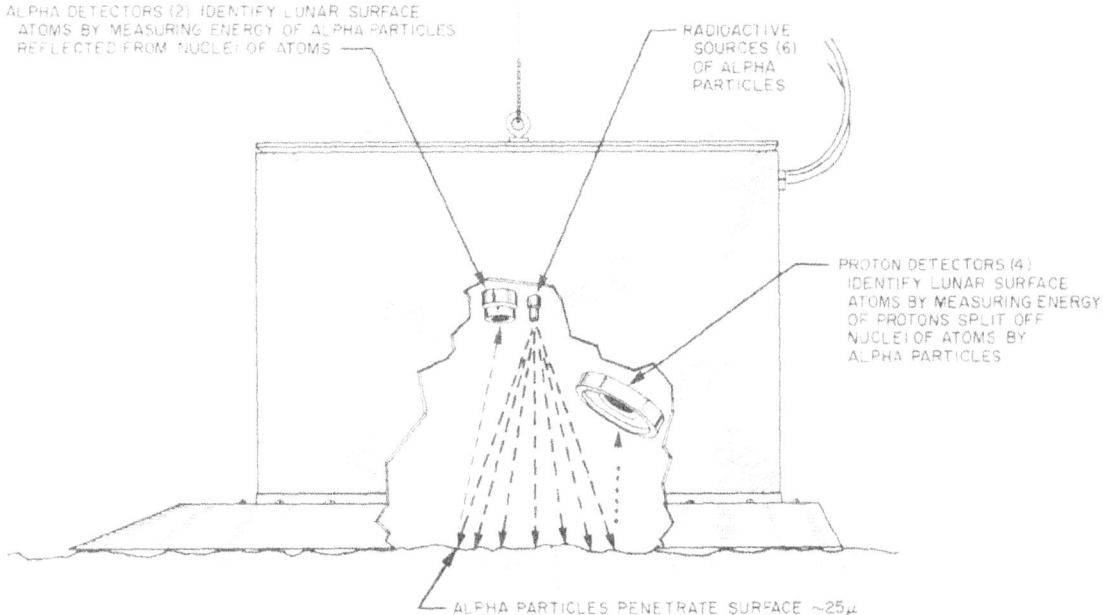

FIGURE 8-10.—Diagrammatic view of the internal configuration of the alpha-scattering sensor head.

that of alpha particles, and because the proton spectra do not vary as seriously with angle as do the alpha spectra, it was possible to locate the proton detectors at a lower angle, closer to the sample, and to make them much larger than the alpha detectors.

The structure and external surfaces of the sensor head were designed for passive thermal control (ref. 8-16). The sides and bottom of the sensor head are gold plated; on the top is a second-surface Vycor-aluminum mirror to promote the radiation of heat from the interior of the sensor head to cold space.

A nylon cord used in the deployment is fastened to an eyebolt on the center of the top of the sensor head. On Surveyor VII, the eyebolt was redesigned as a knob to be grasped by the surface sampler for redeployment of the sensor head to successive samples. The electrical connection to the spacecraft is through a flexible, flat cable attached to the inboard side of the sensor head. To maintain an inert, dry atmosphere inside the sensor head, dry nitrogen was introduced through a short tube on the outboard side of the sensor head during the prelaunch period. The large plate attached to the bottom is circular, with a segment cut out on the side toward the deployment mechanism. The main purposes of this plate were to stabilize the deployment of the sensor head onto irregular or inclined surfaces and to minimize the sinking of the sensor head if the lunar surface were covered with a deep layer of fine dust. The circular segments shown in figure 8-9 are the lugs by which the sensor head was held on the spacecraft during flight.

In addition to the parts already mentioned, the sensor head includes a platinum-resistance thermometer, a 5-watt heater, and an electronic pulse generator for calibration of the instrument.

Sources. Curium-242 is used as the source of alpha particles for the analysis. Its half-life, 163 days, is short enough to allow sources to be prepared with the necessary high activity per unit area and narrow energy distributions, yet long enough that the sources have adequate lifetimes. The main energy of the alpha particles from Cm^{242} is 6.115 MeV, with about one-fourth of the particles 44 keV lower in energy. However, because of internal conversion of this energy, the number of 44-keV gamma photons is only 4×10^{-4} per disintegration. Other, higher energy, photons are emitted to the extent of 6×10^{-5} per disintegration. Most of these are of 102- and 158-keV energy. In addition, Cm^{242} emits 1.6×10^{-7} neutrons per disintegration because of spontaneous fission. The disintegration rate of the daughter product, Pu^{238}, builds up slowly because of its much longer half-life.

Curium is easily handled in solution and in its oxide deposit. Cm^{242} is readily prepared in curie quantities by irradiation of Am^{241} with neutrons in a nuclear reactor, followed by chemical separation of the curium from americium and fission products (ref. 8-17). Excessive exposure of the Am^{241} to neutrons is to be avoided to minimize the amount of long-lived Cm^{243} and Cm^{244} produced.

Flight sources are prepared by vacuum volatilization of the curium onto stainless-steel plates. Each 3.2-millimeter-diameter plate contains 20 to 70 millicuries of curium in an area 6.6 mm². Each plate is enclosed in a stainless-steel capsule with a collimator opening that restricts the egress of the alpha particles from the plate to those that pass through the sample port (about 2.2 percent of the total alpha particles emitted). Details of the construction of the capsule are shown in figure 8-11. The collimator opening is covered with a film of aluminum oxide plus VYNS (polyvinylstyrene), with a total thickness of about 1000 Å, to prevent the contamination of the sample and the interior of the sensor head with radioactive material. A secondary film, mounted on a perforated plate, is placed over the six source openings to prevent contamination in case of failure of one of the primary films. This plate, barely discernible in figure 8-9, fills the area between the alpha detectors. These protective films are thin enough so that the energy of alpha particles passing through them is not seriously reduced.

The source capsules contain breather holes in the sides to permit equalization of the pressure on the two sides of the collimator films, particularly during launch. The passageways from these breather holes lead to the upper part of the sensor head to prevent contamination of the sample chamber. Also, as a precaution against

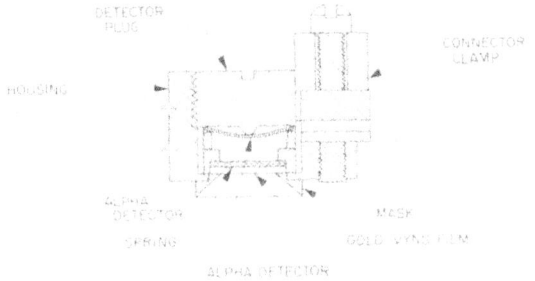

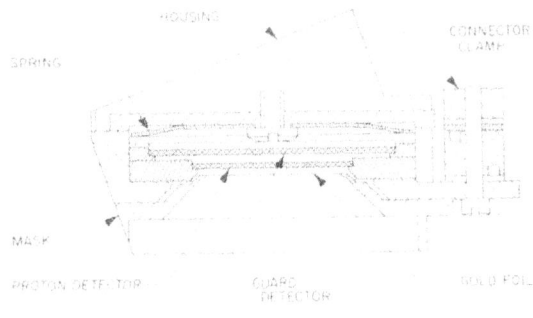

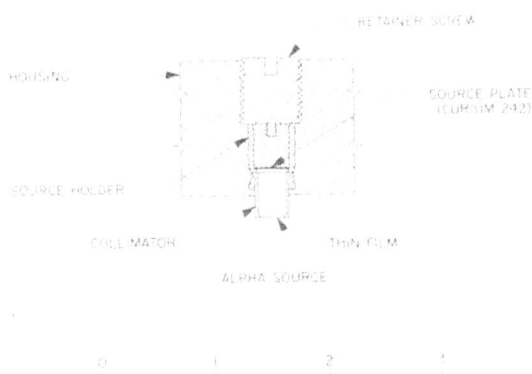

FIGURE 8-11.—Diagram showing some of the details of the source and detector assemblies used in the alpha-scattering sensor head.

possible contamination, the source capsules are loaded into the source block from above, with the secondary film in place.

It was discovered after the Surveyor V mission that sufficient curium can collect on the collimator films by aggregate recoil to become a secondary source of alpha particles. Because these are not collimated, some of them are scattered by the gold plating on the inside bottom of the sensor head. The result is a higher background for the experiment, one which gradually increases, thus reducing the sensitivity of the technique for heavy elements. This effect was quite small in the Surveyor V spectrum, but was considerably larger on Surveyor VI. (See "Background Measurements" in "Results.") For the Surveyor VII mission, the sources were coated with carbon by sputtering from a carbon arc before installation into their capsules. The set for instrument F-3, used as a backup, was the more heavily coated, and showed very little aggregate recoil in preflight tests at Cape Kennedy. In the flight sources actually used in the F-2 instrument on Surveyor VII, more aggregate recoil was found than in the spare sources, but it was considerably less than would be expected without the carbon coating.

In the course of the development of the alpha-scattering method, it was found that sources with activity densities greater than 0.1 curie/cm^2 undergo rapid degradation when exposed to atmospheric moisture. The width at half the maximum height of the peaks in the energy spectra of the flight sources is on the order of 2 percent of the total energy. This width is doubled if the sources are stored in an atmosphere of low humidity for about 1 week. The sources degrade this amount in less than 1 day at high relative humidity. In the laboratory, the sources are stored in desiccators and are shipped in a closed container filled with dry nitrogen and surrounded with magnesium perchlorate, a desiccant. The quality of the sources also had to be preserved during the prelaunch period at Cape Kennedy (up to 2 weeks) when the instrument was on the spacecraft. It was found, in a series of tests with the instrument in simulated prelaunch conditions, that the sources could be protected by a continuous purge of the sensor head, at a rate of 0.3 m^3/hr, with nitrogen containing less than 25 ppm water. To accomplish this, tubing leading from the supply of dry nitrogen was arranged to go to the top of the launch vehicle, through the aerodynamic shroud, and into the instrument sensor head via the short tube on the outboard side. The tubing was disconnected from the shroud just before launch. The connection between the shroud and the short tube was broken after launch when the shroud separated.

The scheduling of source preparation for the Surveyor missions had to be properly fitted into the flight schedules of Surveyors V, VI, and VII, which were launched at 2-month intervals. The first step in source fabrication, the preparation of the Am^{241} oxide for neutron irradiation by pressing with aluminum powder and encapsulation, could be carried out well in advance, because the half-life of the isotope is relatively long (458 years). The irradiation of the Am^{241} (about 3 weeks) and the chemical separation and purification of the curium, however, had to be performed not too long before sources were needed for Surveyor V if the curium from one irradiation were to suffice for all three missions. Otherwise, loss of curium caused by decay would be great, and extensive repurification of the material would be required because of the growth of Pu^{238} activity and the increase of nonradioactive impurities from radiation damage of the containing vessel. Too early preparation of the actual flight sources would result in an undesirable loss in source strength, since the source plates were loaded with as much curium as compatible with the required narrow spectral width of the Cm^{242} energy peak. On the other hand, source plates for a mission were required to be prepared about 2 months before launch to allow adequate time for qualification of the sources, instrument calibration with the sources, and tests involving the complete instrument, spacecraft, and rocket. An additional complication was the requirement of two sets of sources for each mission, one for the actual flight instrument, the other for a backup instrument (F-3) which was prepared on each mission to substitute for the primary instrument if necessary.

The typical sequence of events for each source of a set was as follows: Chemically purified curium was volatilized in vacuum from a filament at white heat onto a stainless-steel plate mounted behind a mask about 1 millimeter from the filament. Seven sources were prepared for each set. Preliminary assays of intensity and quality were obtained on these plates. The plates, on Surveyor VII only, were then coated with carbon from a carbon arc in vacuum and assayed again. The assays of the quality of these sources yielded a detailed energy distribution of the alpha particles emitted by the source. They were obtained by measuring the sources with high-resolution silicon detectors and 512-channel pulse-height analyzers.

The source plates were enclosed in their capsules, collimators with films whose thickness had been measured were installed, and then final certification of the sources was performed. This involved a final assay of the intensity and pulse-height analyses of the energy distribution of the particles from each source. Two of these pulse-height analyses were at normal direction from the source, but covered different energy regions to provide detailed information, not only of the characteristics of the main groups of alpha particles from Cm^{242} but also data on the tiny amounts of other radioactive species present, as well as on the relatively low-energy, continuous distribution always present in small amounts in such sources. Two other pulse-height analyses were made 18° away from the normal direction to obtain this same information in at least one off-normal direction. The set of sources was vibration tested and shipped to Cape Kennedy for installation into the instruments.

This procedure took about 2½ weeks per set of sources, and because two sets were needed per mission, about 5 weeks per mission. The sources were installed into the spare instrument 3 or 4 weeks before the launch; they were installed into the flight instrument 2 or 3 weeks before the launch. Thus, the 2-month interval between launches was barely sufficient for the source preparation schedule.

Particle detectors. All the particle detectors were developed, built, and tested especially for the Surveyor instruments (ref. 8–18). Two of the detectors were designed to register and measure the energy of alpha particles scattered back from the sample. These alpha detectors are located very close to the sources (see figs. 8–9 and 8–10), 7.0 cm above the sample port. The average scattering angle for alpha particles is $172° \pm 4°$.

These alpha detectors are of the silicon surface-barrier type, with an evaporated-gold front surface. Details of the detector assembly are shown in figure 8–11. Removable masks with 0.2-cm² openings are placed in front of the detectors so that alpha particles will be registered only in the central, uniformly sensitive, part of the detector. On these masks, thin VYNS films sputtered with gold are mounted to protect the detectors from radioactive contamination, dust, and excessive light. The detectors were reverse biased (4 to 7 volts) to produce a sensitive depth of approximately 50 microns. (These detectors have some sensitivity at greater depths.) This is large enough to register the full energy of an 8.2-MeV alpha particle, but only a fraction of that of most protons or electrons.

The protons from (α, p) reactions in the sample are detected at smaller angles (about 135°) than the scattered alpha particles. Figure 8–10 shows a diagrammatic view of the source-sample-detector relationships in this case. There are four proton detector assemblies situated symmetrically about the centerline of the instrument, as can be seen through the sample port in figure 8–9.

The proton detectors are of the lithium-drifted silicon type (ref. 8–18) operated with a collection voltage of 45 volts. The sensitive depth is approximately 350 microns, sufficient to register the full energy of a 6.4-MeV proton. Figure 8–11 shows some details of the assembly of these detectors. The detection area of each system is defined by a mask in front with an opening of 1.18 cm². Gold foils, approximately 11 microns thick (about 21 mg/cm²), are mounted on these masks and completely absorb any scattered alpha particles, allowing protons of energy greater than 1.5 MeV, entering normally, to get through.

Isotopic alpha sources (Pu^{239}, Cm^{244}) were used continually to check the characteristics of the detectors. In the case of proton detectors covered with gold foils, a Th^{228} source, in equilibrium with its decay products that include an 8.78-MeV alpha emitter, was used for this purpose.

Very small amounts (about one disintegration/min) of einsteinium (Es^{254}, $t_{1/2}=276$ days, $T_a=6.44$ MeV) are placed on the inside of the gold foil of each proton detector and on

CHEMICAL ANALYSIS

the VYNS film over each alpha detector. The high-energy alpha particles from this nuclide are used as an internal standard for the energy calibration of both alpha and proton systems of the instrument. They appear in the spectra as peaks at approximately channel 110. (See "Background Measurements" in "Results.") The event rate of these alpha particles served also as a crude live-time monitor for the instrument.

Because it was expected that the rate of proton events from a sample would be relatively low and that the contribution of solar and cosmic-ray protons on the Moon to the background would be high, the proton detectors were backed by guard detectors. (See fig. 8–11.) These are also lithium-drifted silicon detectors, somewhat larger (3.08 cm^2) than the proton detectors, and placed as close as possible above them. The sensitive depth of these detectors is 400 microns; they are operated at a bias of 40 volts. (Three of these detectors on the P–4 instrument were biased at 20 volts.) The electrical circuit is such (refs. 8–13 and 8–14) that proton detector events that are coincident with guard events are not registered as proton events in the output of the instrument. This appreciably cuts down the background caused by high-energy protons coming from above, although not affecting the detection of protons from the sample unless the guard event rate is inordinately high. The total event rate in the guard detectors is measured by a ratemeter whose response in volts, as a function of event rate, is shown in table 8–3.

TABLE 8–3. *Guard ratemeter response*

Events per second	Voltage, V
10	0.080
30	.300
100	.900
300	1.800
1000	2.500

Electronics. The basic electronic circuitry in the sensor head is described in references 8–13 and 8–14. It consists essentially of preamplifiers for the individual detectors; a mixing circuit and postamplifier for combining, in parallel, the outputs of the appropriate detectors; a threshold gate (set at approximately 0.6 MeV for events in the alpha and proton detectors and at 0.3 MeV for events in the guard detectors); and a height-to-time converter that produces a signal whose duration is proportional to the energy (less the threshold energy) deposited in a detector by a particular event. It is this time-analog signal that is transmitted along a cable from the sensor head to the electronics package on the spacecraft. There are separate systems for the alpha and proton modes of the instrument. The outputs of any alpha preamplifier or proton system (proton preamplifier and the associated guard system) can be blocked by Earth command.

A block diagram of the electronic circuitry in the sensor head of the Surveyor instruments is shown in figure 8–12. It differs from that described in reference 8–13 because of the inclusion of a mixing circuit with more stable performance (less dependent on the number of detectors included and with a smaller temperature coefficient), an internal pulse generator that can be turned on and off by Earth command, and a guard ratemeter. The pulse generator introduces electrical pulses of two known magnitudes (corresponding to approximately 2.5 and 3.5 MeV) at the detector stages of each of the alpha and proton systems. The pulses, at a rate of approximately 10 per second, are produced by an electromechanical relay. Further details on the design, construction, and performance of this pulse generator can be found in reference 8–14.

Thermal control. A severe limitation on lunar-surface operation was expected and found to be the large range of temperatures that could prevail. In early morning and late evening, and in the shadow of local terrain or parts of the spacecraft, temperatures could be much below $-50°$ C; on the other hand, near local noon, without some temperature control, an object on the lunar surface could easily reach temperatures in excess of $+150°$ C. The sensor head was designed to survive temperatures as high as $+75°$ C and to operate between

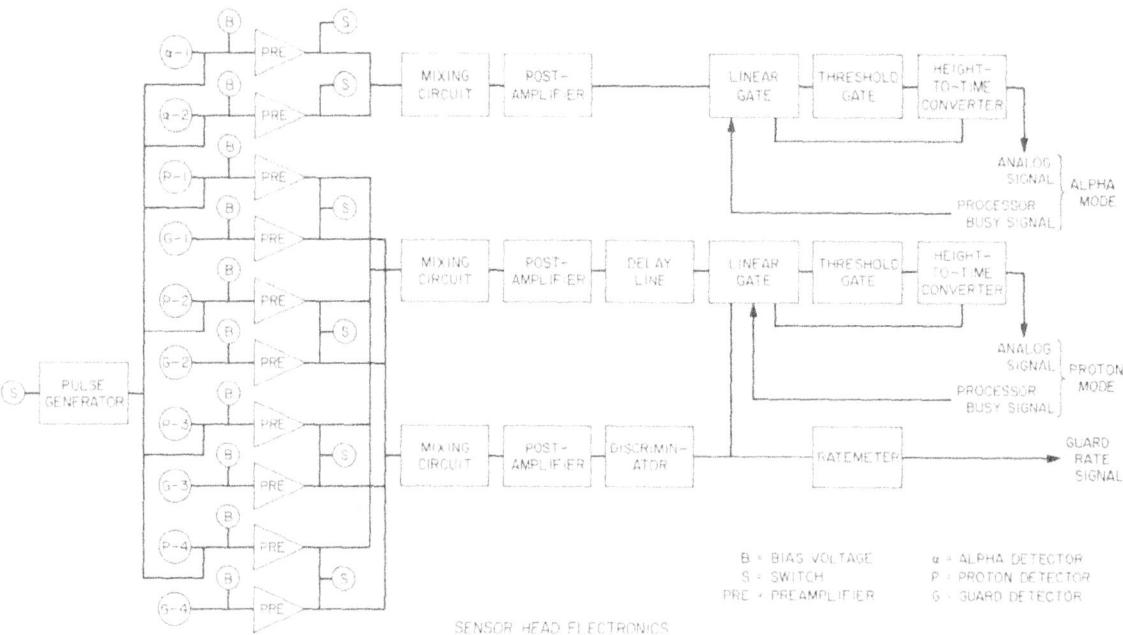

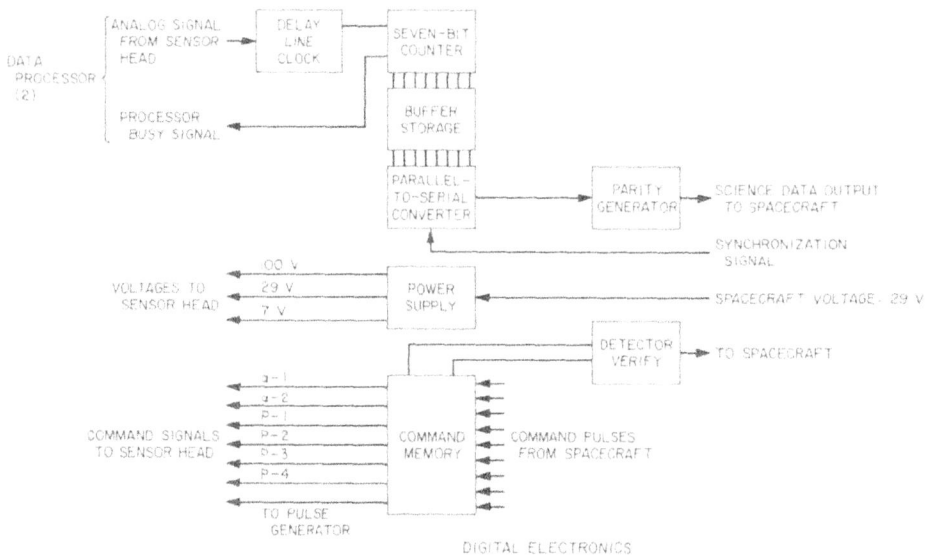

FIGURE 8-12.—Block diagram of the electronic circuitry contained in the alpha-scattering instrument.

−40° and +50° C. The upper operating temperature limit is determined by the detectors whose characteristics degraded rapidly above this point.

This thermal control is achieved by providing a reflecting gold-plated surface for the sides and bottom of the sensor head to reduce the absorption of infrared radiation from the hot lunar surface. These parts are thermally isolated as much as practical from the top and interior parts that contain the detectors, sources, and electronics. Rigid mechanical attachment (with minimum thermal coupling) between the two sections is achieved by using titanium struts of small cross section. The top surface, a second-surface Vycor mirror, is thermally coupled to the interior of the sensor head. It is designed to reflect radiation in the visible region from the Sun and to radiate infrared from the instrument to cold space. A commandable 5-watt heater is used for control at low temperatures. Isothermal conditions for the interior of the instrument, including the detectors, sources, and electronics, were promoted by use of indium gaskets. More details on the thermal design and tests can be found in reference 8-16.

Electronics Package of the Instrument

The electronics associated with the alpha-scattering instrument (other than that in the sensor head) consist of the digital electronics unit and the electronic auxiliary. They are contained in an insulated box (compartment C), which is attached to the upper part of the spaceframe. (See fig. 8-8.) For passive control of temperatures at high Sun angles, the top of this compartment is painted white to promote radiation to space. A 10-watt heater, commandable from Earth, provides some control at low temperatures.

The digital electronics unit (see fig. 8-12) converts the time-analog signals from the sensor head into nine-bit binary words. Separate signals identify the energies of alpha and proton events. Seven bits of each word identify the channel (out of 128) that represents the energy of the registered event. Two extra bits are added before transmission, one to identify the start of the word and one at the end of each word, as a parity check on transmission errors. Buffer registers provide temporary storage of the energy information for readout into the spacecraft telemetry system. The transmission rates are 2200 bits per second for the alpha mode and 550 bits per second for the proton mode. Events of energy greater than the range of the 128-channel analyzers are routed to channel 126 (overflow channel). The digital electronic unit also contains power supplies, a temperature sensor, and the logical electronic interfaces between the instrument and the spacecraft.

The required electrical interfaces among the sensor head, digital electronic, and spacecraft circuits are provided by an electronic auxiliary that provides command decoding, signal processing, and power management. Basic spacecraft circuits that interface directly with the sensor head and digital electronics are the central signal processor, which provides signals at 2200 bits per second and 550 bits per second for synchronization of instrument clocks; and the engineering signal processor, which provides temperature-sensor excitation current and commutation of engineering data outputs. The operating temperature range specified for the electronic auxiliary is $-20°$ to $+55°$ C.

The electronic auxiliary provides two data channels for the alpha-scattering instrument. The separate alpha and proton channels are implemented using two subcarrier oscillators with center frequencies of 70 000 and 5 400 hertz, respectively.

In summary, the output of the alpha-scattering instrument is science data (the data stream giving the channel numbers of the alpha and proton events), and engineering data (temperature of the sensor head and the digital electronics unit, voltage on the guard-detector ratemeter, voltages at two points on the instrument, and information on the combination of detectors used).

Test and Calibration of the Instruments

Each flight instrument was extensively tested and calibrated to insure its quality and to establish its characteristics. The testing was performed under more severe conditions in the type approval model (P-3) to insure that no component in the units was near the threshold of failure at the limits of the normal conditions.

The first of these flight-instrument tests was normally performed on completion of construction. Their characteristics (position of pulser peaks and response to standard sources) were established under thermal-vacuum conditions from $-40°$ to $+50°$ C, with the sensor head and digital electronics unit at the same and different temperatures. The approximate instrument temperature coefficients, used to correct the data obtained in near real time during the Surveyor missions, were derived from these tests. Typical data on the Surveyor instruments gave a few percent change in overall gain over a $90°$ C temperature range.

The units then were subjected to the thermal-vacuum part of the flight-acceptance test at the Jet Propulsion Laboratory. This was similar to, but less extensive than, the test conducted on completion of construction.

The units then went into a science-calibration stage. Here, using flight-intensity sources, the response of the instruments to about 15 different samples (pure elements as well as simple chemical compounds) was established. This science calibration provided the library of elemental responses. (See fig. 8-4.)

The sources were then removed and the instruments were exposed to 50- to 150-MeV protons at the University of Chicago synchrocyclotron (cyclotron tests). The purpose of this test was to verify the proper behavior of the guard detectors and of the instrument in rejecting events in the proton detectors that were caused by energetic particles coming from above the instrument (solar and cosmic rays). The units were then subjected to the shock and vibration part of the flight-acceptance test. Several of the instruments were also tested at the Caltech tandem Van de Graaff machine to study the detailed characteristics of the detectors. The instruments were then delivered to the Hughes Aircraft Co. for further testing, mating to the spacecraft, and checking of the entire spacecraft (including the instruments).

Soon after arrival at Cape Kennedy, the sensor head was put through an additional check of its characteristics over a temperature range of 0° to +40° C. The flight sources were then installed, and a selected set of samples was measured before reinstalling the sensor head on the spacecraft for final, prelaunch tests. During the time that flight sources were in the instrument and it was not in vacuum, dry nitrogen was flushed through the unit to prevent deterioration of the sources.

In all instrument tests that involved the spacecraft, the calibrating pulser was invaluable in providing relatively quick, almost complete checks of the state of the instrument.

Because the silicon detectors were important to the success of the experiment, they were subjected to severe qualification tests even before installation in the instruments. (See ref. 8-18.) If a detector became unreliable (usually as evidenced by low channel noise or a broadening of the pulse-generator peaks), this detector or (in the case of the proton and guard detectors) the detector system could be replaced with tested spares. The instrument was then usually subjected to a penalty test to insure that the change had not affected other characteristics or reliability.

The extensive precautions and tests of the instruments and detectors resulted in more than 150 hours of lunar-surface operation under nominal conditions of 3 alpha-scattering instruments (each with 10 detectors). Although two detectors became temporarily noisy at a temperature close to the upper limit for which they had been qualified, the instruments performed well in all other respects.

Deployment Mechanism

The deployment mechanism provided stowage of the sensor head during flight and after landing on the Moon. It also provided deployment to the background position and to the lunar surface. The sensor head is mounted to the deployment mechanism by means of three support lugs on the bottom plate (fig. 8-9). Clamps that engage these lugs are released in the first stage of the deployment operation. Figure 8-13 shows the two stages of this operation. The sensor head is first released from the stowed position to a suspended position 56 cm above the nominal lunar surface. At this time, the mounting platform falls away on the activation of an explosive pinpuller by command from Earth. From the background position, the sensor head is then lowered directly to the lunar surface on command by activation of another explosive pinpuller device. The deployment velocity is controlled by an escapement.

A sample of known composition is attached to the platform on which the sensor head is mounted in the stowed position. This standard-sample assembly covers the circular opening in the bottom of the sensor head during spacecraft flight and landing to minimize the entrance of dust and light. It also provides a means of assessing instrument performance after landing by an analysis of a relatively complex substance under lunar conditions. The

CHEMICAL ANALYSIS

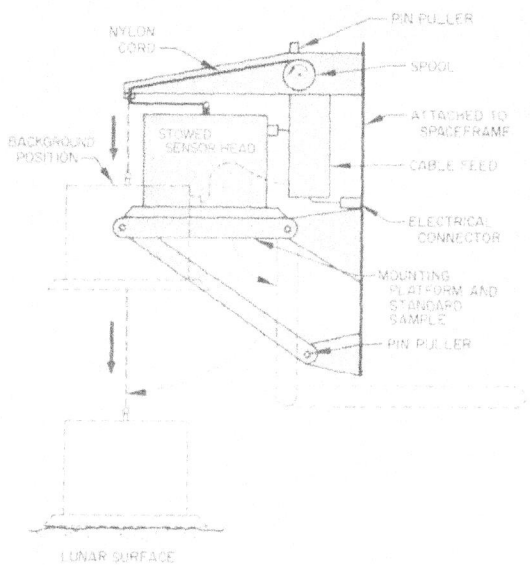

FIGURE 8-13.—Operation of the alpha-scattering deployment mechanism.

standard sample was a piece of glass, the principal constituents of which were oxygen, silicon, magnesium, sodium, and iron. It was partly (20 to 25 percent) covered by a polypropylene grid to provide a response from carbon. Measurements on the actual sample carried on the mission were made by each instrument in the science-calibration stage; during the Cape Kennedy phase, the measurements were repeated on similar samples.

Detailed Characteristics of Various Instruments

Although all the flight instruments were constructed to be essentially identical, there were some small differences among them. Some of the important, detailed characteristics of the instruments and the extent to which they differed are discussed in the following paragraphs.

Table 8-4 lists the dates of completion of various phases of construction and tests of the instruments. Instrument F-3 was not carried through engineering systems tests in conjunction with a spacecraft as were the other three flight instruments. It was loaded with flight sources and carried through final calibration tests at Cape Kennedy for each of the last three Surveyor flights as a backup instrument.

Detailed records of the stability of the instruments' characteristics over the many months of testing are given in reference 8-14. These records indicate that during the missions the channel number corresponding to a given energy event in either the alpha or proton system should be known to better than 0.3 channel at a given temperature of the sensor and electronics unit. It was expected that the calibrating pulser and the Es^{254} would give

TABLE 8-4. *Instrument development events*

Event	P-4	F-1	F-2	F-3
Instrument construction completed	Oct. 1, 1966	Dec. 19, 1966	Mar. 6, 1967	Mar. 30, 1967
Acceptance test [a] completed	Oct. 29, 1966	Jan. 23, 1967	Apr. 19, 1967	May 13, 1967
Science calibration completed	Nov. 22, 1966	Feb. 16, 1967	Apr. 4, 1967	May 2, 1967
Cyclotron tests	Nov. 22, 1966	Feb. 1, 1967	Apr. 4, 1967	May 29, 1967
Unit delivered to Hughes Aircraft Co.	Jan. 18, 1967	Mar. 7, 1967	May 17, 1967	July 7, 1967
Solar-thermal-vacuum tests on spacecraft	May 13, 1967	July 14, 1967	Aug. 4, 1967	NA [b]
Combined system test	June 21, 1967	NA [b]	NA [b]	NA [b]
Radioactive sources installed and final tests started at Cape Kennedy	Aug. 24, 1967	Oct. 20, 1967	Dec. 20, 1967	Aug. 17, 1967 Oct. 10, 1967 Dec. 6, 1967
Launch	Sept. 8, 1967	Nov. 7, 1967	Jan. 7, 1968	NA [b]
Total hours of operation at launch	925	1309	798	NA [b]

[a] Thermal-vacuum part of flight acceptance test.
[b] Not applicable.

more direct, confirming evidence on the characteristics during the mission.

Table 8-5 gives the characteristics of the Cm^{242} sources that were used on the three lunar missions. The total intensity (in terms of disintegrations/min) of the six sources on the date of lunar touchdown was calculated from the last assay of the sources shortly before shipment to Cape Kennedy (a half-life of 163 days was used to correct for decay during the intervening period). The intensities were almost the same on Surveyors V and VI, but were about 70 percent higher on Surveyor VII. The mean peak-energy values are the weighted averages of the peak energies of the individual sources as measured after encapsulation (that is, through the collimator protective films). The uncertainties listed are the standard deviation of the values from the mean. As might be expected from the higher intensity and the presence of the carbon coating, the mean energy of the F-2 (Surveyor VII) sources is the lowest, but the difference from the others is not great. The energy-spread range is a listing of the lowest and highest of the six values of the width of the peak at half its height expressed in percent. Again, the range is shifted only slightly upward for the more intense sources used on F-2.

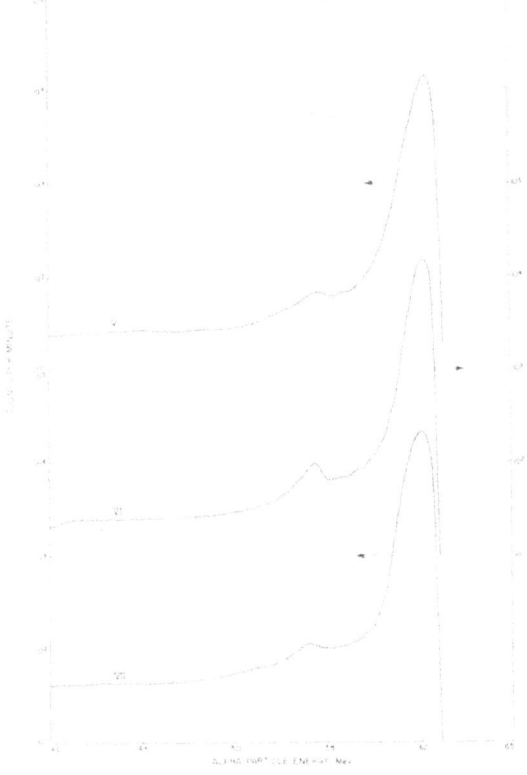

FIGURE 8-14.—Composite energy spectra for each of the three sets (six sources each) of the curium-242 flight sources.

TABLE 8-5. *Principal characteristics of sources used on the lunar missions*

Characteristic	P-4	F-1	F-2
Curium-242 source characteristics:			
Total intensity at touchdown (disintegrations/min)	2.75×10^{11}	2.76×10^{11}	4.70×10^{11}
Date of touchdown	Sept. 11, 1967	Nov. 10, 1967	Jan. 10, 1968
Mean peak energy, through capsule films, MeV	6.04 ± 0.01	6.03 ± 0.02	6.02 ± 0.01
Energy-spread range for 6 sources (FWHM), percent	1.2 to 2.0	1.2 to 1.8	1.3 to 2.2
Characteristics of composite curium-242 source spectra (see fig. 8-14):			
Energy at peak, MeV	6.040	6.041	6.010
Full width at half maximum, percent	1.7	1.5	1.9
Full width at 0.1 maximum, percent	3.3	2.8	3.3
Full width at 0.01 maximum, percent	5.8	4.9	5.3
Thickness of secondary films, energy loss in million electron volts for 6.1-MeV alpha particles, MeV	0.010	0.027	0.026
Energy of alpha particles incident on sample (estimated peak of spectrum), MeV	6.03	6.01	5.98

Figure 8-14 shows the composite energy spectrum of the six curium sources in each of the three sets of sources used on the lunar missions. These spectra were obtained from data on the individual sources after encapsulation. Table 8-5 lists some of the characteristics of these composite spectra, and also lists the thickness of the secondary films protecting against source escape on the three lunar missions. The secondary film on instrument P-4 was much thinner than on the other two instruments. It was made up only of VYNS instead of a combination with aluminum oxide.

Table 8-6 lists some of the principal characteristics of the detector and electronics systems on the instruments. The alpha detectors had an 800-Å layer of evaporated gold on their faces. The energy loss corresponding to this thickness is indicated in table 8-6. The thicknesses of the light-protective films over the alpha detectors, expressed in terms of energy loss for 6.1-MeV alpha particles, is also shown in table 8-6. These films were not well matched in the case of F-1, so that separate values are given for the detectors instead of the average value used in the cases of P-4 and F-2.

The proton detectors had 400 Å of evaporated gold on their front faces. This was negligible, however, compared with the thicknesses of the gold foils in front of the detectors. These foil thicknesses are stated in four different ways: in microns, in milligrams per square centimeter, in energy losses by 6.1-MeV alpha particles, and by 2-MeV protons. F-1 had the thickest foils, while F-2 had the thinnest ones. On a given instrument, the foils on the separate detectors were matched to within 2 percent.

Table 8-6 also contains details concerning the approximate electronic energy scales of the different instruments. These are expressed in terms of a linear equation:

$$N = kE - N^0$$

where N is the channel number corresponding to an energy, E; k is the gain (in channels per million electron volts); and N^0 is the effective zero offset (in channels) of the system. The constants are given separately for the alpha and proton modes and are applicable for both sensor head and electronics at 25° C. They were determined from the response of the system to alpha particles of known energy and from the positions of the two pulser peaks.

Experiment Control and Mission Operations

Alpha-Scattering Analysis and Command

The alpha-scattering experiment on the Surveyor V, VI, and VII missions was controlled

TABLE 8-6. *Principal characteristics of the detectors and electronic systems used on lunar missions*

Characteristic	P-4	F-1	F-2
Alpha detectors:			
Thickness of Au detector surface (energy loss for 6.1-MeV alpha particles), MeV	0.039	0.039	0.039
Thickness of alpha mask films (energy loss for 6.1-MeV alpha particles), MeV	0.029	0.040, 0.029	0.026
Proton detectors:			
Gold-foil thickness, μ	10.8	11.4	10.5
mg/cm^2	20.8	22.0	20.3
Energy loss for 6.1-MeV alpha particles, MeV	5.8	6.1	5.5
Energy loss for 2.0-MeV protons, MeV	1.17	1.24	1.13
Electronics energy scale (at 25° C):			
Alpha:			
k, channels/MeV	19.30	18.94	18.98
N^0, zero-energy channel	−12.0	−10.9	−10.9
Proton:			
k, channels/MeV	19.11	18.86	19.16
N^0, zero-energy channel	−11.65	−10.7	−11.9

from the Space Flight Operations Facility (SFOF), Pasadena, Calif., by means of commands transmitted to the spacecraft from the tracking stations. These commands were chosen on the basis of the analysis of data received from the spacecraft and relayed to the SFOF during the mission. The commands controlled—

(1) Spacecraft power to the instrument
(2) Deployment of the sensor head
(3) Number of detectors used
(4) Electronic calibration pulser
(5) Heater power for thermal control

Table 8–7 lists the command assignments for the alpha-scattering system.

TABLE 8–7. *Command assignments*

Command	Function
3501	Alpha-scattering power on.
3502	Alpha-scattering power off.
3503	Sensor-head heater power on.
3504	Sensor-head heater power off.
3505	Deploy to background position (interlocked with 3617).
3506	Deploy to lunar surface (interlocked with 3617).
3507	Alpha detector 1 on.
3510	Calibration pulser on.
3511	Proton detector 4 on.
3512	Proton detectors 3 and 4 off.
3513	Proton detector 2 on.
3515	Alpha detectors 1 and 2 off.
3516	Proton detectors 1 and 2 off.
3517	Proton detector 1 on.
3520	Calibration pulser off.
3522	Proton detector 3 on.
3523	Alpha detector 2 on.
3617	Interlocked with deployment commands.
0135	Compartment C heater power off.
0136	Compartment C heater power on.

Two types of information relative to the alpha-scattering experiment were transmitted from the spacecraft: engineering data and science data.

The seven engineering measurements monitored are listed in table 8–8. The instrument temperatures were measured to determine whether they were within operating limits, to plan mission operations, and to provide an

TABLE 8–8. *Engineering data*

Engineering commutator	Measurement
AS–3	Sensor-head temperature.
AS–4	Compartment C (digital electronics) temperature.
AS–5	Guard-rate monitor.
AS–6, digital	At least 1 alpha detector on.
AS–7, digital	At least 1 proton detector on.
AS–8	7-V monitor.
AS–9	24-V monitor.

approximate correction to the energy spectra in the real-time data analysis. Power-supply voltages were monitored to determine whether they were within limits and to diagnose possible problems. The two digital measurements gave an indication of detector configuration as a check on the proper receipt of commands. The guard-rate monitor gave a measurement of the event rate in the guard detectors. This provided information on the radiation background as well as on the proper functioning of the detectors.

The science and engineering outputs of the instrument were transmitted to Earth via the spacecraft telemetry system. (Details of this system are described in ref. 8–19.) Figure 8–15 is a simplified block diagram of the spacecraft telemetry system, configured for the transmission of alpha-scattering engineering and science data. In this figure, the numbers above the blocks indicate the commands necessary to activate the various telemetry subsystems. It is seen that the science data and engineering data could be received independently. In particular, science data could be received only when the instrument and presumming amplifiers were activated. Engineering data from the spacecraft were transmitted via one of several engineering commutators or telemetry modes. Alpha-scattering engineering data required use of the mode 4 commutator. Instrument temperatures could be received at all times in mode 4; however, other engineering data could be received only when the instrument power was turned on.

The science data consisted of nine-bit digital

CHEMICAL ANALYSIS

FIGURE 8-15.—Block diagram of part of the spacecraft telemetry system, configured for the transmission of alpha-scattering data.

words that characterized the energy of each of the analyzed alpha particles or protons. These data were generated as separate alpha and proton bit streams and modulated separate subcarrier oscillators (ref. 8-19); they were then combined with the engineering data and transmitted by the spacecraft to Earth. The composite signal from the spacecraft, including non-alpha-scattering engineering data, was recorded on magnetic tape at the tracking stations. These tapes containing the raw data comprise the primary source of the alpha-scattering information that will be used in postmission analysis.

In addition to being recorded for later use, the data were monitored and subjected to computer processing during the mission so that proper control over the experiment could be exercised. This real-time control was necessary for several reasons. Because the outputs of the two alpha and the four proton detectors were summed separately, early in the spacecraft signal processing, a degradation in performance of any of the detectors would have influenced the total output of the data stream generated by that type of detector. It was possible, however, by Earth command to block the output of any detector, or combination of detectors, from contributing to the data stream. Thus, when detector noise was traced to a specific detector, it was possible to remove its contribution and thus suffer no further data degradation other than that corresponding to a lower event rate.

Another need for real-time processing arose from the irreversible sequence of events leading to the deployment of the sensor head from the spacecraft to the lunar surface. Before releasing the instrument from the stowed and background positions, it was necessary that sufficient good data be available to assess both lunar background radiation and the performance of the instrument. This information was needed to analyze the chemical data and to plan other spacecraft operations.

A simplified block diagram of the Earth-based system used in the data-recording and real-time analysis is shown in figure 8-16. In this system, data were processed in two stages: at three Deep Space Stations (Goldstone, Calif.; Robledo, Spain; Canberra, Australia) where the data were received, and at the SFOF where the experiment was controlled.

After demodulation, discrimination, and bit synchronization at each Deep Space Station, the alpha and proton data streams entered an SDS-920 computer where periodic accumulations of the spectral data were made. Each of the three Deep Space Stations was configured to perform the data processing shown in figure 8-17.

For the real-time operations, the alpha and proton data streams were discriminated and conditioned to be acceptable to an SDS-920 computer. This onsite computer performed the following functions:

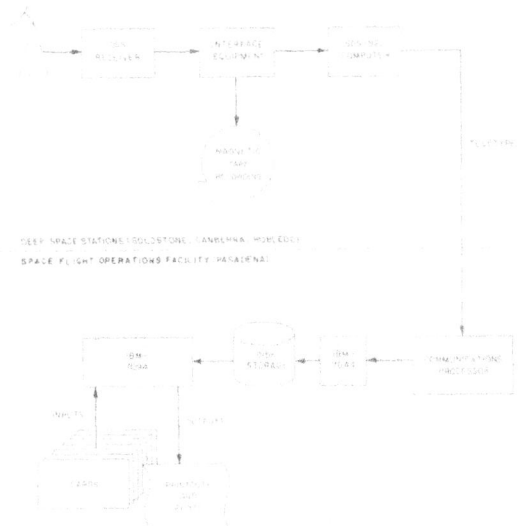

FIGURE 8-16.—Earth-based system used for recording and real-time analysis of alpha-scattering data.

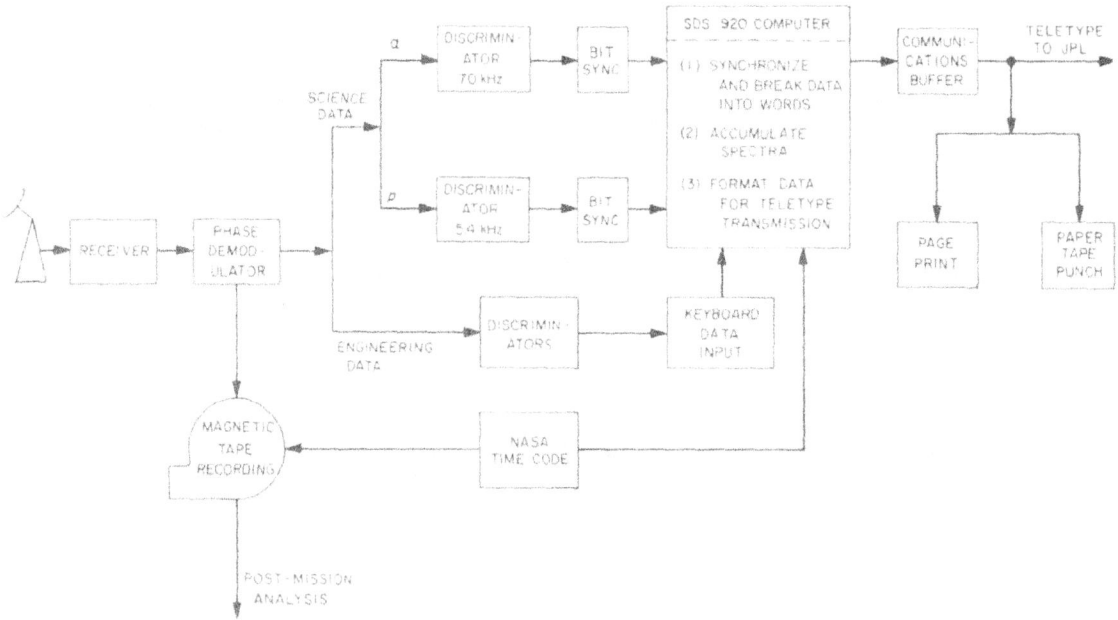

FIGURE 8-17.—Alpha-scattering data processing system at the Deep Space stations.

(1) The two incoming data streams were bit synchronized and formatted into data words representing the energies of the detected alpha particles and protons.

(2) The channel number of each word was determined, and its parity was checked. On the spacecraft, the alpha and proton energies were encoded by the instrument electronics into nine-bit data words (a sync bit, seven information bits, and a parity bit).

(3) On external keyboard command, four spectra representing the number of events detected in each energy channel were simultaneously accumulated. These spectra represented alpha, proton, alpha-parity-incorrect, and proton-parity-incorrect data.

(4) During spectral accumulation, it was possible by keyboard command to interrogate the computer for the current status of any channel in any spectrum.

(5) On keyboard command, a given accumulation was stopped, and the four spectra, as well as other identifiers, were transferred to magnetic tape and/or Teletype lines for transmission to JPL.

The primary function of the onsite processing was a substantial data compression, permitting accumulated spectral batches to be transmitted periodically at Teletype rates to the control center at JPL. Typically, during Surveyor mission operations, spectral accumulations of lunar-surface data lasted 20 to 40 minutes, with a constant transmission time of about 10 minutes for each accumulation.

The batches of raw spectral data accumulated at the Deep Space Stations and transmitted to the SFOF were processed to permit a detailed analysis of instrument behavior. This processing occurred at JPL using the coupled IBM 7044-7094 system; it is outlined functionally in figure 8-18.

After encoding by the 7044, the Teletype data were automatically recorded on the disk for use by the 7094. It was also possible to enter data into the system via cards from the operational control area.

The initial steps in the computer processing in the SFOF involved the correction of the data for transmission errors occurring in the Moon-Earth and Deep Space Station-Jet Propulsion Laboratory transmission links. These corrections were made by using the parity checks included in each transmission. Parity violations were automatically flagged by the

CHEMICAL ANALYSIS

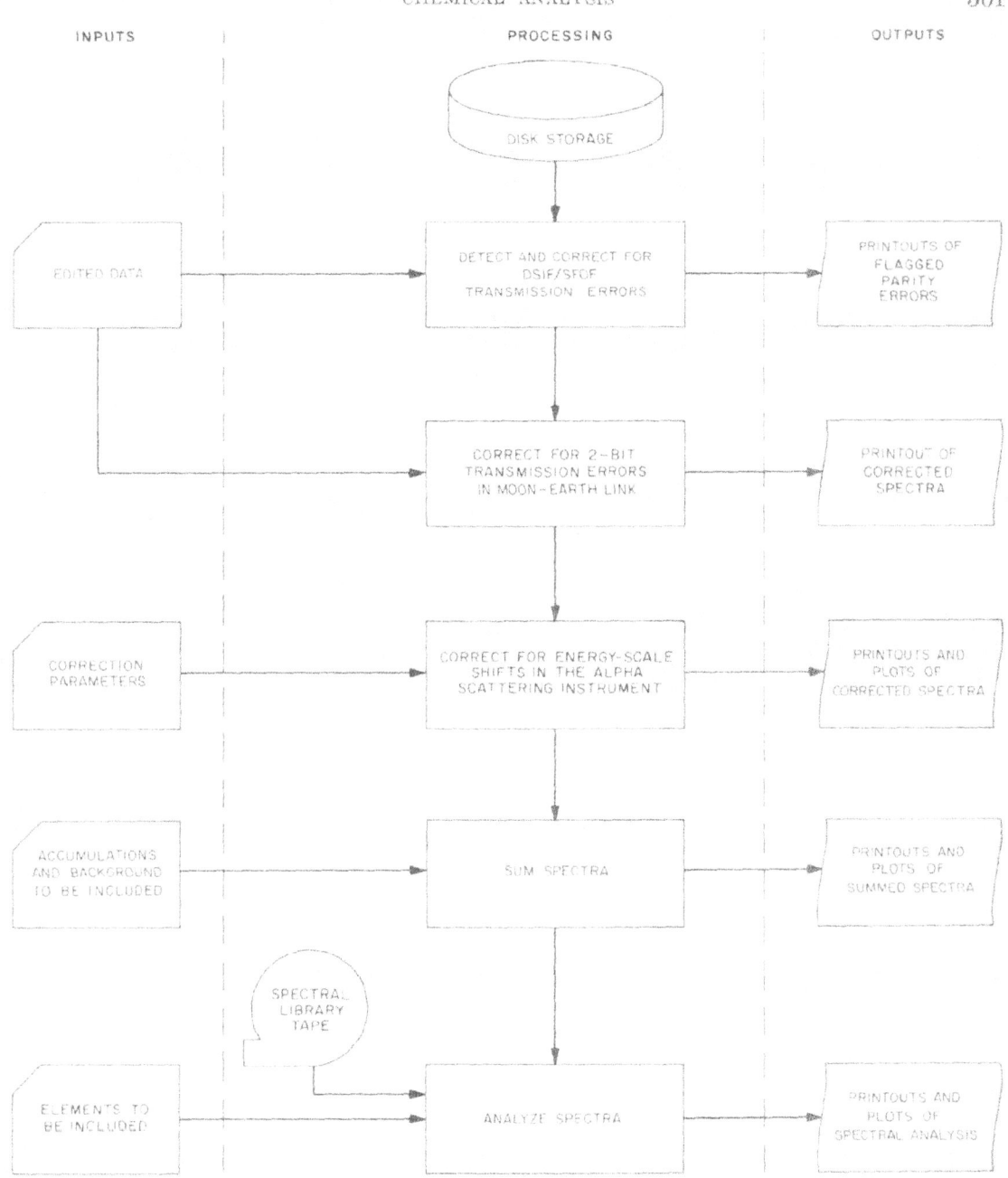

FIGURE 8-18.—Alpha-scattering real-time data processing in the SFOF computer system.

computer; corrections were made automatically using linear interpolation or by manual editing via card input.

The second stage in the real-time computer processing provided a crude correction to the data for temperature-induced shifts in the energy scale of the instrument. Correction parameters, based on premission calibration data, were introduced by card input. The program automatically realined the channel numbers using measured values of temperature included in the Teletype transmission so that any given channel number always corresponded approximately to the same particle-energy interval. In this way, it was possible to combine successive spectral accumulations in terms of a channel-by-channel addition. After transmission and energy corrections were applied, each spectral accumulation was stored on magnetic tape for further processing.

Each incremental accumulation of data represented the spectral results from a period of nominally 30 minutes. The individual accumulations from the various phases of operation were compared with predicted data and with each other to assess the behavior of the instrument and the progress of the measurement. To reduce statistical errors, several individual accumulations occasionally were summed by entering the desired identifying indices for the batches to be summed, as well as indicating an appropriate background (either measured or theoretical) to be subtracted from the sum.

To determine the real-time performance of the instrument in a manner resembling that of postmission data analysis, the spectra of the individual, as well as summed, accumulations were analyzed into their elemental components. A weighted least-squares program was used to find the combination of the elemental spectra of the library that yielded the best fit to the observed spectra. The weighting factor used in this comparison was the reciprocal of the estimated variance in each channel.

In this least-squares program, three different spectra were treated: (1) a selected range of channels in the alpha spectrum only; (2) a selected range of channels in the proton spectrum only; and (3) a spectrum obtained by considering contributions from both the alpha and proton spectra, as though they comprised a single spectrum. By this method, it was possible to determine the elemental composition that best corresponded to the measured alpha and proton spectra. Any significant change in the results of this analysis from one accumulation to the next was an indication of changing experimental conditions; for example, anomalous instrument behavior. The results of this analysis of standard-sample spectra could be compared with premission results to see whether the instrument was behaving normally in the lunar environment.

Calibration spectra obtained with the internal electronic pulser were also handled automatically by the ground data system. After transmission corrections were applied to the data, the peak positions and half widths were determined, and relevant calibration parameters were calculated. At each stage of the spectral processing, the data were plotted.

The entire ground-data-handling system was exercised in a series of tests before the Surveyor V mission. Initial tests included the simulation of flight data using the P–2 instrument. Magnetic-tape recordings of data from this instrument were also made, shipped to each tracking station, transmitted through the onsite equipment, and relayed to JPL for full-scale tests of the system.

The real-time availability of the lunar data and the flexibility of the programs used to process and display the data proved to be very useful during mission operations for the control of the experiment. The real-time data have also served as a source of preliminary results and have been used in the preparation of this and preceding reports (refs. 8–1 through 8–3).

In addition to providing the data for the real-time accumulations of spectra used for performance analysis, the composite signals received from the spacecraft were recorded in analog form on magnetic tape. These recordings contain information on each individual, analyzed event and constitute the prime experimental data to be used in postmission analysis. These tapes have been processed at JPL using a Univac 1219 computer, transferring the alpha-scattering data to digital magnetic tapes for later analysis.

Alpha-Scattering Sequence of Operations

The sequence of operations of the alpha-scattering experiment was planned to obtain information on the performance of the instrument, the background radiation in the lunar environment, and the composition of lunar material. These data were obtained by operating the instrument with the sensor head in each of three positions: stowed, background, and lunar surface.

In the stowed position, the sensor head was supported on a sample of known composition (the standard sample). Data received from this sample and from pulser calibration were compared with prelaunch measurements to give a measure of instrument characteristics in the lunar environment.

At the completion of this phase of operation, the instrument was deployed to the background position by Earth command. The supporting platform and standard sample dropped to one side, leaving the sensor head suspended about 0.5 meter above the lunar surface. In this position, the sensor head responded primarily to cosmic rays, solar protons, and possible surface radioactivity.

When it was determined that sufficient background data had been obtained, the sensor head was lowered directly to the lunar surface on command from Earth. The main accumulation of data was then begun, interrupted only by calibration sequences and other spacecraft operations. The calibration sequence with the pulse generator again was used to check the performance of the individual detectors and their amplifiers by the proper selection of detector on/off commands.

If operations permitted, the television camera was used to view the surface to which the sensor head was deployed and to monitor the deployment of the instrument. Because not all of the deployment area was always visible with the sensor head in the stowed and background positions, an auxiliary mirror was located on the spacecraft to give a more unobstructed view.

A typical alpha-scattering operation planned for the Surveyor missions was as follows:

(1) Television survey: stowed phase (including auxiliary mirror)
(2) Alpha-scattering operations: stowed phase
 (a) Accumulation of data: 2 to 6 hours
 (b) Calibration
(3) Deployment of sensor head to background position
(4) Television survey: background phase
(5) Alpha-scattering operations: background phase
 (a) Calibration
 (b) Accumulation of data: 2 to 6 hours
 (c) Calibration
(6) Lunar surface television survey
(7) Deployment of sensor head to lunar surface
(8) Television survey of sensor head in deployed position
(9) Alpha-scattering operations: lunar surface phase
 (a) Accumulation of data: 1 hour
 (b) Calibration
 (c) Accumulation of data: at least 24 hours
 (d) Calibration as required
 (e) Redeployment of sensor head to subsequent samples (Surveyor VII)

The actual operations conducted on Surveyors V, VI, and VII are described in the following paragraphs.

Description of Missions

Prelaunch operations at Cape Kennedy. Approximately 2 weeks before each launch, the sensor head was removed from the spacecraft at Cape Kennedy and taken to a special facility for final calibration and preparation for flight. The operations performed in this facility were a shortened version of the science calibration conducted on each instrument approximately 9 months earlier. (See "Instrument.")

A special test chamber equipped for thermal-vacuum operation and sample introduction was used for the measurements. The sensor head was calibrated at several temperatures using the electronic pulser and standard test sources of alpha radioactivity; background measurements were taken, and a test of the adequacy of the alpha-detector films in shielding the detectors from excessive light was performed. The Cm^{242} flight sources were installed, and spectra were obtained from a small selection of standard materials including polyethylene, magnesium, aluminum, quartz, iron, and nickel. For each of the three missions, the spare sensor head (F-3) was calibrated by the same procedure before this operation and was stored in another vacuum chamber for potential use.

Based on experience gained during preparations for Surveyor V, improvements were made in the special test facility and in the overall efficiency of operation prior to the Surveyor VI mission. For example, clean-room techniques were used to maintain cleanliness of the sensor head during source-installation procedures (ref. 8-2, first listing).

The thin films covering the alpha detectors and sources were inspected during the final preparations for flight and, if necessary, were changed. The aluminum-oxide films on the source collimators were changed for the Surveyor V mission because of the presence of foreign particles. The films were not changed for the Surveyor VI mission; they were, however, changed for Surveyor VII in order to minimize the amount of alpha activity deposited on them by aggregate recoil before the lunar measurements.

During nominal operations at the special test facility, the sensor heads were calibrated using a nonflight, digital electronic unit because the flight package was already in an inaccessible compartment on the spacecraft. On Surveyor VI, however, the F-1 digital electronic unit had been removed from the spacecraft for modification of the power supply, and thus was available for temperature calibration together with the F-1 sensor head.

The schedule of instrument calibration and source installation on each of the missions was arranged so that the sensor head could be reinstalled on the spacecraft at the appropriate time, about 1 week before launch. The sensor head was continuously purged with dry nitrogen gas during this period. The instrument was tested three times in the final week before launch to insure that it had been properly reinstalled and was still functioning. These tests consisted of short pulser calibration runs and accumulations of alpha and proton spectra. The spectra in the alpha mode showed features characteristic of scattering from nitrogen gas at atmospheric pressure. Figure 8-19 shows alpha spectra from nitrogen taken with the Surveyor VI instrument in the calibration facility 1 week before launch and then on the spacecraft under the shroud of the launch vehicle, a few hours before launch.

Transit operation. Operation of the alpha-scattering instruments while the spacecraft were in flight was desired in order to evaluate the

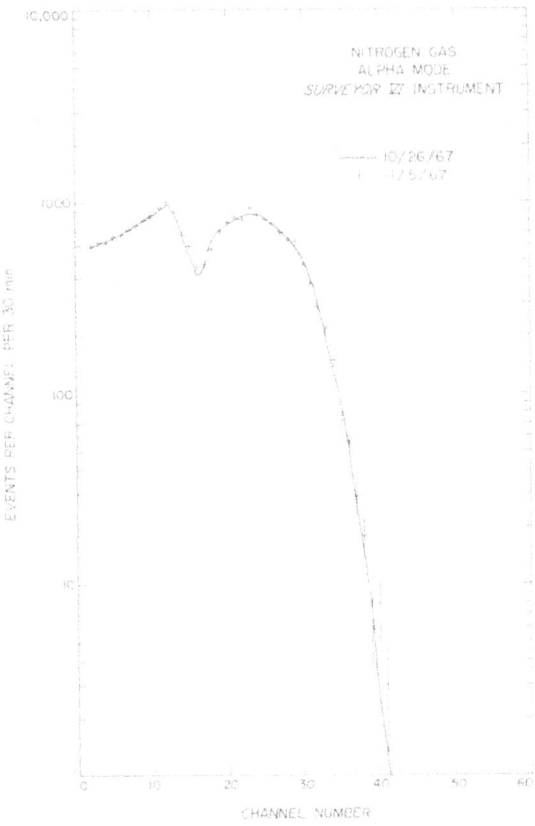

FIGURE 8-19.—Alpha-scattering spectra of nitrogen gas taken during prelaunch tests of the Surveyor VI instrument.

performance of the instruments after the launch, to give a measurement of instrument backgrounds in space, and possibly to shorten the stowed-phase operation after lunar landing. However, spacecraft operations permitted such a measurement during the flight of Surveyor V only.

This initial operation of the instrument occurred several hours after the midcourse velocity correction during the Surveyor V mission. The command to turn alpha-scattering power on was transmitted with the spacecraft 221 000 km from the Earth and 178 000 km from the Moon. The spacecraft, then being controlled from the tracking station at Canberra, Australia, was switched to high power and transmitted alpha-scattering data via an omnidirectional antenna. Two 10-minute accumulations of standard-sample data and seven 2-minute pulser calibration runs were received during less than 1 hour of operation. The data agreed well with prelaunch measurements, and showed that the instrument had survived the launching and midcourse rocket firing. The spectra also showed that the radiation background rates in cislunar space at that time were low enough for useful sample measurements.

Lunar landing. All Surveyors that carried an alpha-scattering instrument (Surveyors V, VI, and VII) landed successfully on the Moon. Certain aspects of the landings are particularly relevant to this experiment. On Surveyor V, for example, the terminal sequence had to be performed much closer to the Moon than had been originally planned. The main retromotor was operated to within 1.6 km of the lunar surface. (The standard end-of-burning distance was about 12 km.) Because aluminum-oxide particles comprise part of the exhaust products of this solid-propellent rocket, the possibility of an effect on the alpha-scattering experiment was investigated and reported in reference 8-1. The percentage of surface area covered by the emitted particles was estimated to be only about 0.04 percent. Moreover, because of the high velocity of impingement, it was considered unlikely that the aluminum-oxide particles came to rest on the very surface of the Moon at the point of impact. The amount of contamination from the more standard descents of Surveyors VI and VII are estimated to be an order of magnitude less than that from Surveyor V.

Because of possible differences in shielding by the Moon between the background and lunar-surface positions, the topography of the local lunar terrain may have to be considered in the detailed evaluation of background data. This may be especially important for Surveyor V which landed on the interior slope of a crater. The sensor head on that spacecraft was deployed to the lunar surface at a level lower than the surrounding terrain.

Postlanding operations. The alpha-scattering operations for Surveyors V, VI, and VII are summarized in figure 8-20. This chart shows, for the first lunar day on each of these missions, the major activities and data-accumulation periods of the experiment. The data-accumulation periods for the various phases of the experiment on each of the missions are given in table 8-9. The following paragraphs describe the operations in more detail.

Stowed position. After spacecraft landing, the instrument was operated with the sensor head in the stowed position, obtaining spectra from a sample of known composition and from calibration tests with the electronic pulser. On each of the three missions, the instrument was found to be operating normally as soon as it was activated. This operation was initiated 2.0, 4.6, and 8.4 hours after spacecraft touchdown on Surveyors V, VI, and VII, respectively.

On Surveyor V, only engineering data (instrument voltages, temperatures, and guard monitor) were received at the start because of other spacecraft operations; however, on Surveyors VI and VII, standard-sample spectra were received at initial turn-on.

After initial receipt of normal engineering data on Surveyor V, the instrument was turned off, and tracking operations were transferred from California to Australia. At this point in operations, television surveys had been planned of the deployment area on the lunar surface. Pictures of the auxiliary viewing mirror were obtained with both wide- and narrow-angle lenses. Only the lower resolution, wide-angle frames included the reflection of the deployment area in the mirror. A picture shown in reference 8-1 indicated that the largest rock in the sample area was smaller than approximately 3 cm. On Surveyors VI and VII, it was possible to view the initial sample area directly during deployment of the sensor head to the background position.

Because operations in the stowed phase during each of the missions occurred early in the lunar day, instrument temperatures remained within normal operating ranges. The performance of the instruments and auxiliary equipment during this phase was excellent, except that when the Surveyor V instrument was turned on for the second time (6.7 hours after touchdown), the calibration pulser started without being commanded on. The pulser was turned off by command 0.4 hour after second turn-on and further operations were normal.

The planned data-accumulation period for the stowed phase was 2 to 6 hours. Because of

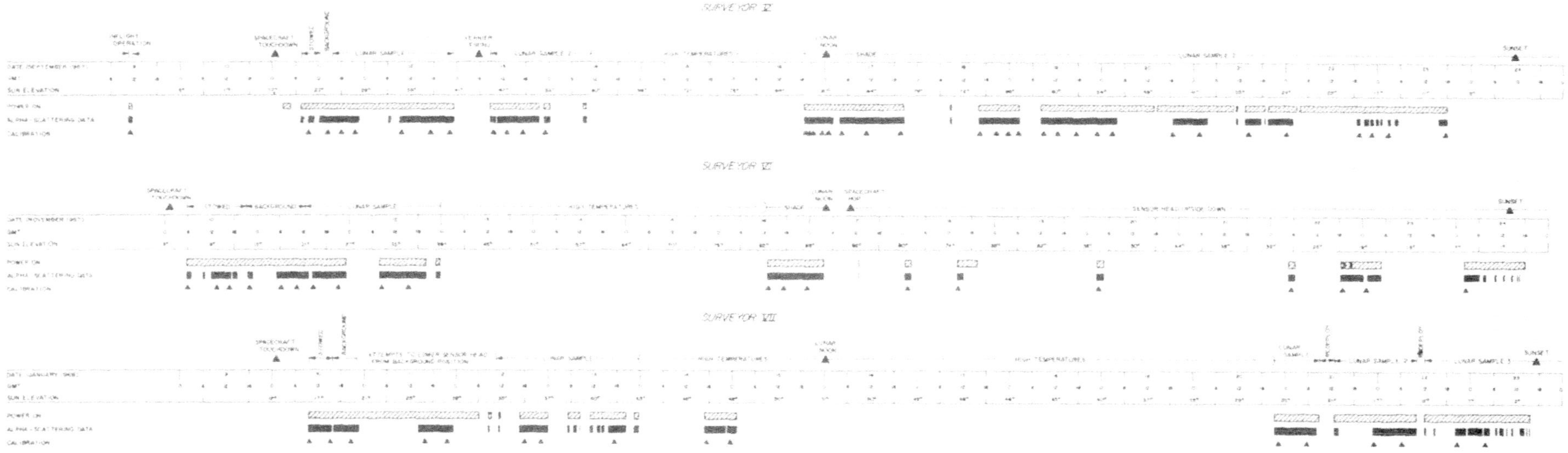

FIGURE 8-20.—Mission-operations summary of the alpha-scattering experiment for the first lunar day of Surveyors V, VI, and VII.

TABLE 8-9. *Alpha-scattering experiment mission operation periods*

Characteristic	Standard sample				Background				Lunar-surface sample measurements					
	Surveyor V transit[a]	Surveyor V	Surveyor VI	Surveyor VII	Surveyor V	Surveyor VI	Surveyor VII		Surveyor V		Surveyor VI	Surveyor VII		
									First	Second	First	First	Second	Third
Total power-on time, hr	0.8	6.6	15.6	6.4	3.4	14.8	38.9		31.6	152.3	39.1	41.4	21.2	27.5
Total time of mode 4 data, hr	.8	2.0	7.9	4.8	2.9	5.9	11.8		16.2	83.3	31.7	31.1	11.4	8.4
Total time of science data accumulation, hr[b]	.3	1.3	5.3	3.2	2.9	7.9	12.6		18.3	71.9	c29.6	c28.0	11.2	7.5
Number of pulser calibrations	1	1	3	2	d7	d2+1	3		5	d28+1	d8+1	7	2	2

[a] These data were accumulated while the spacecraft was in flight to the Moon.
[b] These numbers represent estimates of times that the science data were being recorded on magnetic tape at the tracking stations.
[c] During part of these periods of operation, less than four proton detectors were in use.
[d] Partial calibration.

CHEMICAL ANALYSIS

the schedule of other spacecraft activities, this period was shortened to 1.25 hours on Surveyor V. The duration of the stowed-phase operation was more than 5 hours on the other missions (table 8-9).

Background position. To determine whether the sensor head had deployed successfully to the background position, it had been planned to take television pictures of the instrument just before and after the deployment. Because of the short time available on Surveyor V, the planned television check was not conducted. Instead, the alpha-data bit stream was monitored at the tracking station, using an oscilloscope display of the discriminator output. If the deployment proceeded correctly, it was expected that the alpha-event rate would drop by more than a factor of 10 as the standard sample moved away. The command was sent 11.5 hours after touchdown; after a few seconds, word of the expected decrease in rate, indicative of a successful deployment, was received from the tracking station. Figure 8-21 shows the total alpha-event rate during this period, as derived later from a magnetic-tape recording.

On Surveyors VI and VII, this command was transmitted 20.3 and 14.7 hours after touchdown, respectively. On these missions, a series of television pictures of the deployment operation was planned, primarily to provide a direct view of the lunar surface otherwise obscured by the sensor head. When the sensor head was released from the stowed position, it continued moving for some time like a pendulum and, although the television pictures could not be synchronized with the motion, there was a good chance that at least one of a series of pictures would show a substantial portion of the sample area. This sequence was successful on both missions. Figure 8-22 shows part of the Surveyor VII sequence. In addition to providing views of the sample areas, such a picture series also permits measurement of the period of the sensor-head pendulum under lunar conditions.

The data-accumulation period for the background phase was planned to be about 6 hours. The actual times for the three missions (table 8-9) were 2.9, 7.9, and 12.6 hours. The quality of the data seems to be excellent on each of the missions.

Lunar-surface operations. On Surveyors V and VI, the command to lower the sensor head to the lunar surface was transmitted 14.8 and 35.1 hours, respectively, after spacecraft landing. This operation occurred relatively soon during the Surveyor V mission because of an impending vernier rocket test; it occurred late on Surveyor VI because of other spacecraft operations that restricted alpha-scattering background data accumulations. The deployment operation on each of the missions was monitored by observations of the alpha data stream on an oscilloscope at the tracking station. On Surveyors V and VI, the event rate increased sharply, seconds after sending the deployment command. Figure 8-21 shows the total alpha event rate during this period for Surveyor V. The rate rose from about one event per 10 seconds in the background position to two events per second on the lunar surface. The event rate from the lunar surface can be seen to be comparable to that from the standard sample.

On Surveyor VII, the command to lower the sensor head to the surface was transmitted 20.9 hours after landing. This time, no apparent change in counting rate was observed. The deployment command was retransmitted; again, no increase in counting rate was observed. A 10-minute accumulation of data was consistent with that obtained before the command to lower had been sent, and indicated that the sensor head possibly had not moved.

When tracking operations were transferred

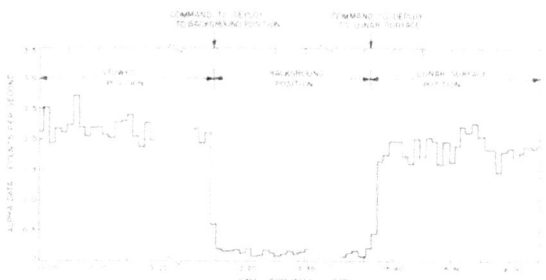

FIGURE 8-21.—Plot of event rate of analyzed alpha particles as a function of time, showing marked decrease when the Surveyor V sensor head was deployed from the stowed to the background position and the increase when lowered from the background position to the lunar surface.

FIGURE 8-22.—Sequence of pictures taken just after Surveyor VII sensor head was deployed from the stowed to the background position. Movement of sensor head is visible in successive pictures (Jan. 10, 1968, 15:49:03, 15:49:12, 15:49:15 GMT).

to the Goldstone, Calif., tracking station, television pictures were taken to help diagnose the problem. The pictures showed that the sensor head was still suspended in the background position, but that the small retaining door (used to prevent premature deployment of the flat electronics cable) had opened correctly. This showed that the deployment command had been properly received by the spacecraft and that the squib-activated pinpuller had operated. This information isolated the problem to the nylon suspension cord or its associated storage spool and escapement mechanism, affording hope that operation of one of the movable parts of the spacecraft would provide enough force to free the sensor head.

After several attempts to free the mechanism, the deployment was accomplished with the aid of the surface sampler, which, in a most ingenious series of operations, improvised and controlled from the SFOF, managed to force the sensor head to the lunar surface (ref. 8-20).

On all missions, the initial operations by the alpha-scattering instrument on the lunar surface were conducted in a period of rising temperatures. This was especially serious on Surveyor VII, because of the relatively late landing (with respect to sunrise), and because of the delayed deployment (57.4 hours after landing). Also, on Surveyor VII, because of the attitude of the spacecraft and the high latitude of the landing site, shading of the instrument during lunar noon by the large panels on the spacecraft was not feasible. Partial shading was provided several times by means of the surface sampler. The high-temperature periods on the three missions are indicated in figure 8–20.

Six lunar samples were analyzed during the three missions. The data-accumulation periods for these samples varied from 7.5 to 71.9 hours for the first-lunar-day operations (table 8–9). The samples are described in the next part of this chapter.

Surveyor V analyzed two lunar samples. The data-accumulation period was 18.3 hours at the initial position of the sensor head on the lunar surface. At 53 hours after spacecraft touchdown, the vernier rocket engines were fired, moving the sensor head about 10 cm farther from the spacecraft. When the instrument was again turned on, it was found to be functioning perfectly, and operations were begun at the new lunar-surface position. By the end of the first lunar day, additional data for 71.9 hours had been obtained. During the second lunar day, the spacecraft was partially reactivated, and the alpha-scattering instrument was operated for several hours. Most of the instrument was found to be in working condition, but digital anomalies, apparently resulting from damage to components during the cold lunar night, prevented obtaining useful science data.

Surveyor VI analyzed only one lunar sample. After a data-accumulation period of 29.6 hours, the vernier engines were fired, moving the spacecraft about 2.4 meters. Subsequent television pictures showed that the sensor head was upside down. Figure 8–23 shows the instrument in this position, as viewed in an auxiliary mirror, with the base plate facing upward. The instrument was found to be functioning after the spacecraft hop, but subsequent data accu-

FIGURE 8–23.—Alpha-scattering sensor head in upside-down position after the hop made by the Surveyor VI spacecraft (Nov. 17, 1967, 13:37:06 GMT).

mulation was restricted to the measurement of radiation from space.

Surveyor VII was the most productive mission from the chemical analysis viewpoint. Because of the presence of the surface sampler on this spacecraft, plans had been made to redeploy the sensor head to additional samples after the initial analysis. In spite of the delayed deployment and high-temperature restrictions, the surface sampler provided three samples for analysis. The first-lunar-day data-accumulation periods were 28.0 hours for sample 1 (undisturbed lunar surface), 11.2 hours for sample 2 (a lunar rock), and 7.5 hours for sample 3 (a trenched area).

During the alpha-scattering operations on Surveyors V, VI, and VII, approximately 75 pulser calibrations of the instrument were performed at a rate of about one calibration per 3 hours of operation.

Description of Lunar Samples

Surveyors V and VI landed in lunar maria, regions of the Moon characterized by their flatness, relative smoothness, and low albedo. Surveyor VII landed in the lunar highlands near the crater Tycho. The region near the landing site of Surveyor VII is topographically complex, but is generally rougher and has a higher albedo than the mare landing sites.

The lunar surface at each of these landing sites consists of weakly cohesive, fine particles; aggregates; and solid fragments. Most of the resolvable fragments are brighter in appearance than the fine-grained matrix material. Of all the five Surveyor landing sites, those of Surveyors V and VI have the least number of resolvable fragments, whereas the Surveyor VII site has the highest number of larger fragments and strewn fields of blocks. The size distribution of small craters (less than a few meters) is about the same at each landing site, but there are fewer larger craters near Surveyor VII than at the mare sites. (See ch. 3 of this report.)

Surveyor V landed in the southwest portion of Mare Tranquillitatis at 23.2° E longitude, 1.4° N latitude. The spacecraft touched down on the southwest wall of a 9- by 12-meter rimless crater, more than 1 meter deep. On the basis of its shape and alinement of small associated craters with other dominant linear features in surrounding areas, the crater has been interpreted as being formed by the drainage of surface debris into a subsurface fissure. The walls of this crater appear to expose parts of the upper meter of this debris layer. Material from the wall of the crater was dislodged during the spacecraft landing and slid into the area analyzed with the alpha-scattering instrument. Figure 8-24 shows this disturbed area, as interpreted by Shoemaker et al. (See ch. 3.) The frequency-size distribution of the lumpy, fragmental material ejected by the footpads during landing is much coarser than that observed on the undisturbed surface.

Only low-resolution television pictures were obtained of the samples analyzed by Surveyor V (ref. 8-1). These pictures indicated that no particles larger than about 3 cm were located in the sample area. Figure 8-25 shows high-resolution-picture mosaics of the sensor head as it conformed to the lunar surface before and after the vernier engine firing. On initial deployment, the outboard edge of the circular base plate can be seen to have become embedded slightly in the surface material. After the vernier firing, that edge of the plate became more deeply embedded, tilting the sensor head further so that the left inboard edge of the plate was raised about 2 cm above the surface. The surrounding surface appears to be relatively smooth, with few fragments large enough to interfere with the operation of the instrument. Operation with individual proton detectors also indicated that the geometry of the sample was relatively flat.

Sinus Medii, the landing site of Surveyor VI, is a relatively small mare plain, about 170 km across, located in the center of the Moon as seen from Earth. The area is bounded by terrae to the north and south. The relatively high density of craters larger than a few hundred meters and the relatively thick debris layer (as suggested by the lack of smaller craters with blocky rims) has been interpreted to mean that the surface material in Sinus Medii is older than that of most of the material exposed at the other mare sites (ref. 8-21). The frequency-size distribution of small craters is closely similar to that found at other mare sites; however, there are fewer fragments larger than a few centimeters at the Surveyor VI and Surveyor V sites. Despite these differences, all of the mare sites are similar in topographic details, that is, in the structure of the surface layer, and in other properties as measured by Surveyor (ref. 8-22).

The area surrounding the Surveyor VI sample (fig. 8-26) can be seen to be free of large fragments. The largest particle within the sample area itself is only about 1.5 cm across. Particles as small as 0.3 cm are distinguishable in the picture.

The Surveyor VII landing site, about 30 km north of the crater Tycho, was much different from all of the previous landing sites. Tycho is a large ray crater located in the southern highlands of the Moon. The ray system and other features of the crater have been interpreted as resulting from hypervelocity impact, although interpretations of its origin involving volcanism have also been given (ref. 8-23). Most investigators agree that Tycho is a relatively young lunar feature and that Surveyor VII landed on material that flowed outward down the rim slope of the crater. According to the impact theory, these flows occurred during or shortly after the formation of the crater and consisted of fluidized masses of particles brought up from several kilometers below the surface. Other interpretations are that the flows are either a result of later "landslides" of rim material or

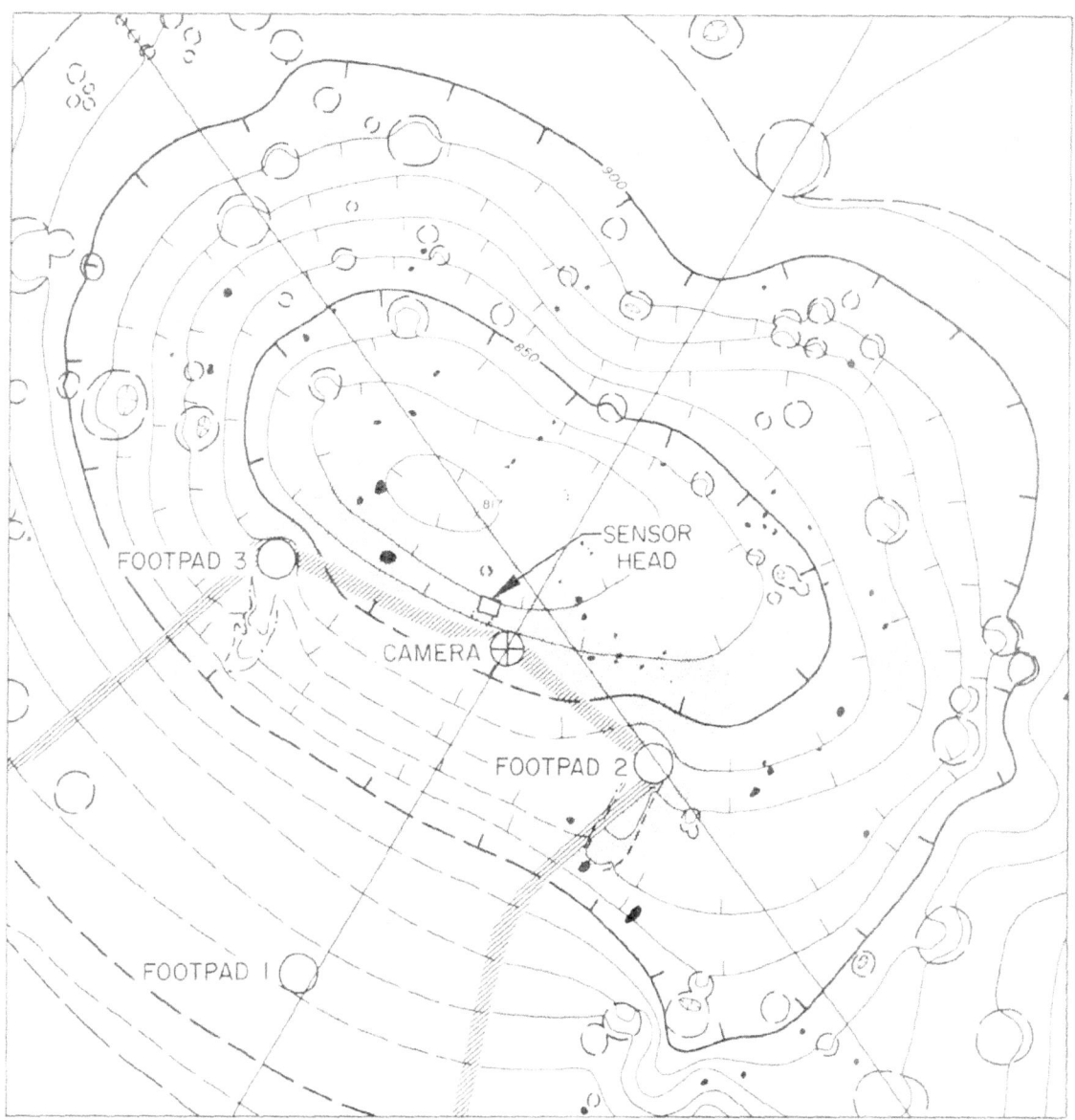

FIGURE 8-24.—Topographic map of part of Surveyor V landing site. The stippled area indicates material ejected by footpads during landing; the sensor head was deployed within this area. Contour interval is 10 cm (topography by R. M. Batson, R. Jordan, and K. B. Larson of the U.S. Geological Survey).

that they are lava flows that originated some time after crater formation.

The terrae in general (and the Tycho region in particular) show a greater diversity in topography and optical properties than do the lunar maria. The Surveyor VII landing site does, however, share with other highland areas their most prominent characteristics of a rougher topography and higher albedo than those of the maria. Per unit area, as seen by the Surveyor camera, more fragments larger than 4 cm were observed at the Surveyor VII site than at any

FIGURE 8-25.—(a) High-resolution mosaic of the Surveyor V sensor head on the lunar surface before vernier firing (Sept. 12, 1967, Catalog 5-MP-24). (b) Mosaic after vernier firing (Sept. 14, 1967, Catalog 5-MP-25).

(b)

FIGURE 8-26.—Wide-angle mosaic of television pictures taken just after the Surveyor VI sensor head was released from the stowed position. The circle shows the area of the lunar surface later analyzed by means of the alpha-scattering experiment (Nov. 10, 1967, 21:17:57 to 21:23:11 GMT).

of the mare landing sites (ref. 8-24). The albedo of the surface material observed with the Surveyor VII television camera was about twice that observed at the Surveyor V and VI sites. This is consistent with Earth-based telescopic observations indicating that the halo region around Tycho, in which Surveyor VII landed,

although darker than the crater or the surrounding terrae, is still brighter than most of the maria.

Figures 8-27 (a) and (b) show mosaics of narrow-angle pictures of part of the surface-sampler area of operation before and after the start of alpha-scattering analyses. Figure

CHEMICAL ANALYSIS

FIGURE 8-27.—(a) Narrow-angle mosaic of Surveyor VII pictures showing area of deployment of the sensor head before the start of surface sampler activities (Catalog 7-SE-4).

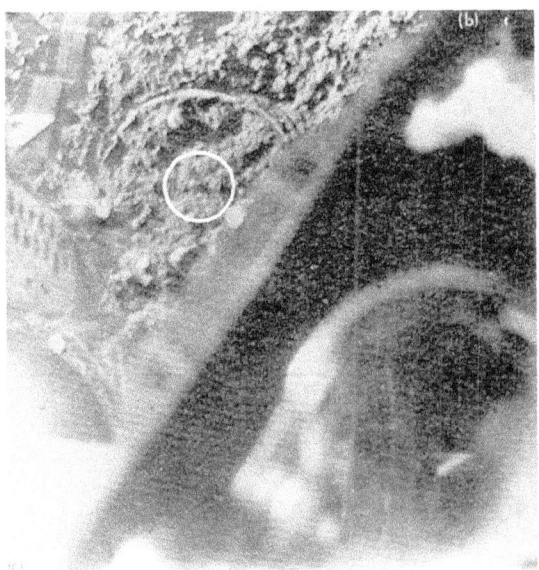

(b) Narrow-angle mosaic of Surveyor VII pictures showing sensor head in place on a lunar rock (sample 2). The surface sampler is seen in the process of digging the second of three trenches at the site of alpha-scattering sample 3. The sample 1 location at the lower

left of the figure is surrounded by a circular impression of the bottom of the sensor head (Catalog 7-SE-16). (c) Surveyor VII lunar sample 1, undisturbed lunar surface, after analysis. The circle shows the actual sample area (Jan. 22, 1968, 18:09:09 GMT). (d) Surveyor VII sample 2, a lunar rock, after analysis. The area outlined by the ellipse (including the rock) shows the size of the sample. (e) Surveyor VII sample 3, an area of the lunar surface trenched by the surface sampler, before analysis. The outline shows the subsequent sensor head and sample positions (Jan. 22, 1968, 10:36:42 GMT).

8-27(c) shows the undisturbed area of the lunar surface, including the first sample analyzed at the Surveyor VII site. Although several fairly large fragments can be seen on the surrounding surface in figure 8-27(c), the central outlined area, comprising the actual sample, can be seen to be relatively smooth. The largest particle in this sample 1 area is about 1.5 cm across. A rock about 4 cm across was located underneath the inboard side of the circular plate of the sensor head during the analysis, causing the sample to be farther than standard from the sources and detectors.

Sample 2 was a lunar rock about 5 by 7 cm on its upper face. This rock was somewhat brighter in appearance than the surrounding surface (see fig. 8-27(a)), and was visible as an exposed object on the lunar surface before the surface-sampler operations. Figure 8-27(b) shows the sensor head in position on the rock. The measured event rate in the alpha mode was about twice that for a sample at the standard distance. This information, together with the television pictures, indicates that during analysis the rock was centered in the sample opening and protruded slightly inside the bottom of the sensor head. During deployment to this sample, the sensor head moved the rock back and forth slightly, but the upper surface of the rock apparently remained on top during this motion. Figure 8-27(d) shows the rock after the sensor head was removed.

The third sample location was a trenched area of the lunar surface previously prepared by the surface sampler (fig. 8-27(b)). Figure 8-27(e) shows an outline of the sensor head as it rested on the third sample, showing that the sample area beneath the sensor head was located primarily within the end of the central trench of three trenches. This indicates that the sample analyzed consisted, at least in part, of subsurface material. Whereas most of the other samples analyzed on the three missions may have consisted of only the top few microns of undisturbed material, this final sample probably consisted of some material originally centimeters below the surface. The actual depth of the trench at the point of analysis cannot be determined by the pictures, however, because the bottom of the trench was in shadow.

Summary of Operations and Performance

Table 8-9 is a summary of operating periods and data accumulation times for the alpha-scattering experiment on Surveyors V, VI, and VII.

The performance of the alpha-scattering equipment and operational system during the three missions was excellent. The semiconductor radiation detectors had been expected to be the least reliable components in the instruments. Of the 30 detectors operated on the Moon, only one detector on Surveyor VI became noisy within the specified temperature limits of $-40°$ to $+50°$ C. A few noise bursts that could not be specifically traced to detectors were observed on Surveyor V in the proton system during a period of rapidly changing temperatures and on Surveyor VII in the guard system for periods of a few seconds. On Surveyors V and VII, detectors survived the lunar night and operated normally after initial periods of noisy behavior.

Data received during each mission showed that the thin films covering the sources and alpha detectors had survived the launch, midcourse maneuver, and lunar landing operations. The quality of the Cm^{242} sources had been preserved during the prelaunch and launch phases of operation, as evidenced by the sharp features observed in the sample spectra.

The electronics, calibration pulsers, and Es^{254} sources performed as expected, as evidenced by agreement of mission data with prelaunch data. Calibration peaks from the pulser were generally sharp, although some low-energy noise was occasionally seen in the pulser data from Surveyors VI and VII, as in prelaunch tests. The guard detector and anticoincidence system worked as designed; guard monitor voltages and proton background spectra agreed well with predicted values.

On the lunar surface, the temperatures of the sensor head were higher than predicted on each of the three missions. The reasons for this discrepancy are not known, but one possibility is that the sides of the box were thermally coupled more strongly with the interior parts than had been expected.

The digital electronic, instrument power supplies, and electronic auxiliaries performed as designed. Circuits used to monitor engineering parameters provided good data. Of the more than 1000 commands transmitted to the instruments, all but two appeared to give the proper response. The first anomaly occurred on Surveyor V. When the instrument was commanded on for the second time, the calibration pulser started operating and had to be turned off by separate command.

The second anomaly associated with commands occurred on Surveyor VII when an explosive squib was blown to allow the sensor head to descend to the lunar surface. The command was received by the spacecraft and the explosive device worked, but the nylon cord or its associated spool and escapement mechanism failed to function. This failure could have been calamitous except for the presence of the surface sampler. All other operations of the deployment mechanisms proceeded flawlessly.

In general, communication links from the spacecraft were excellent; typical bit-error rates encountered were less than 10^{-6}, although much higher levels occurred when the spacecraft were being tracked near the Earth's horizon. The data-handling and computer-processing systems proved to be indispensable for the monitoring of the experiments during the actual missions.

Television support of the alpha-scattering operations was very useful, but viewing of the deployment areas via the auxiliary mirrors did not give good results. The chemical analyses on the Surveyor VII mission would not have been possible without the combined use of the television camera and surface sampler.

Results

The Surveyor V, VI, and VII missions provided the first chemical analyses of lunar-surface material. The Surveyor V and VI analyses were from mare sites on the Moon; the Surveyor VII samples were at a terra site, outside the crater Tycho. Two lunar samples were examined on the Surveyor V mission, one on the Surveyor VI mission, and three on the Surveyor VII mission (see preceding section). On each mission, the first sample examined was the original deployment area of the alpha-scattering instrument. In the last two missions, the first sample was apparently the undisturbed lunar surface. Surveyor V landed just inside a rim of a small crater. During the landing, the spacecraft slid and threw out some subsurface material; thus, the first sample analyzed on this mission may have consisted primarily of this ejected material. The second sample was probably similar material because the instrument was moved only about 10 cm by the static firing of the vernier engines.

Only one sample was examined on the Surveyor VI mission. The second sample examined on the Surveyor VII mission was an exposed lunar rock (see preceding section); the third was a region of the lunar surface previously trenched by the surface sampler and was thus at least partly subsurface material.

All of the chemical analysis results presented in this report are based on spectra relayed by Teletype from the Deep Space Stations. These spectra were sent in essentially real time for purposes of instrument-performance analysis and mission planning (see preceding section). In addition to lack of detailed evidence regarding their reliability, the data have, as yet, been corrected only approximately for nonstandard instrument behavior. The quality of these real-time data was sufficiently high that preliminary science results could be deduced from them. These preliminary results are presented in this report.

The prime data from the experiment were recorded at the Deep Space Stations (often in duplicate). Because of the difficulty of separating them from the mass of other data not relevant to the alpha-scattering experiment, these prime data have become available only recently; thus, it has been possible only to qualify their generally satisfactory format and check on their completeness. The data from the experiment are recorded in two forms:

(1) Science data, a time-tagged record of the channel number (energy) of each event registered by the alpha detectors and by proton detectors. In addition, most commands sent to the spacecraft and checks on the quality of the transmission of the data are recorded.

(2) Engineering data transmitted periodically from the spacecraft. In addition to strictly engineering information, these include the voltage corresponding to the event rate in the guard detectors of the instrument. This rate is determined primarily by the cosmic-ray and solar proton flux in the energy range 50 to 200 MeV. Thus, the guard detectors in the alpha-scattering instrument on the Surveyor missions represent radiation-monitoring devices on the lunar surface.

The actual analytical results reported here are based exclusively on the spectra received in near real time by Teletype. These spectra covered data accumulated over periods of, typically, 20 to 40 minutes. Some of these data have been subjected to preliminary certification to remove any that had suspicious characteristics (high parity errors, anomalous event rates, and so forth). They were then corrected approximately for the temperature coefficients of the instrument using the measured sensor head and electronic temperatures that were available at the time. The spectra were then combined to cover periods of, typically, 100 to 300 minutes and were again examined for consistency. They were then combined to give the data that were subjected to a computer treatment leading to the preliminary analytical results presented here.

Table 8-10 presents, for each phase of operation on each mission and for each sample, the total amount of time during which the certified data were accumulated. These periods are seen to range from 20 minutes for the inflight operation on the Surveyor V mission to 34.5 hours on sample 2 of the Surveyor V mission.

TABLE 8-10. *Total time represented by certified data (used in calculations of preliminary results)*

Phase	Total time, hr		
	Surveyor V	Surveyor VI	Surveyor VII
Inflight	0.33		
Standard sample	1.00	4.3	5.0
Background	2.33	6.10	10.5
Sample 1	15.0	a 13.2	11.2
Sample 2	34.5		9.7
Sample 3			5.9

a Part of the data from this period was recorded with 3 of the 4 proton-detector systems operating.

The near-real-time computer treatment of the data was devised primarily to provide analytical results that would serve as a check on instrument behavior during a mission. For this reason, only approximately correct temperature coefficients were put into the SFOF computer for these calculations. No attempt was made to use the results of the internal electronic pulser calibrations.

The resulting spectra were analyzed using a limited library of responses of the instruments to pure elements. Eight elements were chosen for this library, partially on the basis that they represented contributions in different regions of the alpha and proton spectra and partially on the basis of estimates of the probability of their being present in major amounts in the lunar sample. The elements composing the library were carbon, oxygen, sodium, magnesium, aluminum, silicon, calcium, and iron. The last two elements were meant, at this stage of analysis, to represent elements with mass numbers of about 31 to 47 and about 47 to 65, respectively.

The library used for the missions was taken from the science calibration of the instruments, which had been performed many months before the missions (fig. 8-4 is an example of such a library). At this stage of analysis, little attempt was made to correct this library for slight changes in the characteristics of the instrument during the intervening periods. This meant that slight discrepancies, of the order of one channel, were often present between the energy scales of the instrument at the time of the mission and that of the library used in the analysis.

Standard-Sample Data

The first stage of lunar operations of the alpha-scattering instrument was the measurement on the standard sample. This was usually performed within 8 hours after touchdown and soon after power had been applied to the instrument. On Surveyor V, however, a short period of operation of this type was performed while the spacecraft was in flight to the Moon.

The objective of this stage was to establish that the instrument had survived the flight to the Moon and could perform a chemical analysis of a sample of known composition. Instrument survival was verified by examination of the actual data stream and the positions of electronic pulser peaks. Verification of the lack of radioactive contamination gave assurance that the protective films over the sources were intact. In addition, the data received from the Moon in the standard-sample-measuring phase on the Moon could be compared with those obtained on Earth before launch.

The portion of the Surveyor V inflight data that has been conditionally certified and corrected approximately to standard instrument temperature is presented in figure 8-28. The certified data (table 8-10) for the three missions, obtained on the standard samples after landing, are shown in figure 8-29. In these figures, the number of events per channel in the alpha and

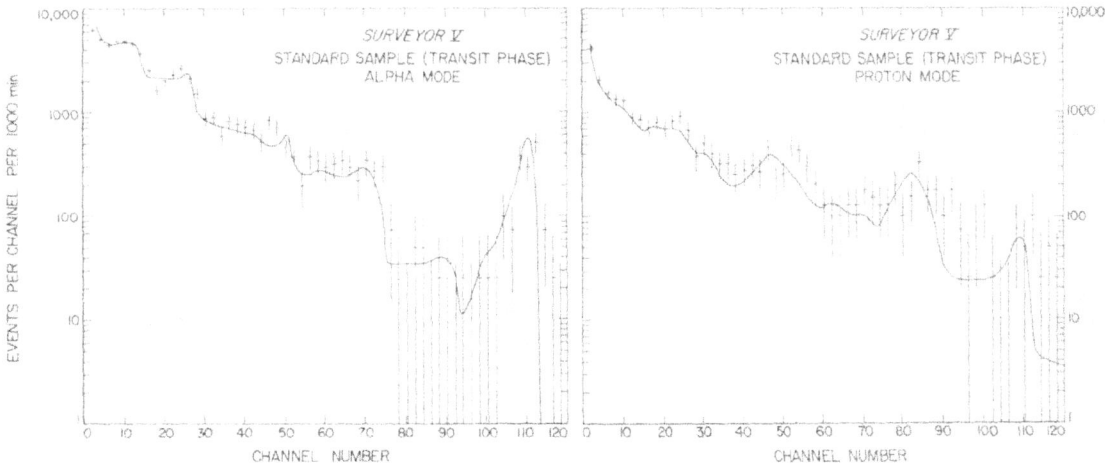

FIGURE 8-28.—Surveyor V inflight data on standard sample. Data taken during a 20-minute measurement while the spacecraft was in transit to the Moon. The experimental points, with statistical errors, are the averages of two channels in order to improve their statistical significance. The smooth curves are derived from data obtained in prelaunch measurements of a similar sample. The peaks at around channel 110 are caused by the Es^{254} located near the detectors as calibration sources.

proton modes of the instrument are plotted on a logarithmic scale as a function of the channel number (energy). The statistical (1 σ) errors are indicated by vertical bars.

In figure 8-28 (the inflight data from Surveyor V), the smooth curves are derived from data obtained on a similar sample during the final calibrations of the sensor head at Cape Kennedy about 2 weeks before launch. The prelaunch data have been corrected for source decay during the intervening period and for the background rates observed later on the lunar surface. It is seen that the data obtained during the short inflight operations are close to those expected from the standard sample.

In figure 8-29 (the standard-sample measurements on the Moon for the three missions), the smooth curves indicate the background levels observed in the following stage of lunar operations. The data presented in these figures are similar to those obtained on standard samples before launch, differing principally as a result of different background levels for the Moon and the Earth.

The lack of excess events in the alpha mode at approximately channels 103 to 104 gave assurance, in each mission, that the thin protective films over the alpha sources had survived the launch, flight, and touchdown conditions and that radioactive contamination of the instrument had not occurred.

A computer analysis of these spectra from the standard samples (after background subtraction) in terms of the library of eight elements reproduces the data quite well (fig. 8-30). The resulting chemical analyses of the standard-sample glass on the three missions are shown in table 8-11. Indicated also, for comparison, are the chemical compositions of the glasses used, as determined before launch, by conventional analytical techniques. These analyses were made on similar but not identical pieces of glass. The agreement, at the present stage of data processing, between the two types of analytical results is certainly not as good as desirable. However, it must be remembered that the lunar standard-sample analyses had short measurement times, that the possibility of the inclusion of tainted data cannot be ruled out by the present certification techniques, and that only approximate corrections have been made so far for nonstandard instrument conditions (for example, for the temperature coefficients and other effects on the energy scales). For example,

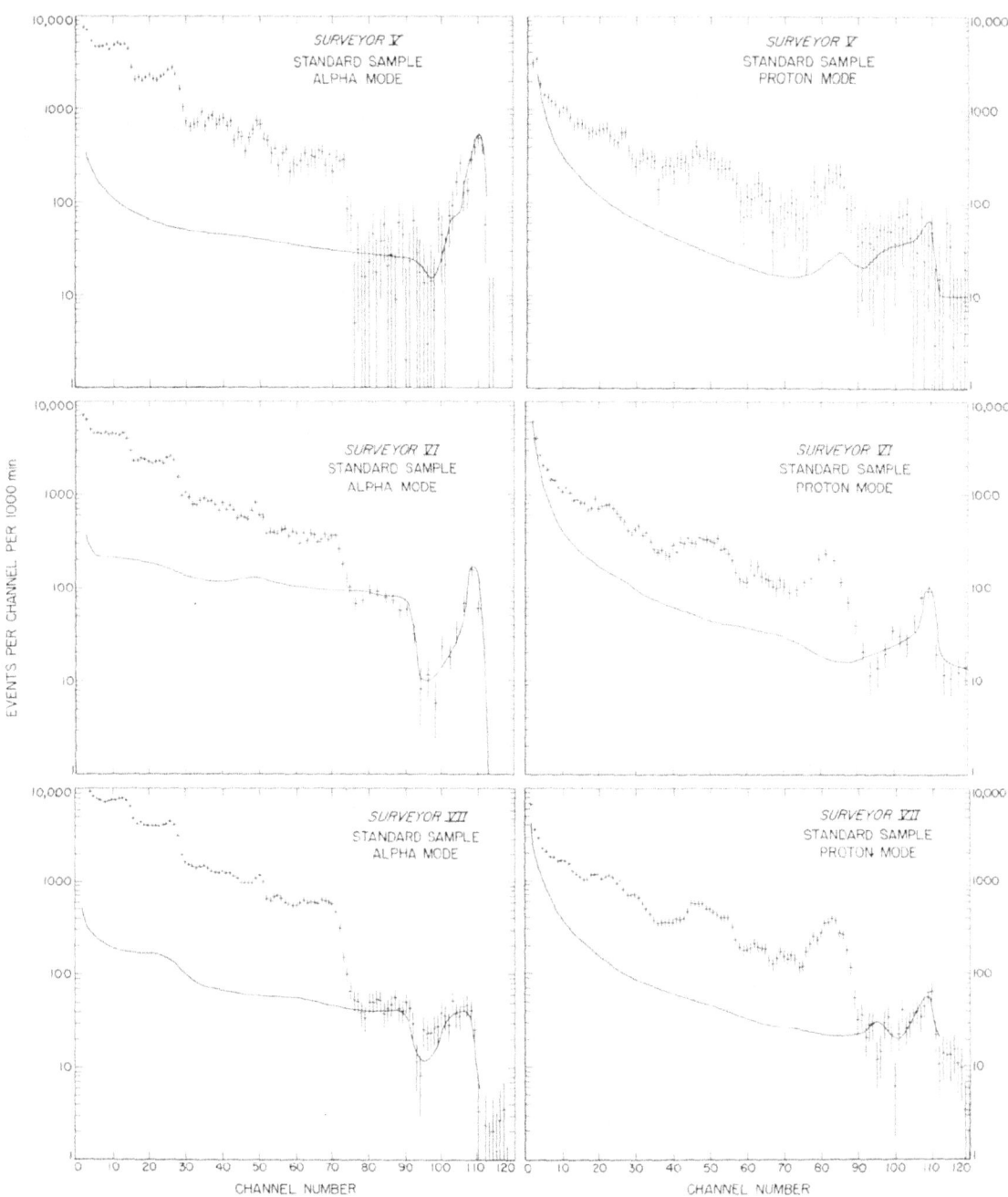

FIGURE 8-29.—Surveyor V, VI, and VII standard sample measurements on the Moon. The experimental points (crosses) are indicated with statistical (1 σ) error bars. The solid curves are smoothed versions of the background spectra observed in the next stage of lunar operations.

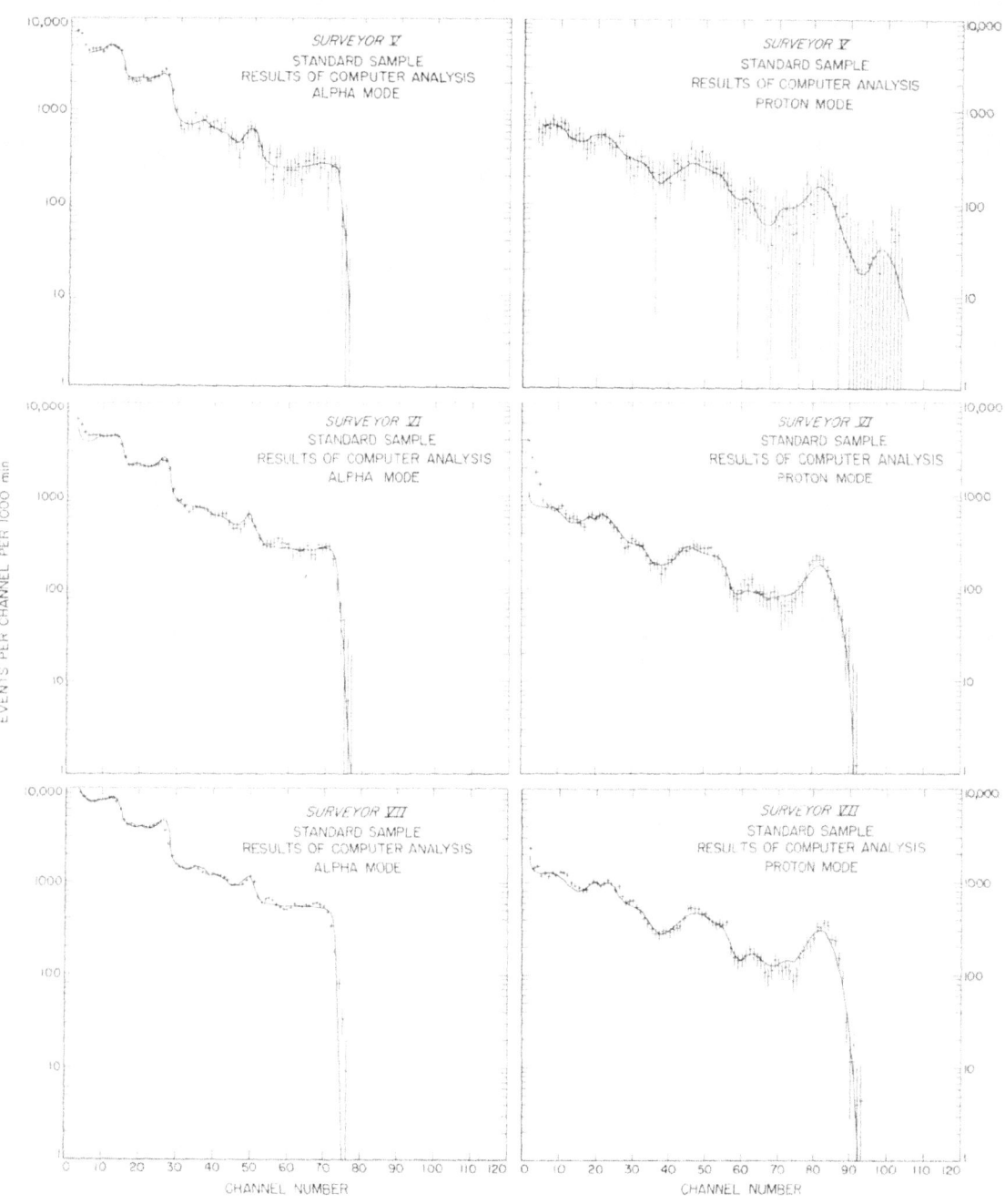

FIGURE 8-30.—Computer analyses of the Surveyor standard sample spectra. Comparison between the calculated spectra (smooth curves), using an eight-element library, and the data obtained on the Moon (after background subtraction).

TABLE 8-11. *Analyses of standard glass samples on the Moon*

Element	Analysis, atomic percent					
	Surveyor V		Surveyor VI [a]		Surveyor VII	
	Surveyor [b]	Standard [c]	Surveyor [b]	Standard [c]	Surveyor [b]	Standard [c]
Oxygen	56	59	51	58	55	59
Sodium	7	8	11	8	10	8
Magnesium	8	9	11	9	10	9
Aluminum	2	2				
Silicon	20	17	18	17	16	17
"Calcium"	~2		1			
"Iron"	8	7	8	6	9	7

[a] The Surveyor VI results were obtained by shifting the spectra to minimize chi-squared in the least-squares fit with the library. They differ somewhat from the results (reported in ref. 8-2) obtained during the mission using the SFOF real-time computer program.

[b] The results are normalized to 100 percent on a carbon-free basis. The glass was covered by a polypropylene grid that masked about 25 percent of the area.

[c] These are the results of conventional chemical analyses of typical standard glass samples.

it is clear from figure 8-30 that the energy scales of the lunar spectra and computer-calculated spectra are not always identical. It is for these reasons that there appears to be little basis for assigning errors to the Surveyor analyses of the lunar standard sample at this stage of data processing. The comparisons made in table 8-11 are considered adequate, however, to establish the capability of the instrument to perform chemical analyses under lunar conditions.

Another method of examining the standard-sample data is to compare them directly (after background subtraction) with the corresponding data obtained in the final calibration at Cape Kennedy a few weeks before launch. To do this properly, small energy-scale corrections were made to the data because of temperature differences and, in the case of Surveyors V and VII, because a different digital electronic unit had been used for prelaunch calibrations than for the actual mission. In addition, the lunar spectra (alpha and proton) were normalized to the prelaunch calibration spectra in the oxygen region (alpha channels 16 to 23) to correct for source decay and different sample distances.

For each of the three missions, these comparisons are shown in figure 8-31. Comparison of the data at this early stage does not introduce the systematic errors associated with the use of the library. In addition, it lends assurance that the background measured in the next stage of lunar operations (with the instrument suspended over the lunar surface) is the appropriate one to use with a sample under the instrument. In all cases, the comparisons made in figure 8-31 show adequate agreement throughout the energy regions in both alpha and proton modes.

To the extent that they have been analyzed to date, the standard-sample measurements on the Surveyor missions give no reason to suspect the data obtained under lunar conditions or the techniques used to interpret them.

Background Measurements

The second stage of the lunar operations of the alpha-scattering instrument was designed to measure the background levels for the instrument under lunar conditions. Because of the lack of a protective atmosphere on the Moon, these were expected to be much different (especially in the proton mode) from those on Earth. The data were accumulated with the instrument suspended about 0.5 meter above the lunar surface (see preceding section).

Figure 8-32 presents the background meas-

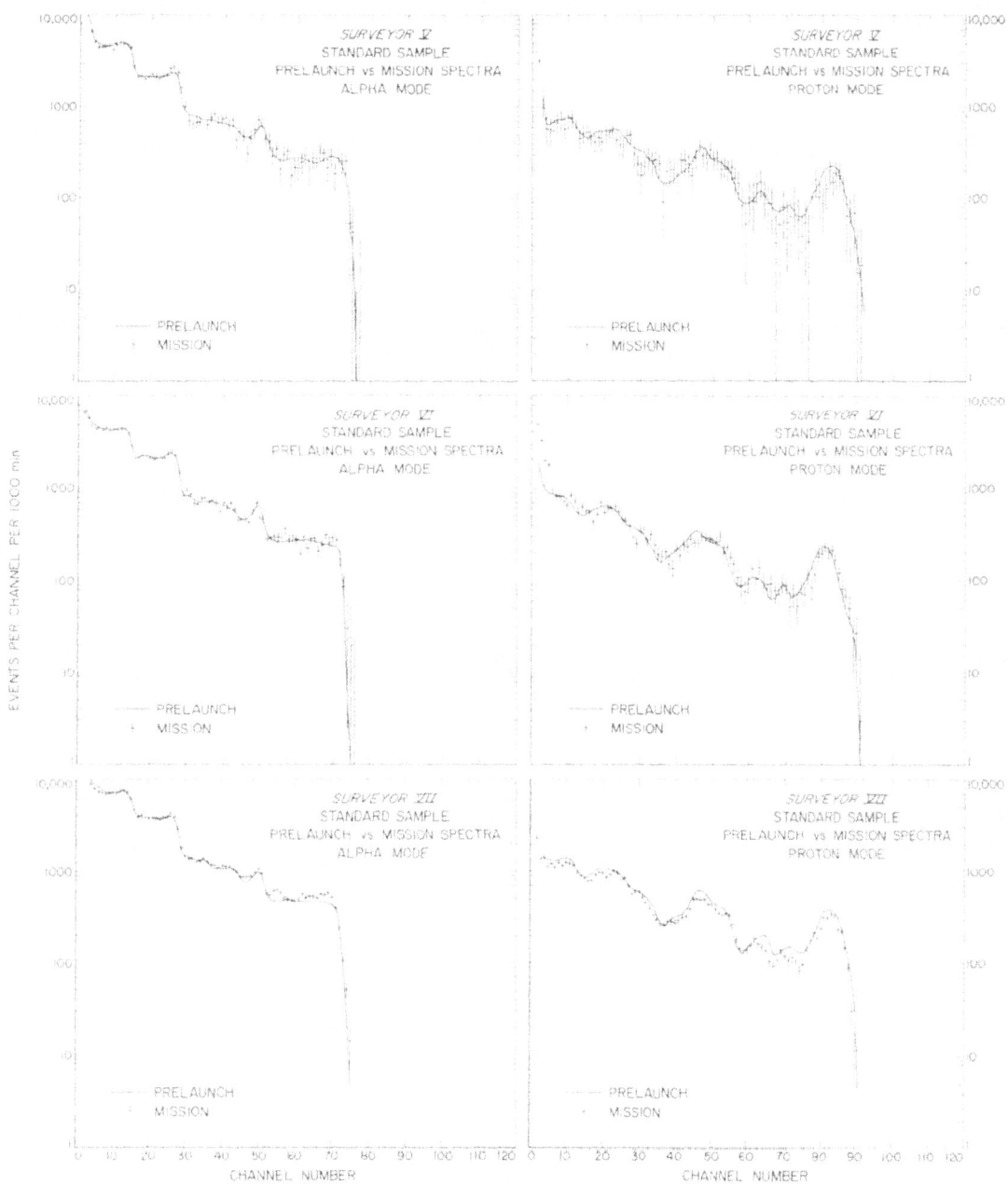

FIGURE 8-31.—Comparison of Surveyor standard sample data with corresponding spectra obtained in prelaunch tests. The experimental points are the data taken during the Surveyor missions, after background subtraction. The smooth curves are derived from data taken on similar samples at Cape Kennedy before each launch.

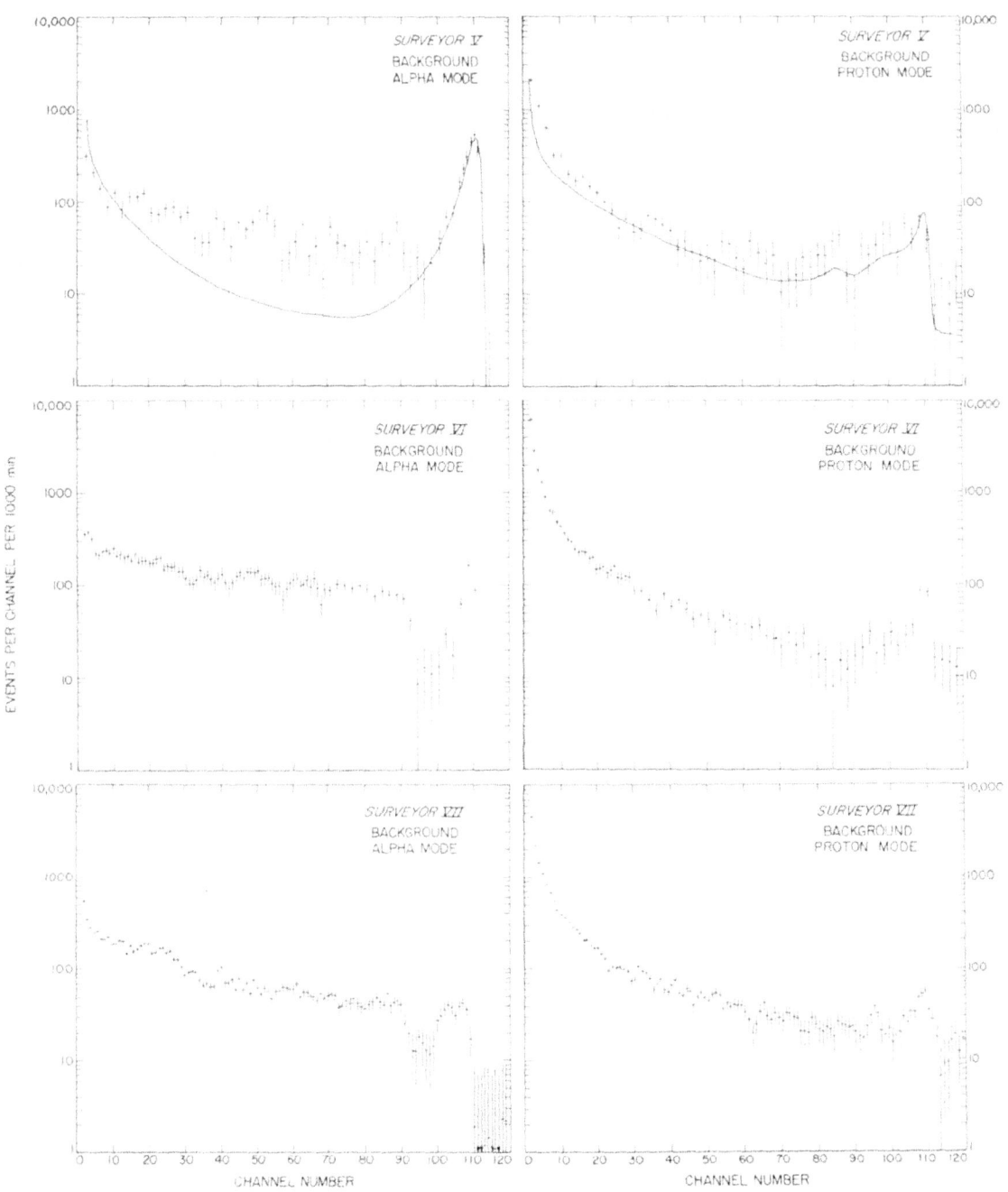

FIGURE 8-32.—Surveyor V, VI, and VII background measurements on the Moon. Parts of the Surveyor V and VI data are the averages of two channels. The smooth curves in the Surveyor V graphs are the premission predictions of the background rates.

urements from the certified spectra (table 8-10) for both alpha and proton modes of the instrument on the three missions. The ordinates are events per channel registered (sometimes averaged over two channels to improve statistics) as a function of channel number (energy). The ordinates are on a logarithmic scale, and the data have been normalized to a 1000-minute measurement time to facilitate comparison with the other figures in this report. The statistical (1 σ) errors in the data are indicated by vertical lines. The original data have already been corrected approximately for the temperature coefficient of the instrument. As is seen from table 8-10, the background measurement of Surveyor V was very short, leading to rather large statistical errors in the data in this mission.

In all of the graphs (alpha and proton), the peak in intensity in the region of channels 110 to 113 is caused by small amounts of Es^{254} ($T_\alpha = 6.44$ MeV) placed very close to each detector. This Es^{254} served as an energy marker as well as a check on detector behavior. Although the intensity and quality of the Es^{254} varied considerably from mission to mission, this alpha source provided assurance, independent of other data, of the correct performance of the instrument.

Before the missions it was anticipated that the main contribution to the background on the Moon, aside from the Es^{254}, would be from the instrument itself in the alpha mode and from cosmic and solar radiation in the proton mode. This difference in response is because the two alpha detectors are so small and have such a small sensitive depth that only 0.6-MeV to about 4-MeV protons, incident normally, would deposit enough energy to be registered above the 600-keV threshold of the electronic system. All such primary particles of the solar and cosmic radiation are absorbed by the nearby components of the instrument. The fraction of originally higher energy particles, which are slowed to the below 4-MeV energy range and the secondaries produced in this energy range at the detectors, was calculated to be small compared to the intrinsic background of the instrument.

On the other hand, the proton detectors were much larger in total area and also had larger sensitive thicknesses, allowing them to register protons of up to 6.4 MeV entering normally. Moreover, because of the gold foil in front of them, these proton detectors were insensitive to radioactive contamination of the instrument. The main contributor to the very low background in the proton mode on Earth (aside from the Es^{254} previously mentioned) was electronic noise and the sensitivity to the small amounts of gamma radiation and neutrons from the Cm^{242} sources.

Because the event rates from the lunar sample in the proton mode were expected to be relatively low, serious attempts were made during instrument design and before the missions to predict the background rates caused by solar and cosmic radiation on the Moon. Early estimates led to the decision to include anticoincidence detectors in back of each proton detector (see "Instrument"). Calculations indicated that these detectors would reduce the predicted background by at least a factor of 5; the remainder of the background would be a result of the radiation entering the sides of the instrument.

The observed backgrounds in the proton mode (fig. 8-32) turned out to be lower than those in the alpha mode over most of the energy range. The smooth curve in figure 8-32 (top right) is the premission prediction of this background spectrum for the proton mode. This is the sum of the prelaunch Cape Kennedy measurements of the instrument background and of the calculated contribution from cosmic and solar rays. It has almost the same shape as that observed and, over most of the energy range, is only about 50 percent lower. In view of the complicated nature of the calculation (involving the use of low-energy solar and cosmic-ray spectra which are known only approximately, and the effect of the complicated geometry and direction-varying amounts of absorbing material between the detectors and space), the agreement between the calculated and observed spectra is considered satisfactory.

The backgrounds observed in the proton mode in the three missions had the same spectral shape (except for the small difference in the Es^{254} region) and were approximately of the same magnitude. For example, the background in the proton mode in the energy region above 6.5 MeV was the same in the Surveyor VI and VII missions, well within the 1-σ statistical error of 5 percent. This means that the flux of 50- to 200-MeV protons (these are the particles which, incident on the instrument from the sides, would contribute to this background) was the same on the Moon at the Tycho site (40° S) in January 1968 as at the Equator in November 1967. This isotropy is consistent with the lack of appreciable magnetic field near the Moon. (A 100 MeV proton in a field of even 30 γ moves in a circular orbit of radius $\sim 10^4$ km.) During the Surveyor V operations, this background was slightly lower, possibly because of shielding by the local terrain, because Surveyor V landed just inside a small crater. The predictability and constancy of the data provide confidence that the nature of the background on the lunar surface in the proton mode of the instrument is adequately understood.

The background observed in the alpha mode of Surveyor V (fig. 8-32, top left), however, provided some surprises. In the intermediate energy range (approximately channels 10 to 90), it was definitely higher (in places more than a factor of 2) than predicted from the last premission measurements at Cape Kennedy. Fortunately, this was still a negligible part (over most of the spectrum) of the response when a sample was in a nominal position under the instrument (fig. 8-29). An explanation of this anomalous behavior was discovered just before the Surveyor VI mission. It proved impractical to eliminate this increase in background on that mission even though this background was much higher on Surveyor VI (fig. 8-32, middle left). However, measurements taken before the Surveyor VII mission were partially successful, so that the background in the alpha mode on that mission (fig. 8-32, bottom left) was lower than that on the Surveyor VI mission.

The clue to the explanation for this higher-than-expected background in the alpha mode lay in the decrease in intensity at approximately channel 93 (fig. 8-32, middle left). Such a behavior is characteristic of the response of the instrument to a gold sample. It is interpreted as a result of a larger-than-previously-encountered recoil behavior of the Cm^{242} flight sources prepared for these missions. Recoiling atoms from the alpha decay of Cm^{242} were transferring radioactive material from the source plate to the protective films which cover the collimating openings of the source capsules. In this position, the radioactivity is not collimated and can strike the gold-plated upper surface of the bottom plate of the instrument. Because of the high cross section for scattering by gold, even a relatively small amount of radioactivity in this position produces a detectable increase in the background in the alpha mode of the instrument.

As can be seen by comparison of the top left with the middle left of figure 8-32, the feature was present but less pronounced in the Surveyor V spectra. This apparently was a result, in part, to the source-protective films on Surveyor V, which had been exposed for a shorter time to this recoil behavior than those on Surveyor VI and, in part, to the very large variability later observed in the efficiency of this recoil-transfer process in individual sources.

For the Surveyor VII mission, a technique of evaporating carbon on the films (enough to reduce the recoil transfer but not enough to reduce the energy of the alpha particles significantly) had been developed. (See "Instruments.") This proved very effective for the sources prepared for the spare instrument for the Surveyor VII mission; it was partially effective for the actual sources flown on this mission. The result was a background in the alpha mode intermediate in level between the two earlier missions, even though the source strength was about 70 percent higher.

The enhanced background in the alpha mode, moreover, gradually increased with time (as more radioactivity was transferred to the films). This could be observed as a slow increase with time of the response of the instrument in an energy region (channels 73 to 90) above the Fe, Co, and Ni endpoints. Although not affecting the results on the principal constituents of the lunar samples examined, this aspect of the experiment will increase the very low limits that could otherwise be placed on the abundance of elements heavier than nickel.

The background measured in the alpha mode in channels above the Es^{254} peak (not shown in the figures) corresponds to events depositing at least 6.5 MeV in the alpha detectors. The rate of such events is very low (54 ± 12 per 1000 minutes on the Surveyor VI mission). It is essentially the same on all the missions and is consistent with the contribution calculated for solar and cosmic ray particles.

The alpha-mode background of the Surveyor VII mission (fig. 8-32, bottom left) shows more structure than is visible on the two preceding missions. This is because, as a result of the higher intensity sources and longer counting time, the effects of the low-probability scattering from the lunar surface, about 0.5 meter away, can be seen. As seen later in "Results of Lunar Sample Data," this structure in the spectrum follows closely the features in the lunar-sample spectrum, although with a very low intensity.

Because only one background measurement was made on each mission, precautions had to be taken in the data analysis to be certain that this background was not changing with time or was changing in a known way. In the alpha mode, the contribution caused by scattering of noncollimated alpha particles was monitored by observing the event rate in channels 73 to 90. The increase in this region was assumed to apply at lower energies as well. It was a significant part of the total event rate only in the region of the alpha spectrum above channel 52.

Changes in solar- and cosmic-ray contributions to the background were monitored in three different ways. The least sensitive method involved the rate of events in the alpha channels with energies greater than 6.5 MeV. Although the rates here were very low (about 54 events/1000 minutes), their constancy, within statistics, provided assurance that the alpha-mode background caused by external solar and cosmic rays was not changing during a mission.

The second method involved the rates of events in the same high-energy region (>6.5 MeV) in the proton mode. Here the rates were higher (about one event/min).

Again, within statistics of about 15 percent there is no indication of changes during the measurements.

The most sensitive monitor of the solar- and cosmic-ray intensity was the guard ratemeter. Preliminary examination of the data indicates no significant changes in intensity over the period of the lunar-sample measurements.

In addition to these internal measurements by the instrument itself, of the constancy of the solar- and cosmic-ray background, data were provided on the charged-particle radiation levels in space from the Imp IV satellite. This satellite was in orbit about the Earth, and readings provided[2] each 4-hour period gave assurance that significant changes in the general level of radiation in space would be detected independently of the alpha-scattering instrument. There were no such changes reported on any of the missions during lunar operations of the alpha-scattering instrument.

In analyzing the data for this report from the standard samples and from the lunar samples on the Surveyor VI and VII missions, the observed backgrounds (increased slightly as appropriate in the alpha mode) were subtracted from the observed spectra before computer processing. The statistical errors associated with the background measurements were considered. In the case of Surveyor V, because of the poorer statistics, a smooth curve was drawn through the background data, and this smoothed version was used in the data analysis.

Results of Lunar-Sample Data

As indicated previously, six lunar samples were examined by the alpha-scattering instrument during the Surveyor V, VI, and VII missions. Three of these were samples of mare material (Surveyors V and VI), and three were from the Surveyor VII terra site. The first sample examined on each mission was the original deployment area of the instrument. On Surveyors VI and VII, this first sample was the relatively undisturbed local surface; on

---

[2] These data were made available on a near-real-time basis by Dr. John Simpson of the University of Chicago and the Small Satellite Tracking Center at Goddard Space Flight Center.

Surveyor V, however, the original deployment area may have been primarily material ejected by the spacecraft footpads on landing. Because the second sample examined during the Surveyor V mission was only about 10 cm away from the first sample, the second sample was probably similar. Only one sample was examined on Surveyor VI. The second sample analyzed on the Surveyor VII mission was a lunar rock that protruded a few centimeters above the originally undisturbed, local lunar surface. With the help of the surface sampler, the alpha-scattering instrument was placed on top of this rock (ref. 8–20 and fig. 8–27). The third sample examined on the Surveyor VII mission was subsurface material; the alpha-scattering instrument was placed on top of a shallow trench previously prepared by the surface sampler.

The relatively raw data from the certified Teletype spectra from each of the samples (table 8–10) are presented in figures 8–33 and 8–34. In each case, the alpha and proton data (in units of events per channel per 1000 minutes) are shown plotted on a logarithmic scale as a function of channel number (energy). The statistical errors (1 σ) are indicated by vertical lines. In each case, the smooth curves are the background rates observed on the Moon in the previous phase of the experiment. In comparing the absolute rates from different samples, it should be remembered that the source strengths were different on different missions and that the average sample-to-detector distance was different. Thus, for example, the average distances of the three samples to the alpha detectors on Surveyor VII were in the ratio of 1 to 0.7 to 0.9.

Examination of these figures indicates that in the alpha mode the response from the samples was adequately higher than the background values over all parts of the spectrum below channel 70. In the proton mode, however, below about channel 10, the background represents a large fraction of the total number of events and is rapidly changing with energy. In most of the higher energy regions, however, the signal-to-background ratio is adequate. As might be expected from the higher source strength on Surveyor VII, the situation is best on the samples examined during that mission.

The qualitative features visible in the raw data of figures 8–33 and 8–34 are the same for all samples in the same mode of the instrument (alpha or proton). These features also resemble those found in many terrestrial rocks examined by Surveyor-type instruments. In the alpha mode, the most prominent features are the sharp drops in intensity at approximately channel 27 (characteristic of the presence of oxygen), the drop often preceded by a slight "bump" at approximately channel 52 (characteristic of the presence of silicon), and the final drop at approximately channel 73 (indicating the presence of elements in the region of iron in the samples). In the proton mode, all samples show the drop in intensity at approximately channel 62 and a broad peak between approximately channels 73 and 100 (characteristic of the presence of aluminum; see fig. 8–4(c), bottom right).

Even though most of the qualitative features are similar in a given mode in the data from all the samples, there are easily visible, slight differences. For example, in the alpha mode above channel 52 (an energy region to which only elements heavier than silicon can contribute), the data from the two samples examined during the Surveyor V mission show no significant breaks before that at approximately channel 73; they merely show a general decrease. The data from Surveyor VI show a definite drop at about channel 63. The drop is much more distinct and goes to lower levels at about the same point in all three samples examined during the Surveyor VII mission.

In view of the unavailability of the primary data in adequate time to be treated properly for this report, the data presented in figures 8–33 and 8–34 have been processed only in an approximate manner using an eight-element library (see "Methods"). After subtraction of the appropriate backgrounds, the data of figures 8–33 and 8–34 have been fitted by a least-squares technique, using the library appropriate to the instrument used.

The results of this computer fitting of the data from the two samples from Surveyor V, the sample from Surveyor VI, and the first sample from Surveyor VII are shown in figures 8–35 and 8–36. The gross features of both alpha

CHEMICAL ANALYSIS

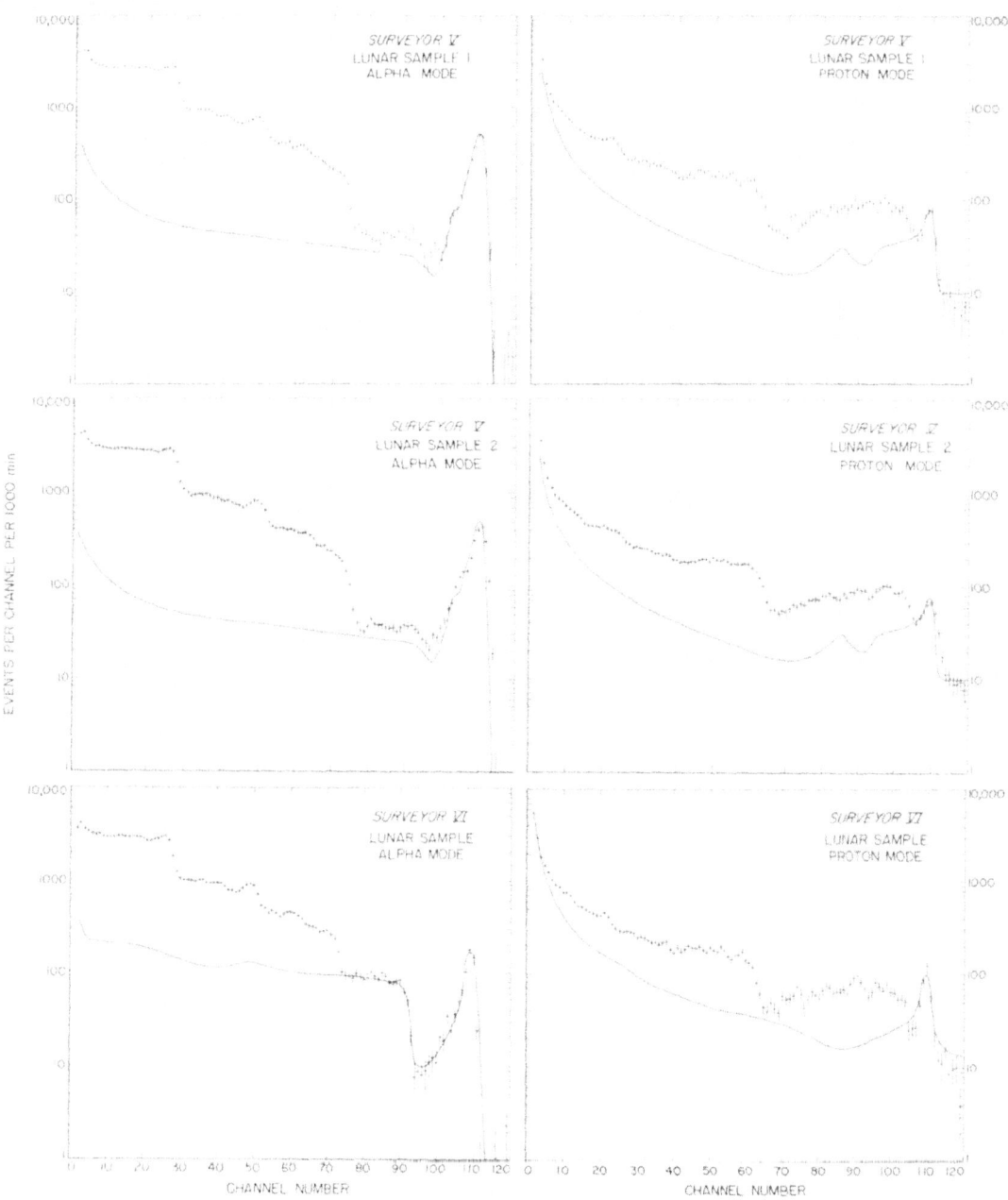

FIGURE 8-33.—Surveyor V and VI lunar samples. The experimental points are indicated with statistical (1 σ) error bars. The solid curve in each case is a smoothed version of the background observed in the previous stage of lunar operations.

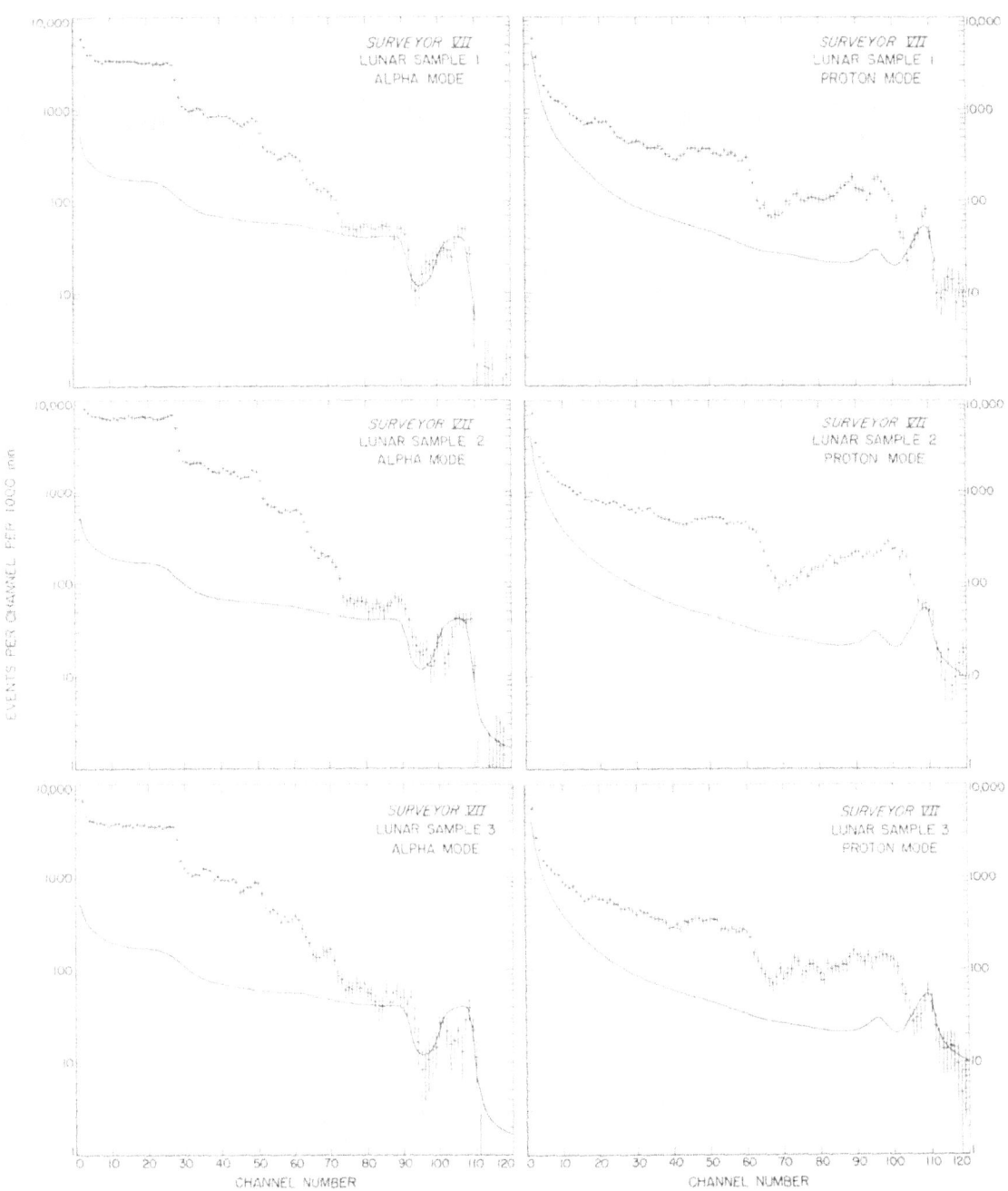

FIGURE 8-34. Surveyor VII lunar samples. The experimental points are indicated with statistical (1 σ) error bars. The solid curve in each case is a smoothed version of the background observed in the previous stage of lunar operations.

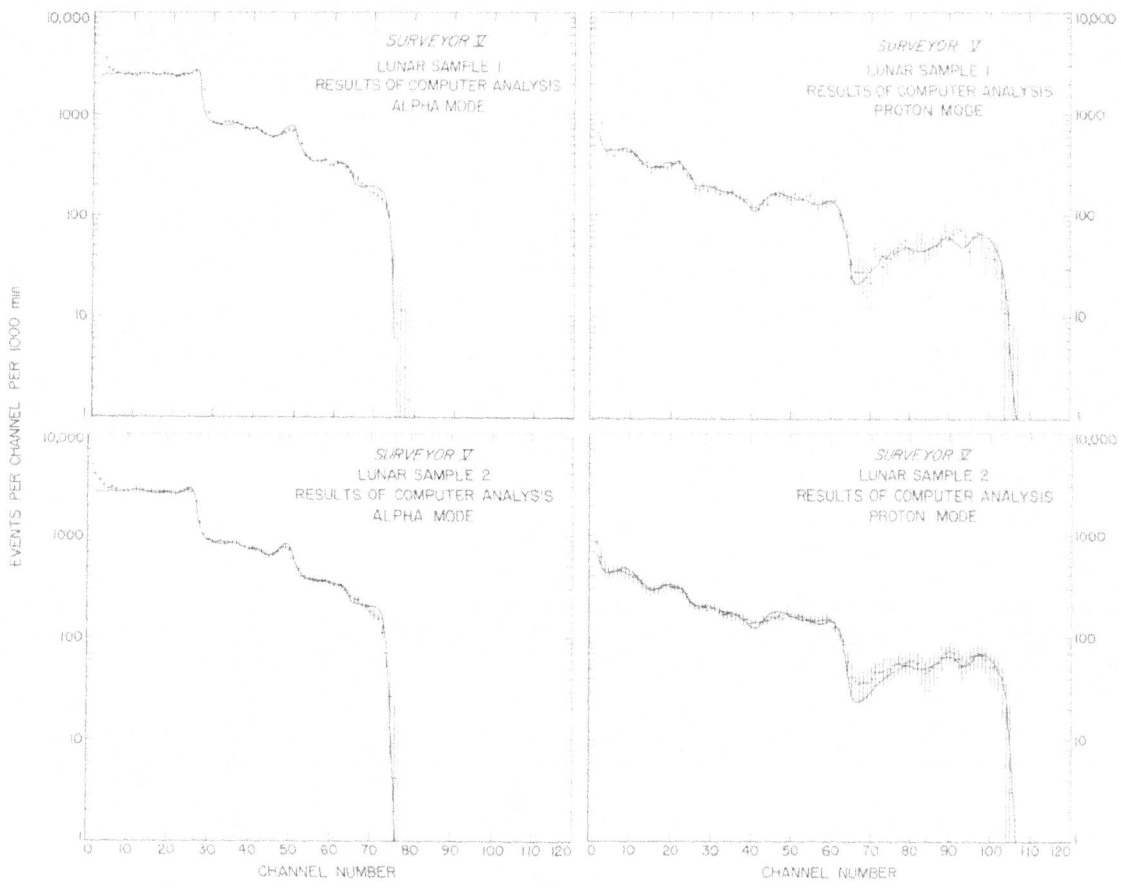

FIGURE 8-35.—Computer analyses of the Surveyor V lunar sample spectra. Comparison between the calculated spectra (smooth curves), using an eight-element library, and the data obtained on the lunar surface (after background subtraction).

and proton spectra for these four samples are well represented by the eight-element library used. The only significant deviations are those explicable by small energy mismatches between the library and lunar data, the tendency for the "bumps" near the oxygen and silicon endpoints of the alpha spectra to be smoothed out in the lunar data, and the inadequacy of a two-component library to fit the alpha data above channel 52 in the two samples studied on Surveyor V. It remains to be seen whether these small discrepancies will persist when the more complete data are processed more rigorously. In spite of these small discrepancies, it is clear that the eight-element library chosen represents the data well. The elements of the library must represent the principal constituents of the lunar soil at the places investigated.

Figure 8-37 is an example of the detailed contribution of the different elements of the library to the net lunar spectra of sample 1 from the Surveyor V mission in both alpha and proton modes. It is seen that the carbon, sodium, and magnesium contributions required are so small that they are untrustworthy at the present stage of analysis.

Computer results are not shown for the second and third samples from Surveyor VII. The far-from-nominal geometries represented by the lunar rock protruding into the instrument opening and by the disturbed-surface (trench) sample produced changes, particularly

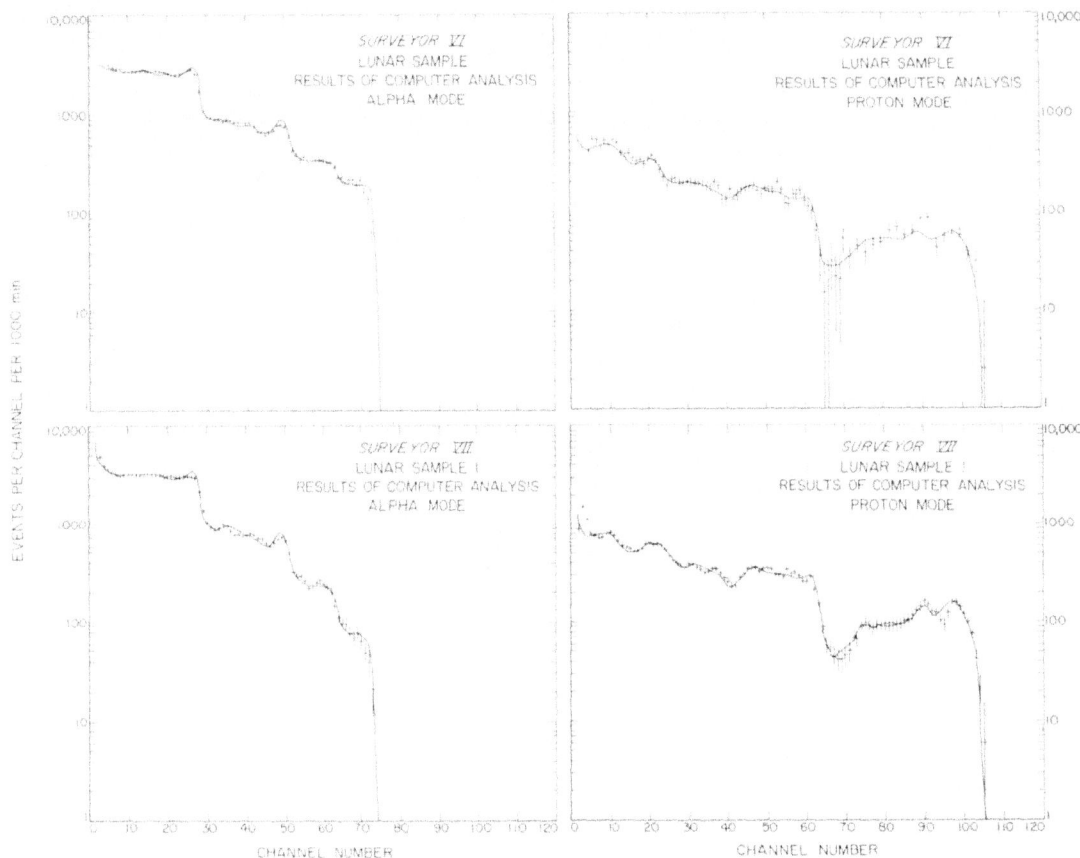

FIGURE 8-36.—Computer analyses of the spectra from the Surveyor VI lunar sample and the Surveyor VII lunar sample 1. Comparison between the calculated spectra (smooth curves), using an eight-element library, and the data obtained on the lunar surface (after background subtraction).

in the proton spectra (in intensity and slightly in shape), which make the use of a standard library less applicable. These effects are qualitatively understood, but will require verification by mockup experiments in the laboratory before the lunar data can be analyzed reliably.

The computer-derived contributions of the separate elements of the library for each sample, after correction for the stopping power of the element (see ref. 8-10 and "Methods" of this report), give the relative number of atoms of the element in the sample. (This treatment assumes that the sample is homogeneous or, if not, that the average atomic stopping power for alpha particles of the components is not too different.) From this relative composition, the atomic fractions of the elements of the library in the samples can be deduced. These are presented in table 8-12. Because of the limitations of the technique, the values in table 8-12 are to be interpreted as chemical compositions normalized to 100 percent excluding elements lighter than beryllium.

The errors quoted in table 8-12 are the present estimates of the reliability of the results. They represent, primarily, systematic errors arising principally from the present lack of information about the reliability of the primary data stream, the approximate nature of the corrections for the small temperature coefficients and other temperature-dependent characteristics of the instrument, slight changes in energy

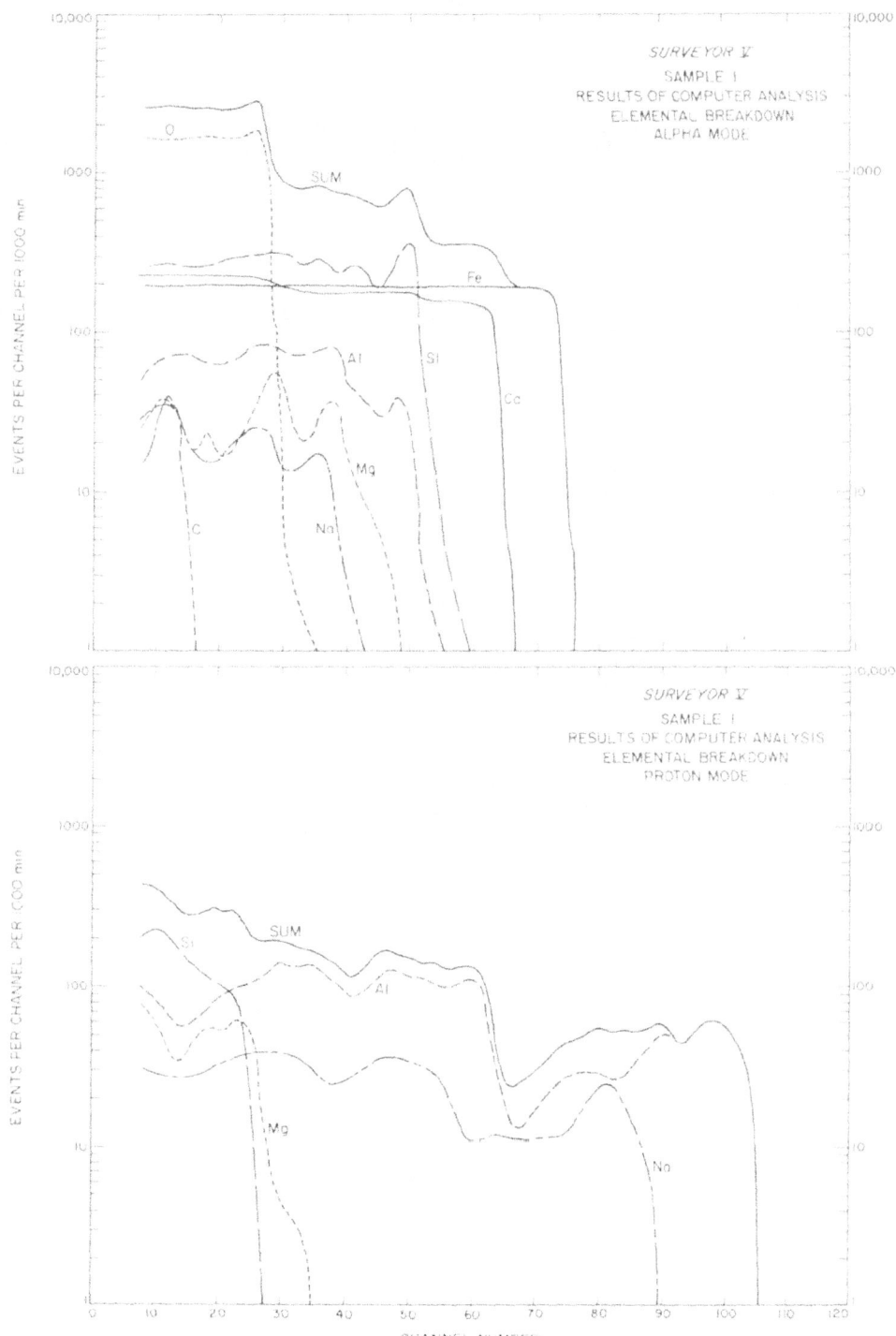

FIGURE 8-37.—Contribution of separate elements to least-squares computer analysis of the Surveyor V spectra from lunar sample 1. The library data for the separate elements are shown in the relative amounts needed to explain the observed lunar data.

TABLE 8-12. *Chemical composition of the lunar surface at the Surveyor V, VI, and VII landing sites, preliminary results*

Element	Chemical composition, atomic percent			
	Mare sites			Terra site
	Surveyor V, sample 1	Surveyor V, sample 2	Surveyor VI	Surveyor VII, sample 1
C	<3	<3	<2	<2
O	58 ±5	56 ±5	57 ±5	58 ±5
Na	<2	<2	<2	<3
Mg	3 ±3	3 ±3	3 ±3	4 ±3
Al	6.5 ±2	6.5 ±2	6.5 ±2	a 9 ±3
Si	18.5 ±3	19 ±3	22 ±4	18 ±4
"Ca" [b]	d 13 ±3	d 13 ±3	6 ±2	6 ±2
"Fe" [c]			5 ±2	2 ±1

[a] The value for aluminum for sample 1 on Surveyor VII has previously been reported as 8±3 percent (ref. 8-3). Additional data analysis indicates the present reported value as being more nearly correct.

[b] "Ca" here denotes elements with mass numbers between approximately 30 and 47 and includes, for example, P, S, K, and Ca.

[c] "Fe" here denotes elements with mass numbers between approximately 47 and 65 and includes, for example, Cr, Fe, and Ni.

[d] At this stage of analysis, the "Ca" and "Fe" groups of elements have not been resolved for the Surveyor V mission. However, a lower limit of 3 percent of "Fe" can be set for each of the samples.

scales of the instrument between the time when the library was determined and the time of the mission, and the use of a limited library. Because usually only a fraction of the eventually usable data has been used, there should also be a significant reduction in the statistical contribution to the error. However, at the present stage of analysis, the contribution of statistics to the quoted error is small for the most abundant elements.

Although the present results are preliminary, an attempt has been made to insure their reliability by the following tests:

(1) Computer analyses of subsets of the data used to obtain the results in table 8-12 have yielded results statistically consistent with the mean values given in that table.

(2) For the first sample on Surveyor V, it has been found that the answers are relatively insensitive to the use of background assumptions, other than the smoothed version shown in figure 8-33. Among the backgrounds tried were the actual background data rather than the smoothed version, and no background.

(3) In several of the cases listed in table 8-12, the inclusion of other elements such as nitrogen, fluorine, and potassium in the library used in the computer has been investigated. It has been found that the results listed in the table are not affected appreciably.

(4) Although the data in the present state hardly justify such refinements, in several of the cases listed in table 8-12, a computer was programed to search for the changes in instrument parameters (gain or zero offset), in both alpha and proton modes, that would improve the fit to the data as judged by a chi-squared test. The results were always a significant increase in goodness of fit as measured by chi-squared, and a better match to such features in the spectra as the oxygen breakpoint. The analytical results were essentially unchanged.

(5) The excellent reproduction of the prelaunch standard-sample spectra by the mission data (fig. 8-31), after subtraction of the corresponding backgrounds, indicates that the lunar background, measured while the sensor head was suspended over the lunar surface, is the appropriate one to use in the analysis of lunar-sample spectra.

While not exhaustive, these tests provide

confidence that the results presented in table 8-12 will not be changed beyond the quoted errors by more refined treatments of the data.

Because of the possibility of geometry-sensitive contributions to the proton spectra of samples 2 and 3 on the Surveyor VII mission, no numerical results are quoted at present for the composition of these samples. The alpha mode of the instrument, however, is rather insensitive to these geometrical effects. Figure 8-38 compares the data obtained in the alpha mode from samples 2 and 3 with those from sample 1. The background has been subtracted in each case, and the data have been normalized so that they match in the oxygen region (channels 8 to 25). Except for the possibly lower values in the region of channels 63 to 70 in the spectrum of sample 2, the data from the three samples agree very well. The conclusion is that there cannot be large differences in the relative amounts of oxygen, silicon, "calcium," and "iron" (the principal contributors to the scattered alpha spectra) in the three samples examined on Surveyor VII. All three sample spectra show a decrease by about a factor of 2 in the "iron" content relative to the values found in the mare samples.

Although no quantitative information is deduced at present from the proton spectra from samples 2 and 3 on the Surveyor VII mission, the data of figure 8-34 (middle and bottom right) show clearly the presence of aluminum in these samples in comparable amounts to that found in the other lunar samples analyzed (figs. 8-33 and 8-34, top). Reliable quantitative results on these two samples will require laboratory simulation studies reproducing the geometrical relationships of the samples and instrument.

The Surveyor VI mission included a hopping maneuver of the spacecraft. This resulted in the alpha-scattering instrument resting upside down on the lunar surface (fig. 8-23), thus ending the possibility of obtaining any more lunar data on that mission. The electronic and detector systems, however, had survived. The data obtained after the maneuver showed a large increase in event rate, particularly around channel 103 in the alpha spectrum. This indicated that at least one of the protective films

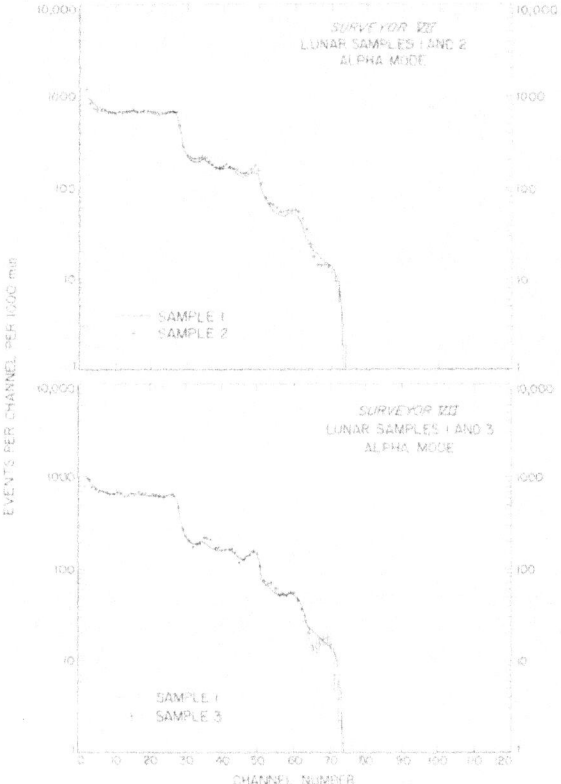

FIGURE 8-38.—Comparisons of Surveyor VII lunar sample spectra. The backgrounds observed in the previous phase of operations have been subtracted and the data have been normalized so that they match in channels 8 to 25.

over the radioactive sources had broken and that recoil contamination of the instrument was taking place.

The rates in the proton mode of the instrument were also appreciably higher than before the hop. Qualitatively, this was understandable in that the proton detectors in the new position of the sensor head were exposed to radiation from space (through the 10.8-cm sample opening of the instrument), with essentially no shielding other than the gold foil over the detectors.

These event rates in the proton mode did not remain constant, however. During the lunar afternoon, they fell significantly, with a larger decrease in the low-energy portion of the spectrum. There are several possible explanations for this behavior. Because of the nonstandard mode of operation, it may be an instrumental effect such as a noisy detector. It may, on the other hand, represent a significant decrease in the number of low-energy protons reaching the surface of the Moon in the late lunar afternoon. This could arise, for example, either as a result of a decrease in the number of low-energy particles following a solar flare or as a result of the shadowing by the Moon of the source of such particles. Detailed study of the primary data, including that from the anticoincidence counters, as well as comparison with the results obtained on satellites in space at the same time, may help to explain this effect.

Discussion

The analytical results deduced from the alpha-scattering experiment on the Surveyor missions represent the first onsite chemical analyses of an extraterrestrial body. There have, of course, been previous speculations about the chemical composition of the Moon. These generally have been based on indirect considerations such as the overall density, thermal and optical measurements, or proposed theoretical relationships of the Moon to either the Sun, the Earth, or meteorites. Until now, the observation most directly related to the chemical constitution of the lunar surface was the gamma-ray experiment on the Russian orbiter Luna 10 (ref. 8-25). The intensities and spectral distributions of gamma rays observed on this satellite were used to set limits on the content of radioactive elements in lunar-surface material. The conclusion was drawn that the upper limit of radioactive material was inconsistent with granitic-type rocks as they exist on Earth. The data appeared more consistent with the amounts of radioactive elements in terrestrial basalts, with the terrae having less, thus possibly being chondritic in composition.

The results obtained as a result of the Surveyor missions and presented in table 8-12 are more direct and comprehensive. It is true they have been obtained at only three lunar sites. Also, at the present stage of data availability and analysis, the limits of error on the analytical results are rather large. However, they would appear to account for at least 90 percent of the atoms present (excluding hydrogen) and so are the most complete analyses of lunar material that are available. They indicate that the most abundant element on the Moon, as on the Earth's surface, is oxygen. More than half of the atoms (not including the undetectable hydrogen) are of this element. Second in abundance, again as in the rocks making up the crust of the Earth, is silicon. Aluminum is very prominent (6 to 9 atomic percent); it is the third most abundant element in the Earth's crust. At present, only upper limits can be placed on the amounts (2 to 3 atomic percent) of carbon and sodium in the samples analyzed by the Surveyors. Thus, inspection of the values given in table 8-12 indicates a gross similarity in chemical composition to that of many rocks found on Earth.

Before proceeding to a more detailed consideration of the results, it is worth recalling some characteristics of the alpha-scattering technique of chemical analysis and aspects of the Surveyor missions that might affect the interpretation of the analyses.

First, the technique provides information on the composition of only the topmost microns of the sample being examined. The possibility that this topmost layer is not representative of the composition of the bulk of the sample, especially under the conditions existing on the lunar surface, must always be kept in mind.

Second, there is the question whether the "undisturbed lunar surface," as exemplified by the first samples examined on the Surveyor VI and VII missions (Surveyor V clearly had been mechanically disturbed), were really "undisturbed." One possibility is that during the landing operation, the spacecraft chemically contaminated the surface, either by reaction with, or deposition of, the retrorocket or vernier rocket exhausts. Surface contamination by the Al_2O_3 from the main rocket exhaust is considered to be negligible, both from theoretical calculations as to the amount and nature of its deposition (see "Description of Missions" under "Experiment Control and Mission Operation") and from the observations of no difference (within presently quoted errors) in the aluminum content of samples which had different exposure to the exhaust.

The possibility of reaction of the topmost

CHEMICAL ANALYSIS 339

layer with the products of the vernier engines (which operated much closer to the surface than the main retrorocket) is made unlikely by the lack of any visible changes in appearance of the surface closest to where their products impinged. Also, preliminary investigation of the presence of nitrogen in the first sample examined by Surveyor V gave amounts below the present detection limits. (The vernier engines operate on dimethyl hydrazine and nitrogen tetroxide.) Moreover, the presence of carbon in any of the Surveyor samples has not been established. These results are consistent with premission tests on powdered terrestrial materials (basalt, granite, and iron), which were exposed for long periods separately to the room-temperature vapor pressure of dimethyl hydrazine and nitrogen tetroxide and showed no detectable effects as measured by a Surveyor-type instrument.

The possibility of some physical removal of the topmost layer of the lunar surface by the vernier-engine blast on landing is harder to exclude. Again, there appears to be no obvious change in physical appearance of the surface near where the vernier engines operated. Moreover, the topmost ($<$1 mm), fragile, higher albedo surface layer seems to be still present at the Surveyor VI and VII sites where the analyses were made and at the Surveyor V site where the throwout material from the footpads did not disturb it. Finally, theoretical estimates of the force exerted on the surface by the vernier blast at the time of cutoff indicate values smaller than those observed to produce visible changes in albedo. These are all arguments against physical removal of the topmost layer by the vernier-engine blast.

Thus, certainly within the presently quoted errors of analysis, several of the samples examined by the Surveyors may reasonably be considered to be characteristic of undisturbed lunar-surface material. These considerations will bear further examination as the more complete data are used to provide more detailed and accurate analysis.

Intercomparison of Results on Different Samples and Surveyor Missions

The limits of error quoted in table 8-12 are large and could accommodate significantly different chemical compositions for the different samples. It has been noted, however, that these estimates of errors can include important contributions from possible systematic errors. Because the different instruments used on the Surveyor missions were identical geometrically and almost identical in detector and electronics characteristics, one can examine for the different samples and different missions to what extent the relatively raw data agree.

For both alpha and proton modes, figure 8-39 shows a comparison between the two samples measured during the Surveyor V mission and comparisons among the first samples measured on the Surveyor V, VI, and VII missions. The original data have been corrected approximately for the temperature coefficients of the respective instruments, and the appropriate backgrounds have been subtracted in each case. The data (both alpha and proton) on all samples have been normalized to the "oxygen" region (alpha channels 8 to 25) to correct differences in source strength, measurement times, and sample distances (approximately). In both parts of the figure, the Surveyor V sample 1 data are represented by solid curves. The experimental points with associated error bars in figure 8-39 (top) are from the second sample measured on the Surveyor V mission. In figure 8-39 (bottom), the dashed curves represent the data from Surveyor VI; the experimental points are the data from the first sample measured on the Surveyor VII mission.

The agreement of these basic data from the two samples measured on the Surveyor V mission is excellent in both alpha and proton modes (fig. 8-39, top). Similarly, the Surveyor VI data agree well with those from sample 1 of Surveyor V (fig. 8-39, bottom). In the latter case, the principal difference is the increased detail previously noted in the scattered-alpha spectrum in the region of channels 53 to 73 in the Surveyor VI data.

The differences among the spectra from these three samples are small, however, compared to those that might be expected from significantly different rock types. For example, figure 8-40 shows spectra taken with a Surveyor

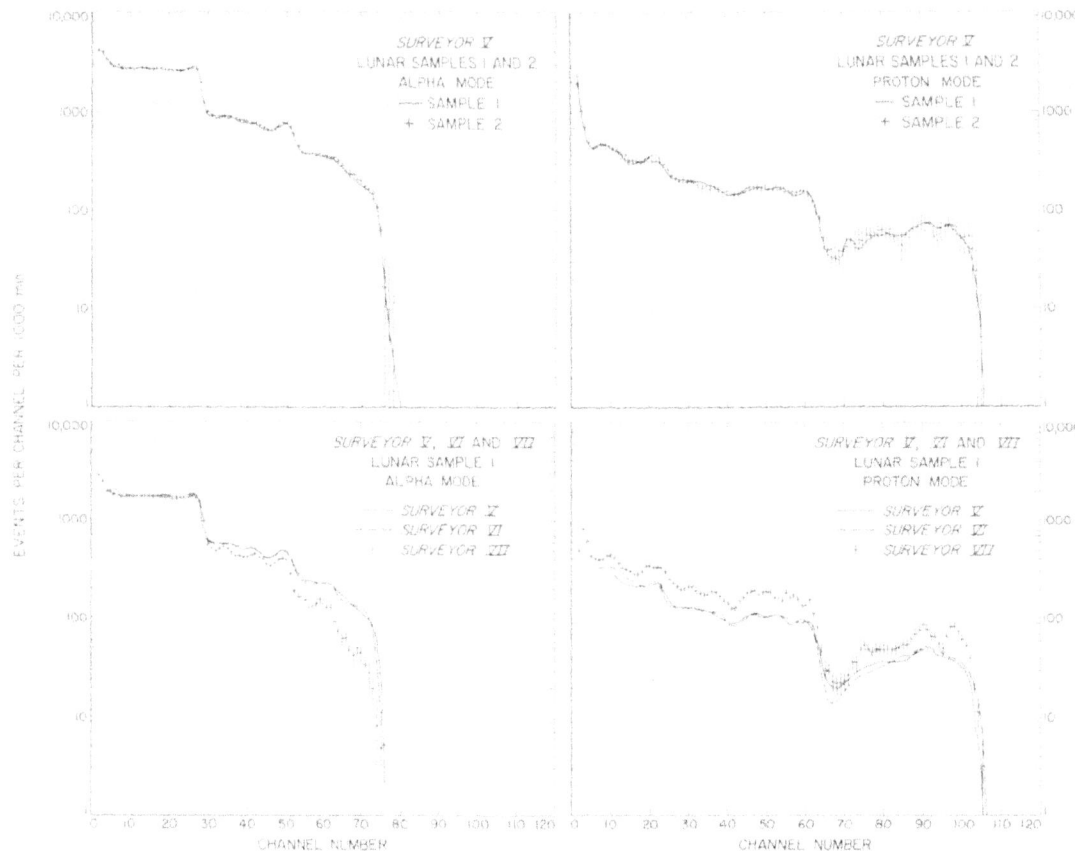

FIGURE 8-39.—Comparison between the spectra of the two Surveyor V lunar samples and among the spectra of the first samples of the Surveyor V, VI, and VII missions. The intensities have been normalized so that they match in channels 8 to 25 of the alpha spectra.

prototype instrument (P-2) from samples of basalt, dunite, and granite.[3] (The construction of this instrument was similar to that of the flight instruments, except that the sensitive depth of the proton detectors of P-2 was shallower than in the Surveyor flight instruments. This resulted in a change of shape of the spectra of the highest energy protons from aluminum.) The data for these three rocks have been normalized in the same manner as for figure 8-39, so that the alpha spectra are matched in the region characteristic of scattering by oxygen. Although there are only small differences between the spectra of basalt and granite, in the proton mode the spectrum is distinctive from dunite (and therefore from chondritic meteorites, which have a composition of proton-producing elements comparable to dunite). In the alpha mode, particularly in the energy regions above channel 40, there are marked differences among the three types of rocks. No such differences in either alpha or proton modes are indicated among the three mare samples. (See fig. 8-39, top, for samples 1 and 2 of Surveyor V, and fig. 8-39, bottom, for samples 1 of Surveyors V and VI.) It must be concluded that relative to oxygen, the amounts of the elements contributing in a major way to the alpha and proton spectra in all three mare samples are the same, probably to within 20 percent. Smaller differences

---

[3] U.S. Geological Survey Standards W-1, DTS-1, and G-2, respectively.

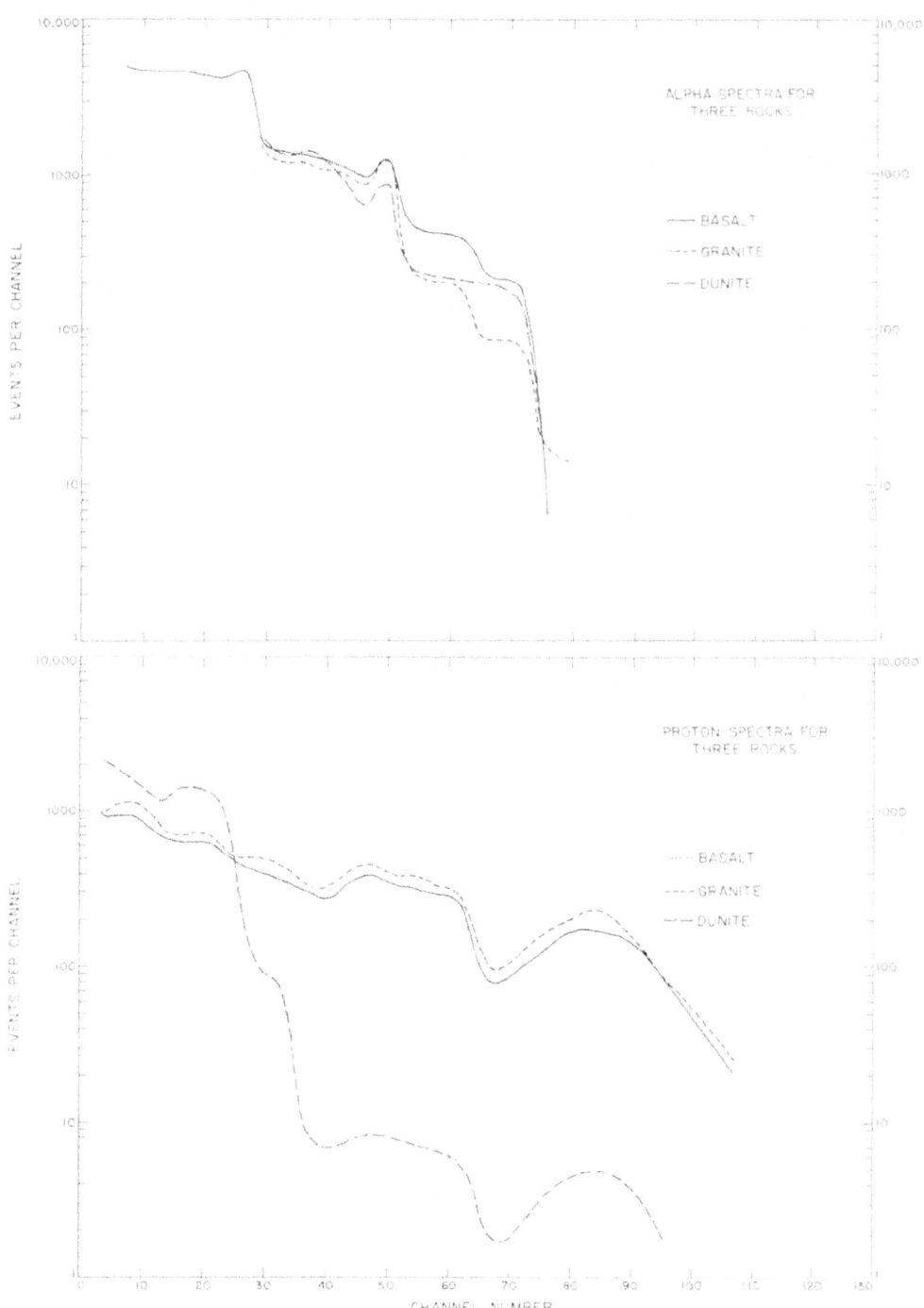

FIGURE 8-40.—Alpha-scattering instrument response from three rocks. Data taken with a Surveyor prototype instrument (P-2) on samples of basalt (USGS-W-1), granite (USGS-G-2), and dunite (USGS-DTS-1). All data from each rock have been multiplied by a factor to make the responses equal in the oxygen region of the alpha mode.

in these major constituents as well as larger differences in minor constituents may become established on more detailed analysis of the data. However, they are not likely to change the conclusion that the three samples from the maria examined on the Surveyor V and VI mission have the same amounts (within 20 percent) of Si, Al, "Ca," and "Fe" relative to oxygen.

Two further remarks can be made pertinent to the generality of these analytical results on mare samples (Surveyors V and VI). (1) Surveyor V landed inside a small crater in Mare Tranquillitatis, and the samples examined were, at least in part, material ejected by the footpads during the landing. Surveyor VI landed in Sinus Medii on relatively flat terrain, and as far as can be determined the sample examined was undisturbed surface material. (2) The two missions differed in the height at which the main retrorocket burned out. In the Surveyor V mission, the end of burning was at 1.6 km; in the Surveyor VI mission, it was at 13.5 km. The retrorocket produced Al_2O_3 which conceivably could affect the analytical results. Estimates made for the Surveyor V mission indicated that even there the effect of the retrorocket exhaust should have been negligible (see app. A of ref. 8-1, first publication). The amount of Al_2O_3 contamination estimated for the Surveyor VI site is an order of magnitude less. The finding that the atomic percent of aluminum in the two samples examined on the Surveyor V mission is the same and indistinguishable from that found at the Surveyor VI site lends strong confirmation to the validity of these estimates.

The close similarity of the analytical results at two mare sites 700 km apart makes it unlikely that this chemical composition is applicable only to the specific landing sites of Surveyors V and VI. It appears much more probable that this composition is representative of large portions of the surface material of lunar maria.

The analyses are distinctly different for the terra samples examined by Surveyor VII. The computer-deduced results (table 8-12) for the chemical composition of the first sample indicate a similar composition to that of the mare samples with primary differences in the higher aluminum and the lower "iron" content of the terra sample. At this stage of analysis, the assigned errors would not exclude the same contents of these two components in the different samples. Examination of figure 8-39 (bottom), however, indicates a distinct difference in the alpha spectrum from the Surveyor VII sample compared with those from the Surveyor V and VI samples. The difference is largest for the highest energies (channels 63 to 73) that record the alpha particles scattered from the "iron" group of elements. The differences at lower energies are primarily a reflection of the lower "iron" contribution. Thus, the basic data indicate a lesser amount of iron-group elements, on the order of a factor of 2, at the terra site than at the mare sites. As mentioned in "Results of Lunar Sample Data" under "Results," the data from the other two samples examined on Surveyor VII (the lunar rock and material in a trench) confirm this lower "iron" content found at the Surveyor VII landing site.

The differences in the proton spectra of the terra site compared with those from the mare sites (fig. 8-39, bottom right) are primarily in magnitude throughout the entire energy region, rather than in spectral shape. These differences could be caused partially by geometrical effects. They will be studied in more detail using laboratory-reproduced geometrical relationships of the sample to the instrument. At present, differences in other than iron content in composition reported in table 8-12 for the terra site, as compared with those of the mare samples, cannot be taken as established.

The Chemical State of Lunar Surface Material

The alpha-scattering experiment provides no direct information about the chemical state of the elements measured. However, chemical experience makes possible an extrapolation from the data of table 8-12 to the probable chemical state of the bulk of Surveyor-type lunar-surface material. Specifically, the large atomic fraction of oxygen, larger than 0.5, suggests that the metals present are in oxide states. The mean values, if taken literally, indicate a slight oxygen deficiency. However, well within the present limits of errors, there is enough oxygen to combine with all elements considered. For example, table 8-13 lists the weight percentages of oxides that would be consistent with the analytical results from the Surveyor missions. Different compositions are presented for the terra samples than for the mare samples, reflecting the mean composition of the latter as presented in table 8-12.

TABLE 8-13. *Oxide compositions of lunar-surface material consistent with the Surveyor analytical results*

Oxide	Weight percent	
	Surveyors V and VI, mare	Surveyor VII, terra
Na_2O	(a)	(a)
MgO	5	7
Al_2O_3	b 14	21
SiO_2	50	50
CaO	15	15
FeO	16	7

[a] The presence of sodium has not been established with certainty. Sodium oxide could be present in amounts up to 3 percent by weight in the mare samples of Surveyors V and VI, and 4 percent by weight in the terra samples of Surveyor VII.

[b] The weight percent of Al_2O_3 is slightly higher than that given in ref. 8-2 (13 percent). The difference arises from rounded-off values in that report.

It must be emphasized that the assigned percentages in table 8-13 are far from unique in representing (within the given error) the analytical results of table 8-12. The limits of variation are hard to estimate at present. However, consistent with the results of the alpha-scattering experiment, the table is an example of the chemical state of the major elements and their relative proportions in the lunar mare and terra surface material examined thus far.

It should be noted that table 8-13 is meant to illustrate the oxide composition of the bulk of the lunar material examined. Minor constituents, adding up to perhaps as much as 10 percent by weight, may be present. In addition, the analytical errors do not exclude some unoxidized metal or radiation-decomposed oxides.

A limit to the amount of metallic iron at the Surveyor VI site has been set by the magnet test at about one-fourth percent by volume. (See "Magnet Data," p. 223.)

It is possible to speculate even further about the chemical state of the lunar material. It is improbable that it exists as a simple mixture of oxides. Rather, these are likely to be combined into more complex minerals. This is the state of material of similar chemical composition in most other natural samples such as terrestrial rocks and meteorites. (In making this additional extrapolation from the basic analytical results, it must be remembered that on the lunar surface the material may be in a non-crystalline form, either as a glass or too radiation damaged to be identified crystallographically.) Even with the present large analytical errors, the types of possible minerals are strongly restricted, although not defined uniquely. For example, the chemical composition of mare material given in tables 8-12 and 8-13 is consistent with most of the material being a mixture of minerals of the feldspar and pyroxene classes. As the analytical errors are reduced and as the amounts of the minor constituents are established, it will be possible to define further the mineral composition of the mare and terra material studied by the Surveyor missions.

Although these interpretations of the analytical results represent an extrapolation from the actual results of the alpha-scattering experiment, they should provide a more secure base from which to predict various other properties of the lunar-mare-surface material than has been available until now. For example, lunar materials in the states postulated should be chemically inert. They should not react with the usual materials of instruments or of structures brought in contact with them. This is consistent with the lack of obvious chemical action of lunar-surface materials with the aluminum-clad footpads of the Surveyor spacecraft. Similarly, because the present analytical results provide information not only on the principal chemical elements but also on their probable chemical state, it should now be feasible to evaluate more confidently the practicality of utilizing the raw materials on the Moon.

Finally, even these preliminary results should make it possible to improve predictions of detailed physical and chemical properties (such as melting point, density, hardness, compressibility, and so forth) of the particles making up lunar-surface material by comparison with terrestrially available substances of similar chemical composition. Because the first returned lunar samples on the Apollo program are likely to be from mare regions, the present analytical results should make possible more definite and economical plans for their investigation. The agreement of the results from Surveyors V and VI also implies that the results of the investigations of even the first returned lunar samples will have more general applicability than could have been expected before the Surveyor results were available.

Comparisons With the Chemical Composition of Various Materials

Although at present the assigned errors to the analytical results of table 8–12 are large, they still allow some significant comparisons to be made with the chemical compositions of various samples of the solar system.

The first such comparison is of the present results on samples of the lunar surface with the chemical composition expected if the Moon were an accumulation of condensed solar-atmosphere material (ref. 8–26). In this case, it may be expected that the volatile elements (hydrogen, noble gases, and so forth) would have escaped as would have those forming volatile hydrides (such as oxygen, sulfur, and so forth). For this reason, the comparison is made only with the metals determined in this work, and silicon is taken as the reference point. This comparison is shown in figure 8–41, using the values found in this work for Na, Mg, Al, (Si), "Ca," and "Fe" in the maria and terra.

It is clear from figure 8–41 that the surface of the Moon, at the places sampled by Surveyor, does not have the chemical composition of condensed solar material. For both mare and terra results, the magnesium values are too low and the aluminum and "calcium" values are too high. In addition, in the terra region near Tycho, the lunar surface does not have quite enough "iron."

The elemental analyses of table 8–12, as well as the representative oxide compositions of table 8–13, suggest silicate rocks such as are common both on the surface of the Earth and in many meteorites. Although an elemental analysis (even one more precise than the present one) can be only a rough indicator of rock type, it is of interest to compare the present results with the chemical composition of some materials that have been considered as constituents of the lunar surface. This is done in figure 8–42, where a comparison of the present results (mare and terra) is made with the analyses of average (refs. 8–27 through 8–30) dunites, basalts, granites, tektites, chondritic meteorites, and basaltic achondrites.

It is seen from figure 8–42 that the lunar surface, at the landing sites of Surveyors V, VI, and VII, cannot consist entirely of material similar to terrestrial ultrabasic rocks such as dunite or to chondritic meteorites. Just as in the comparison with the condensed solar atmosphere, the lunar samples have too much aluminum, calcium, and silicon and not enough magnesium.

At another extreme of the rock spectrum, figure 8–42 also compares the present results

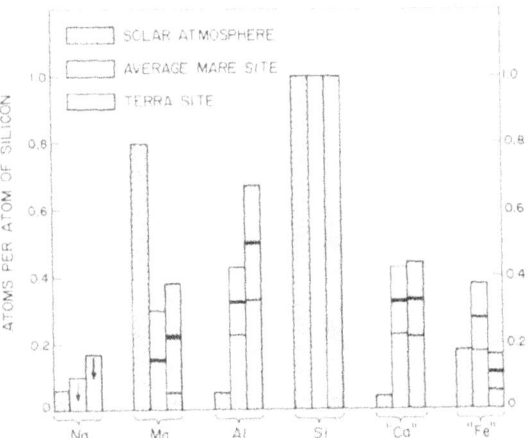

FIGURE 8–41.—Comparison of the observed chemical compositions in the lunar maria (average of Surveyors V and VI) and at the Surveyor VII terra site with that of the nonvolatile elements in the solar atmosphere. The three compositions have been normalized to unity of silicon. The solar values are from reference 8–26.

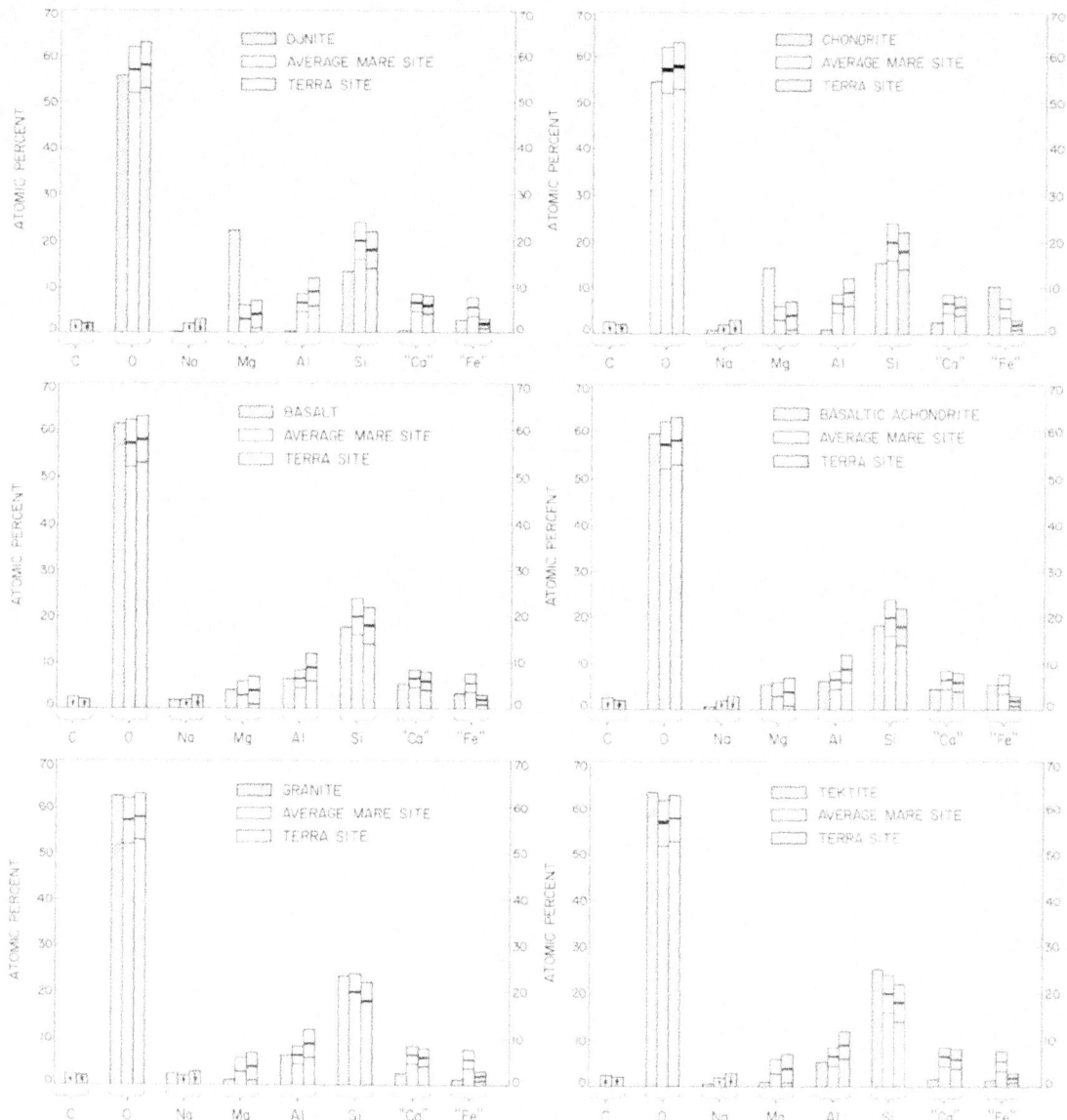

FIGURE 8-42.—Comparison of the observed chemical compositions in the lunar maria (average of Surveyors V and VI) and at the Surveyor VII terra site with that of the average composition of selected materials. The dunite and basaltic achondrite values are from reference 8-27; the basalt values are for Continental Basalts from reference 8-28; the granite values are for the North American Crust from reference 8-29; the chondrite values are the averages of the low-iron group from reference 8-30; the tektite values are those for the Indo-Malayan body quoted in reference 8-27.

with the chemical composition of terrestrial granites and with that of some tektites. Here the lunar results match more closely, especially those from the samples of the terra region.

However, the lunar samples appear to have insufficient SiO_2 and too large an amount of elements of the calcium group.

Of the comparisons made in figure 8-42, the

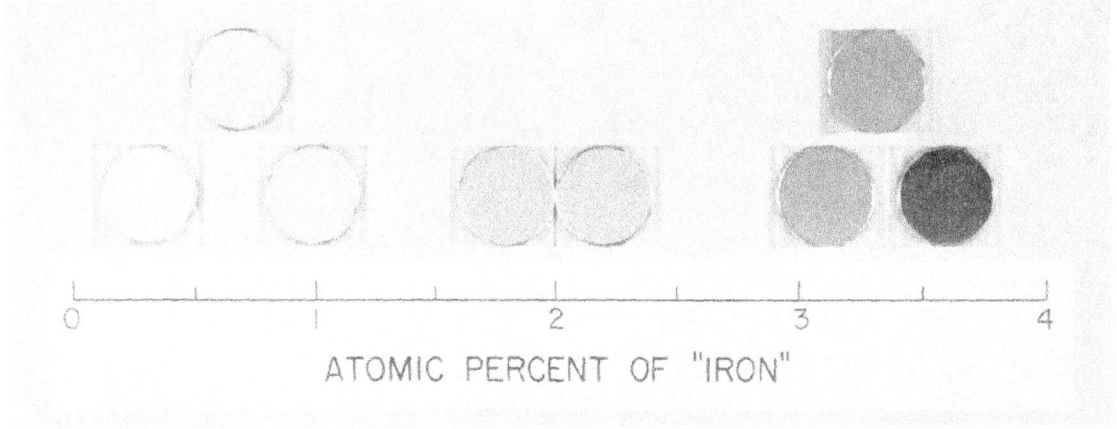

FIGURE 8-43.—A series of powdered terrestrial rocks, arranged according to the concentration of elements of the "iron" group (mass number $47 < A < 65$). The rock powders are all in the 37- to 50-μ particle-size range. In order of increasing concentration of iron-group elements, the rocks are: Mono Crater obsidian, 0.3 percent; Argus granite, 0.7 percent; Half Dome quartz monzonite, 1.0 percent; Black Peak quartz diorite, 1.8 percent; Loomis-88 diorite, 2.2 percent; Little Lake basalt, 3.1 percent; San Marcos gabbro, 3.3 percent; and Pisgah basalt, 3.6 percent. The series of rocks was supplied by Dr. John B. Adams and Dr. Alden Loomis of the Jet Propulsion Laboratory.

composition of the mare samples agrees most closely with the chemical composition of terrestrial basalts and with that of a somewhat rare type of meteorite—the basaltic achondrite. The analysis of the terra sample does not match, within the present errors, the composition of the average terrestrial basalt, although the discrepancy is in the low amount of "iron" only. An intermediate type of rock such as a diorite matches better overall, although still not within the errors quoted. With the present results on only a limited group of elements and with large errors, it may be that comparison with more detailed terrestrial-rock types is not warranted.

In spite of this inability at present to choose a terrestrial-rock type to match the results of the analyses of the samples from the terra site, all of the lunar analyses (mare and terra) are in strong disagreement with that expected for primordial solar system material (whether this is considered condensed solar atmosphere, terrestrial ultrabasic rocks, or chondritic meteorites). The lunar-surface material, where it has been sampled by Surveyors V, VI, and VII, if it originally had such a primordial composition must have undergone cosmochemical or geochemical processing to change the relative amounts of Mg, Al, and Si to those now found on the lunar surface. It is not clear from such arguments alone whether these processes occurred before or after formation of the Moon or whether they are still occurring.

The comparisons of figure 8-42 also make it unlikely that the majority of the meteorites that fall on the Earth (metallic and chondritic) originated on the surface of the Moon. The lunar maria as sources of tektites also appear to be excluded. To the extent that the other terrae have the same composition as that determined by Surveyor VII, they could not originate there either. The carbonaceous chondrites also are ruled out by the analytical results. Thus, on these assumptions, only a small fraction of the meteorites falling on the Earth can now be considered as having an origin at the lunar surface.

Differences Between the Terra and Mare Samples

Although the difference established between the chemical composition of the samples examined by Surveyor VII and the mare samples examined by earlier Surveyors is confined to the lower content of the "iron" group of elements at the terra site, this difference could be significant if it applies generally to the terrae. It should be remembered that the "iron" group, at the present stage of data analysis from the alpha-scattering experiment, includes the elements vanadium, chromium, manganese, iron, cobalt, and nickel. These are elements which, in general, impart color to rocks. Terrestrial rocks that have more of these elements are usually darker and therefore have a lower albedo than rocks with smaller amounts of these elements (fig. 8-43). Thus, although there are several possible reasons for the higher albedo of the terrae of the Moon relative to that of the maria, the lower content of the "iron" group of elements, as found in the Surveyor VII samples, may be a contributing factor.

Similarly, the lower "iron" group content of the Surveyor VII samples, if it is characteristic of terrae in general, probably means that the bulk density of the subsurface rocks of the terrae is less than that of comparable material in the maria. In this case, the very gross topographical relationships of the lunar surface would be similar to those on the planet Earth, where, in general, the continents are composed of material less dense than the basaltic ocean bottoms.

References

8-1. TURKEVICH, A. L.; FRANZGROTE, E. J.; and PATTERSON, J. H.: Chemical Analysis of the Moon at Surveyor V Landing Site: Preliminary Results. Surveyor V Mission Report. Part II: Science Results, Tech. Rept. 32-1246, Jet Propulsion Laboratory, Pasadena, Calif., Nov. 1, 1967, pp. 119-149. Also Chemical Analysis of the Moon at the Surveyor V Landing Site. Science, vol. 158, no. 3801, 1967, pp. 635-637.

8-2. TURKEVICH, A. L.; FRANZGROTE, E. J.; and PATTERSON, J. H.: Chemical Analysis of the Moon at Surveyor VI Landing Site: Preliminary Results. Surveyor VI Mission Report. Part II: Science Results, Tech. Rept. 32-1262, Jet Propulsion Laboratory, Pasadena, Calif., 1968, pp. 127-153. Also Chemical Analysis of the Moon at the Surveyor VI Landing Site: Preliminary Results. Science, vol. 160, 1968, pp. 1108-1110.

8-3. FRANZGROTE, E. J.; PATTERSON, J. H.; and TURKEVICH, A. L.: Chemical Analysis of the Moon at the Surveyor VII Landing Site: Preliminary Results. Surveyor VII Mission Report. Part II: Science Results, Tech. Rept. 32-1264, Jet Propulsion Laboratory, Pasadena, Calif., 1968, pp. 241-266.

8-4. GEIGER, H.; and MARSDEN, E.: On a Diffuse Reflection of the α-Particles. Proc. Roy. Soc., vol. 82A, 1909, pp. 495-500.

8-5. RUTHERFORD, E.: Scattering of Alpha and Beta Particles of Matter and the Structure of the Atom. Phil. Mag., vol. 21, 1911, pp. 669-689.

8-6. DARWIN, C. G.: Collision of Alpha Particles With Light Atoms. Phil. Mag., vol. 27, 1914, pp. 499-506.

8-7. SNYDER, C. W.; RUBIN, S.; FOWLER, W. A.; and LAURITSEN, C. C.: A Magnetic Analyzer for Charged Particles from Nuclear Reactions. Rev. Sci. Instr., vol. 21, 1950, p. 852.

ALLISON, S. K.; and WARSHAW, S. D.: Passage of Heavy Particles Through Matter. Rev. Modern Phys., vol. 25, 1953, p. 779.

8-8. RUBIN, S.; PASSEL, T. O.; and BAILEY, L.: Anal. Chem., vol. 29, 1957, pp. 736-743. Ion Scattering Methods.

RUBIN, S.: Treatise on Analytical Chemistry, Interscience, pt. I, vol. 4, New York, 1957, pp. 2075-2108.

8-9. TURKEVICH, A.: Chemical Analysis of Surfaces by Use of Large-Angle Scattering of Heavy Charged Particles. Science, vol. 134, 1961, pp. 672-674.

8-10. PATTERSON, J. H.; TURKEVICH, A. L.; and FRANZGROTE, E. J.: Chemical Analysis of Surfaces Using Alpha Particles. J. Geophys. Res., vol. 70, no. 6, 1965, pp. 1311-1327.

8-11. RUTHERFORD, E. E.: Collision of Alpha Particles With Light Elements. IV. An Anomalous Effect in Nitrogen. Phil. Mag., vol. 37, 1919, pp. 581-587.

8-12. FRANZGROTE, E.: Compositional Analysis by Alpha Scattering. JPL Space Programs Summary, vol. 37-20, 1963, pp. 186-190.

8-13. TURKEVICH, A.; KNOLLE, K.; EMMERT, R. A.; ANDERSON, W. A.; PATTERSON, J. H.; and FRANZGROTE, E.: Instrument for Lunar Surface Chemical Analysis. Rev. Sci. Inst., vol. 37, no. 12, 1966, pp. 1681-1686.

8-14. ANDERSON, W. A.; LAMPORT, J. E.; TURKEVICH, A. L.; TUZZOLINO, A. J.; PATTERSON, J. H.; and SUDDETH, D. E.: Final Engineering Report Alpha-Scattering Instrument Hardware Fabrication, Test and Support. Laboratory for Astrophysics and Space Research EFI Preprint No. 68-41, Univ. of Chicago, June 1968.

8-15. TURKEVICH, A. L.; KNOLLE, K.; FRANZGROTE, E.; and PATTERSON, J. H.: Chemical Analysis Experiment for the Surveyor Lunar Mission. J. Geophys. Res., vol. 72, no. 2, 1967, pp. 831-839.

8-16. WALKER, J.; WOLF, L., JR.; and KOSTENKO, C.: Thermal Aspects of the Alpha Particle Scattering Device. IITRI Project N6134, Final Rept., Sept. 1965.

8-17. HORWITZ, E. P.; BLOOMQUIST, C. A.; HARVEY, H. W.; COHEN, D.; and BASILE, L. J.: The

Purification of 10 Curies of Cm²⁴². ANL-6998, Argonne National Laboratory, Argonne, Ill., 1965.
8-18. TUZZOLINO, A. J.; KRISTOFF, J. J.; and PERKINS, M. A.: Silicon Detectors Aboard Surveyor V. Nucl. Instr. Methods, vol. 60, 1968, pp. 61-69.
8-19. Surveyor I Mission Report, Part I: Mission Description and Performance. Tech. Rept. 32-1023, Jet Propulsion Laboratory, Pasadena, Calif., 1966, pp. 69-70.
8-20. SCOTT, R. F.; and ROBERSON, F. I.: Soil Mechanics Surface Sampler, Surveyor VII Mission Report. Part II: Science Results. Jet Propulsion Laboratory, Pasadena, Calif., Mar. 15, 1968, pp. 135-185.
8-21. MORRIS, E. C.; BATSON, R. M.; HOLT, H. E.; RENNILSON, J. J.; SHOEMAKER, E. M.; and WHITAKER, E. A.: Television Observations From Surveyor VI. Tech. Rept. 32-1262, Jet Propulsion Laboratory, Pasadena, Calif., Jan. 10, 1968, pp. 9-45.
8-22. JAFFE, L. D.; BATTERSON, S. A.; BROWER, W. E., JR.; CHRISTENSEN, E. M.; DWORNIK, S. E.; GAULT, D. E.; LUCAS, J. W.; NORTON, R. H.; SCOTT, R. F.; SHOEMAKER, E. M.; STEINBACHER, R. H.; SUTTON, G. H.; and TUREKEVICH, A. L.: Principal Science Results From Surveyor VI, Surveyor VI Mission Report. Part II: Science Results. Jet Propulsion Laboratory, Pasadena, Calif., Jan. 10, 1968, pp. 5-7.
8-23. GAULT, D. E.; ADAMS, J. B.; COLLINS, R. J.; KUIPER, G. P.; MASURSKY, H.; O'KEEFE, J. A.; PHINNEY, R. A.; and SHOEMAKER, E. M.: Lunar Theory and Processes. Surveyor VII Mission Report. Part II: Science Results, Tech. Rept. 32-1264, Jet Propulsion Laboratory, Pasadena, Calif., Mar. 15, 1968, pp. 267-313.
8-24. SHOEMAKER, E. M.; BATSON, R. M.; HOLT, H. E.; MORRIS, E. C.; RENNILSON, J. J.; and WHITAKER, E. A.: Television Observations From Surveyor VII. Surveyor VII Mission Report. Part II: Science Results, Tech. Rept. 32-1264, Jet Propulsion Laboratory, Pasadena, Calif., Mar. 15, 1968, pp. 9-76.
8-25. VINOGRADOV, A. P.; SURKOV, YU. A.; CHERNOV, G. M.; KIRNOZOV, F. F.; and NAZARKINA, G. B.: Measurements of the Luna Surface γ-Radiation on the Cosmic Station Luna 10. Geochimiya, no. 8, 1966, p. 891. Also a paper by the same authors presented at the 10th meeting of COSPAR, 1967.
8-26. UREY, H. C.: The Abundance of the Elements With Special Reference to the Problem of the Iron Abundance. Quart. J. Roy. Astron. Soc., vol. 8, Mar. 1967, pp. 23-47.
8-27. PALM, A.; and STROM, R. G.: Space Sciences Laboratory Research Report, series 3, issue 5, Univ. of California, Berkeley, Calif., 1962.
8-28. TUREKIAN, K. K.; and WEDEPOHL, K. H.: Distribution of the Elements in Some Major Units of the Earth's Crust. Geol. Soc. Am. Bull., vol. 72, 1961, pp. 175-192.
8-29. CONDIE, K. C.: Composition of the Ancient North American Crust. Science, vol. 155, 1967, pp. 1013-1015.
8-30. UREY, H. C.; and CRAIG, H.: The Composition of the Stone Meteorites and the Origin of the Meteorites. Geochim. Cosmochim. Acta, vol. 4, 1953, pp. 36-82.

ACKNOWLEDGMENTS

The alpha-scattering experiment on chemical analysis for the Surveyor lunar missions is the culmination of many years' work to which many people have contributed:

At the Enrico Fermi Institute of the University of Chicago, the late Prof. SAMUEL K. ALLISON first suggested to the principal investigator the possibility of using the elastic scattering of charged particles for chemical analysis. Prof. HAROLD C. UREY, while he was still at the institute, inspired people to think of chemical analysis as an important early objective in the exploration of extraterrestrial bodies.

The early work on the alpha-scattering technique at the institute was performed with equipment made available by Prof. HERBERT L. ANDERSON and Dr. T. HINCKS (National Research Council of Canada) and with help from R. R. GABRIEL. Early collaborators of the principal investigator were Dr. ANDY BERETVAS and Dr. KARLFRIED KNOLLE.

The development and design of some of the critical mechanical parts of the instrument were performed by ED BLUME, who supervised their construction at the Central Development Shop of the University of Chicago. WARREN GEIGER and BERND WENDRING of this group were especially involved in this work. In addition, Mr. BLUME and Mr. WENDRING participated in the development and implementation of the hazardous prelaunch source installation operations at Cape Kennedy.

The Laboratory for Astrophysics and Space Research (formerly the Laboratory of Applied Sciences) of the Enrico Fermi Institute of the University of Chicago, under the technical direction of JAMES LAMPORT, had the principal responsibility for the development of the electronics and the construction of the instruments for the Surveyor missions. The design of the electronic system relied heavily on the circuits and techniques developed for the cosmic-ray research programs of Prof. JOHN A. SIMPSON, who provided important early support for this program. BLAINE ARNESON, RICHARD A. EMMERT, ROBERT SAUER, and ROBERT TAKAKI were principally involved in the adaptation of the electronic circuits for Surveyor purposes.

A great deal of the mechanical design work on the instruments was performed by MYRON WEBER. Much of the actual fabrication was performed by outside contractors (ref. 8-14). The instruments were assembled largely by GENE DRAG and TOM WONG. The fabrication and assembly were under the quality-control supervision of WILLIAM SIX. Mr. DRAG also participated in the final preflight tests at Cape Kennedy and in the Mission Operations of the alpha-scattering experiment.

All of the semiconductor silicon detectors used on this Surveyor experiment were developed, built, and tested by the group (Dr. MURRAY A. PERKINS, Dr. JULIUS KRISTOFF, and GEORGE HO) at the Laboratory of Astrophysics and Space Research under the direction of Dr. ANTHONY TUZZOLINO (ref. 8-18). The experiment would have been practically impossible without their dedicated efforts throughout the program to provide detectors that would reliably perform under the extreme temperature conditions on the Moon. At the same laboratory, and at the University of Chicago as a whole, many other people contributed substantially over the years.

At the Illinois Institute of Technology, Dr. G. WALKER (present address: University of Calgary, Calgary, Alberta, Canada) and LOU WOLFE were responsible for the basic design of the passive temperature control of the sensor head of the instrument.

At the Hughes Aircraft Co., we are indebted to the many people who contributed to the design, construction, and testing of the deployment mechanism and of the Electronics Auxiliary; to those who integrated the experiment onto a Surveyor spacecraft designed originally for engineering purposes; and to those who performed the countless tests that assured the experiment's success. They were members of a team led by ROBERT DANKANYIN, and included JOHN N. BUTERBAUGH, HEATON BARKER, and GENE HENDERSON.

At the Argonne National Laboratory of the U.S. Atomic Energy Commission, many members of the Chemistry Division (directed by Drs. W. MANNING, M. MATHESON, F. H. HAGEMANN, and D. STEWART) contributed to the success of this experiment. Some of those who took part in this work are listed here. PAUL R. FIELDS provided laboratory space, equipment, and samples of curium for early work in source preparation. He and other members of the transplutonium group were very cooperative in providing samples of radiochemicals, equipment, and much advice on the chemistry of transplutonium elements throughout the program.

The research apparatus used in the early stages of this program was designed by ALECK P. HRYN of Argonne's Central Shops, and two models were constructed by this department. The electronics were supplied by DALE J. HENDERSON; LEON P. MOORE assisted in the experiments conducted with this equipment.

CAROL A. BLOOMQUIST was responsible for some of the development of the technique for the preparation of the flight sources, and prepared several sets of sources for early prototypes (P-1 and P-2). In cooperation with other members of the Hot Laboratory group under the direction of C. HARRY YOUNGQUIST, she prepared the aluminum capsules containing americium oxide for neutron irradiation in the preparation of curium. Also (under the direction of Dr. E. PHILIP HORWITZ), in collaboration with HOWARD W. HARVEY of the Hot Laboratory group, she performed the separation of the curium from fission products and americium. MICHAEL A. ESSLING developed the method of carbon coating the flight sources, and provided this coating for the two sets of sources for the Surveyor VII mission, as well as assisting in the alpha pulse-height analysis.

JEROME L. LERNER provided a set of test sources by separating and depositing thorium-228 on nickel plates, using an isotope separator. FRED J. SCHMITZ provided general laboratory assistance from time to time. Elemental samples for analytical development were obtained from ROBERT L. McBETH, among others. LOUIS H. FUCHS was very helpful with advice and assistance in geological matters. ARTHUR H. JAFFE was consulted on detector problems.

DALE E. SUDDETH of Argonne National Laboratory's Electronics Division worked with the Laboratory of Astrophysics and Space Research of the University of Chicago in the construction and testing of the flight instruments. He was responsible for the design of the electronic pulse generator, which was incorporated into the instrument. He also participated in the systems testing of the instruments on the spacecraft, the prelaunch operations at Cape Kennedy, and the performance analysis during the missions in Pasadena.

Staff members at the Jet Propulsion Laboratory have also contributed throughout the history of the experiment, which received important early support from Dr. ALBERT E. METZGER, GEORGE O. LADNER, JR., and RICHARD E. PARKER who assisted in the operation and scientific evaluation of prototype instruments and in mission operations. Early contributions to the design of the instrument and test equipment were made by TIMOTHY HARRINGTON and GLENN A. SISK.

The instrument development, construction, and testing was monitored by Dr. DENNIS H. LeCROISSETTE, ROBERT J. HOLMAN, C. DANIEL PORTER, HENRY C. GIUNTA, CHARLES C. FONDACARO, and C. ED CHANDLER; this team was also involved in the final tests and calibration of the flight instruments at Cape Kennedy. The glass used in the standard samples was formulated under the contract to JPL by Dr. O. O. GRAFF of General Electric Co.

CARL R. HEINZEN, MAURICE C. CLARY, FLOYD I. ROBERSON, JAMES J. CARNEGHI, and ROBERT E. IMUS provided support in controlling the instrument during mission operations, under the direction of JACK N. LINDSLEY and DONALD D. GORDON.

Early work by Dr. JACOB I. TROMBKA and RICHARD BIDEAUX contributed to the development of the computer programs used in the control of the experiment during the missions; computational support was provided by CHRISTINE NELSON, SALLY RUBSAMEN, and MARGARET SIMES.

During the program, important support and direction were provided by HOWARD H. HAGLUND, Surveyor

Project Manager; Dr. LEONARD D. JAFFE, Surveyor Project Scientist; and by other staff members of the Surveyor project.

Support was provided by members of the television experiment team, headed by Dr. EUGENE M. SHOEMAKER, on each of the three missions and by members of the Surveyor VII soil mechanics surface sampler experiment team. FLOYD I. ROBERSON of JPL and Dr. RONALD F. SCOTT of Caltech used the surface sampler to deploy the Surveyor VII sensor head to the lunar surface, to shade it during lunar noon, and to redeploy it to subsequent samples.

The entire Surveyor program was conducted under the auspices of the National Aeronautics and Space Administration. The support of the Headquarters Staff of this agency and, in particular, of BENJAMIN MILWITZKY and STEPHEN E. DWORNIK, to this experiment is acknowledged with appreciation.

9. Lunar Theory and Processes

D. E. Gault (Chairman), J. B. Adams, R. J. Collins, T. Gold, G. P. Kuiper, H. Masursky, J. A. O'Keefe, R. A. Phinney, and E. M. Shoemaker

Discussion of Chemical Analysis

D. E. GAULT, J. B. ADAMS, R. J. COLLINS, G. P. KUIPER, H. MASURSKY, J. A. O'KEEFE, R. A. PHINNEY, AND E. M. SHOEMAKER

Preliminary results from the alpha-scattering experiments on Surveyors V, VI, and VII are given in table 9-1. For each of the elemental abundances, an error bar has been given; this error bar involves both the counting statistics and estimates of the uncertainties inherent in this preliminary stage of data reduction. In discussing the analyses, one must consider various compositions that lie within the given error bars. We point out here the problem involved in taking model compositions for which many of the elements lie at the extremes of their permitted ranges. If the likelihood of a single element at an extreme value is, say 0.1, then the joint likelihood that two elements so behave is 0.01, and so on. One may, therefore, ignore model compositions for which several elements are taken near the error limits.

Some rock and meteorite types are given in table 9-1 for comparison with the Surveyor data (refs. 1 to 9). All of these, for one reason or another, are candidates for analogs to the lunar material. The LL chondrite and type 1 carbonaceous chondrite are presented as typical of stony meteorites. The Mg in all chondrites (in the minerals olivine and pyroxene, principally) is too high for any agreement to be possible. Thus, chondritic and carbonaceous chondritic meteorites apparently cannot come from the surface of the Moon, if the analyses are representative. The eucrites (Ca-Fe rich, monomict achondrites) agree better with the Surveyor VI analysis than any other of our analogs. The howardites (Mg rich, polymict achondrites) fail to agree, again by virtue of the high Mg.

The tektites, represented by the Indo-Malayan type, do not fit at all, having too little Ca and excessive Si and O. The granite is not a good analog, although it is possible to find granite compositions that lie within the extreme error bars. The andesite is not as good a fit as some others, having too little Ca and Fe and comparably more Si and O. One of the best fits is an anorthositic gabbro, although Ca and O give marginal comparisons. Because the two mare areas investigated by Surveyors V and VI may be characterized best as basaltic with a high iron content, the simplest characterization of the Surveyor VII composition may be to describe it as basaltic with a low iron content; the precision of the analysis does not seem to warrant a much more detailed statement.

The central scientific questions about the Moon, which might be answered by the compositional data, are:

(1) What is the bulk composition of the Moon? How does this compare with the composition of the Earth and the meteorites?

(2) What are the composition and mode of origin of the lunar crust? (This term is left ill defined, but it includes the surface itself and goes to a depth of at least 2 km, which is the scale of the topography.) Is it derived in ways similar to the terrestrial crust?

(3) What is responsible for the known differences between highlands and maria; the difference in albedo, elevation, crater numbers, etc.?

In the discussion that follows, use is made of terrestrial and meteoritic analogs, both with respect to models of origin and compositional classes. This does not mean that the lunar rocks will be exactly like these analogs; in fact, these rocks are undoubtedly unique in many respects. But in following this approach, it is

TABLE 9-1. *Comparison of preliminary chemical analyses from Surveyors V, VI, and VII with representative rocks*

	Element, atomic percent [a]							
	C	O	Na	Mg	Al	Si	"Ca" [b]	"Fe" [b]
Surveyor V (ref. 9-2)	<3	58±5	<2	3±3	6.5±2	18.5±3	13±3 [c]	
Surveyor VI (ref. 9-3)	<2	57±5	<2	3±3	6.5±2	22.0±4	6±2	5±2
Surveyor VII (sample 1; see ch. 8)	<2	58±5	<3	4±3	9±3	18.0±4	6±2	2±1
Chondrite:								
LL group (ref. 9-4)		58.0	.7	15.2	1.0	16.0	1.0	8.1
Carbonaceous (type 1; ref. 9-4)	6.6	55.4	.6	8.4	.7	8.4	12.3	7.8
Eucrite (ref. 9-6)		60.7	.5	3.6	5.7	18.8	4.2	6.9
Howardite (ref. 9-6)		60.3	.4	7.1	4.6	18.5	3.1	5.8
Dunite (ref. 9-7)		59.0	.1	23.9	.3	14.1	.2	2.3
Peridotite (ref. 9-7)		58.9	.4	19.3	1.9	15.5	1.4	2.5
Anorthositic gabbro (ref. 9-8)		61.4	2.6	1.2	9.4	19.0	4.4	1.7
Basalt (tholeiitic):								
Average oceanic (ref. 9-9)		61.3	1.5	4.1	6.3	18.1	4.5	4.3
Average continental (ref. 9-9)		61.5	1.7	3.2	7.0	18.8	4.3	3.7
Basalt (alkalic):								
Average oceanic (ref. 9-9)		60.8	2.1	3.8	6.7	17.2	4.8	4.3
Average continental (ref. 9-9)		60.8	2.4	3.9	6.8	17.2	4.8	3.9
Andesite (ref. 9-10)		61.2	2.9	.1	6.9	21.1	3.1	3.0
Granite (ref. 9-6)		63.4	2.3	.4	5.0	24.4	2.7	1.0
Tektite (Indo-Malayan; ref. 9-13)		64.0	1.0	1.1	5.4	25.2	3.4	1.5

[a] Excluding elements lighter than beryllium.
[b] "Ca" and "Fe" denote elements with mass numbers between approximately 30 to 47 and 48 to 65, respectively.
[c] "Ca" and "Fe" taken together.

well to remember that we are in a position not unlike the biologist who first tried to describe the fauna of Australia to his colleagues. Furthermore, it must be emphasized that the region in which Surveyor VII landed is found to consist of several flow units that originated from the direction of Tycho. The significance of the chemical analysis by means of the alpha-scattering experiment is clearly dependent on a correct description of the mechanism by which these units were deposited, whether by some volcanic process or by a hot or cold, turbidity-like flow at the time of presumed impact. Nevertheless, this is the only available analysis for the highlands, which constitute more than 80 percent of the lunar surface. We will interpret the analysis, therefore, as being typical in some sense of the composition of these highlands, and discuss the contrasts between the maria and the highlands on the basis of the single Surveyor VII datum and the analyses from the mare sites of Surveyors V and VI. The density and albedo contrasts inferred in this comparison are quite reasonable in terms of the telescopically determined morphological and albedo contrasts. The possibility is accepted, however, that later analyses in this or other highland areas might show the Surveyor VII site to be quite atypical.

Contrasts in Albedo

The low iron content of the material at the Surveyor VII landing site provides a possible explanation of the high albedo of the lunar highlands. Iron is the most abundant of the elements (transition metals) that absorb strongly in the visible part of the spectrum. The change in iron content from the mare sites to the highland site is sufficiently large to account for a distinct change in the opacity and, perhaps in the amount, of the mafic silicate mineral(s). Such a change would, in turn, affect the albedo of the whole-rock powder.

From the present data, it appears unlikely that most of the iron measured by Surveyor VII occurs on the surfaces of the rock particles as free metal. We are not inclined, therefore, to ascribe the albedo contrasts between the highlands and maria to differences in amount of free metal on the lunar surface. Furthermore, low carbon abundances in analyses from the maria and highlands imply that carbon is not a major factor controlling albedo on the Moon.

If it is correct that the iron content of the silicate minerals determines the albedo of large lunar regions, it also is probable that this is not the only factor. For example, the numerous bright craters and rocks in the maria cannot all be intrinsically different in composition from the surrounding darker material. Shoemaker has proposed a "lunar varnish" alteration process (ref. 9-1) to explain these differences in albedo. Adams (ref. 9-2) has emphasized the importance of mean particle size where albedo contrasts are not the result of compositional differences. These ideas have not been tested conclusively by the Surveyor missions. However, the comparisons of analyses (when available) of the undisturbed soil, disturbed soil, and of the rock at the Surveyor VII landing site ultimately may provide evidence on the lunar varnish hypothesis.

Estimated Density of Lunar-Surface Rocks

From the similarity of the atomic abundances in the Surveyor analyses to those of basaltic rocks, it seems reasonable to infer a mineralogy that includes some, or all, of the following:

Mineral		Density
Albite (Ab)	$NaAlSi_3O_8$	2.62
Anorthite (An)	$CaAl_2Si_2O_8$	2.76
Pyroxenes	$(Ca, Mg, Fe)SiO_3$	
Enstatite (En)	$MgSiO_3$	3.20
Diopside (Di)	$CaMgSi_2O_6$	3.28
Hedenbergite (Hd)	$CaFeSi_2O_6$	3.55
Hypersthene (Hy)	$(Mg, Fe)SiO_3$	3.45
Olivines	$(Mg, Fe)SiO_4$	
Forsterite (Fo)	Mg_2SiO_4	3.21
Fayalite (Fa)	Fe_2SiO_4	4.39
Metallic iron	Fe	7.87
Magnetite	Fe_3O_4	5.20
Quartz (Qtz)	SiO_2	2.65

The densities of these minerals and their solid solutions are determined almost entirely by the proportion of iron. Within the pyroxenes, the incidence of high Ca, despite its atomic weight, causes a density decrease. Estimates for the density of the pyroxene present may be made, however, with considerable confidence by ideal weighting in terms of the densities of the end members present. The distinction between the orthopyroxene and clinopyroxene is not significant because hypersthene is used here only to define a particular composition and density. Plagioclase and olivine density may be estimated similarly in terms of two end members.

A series of putative atomic compositions that lie within the error bounds of the Surveyor alpha-scattering analyses is presented in table 9-2. Computed norms for these compositions are given in table 9-3, along with their estimated densities.

In computing the mineralogical norms, the Surveyor analyses do not provide a basis for any confidence in deciding whether or not olivines are present in any amount. Atoms are allocated, therefore, to pyroxene molecules insofar as it is possible. It is apparent from the derived density values that this assumption does not appreciably affect the mean density.

The densities are determined principally by the proportion of plagioclase to total rock and by the "iron" composition of the pyroxenes. The plagioclase proportion is, in turn, determined by the amount of Al in the analysis. Small amounts of free iron, or "iron" as magnetite, affect the density approximately as though the "iron" were in a pyroxene. The computed densities are, however, insensitive to the amount of excess iron, so that the question of whether all the metallic atoms are oxidized does not affect the density analysis.

Comparison of results (table 9-3) for analyses from the highland site (models 1 through 11) with those from the mare sites (models 12 through 15) indicates that reasonable values for the rock grain density for the two regions are approximated by 3.0 ± 0.05 g/cm^3 and 3.2 ± 0.03 g/cm^3, respectively. The difference is significant and reflects the difference in the "iron" content between the two regions. It should be remem-

TABLE 9-2. *Assumed atomic compositions*

Model	Element						
	O	Si	Al	Na	Mg	"Ca"	"Fe"
1	58	18	8	2	4	6	2
2	59	18	8	1	4	6	2
3	59	18	7	1	6	5	2
4	59	18	7	1	5	5	2
5	60	18	8	1	4	5	2
6	60	18	8	1.5	4	5	2
7	60	18	8	2	4	5	2
8	58	18	8	2	4	5	2
9	58	18	7	2	3	5	2
10	58	18	9	1	3	6	2
11	58	18	9	1	5	3	2
Surveyor VII analysis	58 ± 5	18 ± 4	9 ± 3	<3	4 ± 3	6 ± 2	2 ± 1
12	58	20	6.5	2	3	6	5
13	59	20	6.5	1	3	7	6
14	60	18	5.5	1	4.5	5	6
15	60	19	5.5	.5	4.5	4	6
Surveyor VI analysis	57 ± 5	22 ± 4	6.5 ± 2	<2	3 ± 3	6 ± 2	5 ± 2
Surveyor V analysis	58 ± 5	18.5 ± 3	6.5 ± 2	<2	3 ± 3	13 ± 3 [a]	

[a] "Ca" and "Fe" taken together.

bered, moreover, that Turkevich et al. (see ch. 8) state that the rock (sample 2) analyzed at the Surveyor VII site contained about 30 percent less iron than that for the undisturbed lunar surface (sample 1), on which tables 9-2 and 9-3 are based. Thus, differences in rock densities of the highland and mare regions may be even greater than indicated in table 9-3; note that Scott and Roberson (see ch. 5) estimate the density of the rock "weighed" by the surface samples to be 2.4 to 3.1 g/cm³.

Regarding the analyses at the mare sites, the eucrites have been identified as having an atomic composition that falls within all the error bounds for Surveyors V and VI (refs. 9-3 and 9-4). For compositions taken arbitrarily from within the allowed Surveyor V and VI bounds, we find densities between 3.17 and 3.22 g/cm³. The higher densities, especially model 15, are found for compositions that are selected to agree with the most common eucrite compositions. Thus, the estimated densities are in essential agreement with the eucrite densities, although the latter may range up to 3.28 g/cm³.

In short, if the intrinsic density of the mare material is taken as 3.20, then the hypothesis of a eucrite mare composition is not counterindicated. If some of the "Fe" (say about 1 percent) is really Cr or Mn, these elements would be found as impurities in the (already rather non-stoichiometric) pyroxene and plagioclase lattices and will affect the density in a way that cannot be distinguished from the effect of iron.

This analysis has been based on a fairly conventional interpretation of the chemistry. The possibility remains, however, that something rather strange may be masquerading as a basalt or a eucrite. For example, the possibility has not been included that there is 0.5 percent or more K in the "Ca," which would affect the mineralogy. This seems reasonable in view of the indicated low Na values and the usual Na/K ratios of 5 to 10 found in igneous rocks. There has been no discussion of Cl or C in this mineralogical model, nor has consideration been given to insure that the minerals form a stable assemblage.

Many other questions that could be raised about the lunar surface involve effects that are too small for the chemical analysis to pro-

TABLE 9-3. *Estimated densities for assumed normative mineral compositions*

Model	Estimated density, g/cm³	Normative composition [a]								
		Ab	An	Di	En	Hy	Hd	Fe	Fe₃O₄	Other
1	2.95	2	3	3				2		
2	2.99	1	3.5	2.5		1.5				
3	3.03	1	3	2	3	1		1		
4	3.05	1	3	3	1	2				
5	2.99	1	3.5	1.5	1	2				
6	2.98	1.5	3.2	1.8		2				
7	3.05	2	3	2	2				1.5	
8	2.96	2	3	2	2			2		
9	2.92	2	2.5			2.5				
10	2.97	1	4	2		1		1		
11	2.98	1	4	2	2			2		
12	3.17	2	2.3	3.7		.5		4.5		
13	3.20	1	2.7	3			1.3		2.0	2.5 SiO₂
14	3.20	1	2.3		3.2	1.3	2.7	2		
15	3.22	.5	2.5				4.5	1.5		

[a] Units are arbitrary with coefficients proportional to the molar composition.

vide any answers in their preliminary form. It would be desirable to know whether 2 percent or more, by weight, H₂O is present in the surface material as water of hydration. If this were true, the amount of available oxygen for combination with the metals would be reduced by a few percent, thus presumably exacerbating the oxygen deficit. The present error bounds on the chemical analyses, as well as the bounds that must be placed on speculation, permit only the statement that 10 percent water of hydration appears fairly unlikely. The question may also be raised as to how much meteoritic iron is present in the surface soil. From the alpha-scattering experiments, 0 to 4.5 percent, by weight, of the soil could be metallic iron, a result that establishes only an upper limit for the content. On the other hand, the magnet tests seem to indicate less than one-fourth percent magnetic material; this could be all magnetite, if the analogy with terrestrial basalts is at all relevant. The single magnetic object that apparently adhered to the magnets on the surface sampler is spectacular, and may be a fragment of a meteorite; however, it seems inappropriate at this time to base any speculations on a single datum of this kind.

Bulk Composition of the Moon

The Surveyor chemical analyses do not provide any definitive information on the bulk composition of the lunar body. Indeed, the composition of the lunar interior must always remain a matter of inference; evidence will always be circumstantial and remain open to alternate interpretations. At the present time, two questions are crucial to interpretations:

(1) Is the Surveyor VII analysis typical of highland material?

(2) To what depth is the estimated density representative of the lunar "crust"?

These questions obviously cannot be answered until additional highland sites are analyzed. The following circumstances prevail:

(1) The mean density of the Moon (3.34) is very close to the mean density of the Earth's uncompressed mantle material, about 3.35 or slightly higher. Under the pressure and temperature conditions expected in the lunar interior, the mean lunar density may be taken as the true, constant density of the lunar interior with an error less than 0.05 g/cm³, if the estimated density from the Surveyor VII analyses is valid for only a few kilometers of the lunar

"crust" (i.e., no dense core or other structural inhomogeneity). In addition, the composition of the Surveyor VII lunar highland sample agrees closely with that of an anorthositic gabbro and reasonably well with that of oceanic and continental tholeiitic basalts (tables 9-1 and 9-2). Analyses from the Surveyor V and VI sites are similar, but show twice as much iron; thus, the material from the maria also resembles terrestrial basalts, but bears a resemblance to the eucrites, which differ in major element chemistry from the terrestrial basalts in their high iron (as well as having lower alkalies, a matter which cannot be discussed on the basis of the preliminary results from the alpha-scattering experiment). The obvious inference from the similarity between the mantle and Moon densities is that the lunar body and the Earth's mantle are composed of essentially the same substance.

The mantle may be thought of as a mixture of an olivine (80 percent Fo, 20 percent Fa) with a basalt in a ratio of about 5 to 1. Until some strong counterevidence comes from the lunar surface, some heed must be paid to this inference because of the lack of any other obvious candidates for the 3.34 density. For this reason, it is especially interesting that the lunar surface, which we tend to regard as the prime derivative of the lunar body, should have a composition so similar to the basalts, which compose the prime derivative of the terrestrial mantle. The results from the alpha-scattering experiment, therefore, may be viewed as additional circumstantial evidence in favor of the Moon/mantle similarity.

The terrestrial analogy is imperfect, however, and the divergences provide very interesting scientific questions. In the Pacific Ocean, the basin extrusives are andesitic. These two varieties, on the average, show a density difference on the order of 0.1 or 0.2, which is of the same magnitude and sign as the density difference that has been estimated between lunar mare and highland materials. But the sequence of lunar materials is different: The basin deposits on the Earth most resemble the highland deposits on the Moon; the "ferrobasalt" of the lunar basins finds no common terrestrial analog, and the terrestrial andesite has not been found in any of the three Surveyor chemical analyses.

(2) The Surveyor VII analysis bears a close resemblance to a terrestrial anorthositic gabbro (table 9-1), such as may be found in layered basic intrusives, such as the Stillwater, Bushveld, or Skaergaard intrusives. Certainly the possibility should be considered that the layered gabbros and genetically related members are a ubiquitous feature on the lunar surface. In this respect, a geological mapping of the area north of the crater Tycho leads to remarkably consistent agreement that several distinct blankets of material can be identified and stratigraphically placed; only the origin of the various units has been subjected to multiple interpretations. If the units are ejecta deposited as a result of the impact that formed Tycho, then the material around the Surveyor VII site was probably derived from depths of 10 to 15 km. It would not be surprising, therefore, if plutonic igneous rocks were the main constituent of the deposits around the spacecraft. The observation of coarse, light/dark textures in rocks near the spacecraft is suggestive of, and consistent with, such an interpretation. The observations are, however, hardly conclusive. It is equally possible that the material analyzed at the Surveyor VII site is a postcratering volcanic rock. The depth of origin of such material is a matter of speculation, but it seems likely that the source would be very much deeper than for impact ejecta.

Notwithstanding the origin of the Surveyor VII highland samples, if one posits that the density estimated for the material analyzed at the Surveyor VII site is representative of the highland provinces (corresponding to more than 80 percent of the lunar surface) and extends to depths approaching 100 km or more, the mean density of the Moon would require interior densities significantly greater than the mean value. Recent results reported by Lorell and Sjogren (ref. 9-14) from analysis of the Lunar Orbiter tracking data suggest that the Moon has an interior density "moderately higher" than crust density. Because density increases produced by the modest interior pressures of the Moon could be compensated, or even offset, by the effects of increasing temperatures of depth, increased interior density may be interpreted as indicating material that is compositionally different from the Surveyor basaltic chemistry. Ultrabasic composition, high-pressure assemblages, and perhaps even the presence of an embryonic "core" as a result of chemical fractionation of the primordial lunar mass may provide, either individually or collectively, an explanation for higher interior densities. Differentiation within the body of the Moon, however, may not have proceeded as far as terrestrial processes; it is interesting to speculate that the Moon in its present state may represent an evolutionary stage similar to that of a youthful Earth.

On the Thermal Regime in the Moon

The Surveyor chemical analyses are strong circumstantial evidence that melting has occurred in the Moon, and the Lunar Orbiter photographs suggest that this may have been true over a major fraction of the Moon's history. The consequences of such melting in the lunar body are relevant to subsequent discussions and are of intrinsic interest.

It is possible to discuss the heating to be expected in an initially cold Moon by decay of the long-lived radioactivities U^{238}, U^{235}, Th^{232}, and K^{40}. Temperatures estimated in this way are likely to represent the minimum possible temperature, since other effects, such as initial heating, tidal friction, etc., act to raise the temperature. Both time-dependent and steady-state calculations have been made, and all have certain features in common: a nearly constant maximum temperature throughout the interior, decreasing to a nearly constant gradient region near the surface; and a steady increase in the central temperature with time, given by the total heat added to the interior by radioactive decay. By relating the history of heat production to the concentration of heat-producing isotopes, it is possible to investigate whether or not melting in the interior is likely for a given type of material (ref. 9-15). Melting is predicted if the concentration of K_2O exceeds about 0.02 percent.

For oceanic tholeiites (ref. 9-14), K_2O ranges between 0.06 and 0.26 percent; it ranges between 0.04 and 0.22 percent for eucrites (ref. 9-6). Both of these are notable for having the lowest K_2O (and other alkalies) within terrestrial extrusive and stony meteorite groups, respectively. The amount of K_2O in the parent material is less by some factor, which depends on the original proportion of the magma in the parent. Factors of 3 to 6 have been suggested; it is then apparent that the range of uncertainty brackets the critical K_2O value of 0.02 percent. It is probably safer to heed the photographic and chemical evidence in favor of melting, and put a lower limit on the K_2O in the Moon. The values are not very different from the concentrations suggested for the Earth's mantle in discussions of terrestrial heat flow; for that reason, it is convenient to set them equal and compare the steady-state heat flow to be expected. The Moon's volume, and hence its total amount of heat-producing material, is smaller than that of the Earth by $(R_m/R_e)^3$. The area is smaller by $(R_m/R_e)^2$. The heat flow should then be smaller in proportion to the radius, namely by a factor of 4. In all numerical discussions of lunar temperature, the heat flow follows this approximation fairly well, and is insensitive to the transient aspects of the problem. On dimensional grounds then, the near-surface thermal gradient is found to be four times less than the terrestrial gradient. The pressure gradient, away from the center, is about six times less. Thus, approximately, the temperature and pressure gradients in the outer few hundred kilometers of the Moon are expected to be about five times less than on the Earth. The temperature (pressure), T(P), behavior can also be taken over from the terrestrial T(P), but must be scaled by a factor of 5 in depth. That upper portion of the terrestrial crust/upper mantle system that is cool enough to support long-term stresses and not creep is about 50 to 80 km thick, and may be called the lithosphere. The region extending from the lithosphere to at least 180 km is characterized by a low-velocity, high-attenuation zone, and is the locus of primary magma generation. Discussions of temperature indicate that this region is one in which the temperature is near, if not at, the melting temperature of the first melting component and has, in consequence, very little strength (asthenosphere).

If these conditions are "mapped" onto the Moon, the lithosphere must extend to depths of 250 to 400 km, and the remainder of the lunar interior will correspond to the low-velocity asthenosphere. The center of the Moon corresponds to a depth of only 150 km in the Earth.

On Chondritic Meteorites and the Moon

The possibility that some or all varieties of meteorites are derived from the Moon has been a tantalizing prospect for many years (refs. 9–15 and 9–16). However, from even a cursory examination of table 9–1, it is apparent that the chemical composition at the Surveyor V, VI, and VII landing sites in no way resembles the composition of ordinary or carbonaceous meteorites; both types of chondrites have altogether too much Mg and too little Ca and Al; in addition, carbonaceous chondritic meteorites have too much C. The evidence relating to the bulk composition of the lunar body remains circumstantial, however, and can be interpreted in a chondrite framework.

Suppose that ordinary chondrites, with a density of 3.6 to 3.8, comprise a major fraction of the Moon. Two-thirds of this could be fully melted, in a core, without conflicting with present knowledge of the nonequilibrium gravity harmonics of the Moon. Because the average density of such a Moon could not be less than about 3.55, it is necessary to assume that volatiles, as exemplified by the constituents of the low-density (2.9) carbonaceous chondrites, are present in sufficient quantity to bring the mean density down to 3.34. Under the possible conditions of temperature and pressure in the Moon, carbonaceous chondritic material would probably assume a density close to 3.25 when the water was taken into denser phases. The only chondritic Moon that might be arranged to have the correct mean density by this mixture is composed almost entirely of carbonaceous chondrites.

A Moon composed of carbonaceous chondrites in bulk differs principally from terrestrial mantle material in two ways:

(1) The chondrites have significantly more iron, either as metal or in a silicate phase. The effect of this iron to increase the density is offset by the presence of a great deal of water, on the order of 10 percent of the total mass instead of 1 percent or less, as is the case with the mantle.

(2) The chondritic meteorites appear to be enriched in Na, K, etc., with respect to the Earth's mantle.

It is possible to discuss implications of these differences, but not conclusively. From an analysis of the probable pressures and temperatures in the Moon, there are indications that the $T(P)$ is very much like that of the Earth, but with a depth scale about five times greater. From the center of the Moon out to 1200 to 1400 km, temperatures appear to be at, or near, melting conditions for the first melting fraction. Under these conditions, with approximately 1 percent water, the Earth's upper mantle is extremely mobile on geological time scales; this mobility is responsible for drastic displacements of crustal blocks, island arcs, mountain building, etc. No evidence for a similar tectonics of large-scale, lateral displacement is seen on the lunar surface; however, this is compatible with the possibility that the lunar lithosphere (mechanically rigid crust) is six times thicker than the terrestrial lithosphere, a situation which is likely to suppress large-scale displacements. To introduce 10 percent water, however, and retain such stability seems totally unreasonable. Moreover, it seems unlikely that a Moon with a mobile, high-temperature interior could retain 10 percent water against outgassing over times of 10^9 years or more. The circumstance most favorable to a chondritic Moon is, therefore, that of an interior which has remained at temperatures significantly below the melting point; this does not appear to be compatible with the amounts of heat-producing K, U, and Th in chondritic meteorites.

If material of chondritic composition occurs in the lunar interior and does not come to the surface, the chondrites arriving on Earth must have originated elsewhere. Although the possibility cannot be overlooked that chondritic material eluded three Surveyors, the fact remains that chondrites constitute the overwhelming majority of all meteorites, and the ordinary chondrites, high density and all, are still to be explained. (Carbonaceous chondrites are undoubtedly numerically more significant outside the atmosphere; they are easily broken apart and consumed by ablation processes on entering the Earth's atmosphere.) If the Surveyor analyses are typical, it is difficult to see how some, or all, of the chondrites come from the Moon, without conflicting with either the composition or the mean density.

The resemblance to eucrites, shown by analyses from the maria, has been cited in the past on circumstantial evidence in favor of the Moon as an origin for the basaltic achondrites (refs. 9-6 and 9-16). The Surveyor VII analysis does not support the lunar origin for these objects and, in fact, tends to refute the possibility. There are two difficulties: First, the basaltic achondrites constitute about 5 percent of the observed falls and, if they have a lunar origin, are derived from less than 20 percent of the lunar surface covered with mare material. Objects derived from the remaining 80 percent of the lunar surface, the highlands, also should be present in the meteorites arriving on Earth. But there are no known meteorites with a composition similar to that indicated by the Surveyor VII analysis. Either the Surveyor VII analysis is not representative of the highlands or one must invoke the absurd conclusion that most meteorites are "filtered" by some unknown process that excludes all but those from the maria arriving on Earth.

A second pitfall for the Moon/eucrite analogy stems from the observation (ref. 9-6) that eucrites might be genetically related to the howardites and mesosiderites and they, in turn, might be representative of the highlands. It is clear from tables 9-1 and 9-2 that the Surveyor VII analysis does not support such a possibility. It should be noted in passing that, with the potassium argon ages of eucrites 4.5 billion years, it is clear that the surfaces in the maria are either 4.5 billion years old or that the eucrites do not come from the Moon (barring circumstances of surface heterogeneity). Lunar Orbiter photography provides a wealth of morphological and geological detail about mare surfaces; many mare areas are among the stratigraphically youngest places on the Moon.

Some members of this Surveyor working group are inclined to the view that the stratigraphic youth is equivalent to geological youth, with ages of some millions to tens of millions of years. However, others in this working group feel that the stratigraphically youngest areas are 4.5 billion years old. This question of age and eucrite origin should be settled beyond reasonable doubt when lunar samples are available for radiometric dating.

On Tektites

Chemical measurements at the Surveyor VII landing site (see ch. 8) add to the evidence (refs. 9-3 and 9-4) that tektite material is not widely distributed on the lunar surface. The importance of such material in the formation of the mare surface, if any, is clearly not as great as indicated by O'Keefe in reference 9-17.

The analysis of the rock from the Surveyor VII site indicates a material that may have a density of 3.0 or as low as 2.9, in remarkably good agreement with the best estimate for the rock weighed by the surface sampler (see ch. 5). The contrast in density between this rock and the material of the maria, which is much richer in iron and may have a density of 3.2, is conceivably sufficient to account for isostatic differences in the elevations of the two regions. On the Earth, isostatic differences correspond to density differences of 2.7 versus 3.0. It follows that the argument for a silicic rock in the highland parts of the Moon, contrasting with a basaltic rock in the maria, is not securely based.

On the other hand, it is well to keep in mind that large basaltic intrusions in the Earth are normally accompanied by small volumes of silicic rock, the so-called granophyres. It should be expected, therefore, that acidic rock may occur somewhere on the lunar surface; the Surveyor analyses, therefore, do not rule out the possibility that tektitic material may be found in some parts of the Moon.

Solar System Implications

The chemical analysis and the results of the data derived from the magnet test exclude the possibility that the Surveyor VII site is composed of chondritic material. This discovery, coupled with the findings of Surveyors V and VI in the maria, supports the conclusion that the Moon is not the source of chondritic meteorites. This conclusion bears directly on our present knowledge of the chemical composition of the terrestrial planets.

The high density of Mercury (ref. 9-18) and the generally lower (uncompressed) densities of the planets more distant from the Sun have led to the idea that the dispersed material from which the planets accreted was somehow affected by solar irradiation early in the evolution of the solar system.

Urey (ref. 9-15) suggested that chondritic meteorites might come from the Moon. If true, this would mean that the bulk of the meteoritic data applies to a relatively restricted portion of the solar system. If the chondrites are now ruled out by the Surveyor evidence, it appears that most meteorites are samples from outside the Earth/Moon system. The source of the chondritic meteorites is, of course, undetermined. However, the existing chemical and isotopic analyses of meteorites, as compared with terrestrial and lunar data, now become more significant.

The Surveyor analyses raise doubts about whether any primitive lunar material is preserved at the surface. If the basaltic rocks measured by Surveyors are the product of magmatic differentiation, the Moon probably has been extensively modified since accretion. A differentiated Moon would imply that the (larger) terrestrial planets also are likely to be differentiated.

Chemical Observations of Surveyor V

T. Gold

The important observation that the lunar soil at the Surveyor V site is basaltic in composition is taken by many to substantiate the viewpoint, previously widespread, that volcanism formed most of the lunar surface, supplying a differentiated type of rock. The case for this is, however, by no means so simple or so clear cut. The arguments previously voiced against a widespread differentiation on the Moon are now just as strong or, in some cases, even strengthened, by recent observations. For example:

(1) The mean density of the Moon is lower than the density it would have if it were initially composed of the same material as the Earth. A chemical composition different from that of the Earth is, therefore, implied.

(2) The value of $C/(ma)^2$ (where C is the largest moment of inertia, m is the mass, and a is the radius) is now known to be close to 0.4, indicating the absence of any central condensation. Thus, it does not seem likely that very widespread internal melting leading to internal differentiation occurred, since this most probably would have resulted in central condensation also.

(3) The structural strength of the Moon is high enough to allow the persistence of the present large departure from equilibrium in the distribution of its mass. A hot interior is not compatible with such material strength. This is true for the maintenance of both the lowest and the highest harmonics of the gravitational field.

(4) The Moon loses very little gas at the present time. Gas emission from any present-day volcanism on the Moon must be a factor of 10^5 or 10^6 down from that of terrestrial volcanism in order to have escaped detection. The presence of liquid rocks at shallow depths seems to be excluded by the paucity or absence of gas emission.

(5) The lunar material has suffered very little horizontal deformation in the whole of the history depicted by its present surface. Even the oldest craters show as much tendency to circularity as the youngest. There are no chains of folded mountains or any large distortions of the high ground that would be expected if large volumes had been displaced in pouring lava over the low ground.

(6) There is no widespread stratification visible on the Moon, even on the steep slopes of large, fresh craters. Corresponding slopes on the Earth generally would demonstrate stratification, both with respect to color or albedo and to the tendency of erosion to cause terracing.

(7) The large increase in resolution that the Lunar Orbiter photographs give over terrestrial telescopes has led to very little new morpho-

logical information that indicates volcanism. Much of the information was barely detectable with terrestrial telescopes, but is not seen any more clearly by using the photographs. Internally caused visible features may be due to the movements of subterranean ice and water, rather than magma.

There are several other interpretations of the recent findings that do not contradict these points. The meteorites show that various degrees of differentiation have occurred in the solar system in bodies other than those that now exist. We assume that these bodies were shattered by collisions. We may then ask whether the present-day meteorites represent a selection of material left over from these earlier phases, and which type of such material was responsible for building the Moon or for adding the outermost layer to it. If the basaltic achondrites represent this material, the composition would fit, and one may even wonder whether the basaltic layer that covers most of the deep oceans on the Earth has perhaps a similar origin. (The basaltic ocean mounds have, without doubt, been heated in the process of being raised, and the short age determinations may be quite irrelevant for the ages of the seismically determined deep ocean layer.) The observation of the composition of lunar surface material in fine detail and of the deep-ocean basalt would be most revealing.

It may be that there are other ways of reconciling the known evidence. One must search for these and ask the appropriate questions of the lunar exploration program. It would be a disservice to this program if an important point were regarded as settled, despite a mass of conflicting evidence.

The Physical Condition of the Lunar Surface
T. Gold

The Surveyor landings demonstrated that the lunar surface is composed, in general, of very fine, slightly cohesive rock powder. The depth of this material, the particle size, and the ubiquity of this type of surface can still be debated; but very significant constraints can be placed on each.

The small particle size of the material that covers all the Surveyor landing sites can be established from a variety of observations. Many Surveyor pictures clearly show imprints of spacecraft members in the lunar surface material. An outstanding example is the Surveyor III picture of an imprint made by a footpad during the spacecraft's initial landing. The depth of the regular "waffle pattern" cannot be more than 100 microns, which is the maximum deformation that the flat aluminum sheet could have suffered. This waffle pattern, which is clearly visible, would not be seen if the medium contained a significant admixture of, for example, 1/10-mm grains. Experiments show that even a 20-percent, by weight, admixture of 50-micron particles in an otherwise sufficiently fine medium will spoil the precision molding properties observed.

Pronounced changes occurred in the optical scattering law of the surface material as a result of the imprints. The very great brightness of the conical part of the Surveyor III footpad can be understood only in that way. Instead of the normal scattering law of the lunar material with a strong peak in the backscatter direction, evidently there is now a strong forward-scatter component around the direction of specular reflection. A surface of that nature would be said to possess "sheen." The same effect is visible in many other Surveyor pictures, where smooth spacecraft surfaces have been in contact with the powdered rock. One can estimate by theorizing or test by experimenting how small a particle size is required to achieve these effects. The answer in either case is less than 10 microns.

These observations also resolve whether the normal scattering law of the Moon is a consequence of the "fairy castle" structure in which the surface of the powder is normally assembled, or whether it is a consequence of individual particles that have the complex shapes needed to generate this law. It is clear that the latter cannot be the answer, since the backscatter law of the surface then would not have been changed significantly by contact with spacecraft surfaces. Thus, it is clear that the particle size is predominantly less than 10 microns, and that the usual configuration of the particles on the surface indeed involves the type of structure called "fairy castle"; i.e., a complex microscopic structure in which the shadows cast have a profound effect on the optical properties.

Since the optical polarization law of the Moon's surface requires roughness on a scale of 2 or 3 microns or less, it is likely that this places a limit on the dominant particle size, rather than the slightly larger limit that would be allowed by the previous considerations. Most physical processes that generate small particles tend to make rather smooth surfaces on this scale where surface tension is very significant. It is, therefore, most probable that almost all of the Moon is covered with particles no more than a few microns in size.

The Observed Spray Phenomena

Many observed phenomena are best understood as spray of a slightly adhesive powder. The best example was the Surveyor VI mission during which a photometric target, first observed to be quite clean, was almost completely covered with a thick, adhering layer of rock powder after the spacecraft executed a brief takeoff-and-landing maneuver. It is thought that this coating occurred during the liftoff when the vernier engines were fired close to the lunar surface. The photometric target was at a height of about 1 meter above the ground and about 2 meters from the surface area under the nearest vernier engine. In the Surveyor V vernier firing, many clumps of material on the surface were observed to be dislodged and, presumably, one similar clump hit this target. It is noteworthy that a clump striking such a surface will splatter itself and form an adhesive coating. The adhesion must have been sufficiently good to not only survive against the weak force of gravity but also against the much larger inertial forces of the subsequent landing impact (between $1\frac{1}{2}$ and 3 g).

Expressed in more familiar terms, it will mean that the lunar powder can be picked up in clumps, thrown like a snowball and, like a snowball, will tend to cover the target surface with a coating that is not easily shaken off.

A slightly different behavior was observed on Surveyor V, where the vernier engines were fired briefly without moving the spacecraft. Many changes in the detail of the surrounding surface could be seen. Many clumps had evidently been thrown to a distance of more than 3 meters from the vehicle. The sensor head of the alpha-scattering instrument initially had a gold-plated, highly reflecting mirror finish. In the picture taken before the vernier firing, one can see the vertical face of the sensor head giving a clear mirror image of the surrounding ground. After the vernier firing, that same face seemed to be completely matte and dark gray with an albedo similar to that of the lunar surface. No mirror-image effect is observed, and a shadow (of the electrical connecting tape) appears as dark on that surface as on the lunar ground. This implies that the surface was coated substantially with the lunar dust. In this case, the coating appears quite smooth, unlike that on the photometric target; it is not at all probable that the coating was caused by clumps thrown against the instrument. Instead, a fine spray must have been showered over it, and an almost complete monolayer, or more, must have become attached. This is quite consistent with the observations of the photometric target, and it is very probable that if clumps can be thrown and if they adhere, many individual small particles would also be thrown and would attach themselves to the target surfaces. It is observed in the laboratory, even without high vacuum, that small dust grains would indeed readily adhere and coat surfaces. In a better vacuum, this tendency increases.

All these phenomena of clumping and adhering to surfaces again imply a rather small particle size. In the laboratory, particles of 10 microns or smaller behave in that manner; however, 100-micron rock particles behave in a different manner. The cohesion between them cannot be made large enough, even in a good vacuum, for them to clump together; when striking a surface they tend to bounce off rather than to adhere. The observations of these coated surfaces are, therefore, in accord with the previous deduction that the particle size is predominantly below 10 microns.

On Surveyor III, another particle-adhesion phenomenon occurred. The camera mirror lost its optical quality over approximately half of its area and scattered light diffusely. This can again be attributed to particles sprayed up from the surface during the landing maneuver executed by Surveyor III. The vernier engines continued to fire during the first and second contact with the ground and, no doubt, some lunar dust also was stirred up on these occasions. The part of the mirror that seems to have been obliterated was the region that would have been exposed, and not protected by the camera housing.

It has been argued that some of these phenomena may not have been due to adhesion of small particles, but to sandblasting effects and to the destruction of the surfaces. This would require a much larger particle size than 10 microns before any sandblasting effects became optically significant. It would also require very high particle velocities. There is no question that particles adhered to the photometric target; therefore, it is clear that, in general, particles are capable of such adhesion. Adhesion is easily demonstrated in the laboratory, and occurs with almost all small particles and all surfaces. Therefore, adhesion is considered a more probable explanation than sandblasting for the observations on the two mirror surfaces.

The fact that many other surfaces were observed to remain clean during the same events that coated the surfaces mentioned merely indicates that the rocket spray from the ground tends to be in a few jets rather than a diffuse shower.

The depth to which the powder layer extends cannot be established from any of the present observations. It is evidently more than a few meters in most areas, for many craters of that depth have been seen that show no discontinuity in their interior. The larger craters visible in Surveyor pictures are often surrounded by many angular rocks. It is not known whether these objects are hard rocks or a variety of aggregates with varying degrees of firmness. The depth that craters must have before blocks can be seen around them is different in different regions. There is either a subsurface layer of a stronger rock at different depths, or the degree of compaction with depth is regionally different. Material ejected from very deep craters such as Tycho, previously at a depth or more than 4 kilometers, for the most part, must be hard. The material will have been compressed by the initial overburden under which it would lie for a very long time, and then by the intense shock waves necessary to excavate the crater.

Because the powder layer in most areas is at least a few meters thick, it is probably the layer responsible for the low radio reflectivity. If this is true, it would have to extend in most areas to a depth of at least 10 meters to account for the radar observations. The alternative would be to suppose that another different kind of underdense rock exists beneath a few meters of powder. There is no suggestion of this in any of the pictures, and it is, therefore, more probable that the powder continues to the depth required by the radio measurements, though quite possibly with gradually increasing density of packing, and increasing cohesive strength.

Surveyor VII landed on highland ground, and although a great many more rocks were found in the area, the consistency of the soil still seemed much the same. In addition, several tracks of rolling boulders can be seen in Lunar Orbiter photographs; these tracks imply penetration to a depth of several meters. Thus, the material of the highlands and maria appear to be covered with a thick layer of fine rock powder. The thermal and radar anomalies for certain regions on the Moon can be accounted for by an increase in the number density of rocks, as observed by Surveyor VII in the vicinity of Tycho.

Importance for Future Technology

For purposes of future lunar technology, the findings discussed would be important in the following respects:

(1) Digging should be easy. It should be remarkably easy to dig down a few meters with a large shovel. The weight of the material in lunar gravity is only a little more than snow is on Earth. The material can be cut easily; it is cohesive enough to be lifted in large blocks, more or less like snow, and the sides of a hole should be fairly firm and not easily caved in. To investigate the lunar subsurface at a depth of about 1 meter should be considered the equivalent of digging in snow on Earth, about 1 minute of work, rather than digging in terrestrial soil or rock.

(2) Similarly, pushing rods into the surface should be easy, especially if they are pointed, thin, and possess a surface that will slide comparatively freely against rock powder. Thermal measurements obviously can benefit from this convenient property. An early test with thin, long rods coupled together should determine the circumstances under which a drill would be required.

(3) Spray and adhesion of lunar powder are likely to present problems in many situations, e.g., during the landing maneuver of a large vehicle, when visibility of the ground may be seriously impeded, and when surfaces on the vehicle may become coated. The experience with the photometric target shows that splashing does occur, and surfaces that are not in direct-line access from the originating area of the spray may still be coated as a result of impact and deflection of some material. The boom holding the photometric target was clearly in the way of the spray and, although a reduced coating is seen in its spray "shadow," this shadow is not complete. It is, therefore, not quite certain what surfaces on the landing vehicle may be affected, and one should consider the problems raised by a splash that coats any part of the vehicle's surfaces.

The proposed lunar flying machines that propel an astronaut off the surface will create a hazard. Both the nearby LM and the astronaut's visor may provide surfaces where contamination would be damaging or disastrous, yet a strong blast is required close to the ground.

Adhesion of powder, though in general apparently not very severe, may still be a problem in several astronaut activities. Door mechanisms and gaskets, plugs, and other accurately fitting surfaces would clearly be affected by adhesion such as that on the alpha-scattering instrument or on the photometric target. Some dust necessarily will be brought into the LM by the astronauts and will later float in the interior.

Postsunset Horizon "Afterglow"

D. E. GAULT (CHAIRMAN), J. B. ADAMS, R. T. COLLINS, G. P. KUIPER, J. A. O'KEEFE, R. A. PHINNEY, AND E. M. SHOEMAKER

Observations of the western horizon shortly after sunset during the Surveyor VII mission

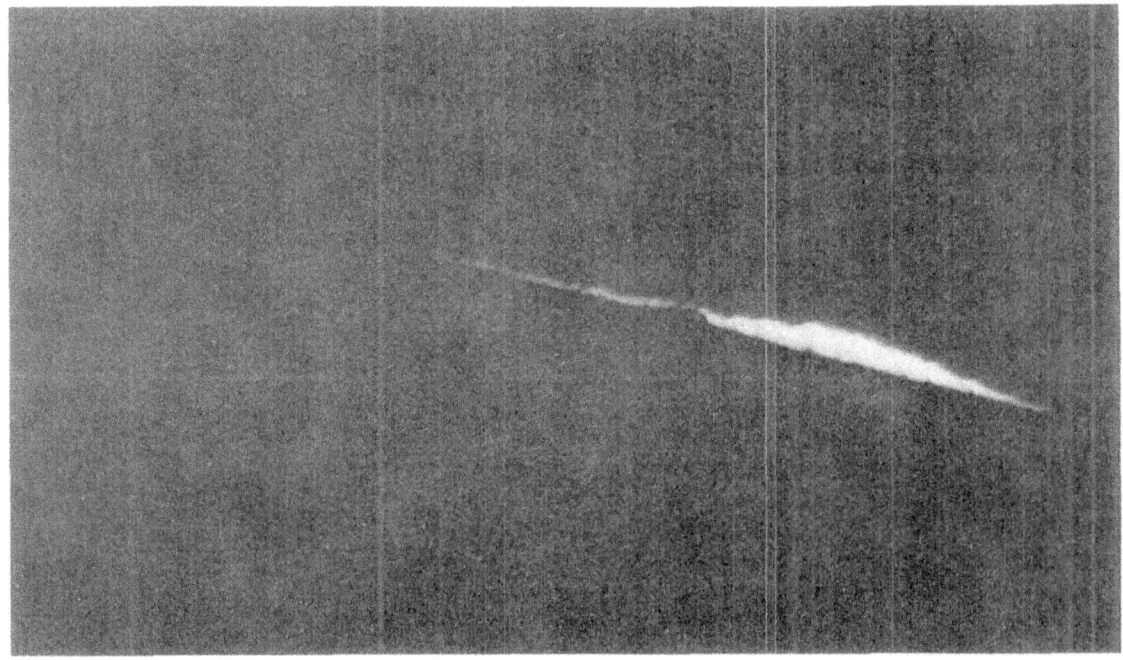

FIGURE 9-1.—Illumination along western horizon about 15 minutes after local sunset. Second disk; exposure time: 0.2 second (Jan. 23, 1968, 06:18:32 GMT).

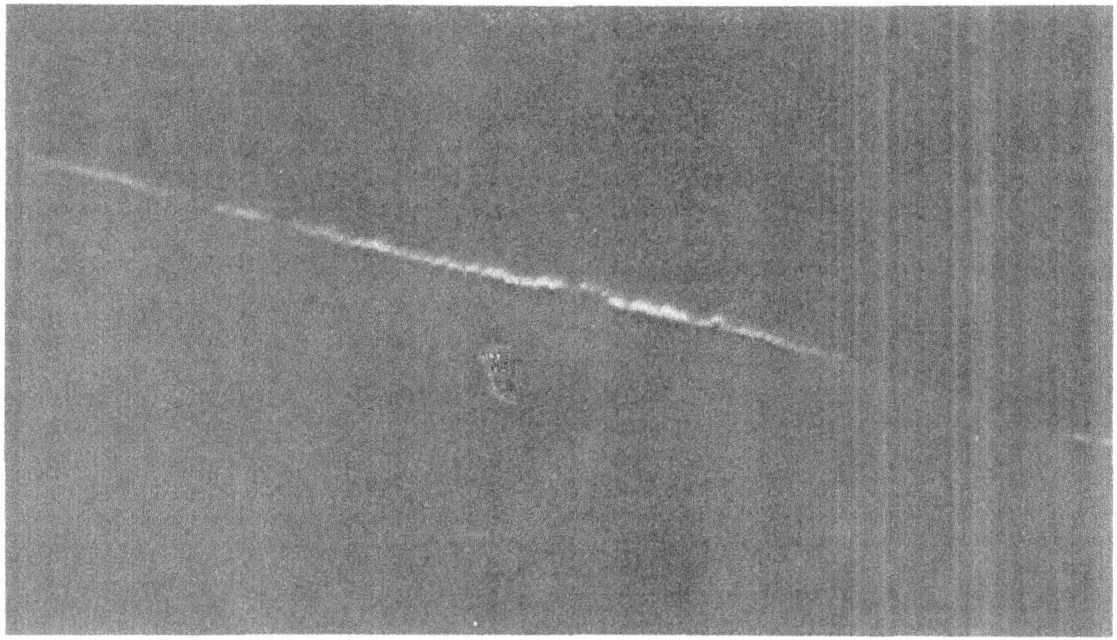

FIGURE 9-2.—Illumination along western horizon about 90 minutes after local sunset. Second disk; exposure time: about 1.2 seconds (Jan. 23, 1968, 07:32:49 GMT).

revealed, along the crest of the horizon, a bright line of light similar to that previously reported for the Surveyor V and VI missions (refs. 9-16 and 9-19). Though not sufficiently well defined to be recognized at the time, the phenomenon also occurred during the Surveyor I mission. Although no sunset observations were made on Surveyor III, it appears that this postsunset phenomenon along the western horizon (and probably the eastern horizon at sunrise) is not an unusual event, but occurs regularly as the natural consequences of some aspect of the lunar environment.

The light was observed for periods of time up to about 2 hours after sunset. The center of the solar disk, therefore, is about 1.25° below the horizon when the "afterglow" either stops or the intensity falls below the limits of detection. Pictures of the light from the Surveyor VII missions are shown in figures 9-1 and 9-2 when the Sun was centered approximately 0.4° and 1.0°, respectively, below the horizon. In figure 9-1, the light intensity permitted normal shutter operation (exposure time, 0.15 second); the bright line appears to extend only about 2° along, and ⅛° above, the horizon. The light intensity decreased rapidly; about 1½ hours later, a nominal 1.2-second exposure (fig. 9-2) showed a faint line of illumination extending at least 4° along the horizon. A 40-second exposure (fig. 9-3), taken about 2 hours 40 minutes after sunset, showed no edge of light along the horizon. This last picture (illumination provided by light backscattered from the ridges east of the spacecraft, and by earthlight) provides a valuable comparison of the rocks and horizon geometry with the shape of the bright regions in figures 9-1 and 9-2. A particularly striking facet of the phenomenon is the "mapping," or shadows, in the edge of light, apparently caused by the rocks extending along and above the lunar horizon line.

Although no complete explanation can be offered at this time, the relative intensities of the light on Surveyors VI and VII suggest that scattering by small particles above the lunar surface is not the mechanism for the phenomenon. This conclusion is drawn from the fact

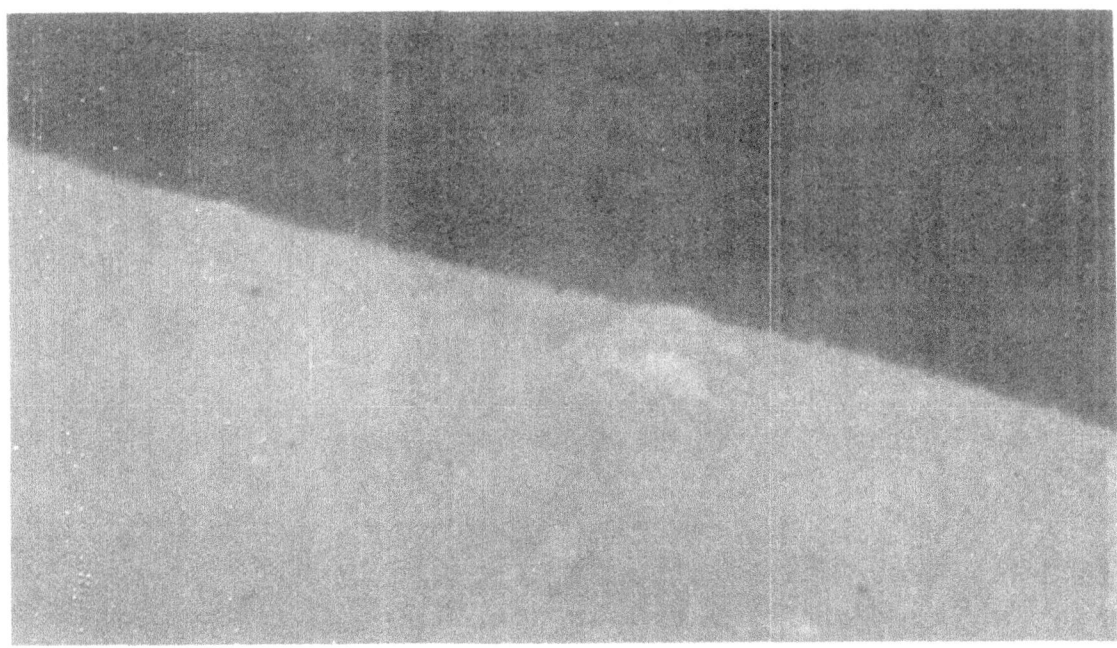

FIGURE 9-3.—Same field of view of western horizon as figures 9-1 and 9-2 about 160 minutes after local sunset. Second disk; exposure time: about 40 seconds (Jan. 23, 1968, 08:46:56 GMT).

that, while the intensity of the bright edge appears to be greater for Surveyor VII than for Surveyor V or VI, the distance to the horizon and the path length of the light immediately above and along the surface is probably shorter. For equal spatial density of the particles above the surface, the longer path length, contrary to observations, should have produced a pattern of greater brightness. Alternatively, diffraction by small particles on the lunar surface, as discussed by O'Keefe et al. (ref. 9-16), may provide a mechanism for producing the phenomenon; however, further study is required before any explanation is considered firm.

References

9-1. SHOEMAKER, E. M.; BATSON, R. M.; HOLT, H. E.; MORRIS, E. C.; RENNILSON, J. J.; and WHITAKER, E. A.: Principal Scientific Results of the Surveyor III Mission. Surveyor III Mission Report. Part II: Scientific Results, Tech. Rept. 32-1177, Jet Propulsion Laboratory, Pasadena, Calif., June 1, 1967, pp. 9-67.

9-2. ADAMS, J. B.: Lunar Surface Composition and Particle Size: Implications from Laboratory and Lunar Spectral Reflectance Data. J. Geophys. Res., vol. 72, no. 22, 1967, pp. 5715-5720.

9-3. TURKEVICH, A. L.; FRANZGROTE, E. J.; and PATTERSON, J. H.: Chemical Analysis of the Moon at the Surveyor V Landing Site: Preliminary Results. Surveyor V Mission Report. Part II: Science Results, Tech. Rept. 32-1246, Jet Propulsion Laboratory, Pasadena, Calif., Nov. 1, 1967, pp. 119-149.

9-4. TURKEVICH, A. L.; FRANZGROTE, E. J.; and PATTERSON, J. H.: Chemical Analysis of the Moon at the Surveyor VI Landing Site: Preliminary Results. Surveyor VI Mission Report. Part II: Science Results, Tech. Rept. 32-1262, Jet Propulsion Laboratory, Pasadena, Calif., Jan. 10, 1968, pp. 127-153.

9-5. KEIL, K.: Meteorite Composition. Handbook of Geochemistry. Berlin, Heidelberg, New York, Springer-Verlag (in press).

9-6. DUKE, M. B.; and SILVER, L. T.: Petrology of Eucrites, Howardites, and Mesosiderites. Geochim. et Cosmochim. Acta, vol. 31, 1967, pp. 1637-1665. See also UREY, H. C.; and CRAIG, H.: The Composition of the Stone

Meteorites and the Origin of the Meteorites. Geochim. et Cosmochim. Acta, vol. 4, 1953, pp. 36–82.

9-7. CLARK, S. J., ed.: Handbook of Physical Constants. Geological Society of America, New York, 1966, p. 2.

9-8. TURNER, F. J.; and VERHOOGEN, J.: Igneous and Metamorphic Petrology. McGraw-Hill Book Co., Inc., 1960, p. 325.

9-9. MANSON, V.: Geochemistry of Basalts: Major Elements. Basalts, The Poldervoart Treatise on Rocks of Basaltic Composition, vol. 1, Interscience Publishers, 1967, pp. 223–225.

9-10. HATCH, F. H.; WELLS, A. K.; and WELLS, M. K.: The Petrology of Igneous Rocks, Thomas Murby & Co., London, 1952, p. 271.

9-11. PALM, A.; and STROM, R. G.: Space Sciences Laboratory Research Report, series 3, issue 5, Univ. of California at Berkeley, 1962.

9-12. LORELL, J.; and SJOGREN, W. L.: Lunar Gravity: Preliminary Estimates from Lunar Orbiter. Science, vol. 159, no. 3815, 1968, pp. 625–627.

9-13. PHINNEY, R. A.; and ANDERSON, D. L.: Present Knowledge About the Thermal History of the Moon. Physics of the Moon, AAS Science and Technology Series, vol. 13, Am. Astronaut. Soc., Washington, D.C., 1967, pp. 161–179.

9-14. ENGEL, A. E. J.; ENGEL, C. G.; and HAVENS, R. G.: Chemical Characteristics of Oceanic Basalts and the Upper Mantle. Geol. Soc. Amer. Bull., vol. 76, 1965, pp. 719–734.

9-15. UREY, H. C.: Primary and Secondary Objects. J. Geophys. Res., vol. 64, 1959, pp. 1721–1737. See also The Origin of Some Meteorites From the Moon. Die Naturwissenschaften, vol. 55, no. 2, 1968, pp. 49–57.

9-16. O'KEEFE, J. A.; ADAMS, J. B.; GAULT, D. E.; GREEN, J.; KUIPER, G. P.; MASURSKY, H.; PHINNEY, R. A.; and SHOEMAKER, E. M.: Theory and Processes Relating to the Lunar Maria From the Surveyor Experiments. Surveyor VI Mission Report. Part II: Science Results, Tech. Rept. 32-1262, Jet Propulsion Laboratory, Pasadena, Calif., Jan. 10, 1968, pp. 171–176.

9-17. O'KEEFE, J. A.: Origin of Tektites. Space Science Reviews, vol. 6, 1966, pp. 174–221.

9-18. ASH, M. E.; SHAPIRO, I. I.; and SMITH, W. B.: Astronomical Constants and Planetary Ephemerides Deduced from Radar and Optical Observations. Astron. J., vol. 72, 1967, pp. 338–350.

9-19. NORTON, R. H.; GUNN, J. E.; LIVINGSTON, W. C.; NEWKIRK, G. A.; and ZIRIN, H.: Astronomy. Surveyor V Mission Report. Part II: Science Results, Tech. Rept. 32-1246, Jet Propulsion Laboratory, Pasadena, Calif., Nov. 1, 1967, pp. 115–118.

10. Surveyor Posttouchdown Analyses of Tracking Data

F. B. Winn

Lunar posttouchdown tracking data (coherent two-way doppler) were collected intermittently throughout the first lunar day of the Surveyor I, III, V, VI, and VII missions. These data were used to determine the selenocentric locations of the Surveyors and the geocentric locations of the Deep Space Stations (DSS) involved in the tracking operations. During the Surveyor data reductions, model limitations were apparent. The "observed minus computed" (O−C) residuals emanating from the data fits exhibited long-term periodicities and diurnal signatures. The long-term periodicities are lunar ephemeris dependent. The diurnal signatures result from tropospheric, ionospheric, and lunar motion-modeling defects. High correlations between the lunar radius and selenocentric longitude do exist in the solutions of the Surveyor I and III data. A high correlation between the probe's selenocentric distance and latitude were experienced in the Surveyor VII data fits. This necessitated the adoption of the Aeronautical Chart and Information Center (ACIC) lunar radii for these solutions. The lunar radius is a valid parameter in the Surveyor V and VI data reductions.

The Surveyor positions, as deduced from Lunar Orbiter photographs (ref. 10–1), from inflight data fits (ref. 10–2), and posttouchdown data fits, are presented for comparison in table 10–1.

Long-Term Periodicities

Three lunar ephemerides were used in the processing of the Surveyor posttouchdown tracking data. The ephemerides employed were Lunar Ephemeris 4 (LE 4, ref. 10–3), LE 5 (refs. 10–4 and 10–5), and LE 8 (a recent construction, no reference available at this time).

LE 4 can be regarded as the modern, evolved Brown's lunar theory. It recently has been discovered that LE 4 has radial position and velocity components that deviate from observations (refs. 10–6 and 10–7; see figs. 10–1 and 10–2).

LE 5 is a numerical integration of the equations of motion, which uses LE 4 positions as input observables. Essentially, this amounts to a smoothed LE 4, which is gravitationally consistent.

TABLE 10–1. *Surveyor selenocentric coordinates (in degrees)*

Spacecraft	Lunar Orbiter photographs (ref. 10–1)		Inflight data fit (ref. 10–2)				Posttouchdown data fit			
	λ	ϕ	λ	σ_λ	ϕ	σ_ϕ	λ	σ_λ	ϕ	σ_ϕ
I [a]	316.79	2.46S	316.665	0.069	2.439S	0.084	316.676	0.029	2.501S	0.005
III [a]	336.66	3.01S	336.59	.02	3.01S	.11	316.682	.011	3.055S	.005
V			23.20	.04	1.49N	.16	23.201	.026	1.415N	.006
VI	358.62	.47N	358.624	.006	.419N	.064	358.629	.018	.460N	.006
VII [a]	348.55	40.92S	358.59	.06	41.01S	.07	348.653	.008	40.971S	.010

[a] Posttouchdown solutions constrained to ACIC lunar radius value.

λ = selenocentric longitude, deg.
ϕ = selenocentric latitude, deg.

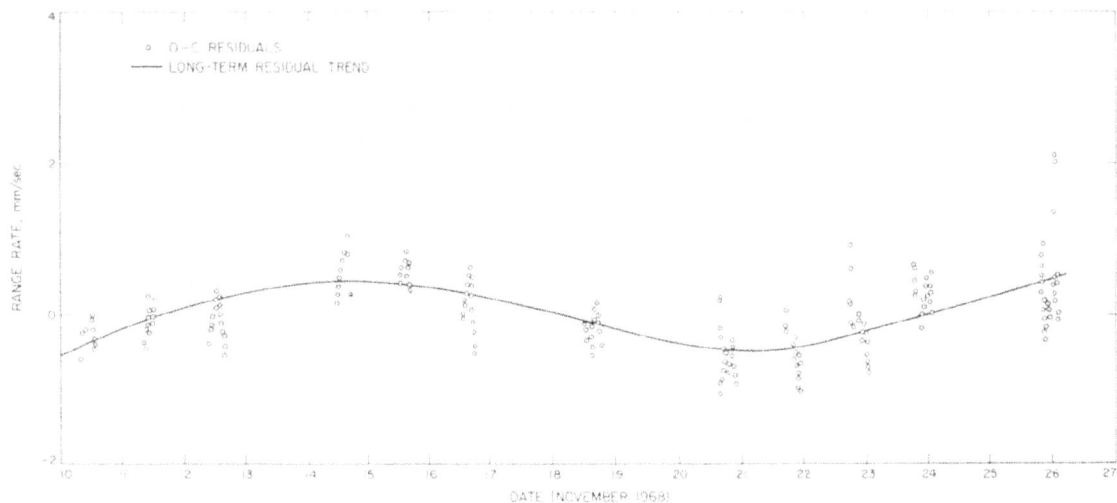

FIGURE 10-1.—Surveyor VI, DSS 42. First lunar day, O−C residual set (compressed five data points/min). LE 4 was used. Note periodicity.

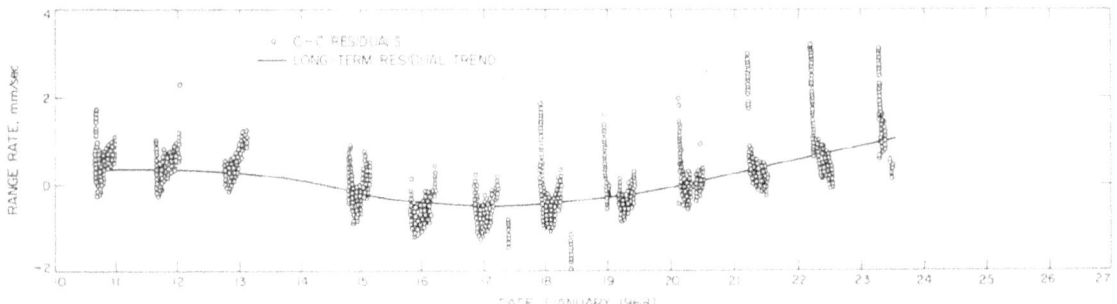

FIGURE 10-2.—Surveyor VII, DSS 61. First lunar day, O−C residual set (one data point/min). LE 4 was used.

LE 8 is analogous to LE 5 in that LE 8 is a numerical integration of the equations of motion using "refined" LE 4 positions as input observables. The "refinement" of LE 4 is in two parts: (1) the J_2 defect (the incorrect coefficient of the second harmonic term of the harmonic series used to describe the Earth's gravitational potential) was changed to be consistent with the definition by the International Astronomical Union (IAU in 1964); (2) suspected faulty fitting to the Brown lunar theory (ref. 10-8) was corrected to one significant place.

The O−C residual, long-term periodicities represent lunar motion modeling errors. All three ephemerides exhibit this trait. The LE 4 (nonintegrated ephemeris) induced, long-term periodicities are more pronounced than those associated with the integrated ephemerides, LE 5 and LE 8. The O−C residual, long-term trends, LE 4 dependent, are shown in figures 10-1 and 10-2. The LE 5-associated residual, long-term periodicity shown in figure 10-3 depicts an approximate period of one lunation with an amplitude of 0.7 mm/sec. In retrospect, all of the Surveyor O−C residual sets demonstrate LE 5-induced, long-term periodicities (figs. 10-4 and 10-5).

LE 8-dependent, long-term residual trends

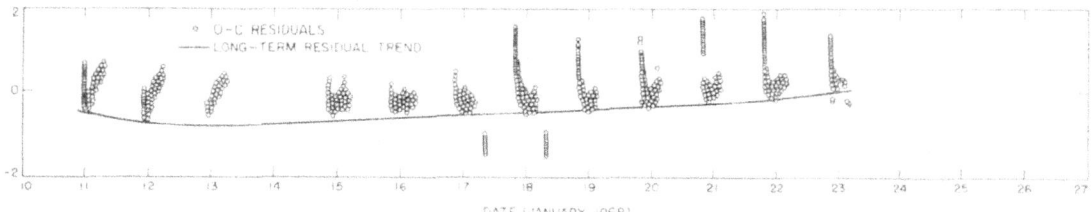

FIGURE 10-3.—Surveyor VII, DSS 61. First lunar day, O−C residual set (one data point/min). LE 5 was used.

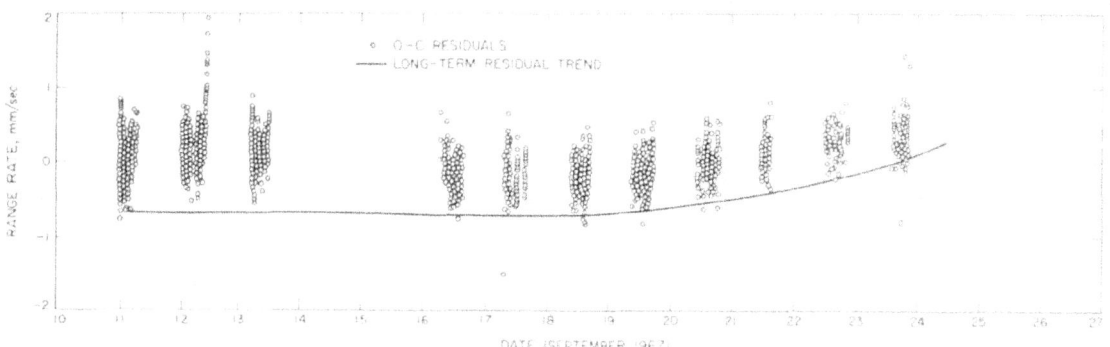

FIGURE 10-4.—Surveyor V, DSS 61. First lunar day, O−C residual set (one data point/min). Refractivity index: $N_{01}=270$. LE 5 was used.

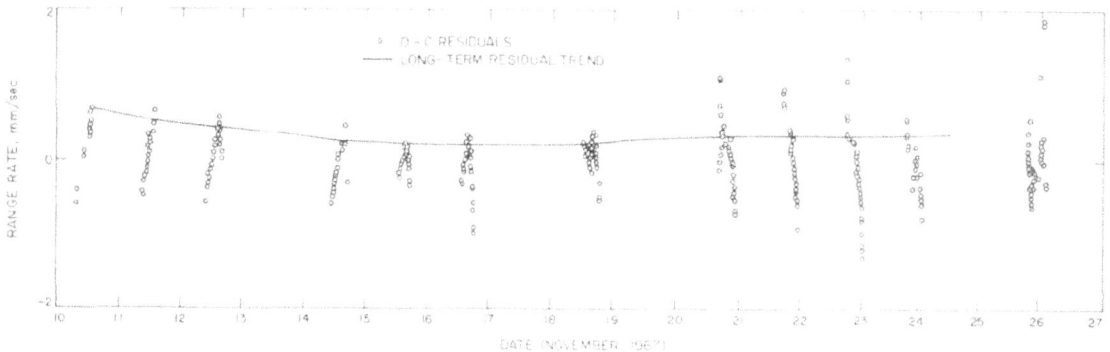

FIGURE 10-5.—Surveyor VI, DSS 42. First lunar day, O−C residual set (compressed five data points/min). LE 5 was used.

are apparent in figure 10-6. LE 8 was used only in the Surveyor VII analysis. Although LE 8 is of such recent origin that, as yet, its detailed character has not been assessed, it is valid to say that LE 5 fits Surveyor VII observations better than LE 8.

Diurnal Signature

The O−C residuals for Surveyors I, III, V, VI, and VII exhibited diurnal signatures. The character of the diurnal signature is governed by that portion of the pass used for tracking data collection. During the operational lifetimes of Surveyors I, III, and V, spacecraft control was frequently transferred as soon as possible to DSS 11 (Goldstone, Calif.) by DSS 42 (Canberra, Australia) and DSS 61 (Robledo, Spain). This was desirable from the standpoint

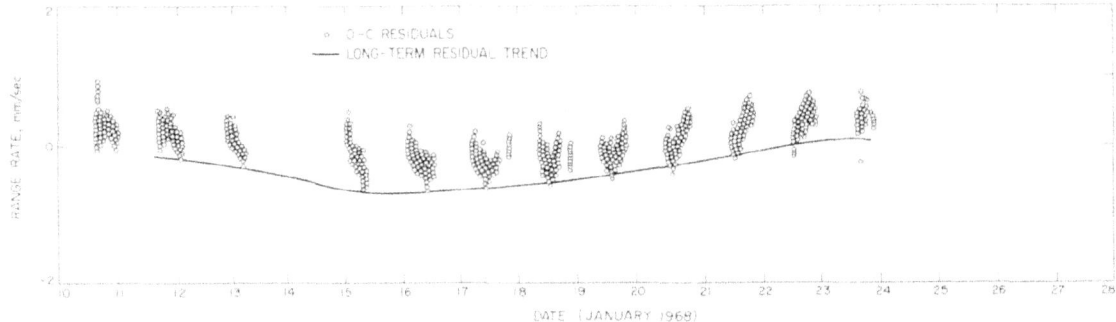

FIGURE 10-6.—Surveyor VII, DSS 61. First lunar day, O−C residual set (one data point/min). LE 8 was used.

of video research operation. The lunar rise over DSS 42 and the lunar set over DSS 61 were infrequently observed. As a consequence, the daily residual traits are deceptive (see figs. 10-7 through 10-9; also see fig. 10-4). The diurnal signature is descriptive of DSS longitude and/or timing errors, lunar longitude errors, tropospheric refraction modeling errors, DSS latitude, or spin-axis distance errors.

To maximize the effectiveness of the tracking data samples of Surveyors VI and VII, the following data acquisition policy was requested:

(1) All tracking data collection periods to be a minimum of 30 minutes.

(2) Tracking data collected during 1 lunar day to be equally distributed throughout the mean lunar pass, rather than collected at the same points or positions during each pass.

An extensive effort was made to create these data characteristics. Because of the acquisition of low-elevation tracking, a more complete picture of the diurnal signature is available (see figs. 10-7 through 10-9). This signature can be attributed to tropospheric refraction (deficient modeling), ionospheric charged-particle effects (not modeled), and/or lunar latitude errors (suspected lunar ephemeris defect).

Recent investigations by Liu (ref. 10-9) and Mulholland (ref. 10-10) have tentatively identified and ordered the three diurnal errors presently incorporated into the Single Precision Orbit Determination Program (SPODP, ref. 10-11) residual sets:

(1) Tropospheric refraction: ~33 mm/sec/100 N at 0° elevation (maximum).

(2) Suspected lunar ephemeris error functions: ~1.0 mm/sec (maximum).

(3) Ionospheric charged-particle effects: ~0.5 mm/sec (maximum).

Tropospheric Refraction

The refraction signature has been empirically determined and programed into the SPODP (ref. 10-11). The empirical refraction function is

$$\Delta r \dot{\rho} = \frac{C_1}{\tau} \left\{ \frac{1}{\left[\sin\left(\gamma + \frac{\dot{\gamma}\tau}{2}\right) + C_2 \right]^{C_3}} - \frac{1}{\left[\sin\left(\gamma - \frac{\dot{\gamma}\tau}{2}\right) + C_2 \right]^{C_3}} \right\} \frac{N}{340.0}$$

where C_1, C_2, and C_3 are empirically determined constants ($C_1 = 0.0018958$, $C_2 = 0.06483$, and $C_3 = 1.4$) and

$\Delta r\dot{\rho}$ = refraction correction applied to the SPODP-calculated data types, hertz
τ = doppler count interval, seconds
γ = elevation angle, radians
$\dot{\gamma}$ = rate of elevation-angle change, radians per second
N = refractivity index

The tropospheric refraction indices, N, used in the initial SPODP solution for the Deep Space Station locations are all set at $N = 340.0$. Recent research by A. Liu (ref. 10-9) has provided evidence that the following values for N are more precise:

DSS 11: $N_{11} = 240.0$

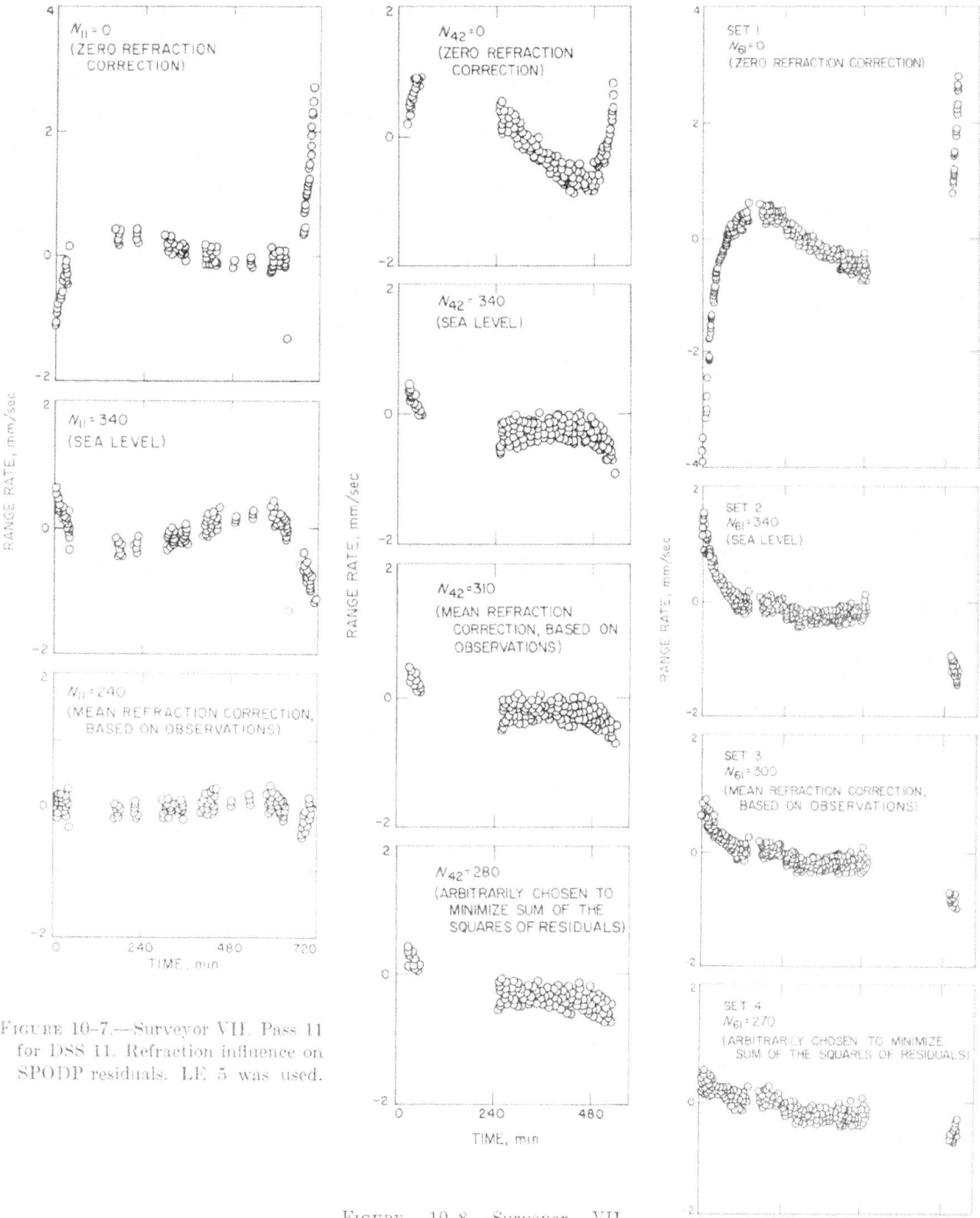

FIGURE 10-7.—Surveyor VII. Pass 11 for DSS 11. Refraction influence on SPODP residuals. LE 5 was used.

FIGURE 10-8.—Surveyor VII. Pass 11 for DSS 42. Refraction influence on SPODP residuals. LE 5 was used.

FIGURE 10-9.—Surveyor VII. Pass 11 for DSS 61. Refraction influence on SPODP residuals. LE 5 was used.

DSS 42: $N_{42} = 300.0$
DSS 61: $N_{61} = 310.0$

This influence of tropospheric refraction is primarily a phase retardation plus a bending, and a consequential lengthening, of the ray path. By using Liu's formulation, an error of 100 units of N generates O−C residuals of 0.5 Hz (33 mm/sec) for horizon range-rate observations. The refraction-induced O−C residual signature contained in reference 10-9 greatly resembles the O−C residual characteristics of the Surveyor passes. An examination of figures 10-7 through 10-9 (Surveyor VII, DSS 11 residuals, pass 11, with varying values of N) shows significant elevation-dependent O−C residual biases that correlate remarkably well with the computed refraction-error functions. Examination of figures 10-8 (DSS 42) and 10-9 (DSS 61) reveals evidence of like influences.

Lunar Ephemerides (Diurnal O−C Residual Contribution)

Although the troposphere is an acknowledged major, but unevaluated, error source that warrants the evaluation efforts in process, there are other model limitations such as the lunar ephemeris.

J. D. Mulholland (ref. 10-10) has provided tentative evidence of lunar ephemeris defects (one significant place) that have approximately a daily influence. A correlation study of Surveyor O−C residuals and the suspected error functions is inconclusive (figs. 10-10 through 10-12). A new lunar ephemeris (LE 8) that incorporates these error functions has been constructed, and will be used in future analyses.

Ionospheric Charged-Particle Effects

Ionospheric charged-particle effects have been omitted from model considerations to the present. The ionospheric influence on the coherent, continuous, two-way doppler O−C residuals is a function of effective electron density, which is dependent on elevation angle, elevation range of change, the Sun's local hour angle, solar activity, and transmitter frequency. The residual signature resulting from this SPODP model omission is similar to the tropospheric refraction-error function signature; however, the ionospheric influence on the tracking data is of a lesser magnitude. Accordingly, the ionospheric effect can be removed or diminished from the O−C residuals by a slightly erroneous value of N in the tropospheric model.

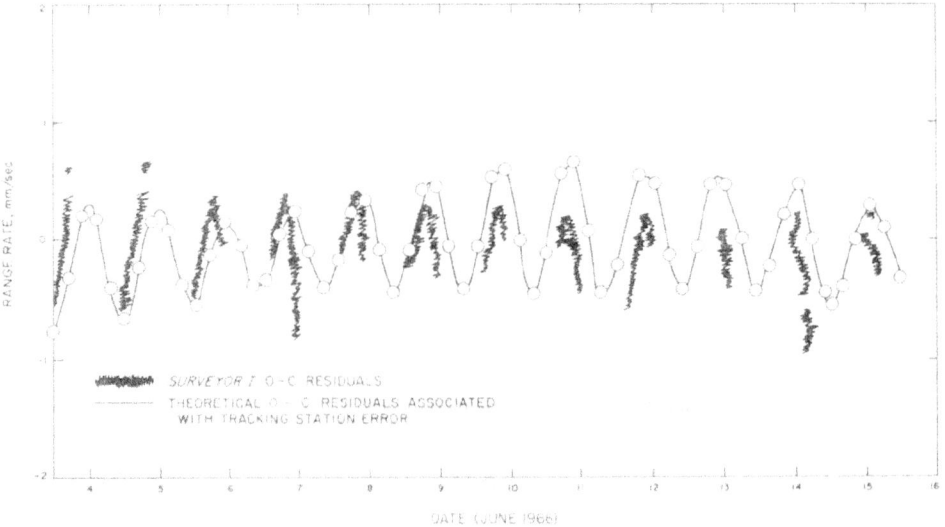

FIGURE 10-10.—Surveyor I, DSS 42. First lunar day, O−C residual set (compressed five data points/min). LE 5 was used.

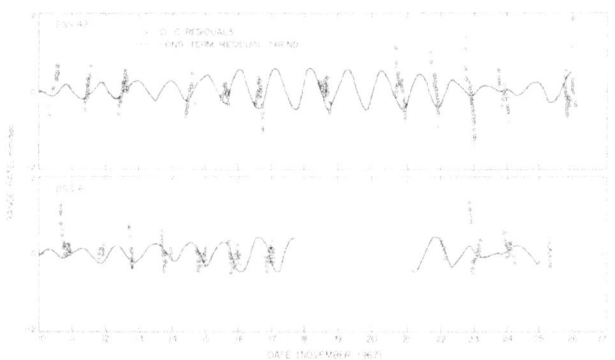

FIGURE 10-11.—Surveyor VI, DSS 42 and DSS 61. Two-way doppler O−C residual set versus suggested ephemeris-dependent tracking station position error. LE 5 was used.

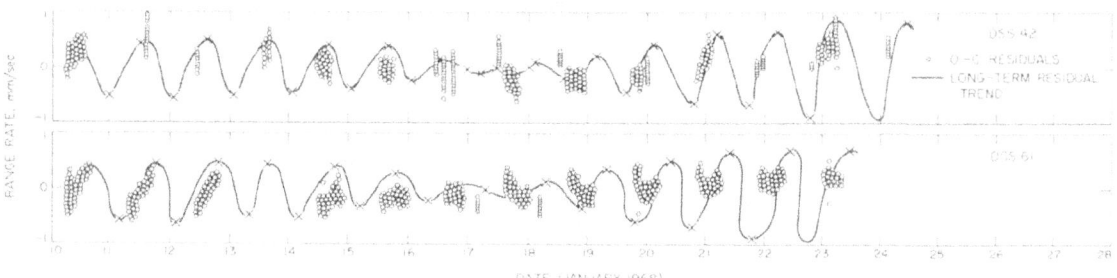

FIGURE 10-12.—Surveyor VII, DSS 42 and DSS 61. Two-way doppler O−C residual set versus suggested ephemeris-dependent tracking station position error. LE 5 was used, coupled with refraction refinement.

A history of ionosphere activity for the first lunar day of all successful Surveyor missions is being compiled.[1] Once this information is available, the correlation of the tropospheric refraction and ionospheric charged-particle influences with the O−C residuals will be more fully investigated.

One Combinational Parameter

The three diurnal components that comprise the diurnal signature are highly correlated. The use of one "combinational" parameter, as a means of fitting out of the O−C residual, sets the influences resulting from tropospheric refraction, ionospheric charged-particle effects, and lunar ephemeris defects, is the only available approach at this time.

The results from the use of this procedure are most striking. The preponderance of the diurnal signature has been removed by the manipulation of the refraction indices. Figures 10-7 through 10-9 show that the O−C residuals, emanating from the several SPODP Surveyor tracking data reductions, used the following refraction indices:

(1) $N_{11} = N_{42} = N_{61} = 0$ No refraction correction.

(2) $N_{11} = N_{42} = N_{61} = 340.0$ Sea-level refraction correction.

(3) $N_{11} = 240$; $N_{42} = 310$; $N_{61} = 300$ Refraction correction based on observations of Lunar Orbiter II (ref. 10-12).

(4) $N_{11} = 240$; $N_{42} = 280$; $N_{61} = 270$ Arbitrarily chosen to minimize the sum of the square of the residuals.

---

[1] M. Davis, Stanford University Electronics Laboratories, Calif.

The influence of the "combinational" parameter on the parameter list is presented in the specific discussions for each Surveyor.

It is not the purpose of such an arbitrary procedure to evaluate numerically any of the parameters under discussion. The sole purpose is to demonstrate the nature of the error sources.

High Parameter Correlation

C. N. Cary (ref. 10-13) has demonstrated the insensitivity of the range-rate measurement in the determination of the selenocentric distance of a Surveyor spacecraft (lunar radius). Because of the lunar rotation and revolution rates relative to Earth, and the coincidence of the Moon's orbital plane and equator, the Earth's motion in the selenographic coordinate system is a small oscillatory motion. Cary has shown that the error ellipsoid resulting from a recursive least-squares fit of range-rate measurement is extremely elongated in the mean direction of the Earth (fig. 10-13(a)).

If a Surveyor spacecraft should be situated on the lunar surface such that the mean Earth direction is orthogonal to the surface, the correlation between selenocentric latitude, longitude, and radius is minimal. As the obliqueness of the mean Earth direction increases relative to the lunar surface, the surface parameters become more highly correlated with lunar radius (fig. 10-13(b)).

It is this latter situation that necessitated the adoption of the ACIC lunar radii to effect solutions for the surface parameters associated with Surveyors III and VII.

A Priori Parameter Constraints

It is possible to constrain the terrestrial tracking station position parameters in the SPODP tracking data reduction to those of some previous determination. However, such a constrained solution can lead to systematic distortion. There are many time-dependent variables incorporated in the theoretical model (e.g., Brown's lunar theory, lunar librations,

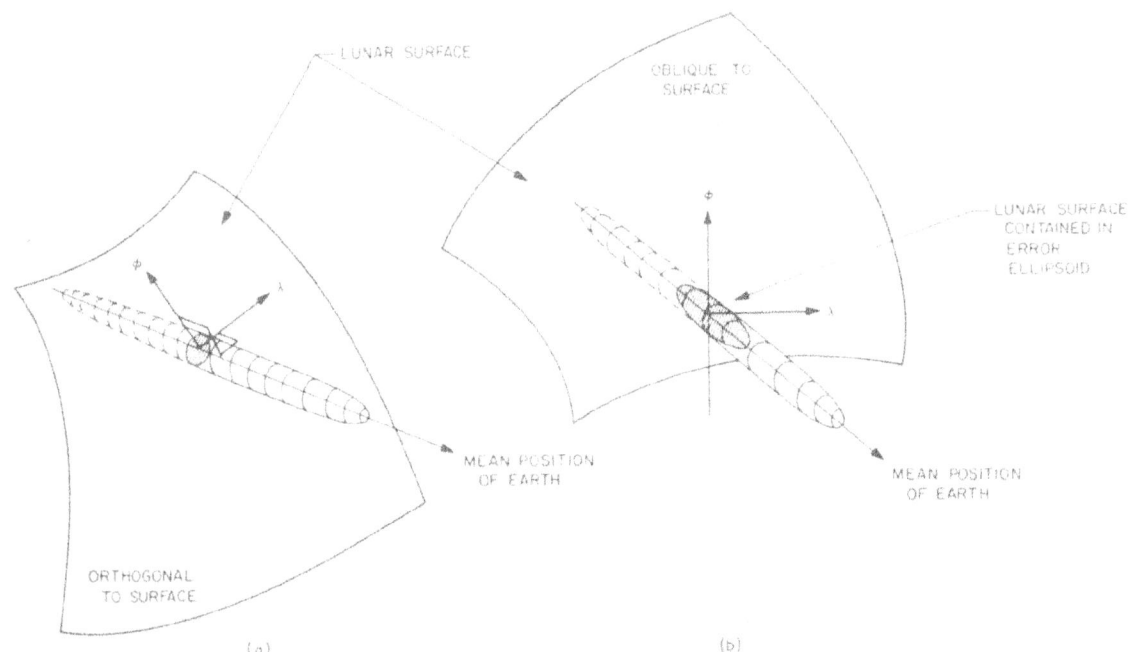

FIGURE 10-13.—(a) Surveyors III, V, and VI were in sufficiently close proximity to the selenographic X-axis to approximate an orthogonal projection of the error ellipsoid on the lunar surface. Thus, the correlations between the surface coordinates and the radius remain small. (b) Surveyors I and VII did not touch down close to the selenographic X-axis. Surveyor I was displaced in longitude 44° W, Surveyor VII at 41° S latitude.

diurnal rotation, ionosphere, space plasma effects, etc.). There is a series of models used to provide values for some of these time-dependent parameters; some are not modeled at all. Thus, there is always the danger of introducing systematic errors into a tracking data fit by constraining to the previously determined terrestrial tracking station positions. The a priori standard deviations associated with the parameters are:

(1) Surveyor III, VII selenocentric distance: 0.1 km; Surveyor I, V, VI selenocentric distance: 10.0 km

(2) Surveyor selenocentric latitude: 4.0° (150 km)

(3) Surveyor selenocentric longitude: 5.0° (150 km)

(4) DSS 11 geocentric distance: 300 meters

(5) DSS 11 geocentric longitude: 0.005° (0.5 km)

(6) DSS 42 geocentric distance: 300 meters

(7) DSS 42 geocentric longitude: 0.005° (0.5 km)

(8) DSS 61 geocentric distance: 300 meters

(9) DSS 61 geocentric longitude: 0.005° (0.5 km)

Surveyor I

Surveyor I landed on the lunar surface on June 3, 1966; intermittently, coherent two-way doppler data were acquired until lunar night on June 17. Posttouchdown data for Surveyor I were analyzed by C. N. Cary. The density of the data collected by DSS 11 and DSS 42 can be visualized by inspecting figures 10–14 and 10–15.

The $O-C$ residuals contained in figures 10–14 and 10–15 exhibit diurnal signatures and long-term periodicities. The diurnal signatures associated with DSS 42 reflect the consistent observation of the premeridian portion of the passes. The diurnal nature of the DSS 11 residuals signify a scattered distribution throughout the passes. The nature of this phenomenon is discussed in the section "Diurnal Signature."

The long-term trend of the Surveyor I residual sets results from the lunar motion modeling errors of LE 5.

The solution parameters that result from the Surveyor I data fits are presented in table 10–2. The high correlation anticipated between the lunar radius and longitude is in evidence in the correlation matrix based on the normal equation matrix (table 10–3). Table 10–3 depicts the correlation between parameter pairs: two-by-two correlations independent of other parameter influences. A correlation based on the variance/covariance matrix depicts all N parameter correlations considered. Table 10–4 is such a correlation matrix for Surveyor I position parameters. The absence of the high

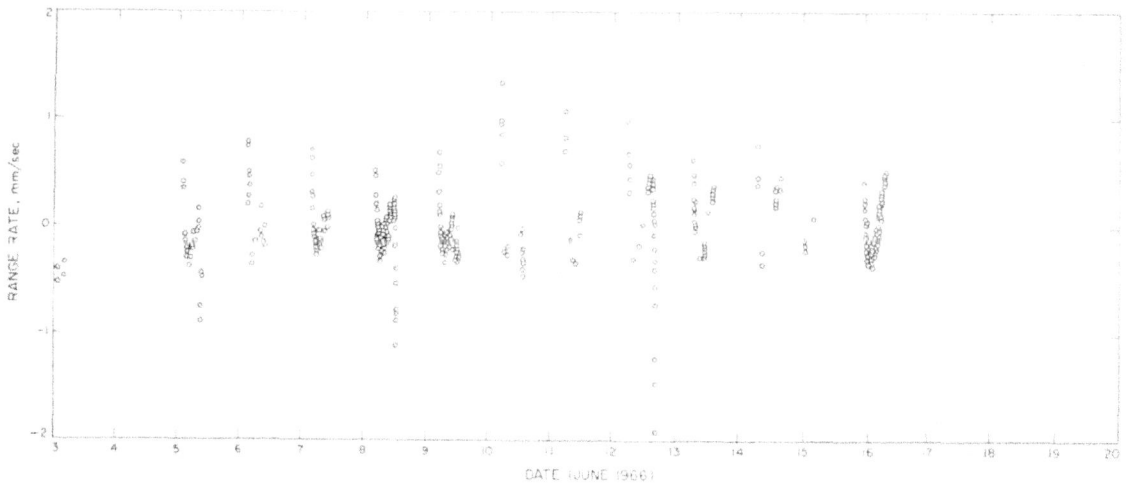

FIGURE 10–14.—Surveyor I, DSS 11. First lunar day, two-way doppler $O-C$ residual set (one data point/min). LE 5 was used.

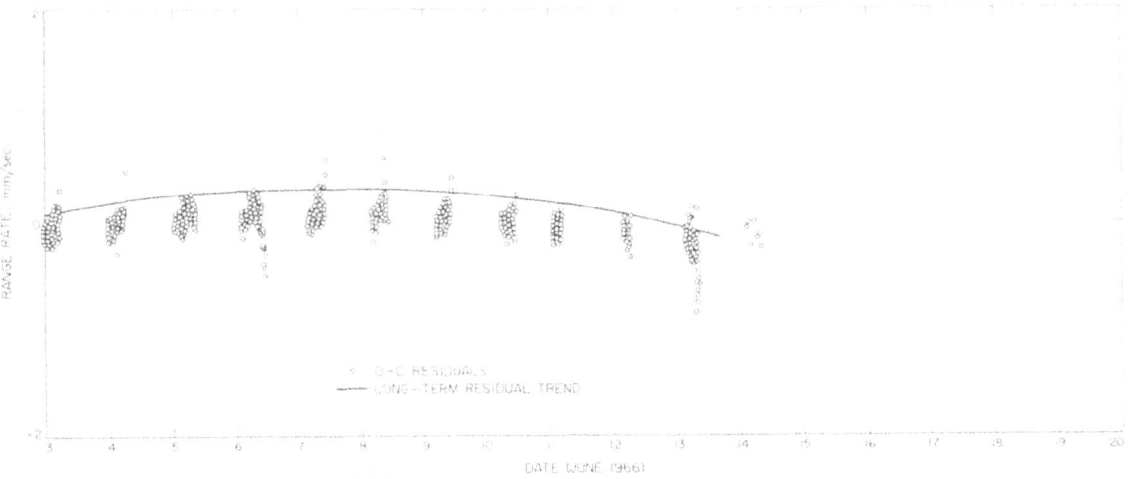

FIGURE 10-15.—Surveyor I, DSS 42. First lunar day, two-way doppler O—C residual set (four data points/5 min). LE 5 was used.

TABLE 10-2. *Surveyor I parameter solutions*

Parameter [a]	A priori parameters	A priori constraints	Solution vector	Standard deviation
RADS	1735.6	10^5	1735.474	1.338
LATS	2.4118	10^5	2.5028	.091
LONS	316.654	10^5	316.676	.061
DSS 11:				
RI	6372.0188	0.500	6372.0071	.004
LO	243.15067	0.005	243.15099	.0002
DSS 42:				
RI	6371.6881	0.500	6371.694	.004
LO	148.98141	0.005	148.98172	.0002

[a] The parameters are defined as:
  RADS = selenocentric distance, km
  LATS = selenocentric latitude, deg
  LONS = selenocentric longitude, deg
  RI = geocentric distance of Deep Space Station, km
  LO = geocentric longitude of Deep Space Station, deg

correlation in table 10-4 between the selenocentric radius and longitude indicates that multiple parameter correlations have disguised the known dependence.

Future analyses of Surveyor I will explore the relative correlations between parameters in the selenocentric coordinate system. It might be advantageous to use a different reference frame, or to simply constrain the lunar radius to the ACIC determination.

Surveyor III

Like Surveyor I, Surveyor III was displaced in longitude from the selenographic X-axis; accordingly, Surveyor III selenocentric distance and longitude are highly correlated (table 10-5). This fact is disguised by the influence of the remaining parameter list (table 10-6). By constraining the lunar radius to the accepted ACIC value, this statistical weakness was

TABLE 10-3. *Surveyor I position parameter correlations, based on normal equation matrix*

Parameter [a]	RADS	LATS	LONS
RADS	1.0	0.555	0.996
LATS		1.0	-.578
LONS			1.0

[a] The parameters are defined as:
RADS = selenocentric distance, km
LATS = selenocentric latitude, deg
LONS = selenocentric longitude, deg

TABLE 10-4. *Surveyor I position parameter correlations, based on variance/covariance matrix*

Parameter [a]	RADS	LATS	LONS
RADS	1.0	-0.331	0.553
LATS		1.0	-.151
LONS			1.0

[a] The parameters are defined as:
RADS = selenocentric distance, km
LATS = selenocentric latitude, deg
LONS = selenocentric longitude, deg

TABLE 10-5. *Surveyor III position parameter correlations, based on normal equation matrix*

Parameter [a]	RADS	LATS	LONS
RADS	1.0	-0.1058	0.9820
LATS		1.0	.2808
LONS			1.0

[a] The parameters are defined as:
RADS = selenocentric distance, km
LATS = selenocentric latitude, deg
LONS = selenocentric longitude, deg

TABLE 10-6. *Surveyor III position parameter correlations, based on variance/covariance matrix*

Parameter [a]	RADS	LATS	LONS	Standard deviation
RADS	1.0	0.8754	0.8252	2.828
LATS		1.0	.4726	.022
LONS			1.0	.045

[a] The parameters are defined as:
RADS = selenocentric distance, km
LATS = selenocentric latitude, deg
LONS = selenocentric longitude, deg

overcome. The resulting parameter solutions and the related correlation matrix are presented in table 10-7.

Diurnal and longer termed characteristics are evident in the Surveyor III O-C residual sets (figs. 10-16 and 10-17).

Surveyor V

The Surveyor V O-C residual sets exhibit the long-term periodicities and diurnal signatures familiar to the other Surveyor residual sets. The long-term periodicities that are lunar ephemeris dependent are shown in figures 10-5 and 10-18 through 10-20. There is one interesting aspect to the Surveyor V daily residual character: The use of the tropospheric refractivity indices derived from the Surveyor VII data reduction for the Deep Space Stations did not diminish the magnitude of the DSS 11 and DSS 61 residual sets (see fig. 10-21). In fact, for these two stations, it appears that overcorrection has occurred. The significance of this "combinational" parameter's variation is not understood at this time. It is not known whether this variation is refraction dependent, charged-particle influenced, lunar ephemeris dependent, or a combination. The DSS 42 response was more as anticipated. The diurnal signature was dramatically diminished. The performance of the "combinational" parameter on the daily variations of DSS 42 suggests that nonlunar ephemeris-dependent influences are responsible for the "combinational" parameter variational.

Future Surveyor V data reductions will determine the value of N for DSS 11 and 61; this determination will permit a more detailed analysis of this parameter.

The parameter solutions that resulted from the use of sea-level ($N=340$) refractivity and the Surveyor VII computed values for N are

TABLE 10-7. *Surveyor III parameter solution vector correlation matrix*

Parameter[a]	A priori parameter	A priori constraints	Solution vector	Standard deviation	Correlation matrix								
					Surveyor III			DSS 42		DSS 61	
					RADS	LATS	LONS	RI	LO	RI	LO
RADS	1736.1	0.1	1736.106	0.314	1.0	0.273	0.552				
LATS	−2.940	5.0	−3.055	.005		1.0	−.264				
LONS	336.6	5.0	336.68251	.011			1.0				
DSS 42:											
RI	6371.6881	.500	6371.7019	.002				1.0	−0.395	0.048	−0.219
LO	148.98140	.005	148.98139	.0003					1.0	−.137	.817
DSS 61:											
RI	6370.0005	.500	6370.0053	.003						1.0	−.129
LO	355.75114	.005	355.75113	.0003							1.0

[a] The parameters are defined as:
    RADS = selenocentric distance, km (constrained to ACIC determination)
    LATS = selenocentric latitude, deg
    LONS = selenocentric longitude, deg
    RI = geocentric distance of Deep Space Station, km
    LO = geocentric longitude of Deep Space Station, deg

POSTTOUCHDOWN TRACKING DATA ANALYSES 381

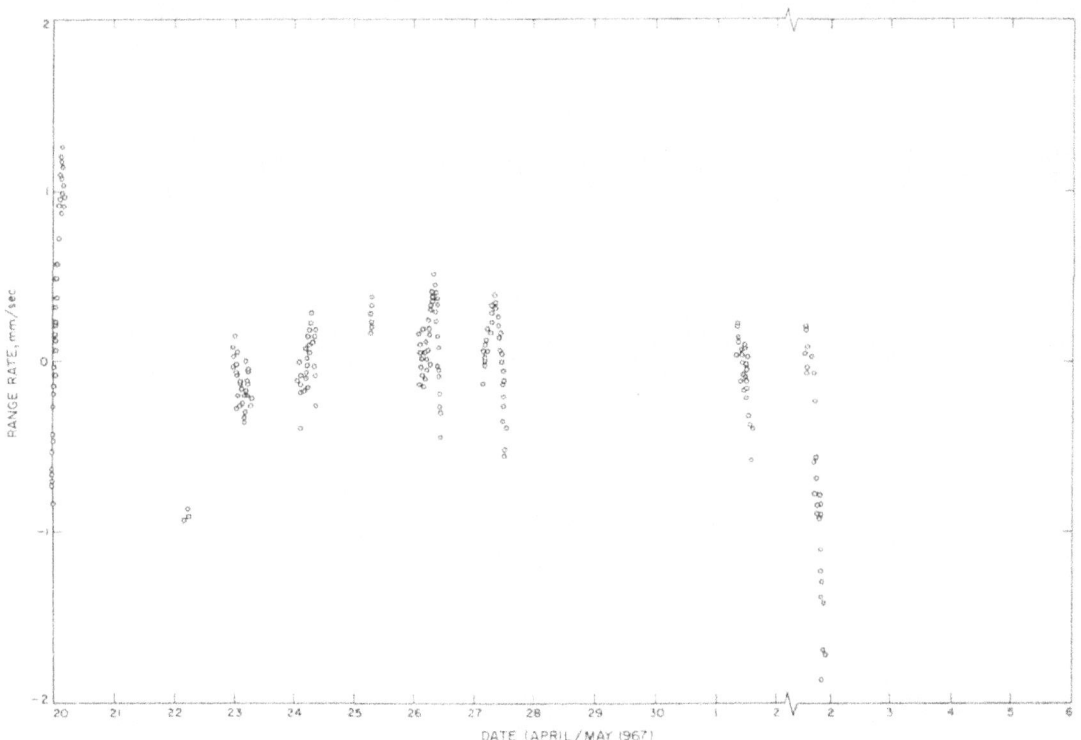

FIGURE 10-16.—Surveyor III, DSS 42. First lunar day, two-way doppler O − C residual set.

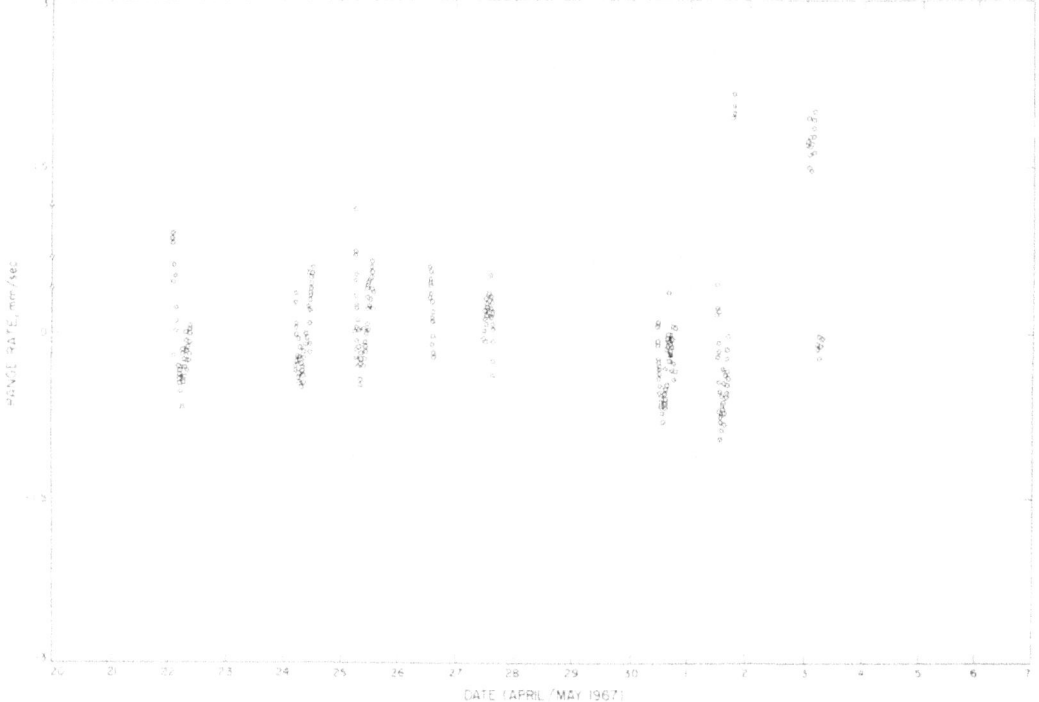

FIGURE 10-17.—Surveyor III, DSS 61. First lunar day, two-way doppler O − C residual set.

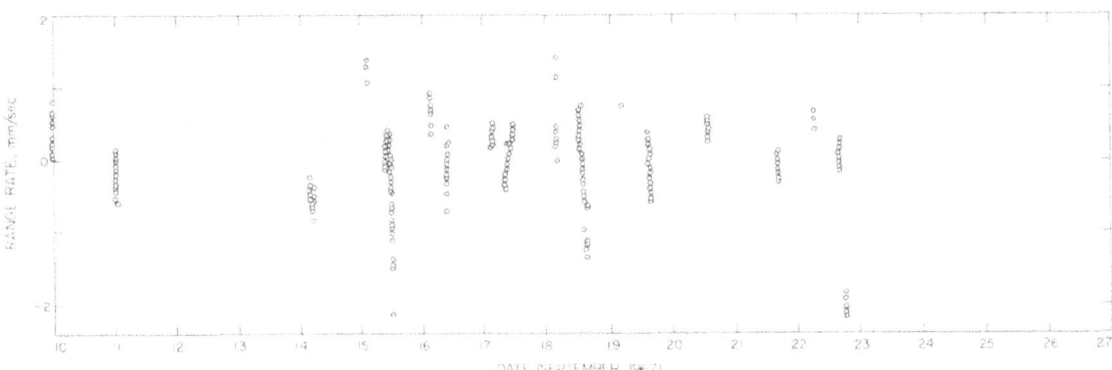

FIGURE 10-18.—Surveyor V, DSS 11. First lunar day, O−C residual set (one data point/min). Refractivity index: $N_{11}=340$. LE 5 was used.

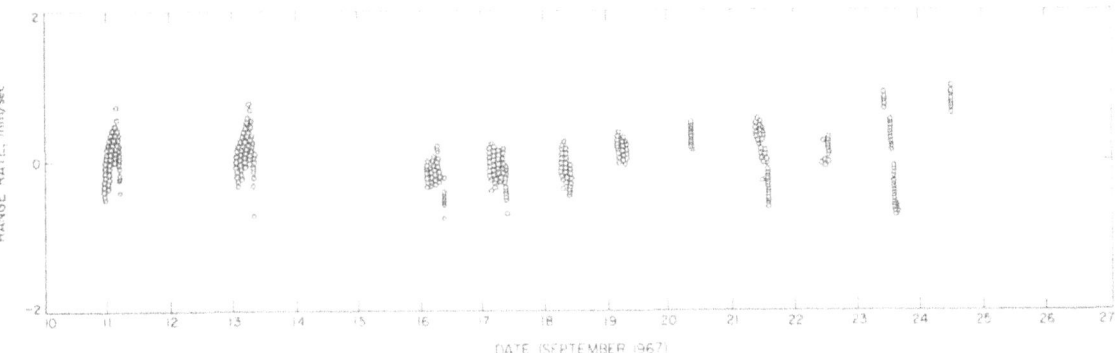

FIGURE 10-19.—Surveyor V, DSS 42. First lunar day, O−C residual set (one data point/min). Refractivity index: $N_{42}=340$. LE 5 was used.

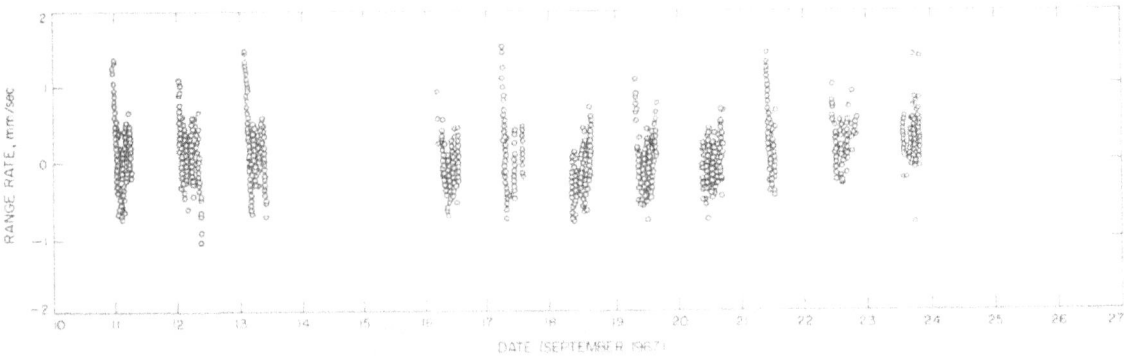

FIGURE 10-20.—Surveyor V, DSS 61. First lunar day, O−C residual set (one data point/min). Refractivity index: $N_{61}=340$. LE 5 was used.

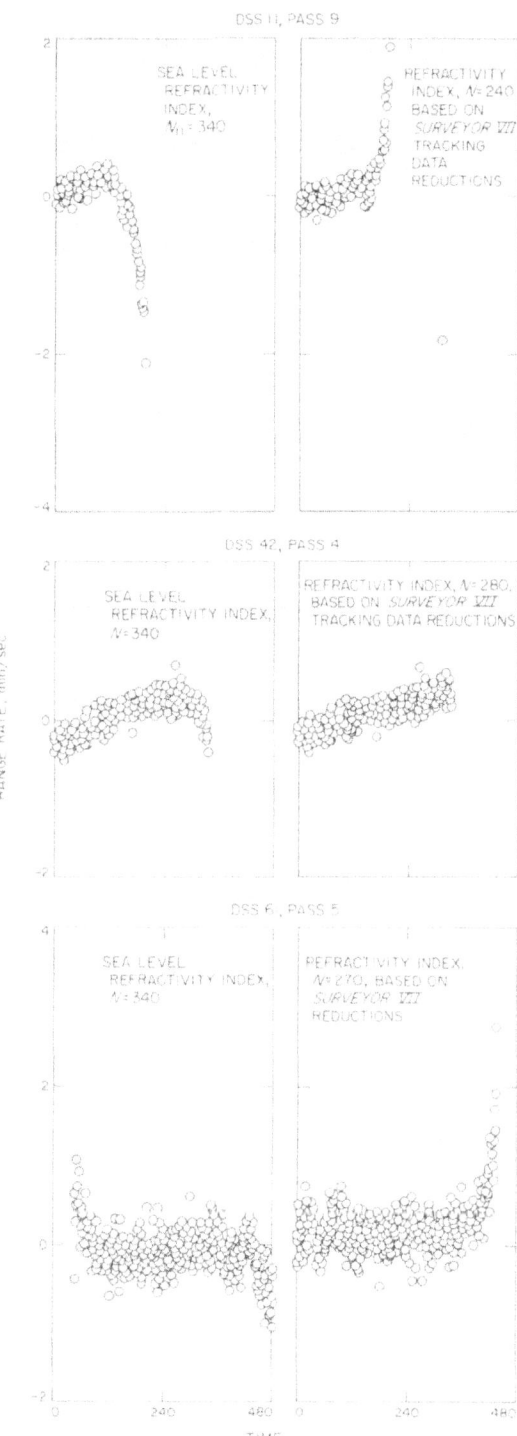

FIGURE 10-21.—Surveyor V. DSS 11, pass 9; DSS 42, pass 4; and DSS 61, pass 5. Refraction influence on SPODP residuals.

presented in table 10-8. The correlations among the parameters are presented in table 10-9.

Unlike Surveyor I and III, the Surveyor V selenographic location does not create an excessively high correlation between the selenocentric distance and longitude parameters (see table 10-10); thus, the radial determination is valid.

Surveyor VI

Parameter Solution Vectors

Three Surveyor VI postlanded data reductions are presented. Two of the solutions employ compressed data (five data points/min) and a third fit uses scrubbed, uncompressed data (one data point/min). LE 4 is used in one of the compressed data fits; LE 5 is used in the other two solutions.

Surveyor VI position parameters and Deep Space Station locations, determined by the three solutions, are presented in table 10-11 with formal standard deviations specified.

The Surveyor VI position error ellipsoids derived from the use of LE 5 and LE 4, coupled with identical, compressed data samples, have the same respective dimensions. There is a large relative displacement of the error ellipsoid centers; this displacement is attributed to the different ephemerides used. The relative metric displacements in selenocentric components are:

Radius: 8000 meters
Lunar latitude: 550 meters
Lunar longitude: 1800 meters

LE 4 and LE 5, coupled with identical data samples, were used to reduce Surveyor I posttouchdown tracking data. A similar, relative displacement of the two Surveyor I positions was the result (ref. 10-6).

The two LE 5 fits, one used uncompressed data (one data point/min) and the other solutions used compressed data (five data points/min), also produced different probe locations. However, this small, relative displacement reflects data and computer noise and is acceptable because it is within the specified confidence levels. The separation of the two Surveyor VI position determinations is:

TABLE 10-8. *Surveyor V parameter solutions*

Parameter [a]	A priori parameters	A priori standard deviation	Using LE 5 and sea-level refraction indices (N=340)	Standard deviation	Using LE 5 and solved for refraction indices
RADS	1734.9	0.3	1735.114	0.298	1735.113
LATS	1.49N	5.0	1.406N	.006	1.415N
LONS	23.2	5.0	23.217	.025	23.201
DSS 11:					
RI	6371.999	.3	6371.997	.004	6372.009
LO	243.15070	.005	243.15101	.0001	243.15101
DSS 42:					
RI	6371.688	.3	6371.696	.003	6371.699
LO	148.98140	.005	148.98169	.0001	148.98174
DSS 61:					
RI	6370.000	.3	6369.998	.003	6370.005
LO	355.75114	.005	355.75143	.0001	355.75150

[a] The parameters are defined as:
  RADS = selenocentric distance, km ($N_{11}=240.0$)
  LATS = selenocentric latitude, deg ($N_{42}=280.0$)
  LONS = selenocentric longitude, deg ($N_{61}=270.0$)
  RI = geocentric distance of Deep Space Station, km
  LO = geocentric longitude of Deep Space Station, deg

TABLE 10-9. *Surveyor V correlation matrix of parameters*

Parameter [a]	Standard deviation	Surveyor V			DSS 11		DSS 42		DSS 61	
		RADS	LATS	LONS	RI	LO	RI	LO	RI	LO
RADS	0.298	1.0	0.018	−0.313	0.009	0.186	0.255	0.179	−0.209	0.185
LATS	.006		1.0	−.301	−.162	.185	−.242	.207	−.559	.226
LONS	.026			1.0	−.271	−.949	.127	−.969	.412	−.982
DSS 11:										
RI	.004				1.0	.151	−.172	.284	−.129	.285
LO	.0001					1.0	−.105	.947	−.414	.956
DSS 42:										
RI	.003						1.0	−.208	.044	−.117
LO	.0001							1.0	−.422	.977
DSS 61:										
RI	.003								1.0	−.403
LO	.0001									1.0

[a] The parameters are defined as:
  RADS = selenocentric distance, km
  LATS = selenocentric latitude, deg
  LONS = selenocentric longitude, deg
  RI = geocentric distance of Deep Space Station, km
  LO = geocentric longitude of Deep Space Station, deg

Radius: 1 meter
Lunar latitude: 30 meters
Lunar longitude: 90 meters

The relative displacements of the Surveyor VI selenocentric position error ellipses are shown in figure 10-22. The position solutions from Lunar Orbiter IV photographs and from the Surveyor VI cruise data fit are given in table 10-12.

TABLE 10-10. *Surveyor V position parameter correlation, based on normal equation matrix*

Parameter [a]	RADS	LATS	LONS
RADS	1.0	−0.6746	−0.8657
LATS		1.0	−.7737
LONS			1.0

[a] The parameters are defined as:
  RADS = selenocentric distance, km
  LATS = selenocentric latitude, deg
  LONS = selenocentric longitude, deg

The statistical dependence of one parameter, relative to other parameters within a recursive least-squares fit, can be inferred from the correlations of the parameter in question and the remaining parameter list. The small magnitudes of the parameter correlations in the correlation matrices (table 10-13) indicate the relative statistical independence of the parameters. A model weakness to be noted is the high correlation exhibited among all selenocentric and geocentric longitude determinations.

O − C Residuals

The thoroughness of the data-validity testing techniques, in conjunction with the sophistication of the SPODP model, did not produce good data fits. In terms of past experience, the solutions are relatively good; but in an absolute sense, the model has not evolved far enough. Diurnal periodicities and longer termed patterns, coupled with high-frequency data and computer noise, are the obvious characteristics of the data fits.

TABLE 10-11. *Surveyor VI parameter solutions*

Parameter [a]	A priori		Compressed data				Uncompressed data	
	Parameter estimates	Standard deviation	LE 4 parameter solution	Standard deviation	LE 5 parameter solution		LE 5 parameter solution	Standard deviation
RADS	1736.0	10.0	1744.027	0.894	1736.439		1736.528	0.840
LATS	.437	5.0	.474	.007	.459		.460	.006
LONS	358.630	5.0	358.65102	.019	358.63229		358.62941	.018
DSS 11:[b]								
r_s	5206.333	.24	5206.332	.002	5206.332		5206.332	.002
LO	243.15070	.005	243.15079	.001	243.15081		243.105082	.001
DSS 42:[b]								
r_s	5205.348	.24	5205.342	.002	5205.342		5205.342	.002
LO	148.98140	.005	148.98153	.001	148.98156		148.98157	.001
DSS 61:[b]								
r_s	4862.601	.23	4862.606	.002	4862.606		4862.606	.001
LO	355.75114	.005	355.75126	.001	355.75126		355.75128	.001

[a] Parameters are defined as:
  RADS = selenocentric distance, km
  LATS = selenocentric latitude, deg
  LONS = selenocentric longitude, deg
  r_s = spin-axis distance of Deep Space Station, km
  LO = geocentric longitude of Deep Space Station, deg
[b] Terrestrial tracking station location referenced to 1903.0 pole.

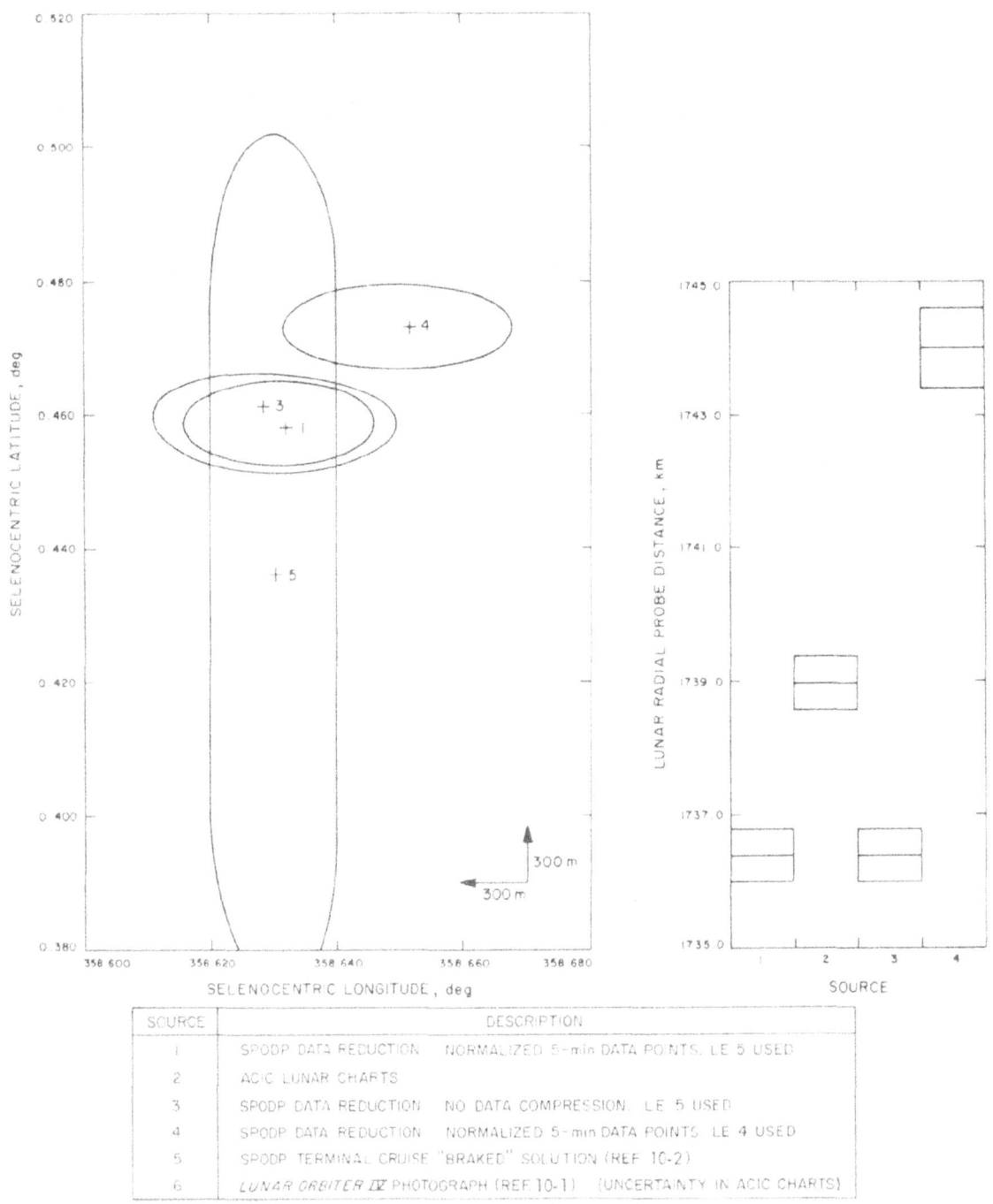

FIGURE 10-22.—Surveyor VI. Relative displacements of selenocentric position error ellipses.

TABLE 10-12. *Surveyor VI location determinations*

Source	Selenographic latitude, ϕ, deg	Standard deviation of ϕ	Selenographic longitude, λ, deg	Standard deviation of λ	Lunar radius, R	Standard deviation of R
1 [a]	0.49N		358.60			
2 [b]	.437N	0.065	358.630	0.006		
3 [c]	.474N	.007	358.651	.019	1744.04	0.89
4 [d]	.459N	.007	358.632	.019	1736.44	.89
5 [e]	.460N	.006	358.629	.018	1741.468	.84

[a] Lunar Orbiter IV photograph (see ref. 10-10).
[b] Terminal cruise SPODP position (cruise data; see ref. 10-2).
[c] SPODP data reduction using LE 4 (compressed postlanded Surveyor data).
[d] SPODP data reduction using LE 5 (compressed postlanded Surveyor data).
[e] SPODP data reduction using LE 5 (uncompressed postlanded Surveyor data).

TABLE 10-13. *Surveyor VI correlation matrix using LE 5 (compressed data fit)*

Parameter [a]	Standard deviation	Surveyor VI			DSS 11		DSS 42		DSS 61	
		RADS	LATS	LONS	RI	LO	RI	LO	RI	LO
RADS	0.894	1.0	0.795	-0.698	0.322	0.698	0.392	0.691	0.047	0.705
LATS	.007		1.0	-.817	.270	.793	.347	.796	.101	.793
LONS	.019			1.0	-.358	-.981	-.396	-.991	.131	-.988
DSS 11:										
RI	.002				1.0	.308	.164	.357	-.061	.360
LO	.001					1.0	.393	.973	-.135	.971
DSS 42:										
RI	.002						1.0	.352	-.038	.397
LO	.001							1.0		
DSS 61:										
RI	.002								1.0	-.092
LO	.001									1.0

[a] The parameters are defined as:
 RADS = selenocentric distance, km
 LATS = selenocentric latitude, deg
 LONS = selenocentric longitude, deg
 RI = geocentric distance of Deep Space Station, km
 LO = geocentric longitude of Deep Space Station, deg

The standard deviation of the high-frequency noise associated with the one data point/min sample (with resolver data) is 0.002 Hz (0.13 mm/sec).

The LE 4-dependent, longer term periodicities demonstrated by the residual sets (see figs. 10-23, 10-24, and 10-1) are as anticipated. The residual sinusoidal pattern is descriptive of the range-rate differences between LE 5 and LE 4 (ref. 10-6) after least-squares minimization has been attempted (see fig. 10-6). LE 5 is a better model of lunar motion than LE 4. A graphical representation of the LE 5-LE 4 differences and the DSS 42 O-C residual set assist in the comparison of the two functions (see fig. 10-25); a high correlation is evident.

The absence of any detectable long-term pattern in the LE 5 O-C residual sets indi-

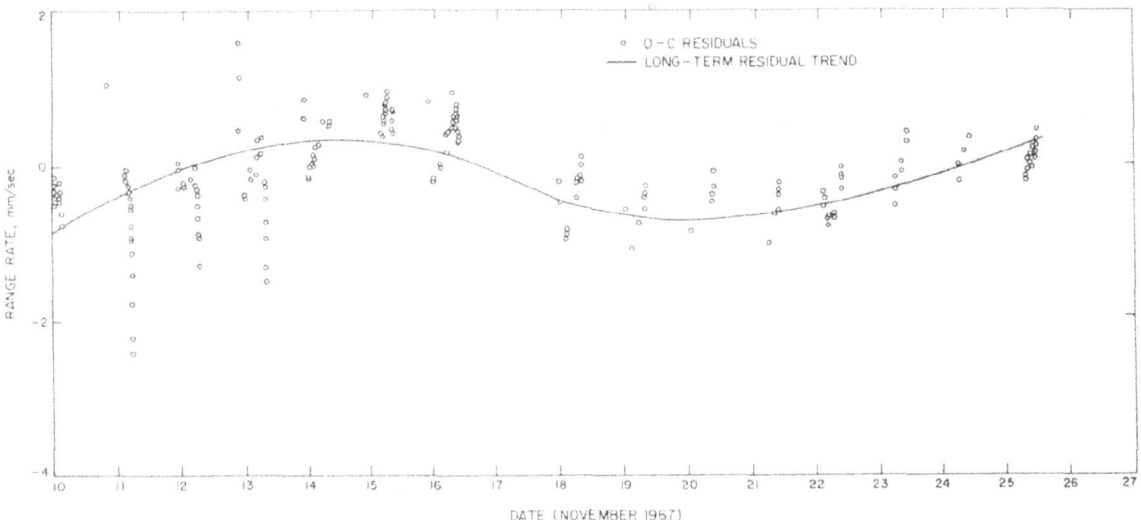

FIGURE 10-23.—Surveyor VI, DSS 11. First lunar day, O−C residual set (compressed five data points/min). Note the long-term periodic nature. LE 4 was used.

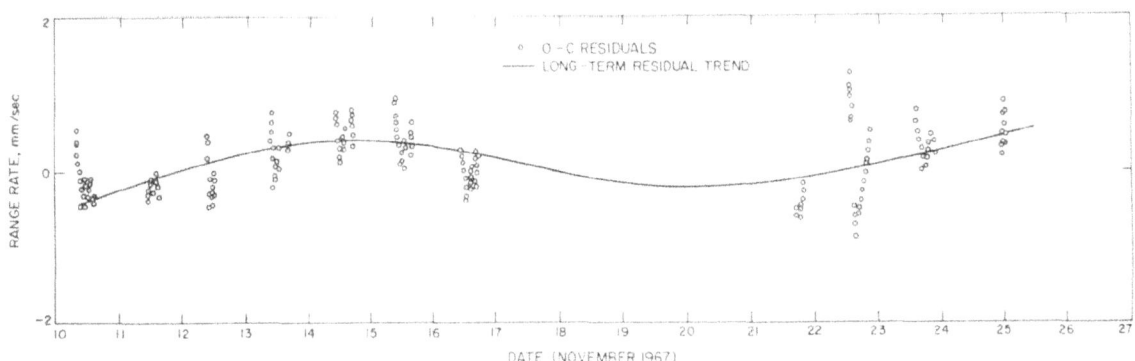

FIGURE 10-24.—Surveyor VI, DSS 61. First lunar day, O−C residual set (compressed five data points/min). Note periodicity. LE 4 was used.

cates the ability of LE 5 to model the lunar motion (see figs. 10-26, 10-27, and 10-5).

The diurnal nature of the Surveyor VI O−C residuals is the same as the daily variations identified with the O−C residuals of Surveyors I, III, V, and VII (see figs. 10-28 through 10-30).

Surveyor VII

Because of the selenocentric error ellipsoid orientation (fig. 10-14) resulting from the high lunar latitude of Surveyor VII, it is reasonable to expect a high correlation between selenocentric distance and latitude (ref. 10-13), as verified by an examination of table 10-14. To avoid this limitation, the lunar radius was not treated as a parameter in any of the reported Surveyor VII reductions. Under the influence of a constrained ACIC lunar radius, the correlation between parameter pairs can be viewed in table 10-15.

The parameter determinations that result from lunar ephemeris variations and "combinational" parameter values are presented in table 10-16, and plotted in fig. 10-31. The associated parameter correlations are contained in table 10-17.

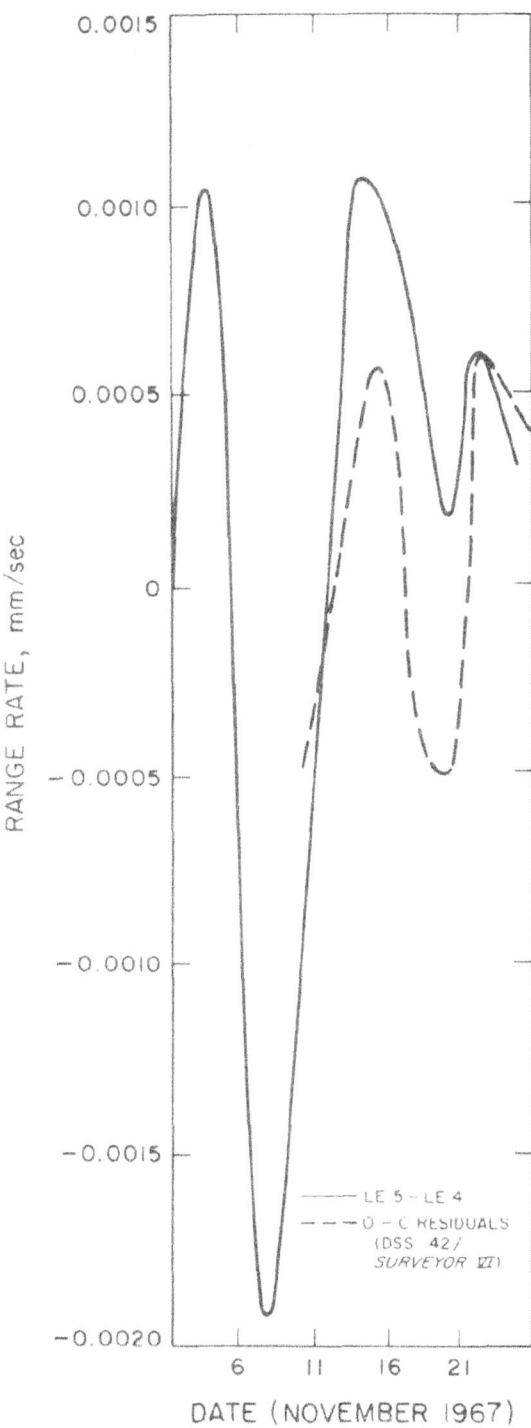

FIGURE 10-25.—LE 5–LE 4 range-rate differences versus Surveyor VI DSS 42 O–C residuals.

TABLE 10-14. *Surveyor VII correlations, based on normal equation matrix*

Parameter [a]	RADS [b]	LATS	LONS
RADS [b]	1.0	0.9642	0.3954
LATS		1.0	.1526
LONS			1.0

[a] The parameters are defined as:
RADS = selenocentric distance, km
LATS = selenocentric latitude, deg
LONS = selenocentric longitude, deg
[b] RADS unconstrained.

TABLE 10-15. *Surveyor VII correlations, based on normal equation matrix*

Parameter [a]	RADS [b]	LATS	LONS
RADS [b]	1.0	0.8302	0.0400
LATS		1.0	.4588
LONS			1.0

[a] The parameters are defined as:
RADS = selenocentric distance, km
LATS = selenocentric latitude, deg
LONS = selenocentric longitude, deg
[b] RADS constrained to ACIC determination.

The deficiencies of the data fits are demonstrated by the O–C residuals. In terms of past experience, these solutions are good. A residual feature is a discontinuity in the O–C residual of DSS 61 at 00:00 GMT on 4 consecutive days (figs. 10-7 through 10-9). The reason is not known at this time.

The standard deviation of the high-frequency noise associated with the residual sets is 0.002 Hz (0.13 mm/sec).

The LE 4-dependent, longer term periodicities demonstrated by the residual sets (see fig. 10-12) are as anticipated. The residual sinusoidal pattern is descriptive of the range-rate differences between LE 5 and LE 4 (ref. 10-13) after least-squares minimization has been attempted (fig. 10-25). LE 5 is a better model of lunar motion than LE 4.

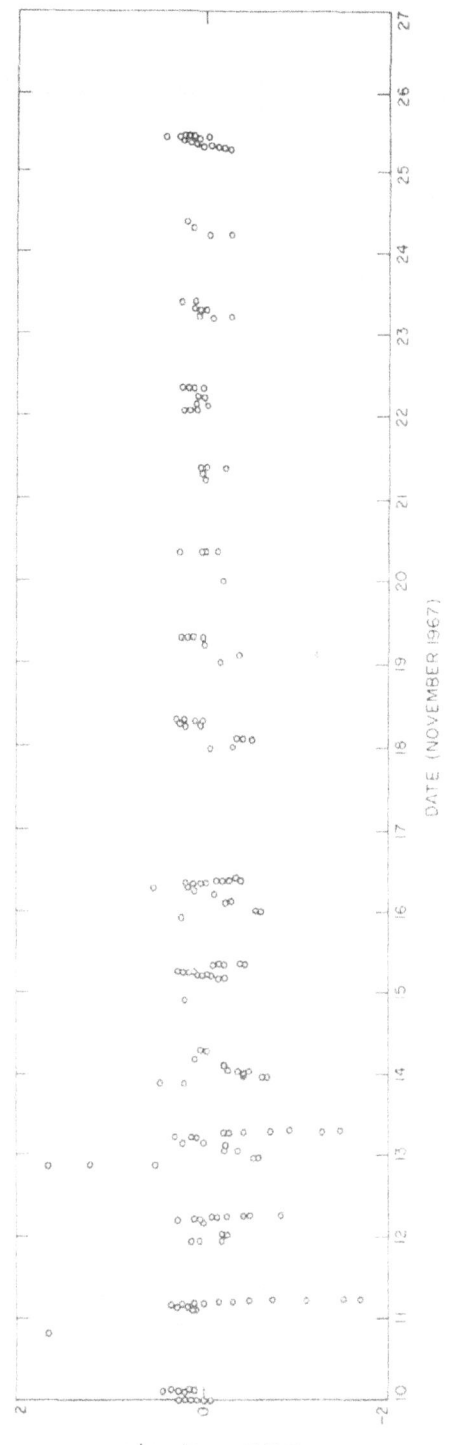

FIGURE 10-26.—Surveyor VI, DSS 11. First lunar day, O — C residual set (compressed five data points/min). LE 5 was used.

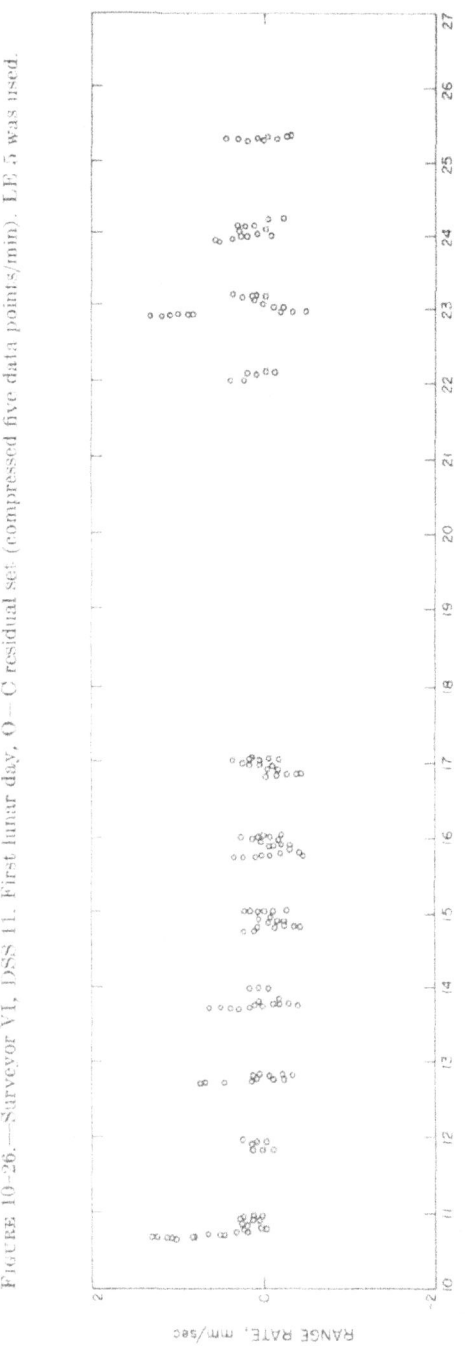

FIGURE 10-27.—Surveyor VI, DSS 61. First lunar day, O — C residual set (compressed five data points/min). LE 5 was used.

POSTTOUCHDOWN TRACKING DATA ANALYSES 391

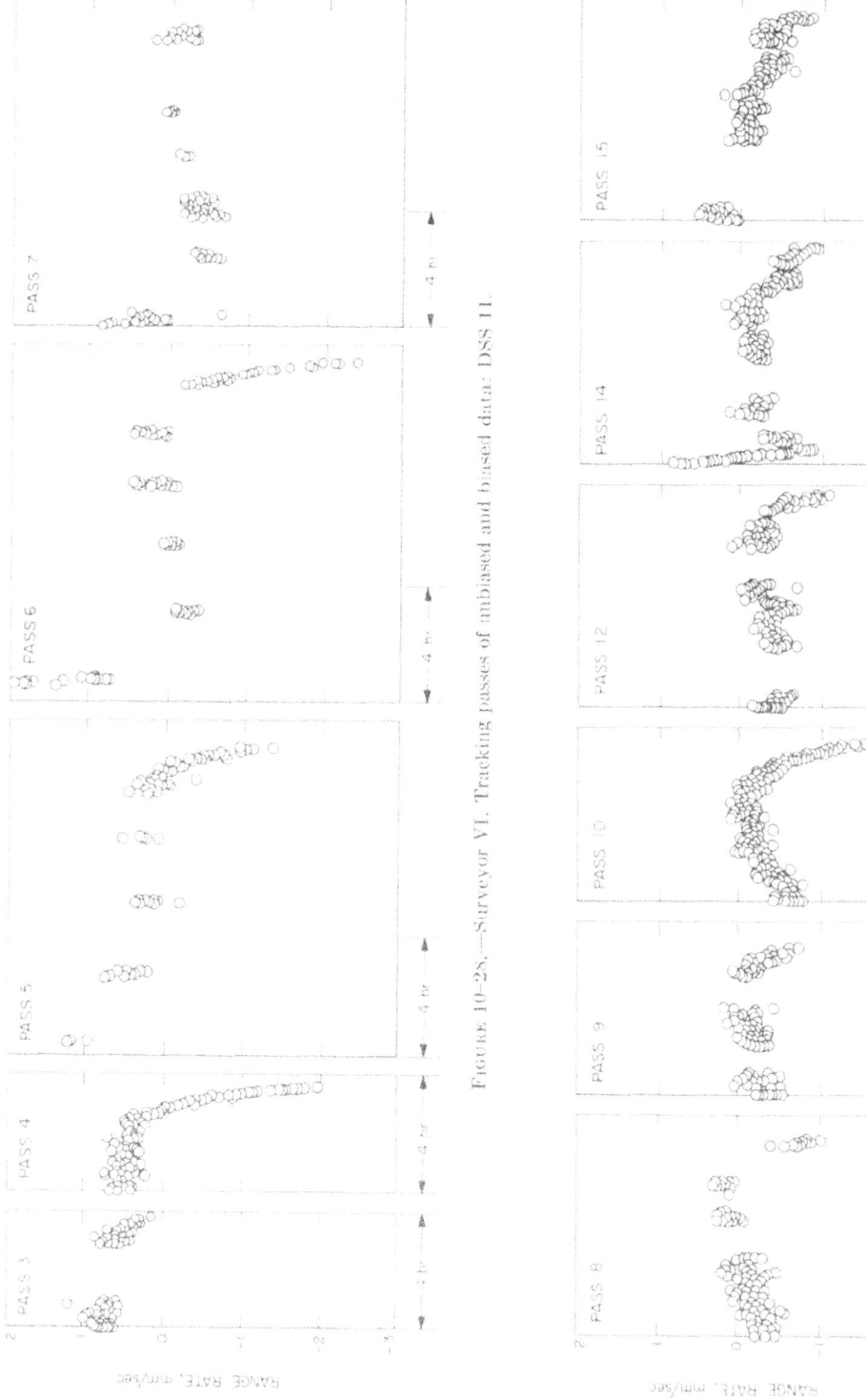

FIGURE 10-28.—Surveyor VI. Tracking passes of unbiased and biased data: DSS 11.

FIGURE 10-29.—Surveyor VI. Tracking passes of unbiased and biased data: DSS 42.

392 SURVEYOR: PROGRAM RESULTS

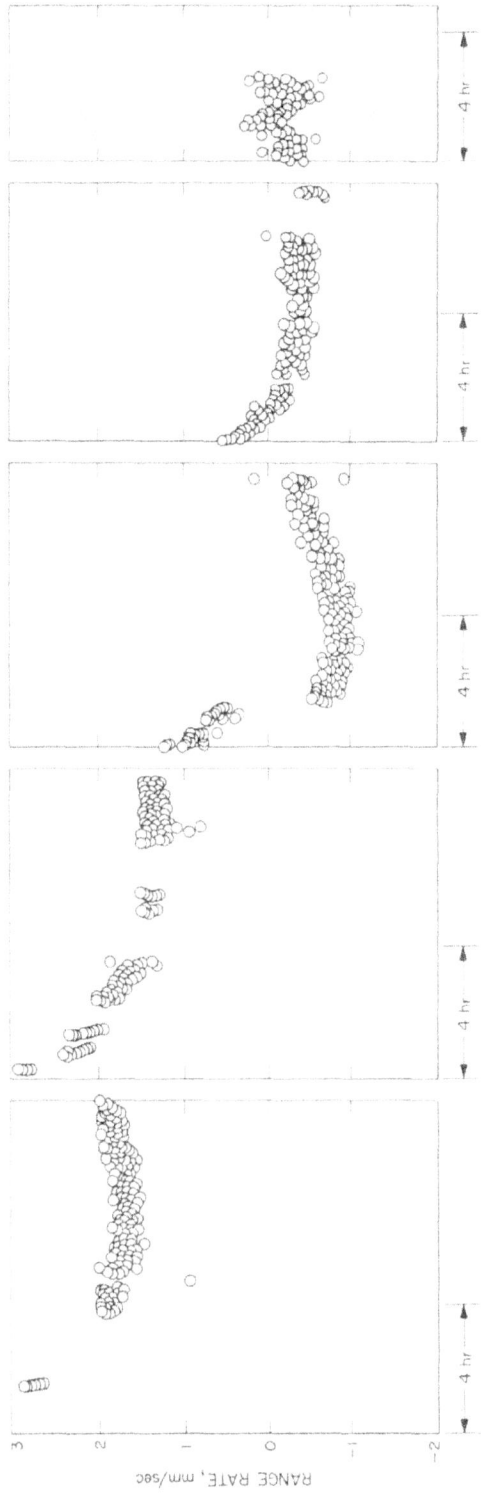

FIGURE 10-30.—Surveyor VI. Tracking passes of unbiased and biased data: DSS 61.

Surveyor VII O−C residuals that result from recursive least-squares fits utilizing the LE 5 lunar motion model exhibit long-term periodicities with an approximate period of one lunation and an amplitude of about 0.7 mm/second (fig. 10-3).

The new integrated ephemeris, LE 8, induced long-term trends into the O−C residuals (fig. 10-6) which are larger than those of LE 5.

The diurnal nature of the Surveyor VII O−C residuals is the same as the daily variations identified with the O−C residuals of Surveyors I, III, V, and VI. However, because of the acquisition of Surveyor VII low-elevation tracking data, a more complete picture of the residual behavior was made available. The

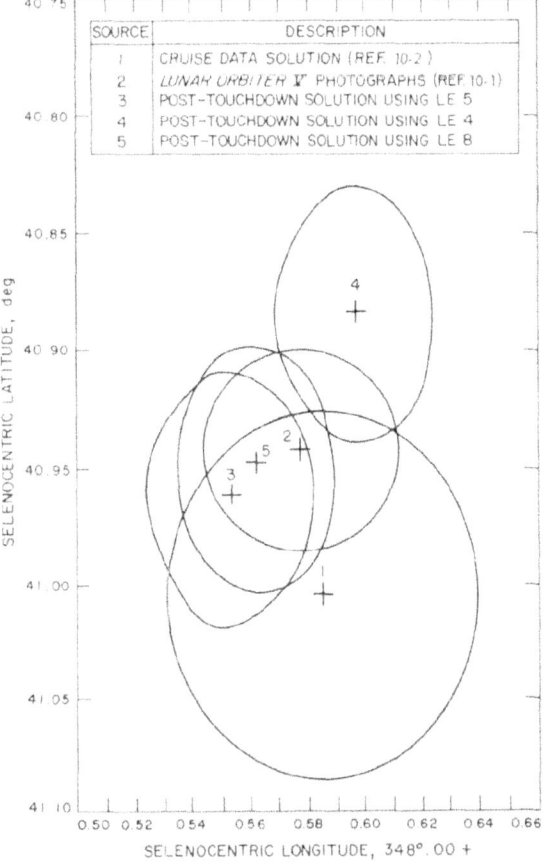

FIGURE 10-31.—Surveyor VII. Relative displacements of selenocentric position error ellipses.

TABLE 10-16. *Surveyor VII parameter[a] solutions*

A priori

DSS	RI	Standard deviation of RI	LO	Standard deviation of LO	RADS	Standard deviation of RADS	LATS	Standard deviation of LATS	LONS	Standard deviation of LONS
11	6371.9989	0.3	243.15076		1736.0	0.1	41.1	5.0	348.56	5.0
42	6371.6881	.3	148.98140							
61	6370.0085	.3	355.75114							

Solution

LE	DSS	N	RI	Standard deviation of RI	LO	Standard deviation of LO	RADS	Standard deviation of RADS	LATS	Standard deviation of LATS	LONS	Standard deviation of LONS
	11	240	6371.9998	0.002	243.15067	0.0003	1739.051	0.311	40.9758	0.010	348.563	0.008
5	42	280	6371.6861	.003	148.98145	.0003						
	61	270	6370.0049	.002	355.75117	.0003						
	11	240	6371.991	.002	243.15067	.0003	1739.059	.311	40.9758	.010	348.561	.008
5	42	310	6371.6795	.003	148.98146	.0003						
	61	340	6370.0001	.002	355.75116	.0003						
	11	240	6372.0102	.002	243.15065	.0003	1739.171	.311	40.8988	.010	348.605	.008
4	42	280	6371.6534	.003	148.98142	.0003						
	61	270	6370.0101	.002	355.75115	.0003						
	11	340	6372.0039	.002	243.15068	.0003	1739.179	.311	40.9028	.010	348.605	.008
4	42	340	6371.6868	.003	148.98143	.0003						
	61	340	6370.0853	.002	355.75114	.0003						
	11	240	6372.0021	.002	243.15068	.0003	1739.030	.311	40.9558	.010	348.568	.008
8	42	280	6371.6568	.002	148.98146	.0003						
	61	270	6370.6032	.002	355.75118	.0003						
	11	340	6371.8077	.002	243.15068	.0003	1739.038	.311	40.9598	.010	348.56595	.008
8	42	310	6371.6802	.003	148.98147	.0003						
	61	340	6370.0085	.002	355.75118	.0003						

[a] The parameters are defined as:

RI = geocentric distance of Deep Space Station, km
LO = geocentric longitude of Deep Space Station, deg
RADS = selenocentric distance, km
LATS = selenocentric latitude, deg
LONS = selenocentric longitude, deg

TABLE 10-17. *Surveyor VII correlation matrix of parameters*

Parameter [a]	Standard deviation	Surveyor VII			DSS 11		DSS 42		DSS 61	
		RADS	LATS	LONS	RI	LO	RI	LO	RI	LO
RADS	0.3	1.0	0.8665	0.3283	0.0656	0.0025	0.0172	0.0112	0.0470	−0.0062
LATS	.010		1.0	.9434	.1619	.2550	.1213	.2538	.1558	.2837
LONS				1.0	.0080	−.7298	−.1334	−.7402	.0371	−.8456
DSS 11:										
RI	.002				1.0	.0811	.0408	.0534	.0615	.0653
LO	.00003					1.0	.1496	.6489	.0381	.7416
DSS 42:										
RI	.003						1.0	−.1828	.0415	.1716
LO	.00003							1.0	.0315	.7481
DSS 61:										
RI	.002								1.0	.1744
LO	.0003									1.0

[a] The parameters are defined as:
RADS = selenocentric distance, km
LATS = selenocentric latitude, deg
LONS = selenocentric longitude, deg
RI = geocentric distance of Deep Space Station, km
LO = geocentric longitude of Deep Space Station, deg

diurnal signature in evidence is characterized by figures 10-7 through 10-9. This signature can be attributed to tropospheric refraction (from deficient modeling), ionospheric charged-particle effects (not modeled), and/or station spin-axis distance, latitude error (suspected lunar ephemeris defect). Because of the correlation between these variables, most of the influence of these combined errors in the O−C residuals can be effectively removed by incorporating any one of the three variables into the SPODP as a solution parameter. It is not the intent of such a procedure to evaluate numerically, in a physically meaningful manner, any one of the three parameters; this approach provides a means of increasing the accuracy of the data fit by using a combination parameter (figs. 10-7 through 10-9).

The use of this "combinational" parameter produces dramatic improvements to the data fits of Surveyor VII. It should be noted, however, that when the same "combinational" parameter values ascertained from the Surveyor VII data reductions were used in the Surveyor V data reductions, DSS 11 and DSS 61 appeared to have been grossly undercorrected for tropospheric refraction. The implications of this influence are not known at this time.

LE 8 was used in the Surveyor VII data reduction with the intention of diminishing the diurnal signatures associated with Surveyor VII two-way doppler residuals. The use of LE 8 did not produce this desired effect uniformly. That is, the phase and period of the error functions incorporated into LE 8 do not correlate with Surveyor diurnal signatures (see figs. 10-10 through 10-12). As a consequence, the ability of LE 8 to reduce the daily residual excursions is time dependent.

The ephemeris influence on Surveyor VII data fits can be demonstrated by the sum of the squares of the weighted residuals.

Ephemerides used:	LE 5	LE 8	LE 4
Sum of the squares:	57	86	114
Refractivity value:	$N_{11}=N_{42}=N_{61}=340$		

Ephemerides used:	LE 5	LE 4	LE 8
Sum of the squares:	51	78	83
Refractivity value:	$N_{11}=240;\ N_{42}=280;\ N_{61}=270$		

By this standard, LE 5 is the most accurate lunar ephemeris currently in existence.

Summary

(1) Although LE 5 demonstrates a long-term periodicity with a period of one lunation and amplitude of 0.7 mm/sec, it is the most accurate lunar ephemeris known to exist.

(2) The selenocentric distance, latitude, and longitude for Surveyors V and VI have been determined and presented.

(3) A high selenocentric distance/selenocentric latitude and longitude correlation permitted a solution for only the lunar-surface coordinates of Surveyors VII and III, respectively (based on ACIC lunar radius determination).

(4) The Surveyor I position determination is subject to review in that a high correlation between selenocentric longitude and distance has distorted the solution.

(5) The new integrated ephemeris, LE 8, demonstrates a time-dependent ability to diminish the diurnal signature of Surveyor VII residuals.

(6) A "combinational" parameter was used to maximize the data fits of Surveyors V and VII. This parameter took the form of refractivity indices.

References

10-1. MORRIS, E. C.; BATSON, R. N.; HOLT, H. E.; RENNILSON, J. J.; SHOEMAKER, E. M.; and WHITAKER, E. A.: Television Observations from Surveyor VI. Surveyor VI Mission Report. Part II: Science Results. Tech. Rept. 32-1262, Jet Propulsion Laboratory, Pasadena, Calif., Jan. 10, 1968, pp. 9–45.

10-2. Surveyor VII Mission Report. Part I: Mission Description and Performance. Tech. Rept. 32-1264, Jet Propulsion Laboratory, Pasadena, Calif. (to be published).

10-3. LAWSON, C. L.: Announcement of JPL Developmental Ephemeris No. 19. Tech. Memo. 33-162, Jet Propulsion Laboratory, Pasadena, Calif., Apr. 13, 1967.

10-4. MULHOLLAND, J. D.; and BLOCK, N.: JPL Lunar Ephemeris Number 4. Tech. Memo. 33-346, Jet Propulsion Laboratory, Pasadena, Calif., Aug. 1, 1967.

10-5. DEVINE, C. J.: Description of a Numerical Integration of Lunar Motion Employing a Consistent Set of Constants. Supporting Research and Advanced Development. Space Programs Summary 37-48, vol. III, Jet Propulsion Laboratory, Pasadena, Calif., Dec. 31, 1967.

10-6. SJOGREN, W. L.; and CARY, C. N.: Lunar Ephemeris Errors Confirmed by Radio Observations of Lunar Probes. The Deep Space Network, Space Programs Summary 37-48, vol. II, Jet Propulsion Laboratory, Pasadena, Calif., Nov. 30, 1967, pp. 4–7.

10-7. STURMS, F. M., JR.: An Integrated Lunar Ephemeris. The Deep Space Network, Space Programs Summary 37-48, vol. II, Jet Propulsion Laboratory, Pasadena, Calif., Nov. 30, 1967, pp. 7–12.

10-8. VAN FLANDERN, T. C.: A Preliminary Report on a Lunar Latitude Fluation. Proceedings of the JPL Seminar on Uncertainties in the Lunar Ephemeris, Tech. Rept. 32-1244, Jet Propulsion Laboratory, Pasadena, Calif. (to be published).

10-9. LIU, A.: Recent Changes to the Tropospheric Refraction Model Used in the Reduction of Radio Tracking Data From Deep Space Probes. The Deep Space Network, Space Programs Summary 37-50, vol. II, Jet Propulsion Laboratory, Pasadena, Calif., Mar. 31, 1968.

10-10. MULHOLLAND, J. D.: A Possible Explanation of Landed Surveyor Residuals. The Deep Space Network, Space Programs Summary 37-49, vol. II, Jet Propulsion Laboratory, Pasadena, Calif., Jan. 31, 1968.

10-11. WARNER, M. R.; and NEAD, M. W.: SPODP—Single Precision Orbit Determination Program. Tech. Memo. 33-204, Jet Propulsion Laboratory, Pasadena, Calif., Feb. 15, 1965.

10-12. MOTTINGER, N. A.: Effect of New Tropospheric Corrections on Lunar Orbiter II Cruise Solution Vector. The Deep Space Network, Space Programs Summary 37-50, vol. II, Jet Propulsion Laboratory, Pasadena, Calif., Mar. 31, 1968.

10-13. CARY, C. N.: Spacecraft on the Lunar Surface. Part I: Range-Rate Tracking. The Deep Space Network, Space Programs Summary 37-47, vol. II, Jet Propulsion Laboratory, Pasadena, Calif., Sept. 30, 1967.

11. Laser Beam Pointing Tests

C. O. Alley (Chairman) and D. G. Currie

An opportunity to verify the ability of Earth stations for directing very narrow laser beams to a specific location on the lunar surface was provided by the detection sensitivity of the Surveyor VII vidicon camera operating in its integration mode. Such tests were of interest primarily because of a planned Apollo lunar surface experiment in which an astronaut will emplace a corner reflector array to provide a fixed point for very precise laser ranging. The successful monitoring of point-to-point Earth-Moon distances to the expected accuracy of ±15 cm would provide: (1) a definitive test of the conjectured slow decrease of the gravitational constant; (2) an experimental study of whether continental drift is occurring now; (3) new knowledge on the physical librations, size and shape, and orbital motions of the Moon; and (4) new information on the rotation of the Earth (refs. 11–1 through 11–4). An additional factor in testing narrow laser beam pointing and tracking techniques lies in their potential use in space communications systems.

The idea of using a Surveyor television camera for such tests occurred during a discussion on whether an astronaut could see the pulsed ruby laser beam planned for the retroreflector ranging experiment. Measurements on the wavelength sensitivity of the vidicon surface were conducted in November 1967 at the Jet Propulsion Laboratory (JPL) and indicated a decrease from the peak sensitivity by a factor 1/300 for the ruby laser wavelength of 6943 Å, making detection marginal for existing and planned ruby laser systems. However, the availability of argon-ion lasers operating in the blue-green (main wavelengths at 4880 and 5145 Å), within the peak of the vidicon sensitivity with average powers of a few watts, suggested their use for the tests. The pointing and tracking techniques would be similar to those used with pulsed lasers.

Estimates of the power density on the Moon of a 10-watt (transmitted) argon-ion laser beam contained within a divergence cone angle (half) of 10 seconds of arc yielded a value 2.25 times the power density of a magnitude 0 star, or nearly magnitude −1. The power density would scale directly as the power transmitted and inversely as the square of the beam angle. Experience with star observations on previous Surveyor missions (p. 15 of ref. 11–5) indicated that the laser beams could be easily observed if they were directed to illuminate the spacecraft. The diameter of the illuminated area on the Moon is about 2 km per arc second of divergence.

Laser Transmitting Stations

Six transmitting stations were established; each consisted of an argon-ion laser with a suitable optical system for collimating and aiming the laser beam. All six stations used the technique of directing the laser beam backward through a telescope to reduce the beam divergence. However, each station used a different method for aiming the laser beam. A brief description of each station is given below.

(1) *Kitt Peak National Observatory, Tucson, Ariz.*—The McMath solar telescope (60-in., f/60, heliostat configuration) and a 2-watt laser were used. The telescope was used in the normal direction for aiming. The guide beam and the laser beam were separated by a specially constructed, divided-mirror beam splitter placed near the telescope focal plane. A reticle, designed for the purpose and permitted offset guiding from nearby lunar features, and a field lens were placed in the focal plane.

(2) *Table Mountain Observatory, Wrightwood,*

Calif.—The JPL 24-inch telescope, utilized at its $f/36$ Coudé focus, and a 2-watt laser were used. A beam splitter with a pinhole was placed in the telescope focal plane to separate the guide beam from the laser beam. A 2.5 magnification microscope with a crosshair reticle was used as a viewing eyepiece.

(3) *Wesleyan University, Raytheon Research Laboratory, Waltham, Mass.*—A 6-inch, two-mirror coelostat directed the beam from a specially constructed 4-inch, $f/15$ telescope toward the Moon; a 60-watt laser was used. The guide beam and the laser beam were separated using a clear pellicle beam splitter located ahead of the primary focal plane. The use of an appropriate glass filter over the eyepiece permitted continuous viewing of the crosshair reticle.

(4) *Lincoln Laboratories, Lexington, Mass.*—A beam from a 3.5-watt laser collimated with a 3-inch telescope was directed using a special servo-driven az-el flat mirror. Guiding was accomplished using a second 3-inch telescope, which was boresighted to the first telescope.

(5) *Goddard Space Flight Center, Greenbelt, Md.*—An existing mobile laser satellite ranging system was used; the pulsed ruby laser was replaced by a 10-watt argon-ion laser. A series of mirrors guided the beam along the rotation axes of the az-el mount through a 5½-inch output aperture. Viewing of the Moon was accomplished by an image orthicon television display from a boresighted 16-inch telescope.

(6) *Perkin-Elmer Corp., Norwalk, Conn.*—A portable 2-watt laser was attached at the Cassegrain focus of a 24-inch telescope. Aiming was accomplished by the 6-inch guide telescope, which was boresighted to the main telescope.

To aid in locating Surveyor VII on the lunar surface, Lunar Orbiter photographs of the region around Tycho and ACIC Lunar Chart LAC 112 were supplied to all stations. The initial estimates of the landing coordinates, as well as the accurate location of Surveyor VII, were communicated with respect to both the Lunar Orbiter photographs and the lunar chart.

Lunar Schedule of Tests

The heavy demands on the Surveyor camera resulted in the initial allocation of only one 10-minute block of laser observing time on each of four different nights. By combining the laser observations with the planned earthlight polarization observations, it was possible to increase the length of observing periods and to have a second period on January 20, 1968. Time windows were chosen so that stations on both east and west coasts could be observed simultaneously during control of the spacecraft by the Goldstone Deep Space Tracking Station; the primary constraint was that no station be too close to the terminator. During the window, the laser stations were responsive to the availability of the television camera. Communication was handled by a telephone network connecting all stations with the JPL Space Flight Operations Facility.

The first few days after touchdown were needed for other Surveyor activities and were used for final preparations at the stations. With the exception of the Norwalk and Greenbelt stations, the first test period was held at 04:30 GMT on January 14, 1968. It was necessary to interrupt the tests during the period near lunar noon because of glare in the camera caused by the proximity of the Earth and Sun. This time was used to modify techniques at some of the stations on the basis of the first test. Test periods were resumed on January 19, and continued on January 20 and 21. The time on January 21 was chosen to maximize the probability of observing stations on the east coast by having them far from the terminator even though, for stations on the west coast, it placed the Moon very low in the sky.

During each test period, modes of operation for the stations were prescribed with definite on-and-off sequences to identify stations that were geographically close together. The aperture and exposure time were varied to produce on the A-scope display approximately one-half the saturation voltage level in the dark part of the Earth crescent where laser beams were being transmitted, as this maximized the sensitivity. With this setting, repeated exposures were taken while the stations were directed to follow the above modes.

Detection of Laser Beams

Detection was achieved visually during the

first observing period on January 20 for both Tucson and Wrightwood. Suspected laser beam spots with the correct locations, as shown in figure 11-1, were observed at the JPL Space Flight Operations Facility. Further confirmation resulted when the Earth image was shifted 3° within the 6.5° narrow field of view, the two spots shifting with it. The on-off sequencing

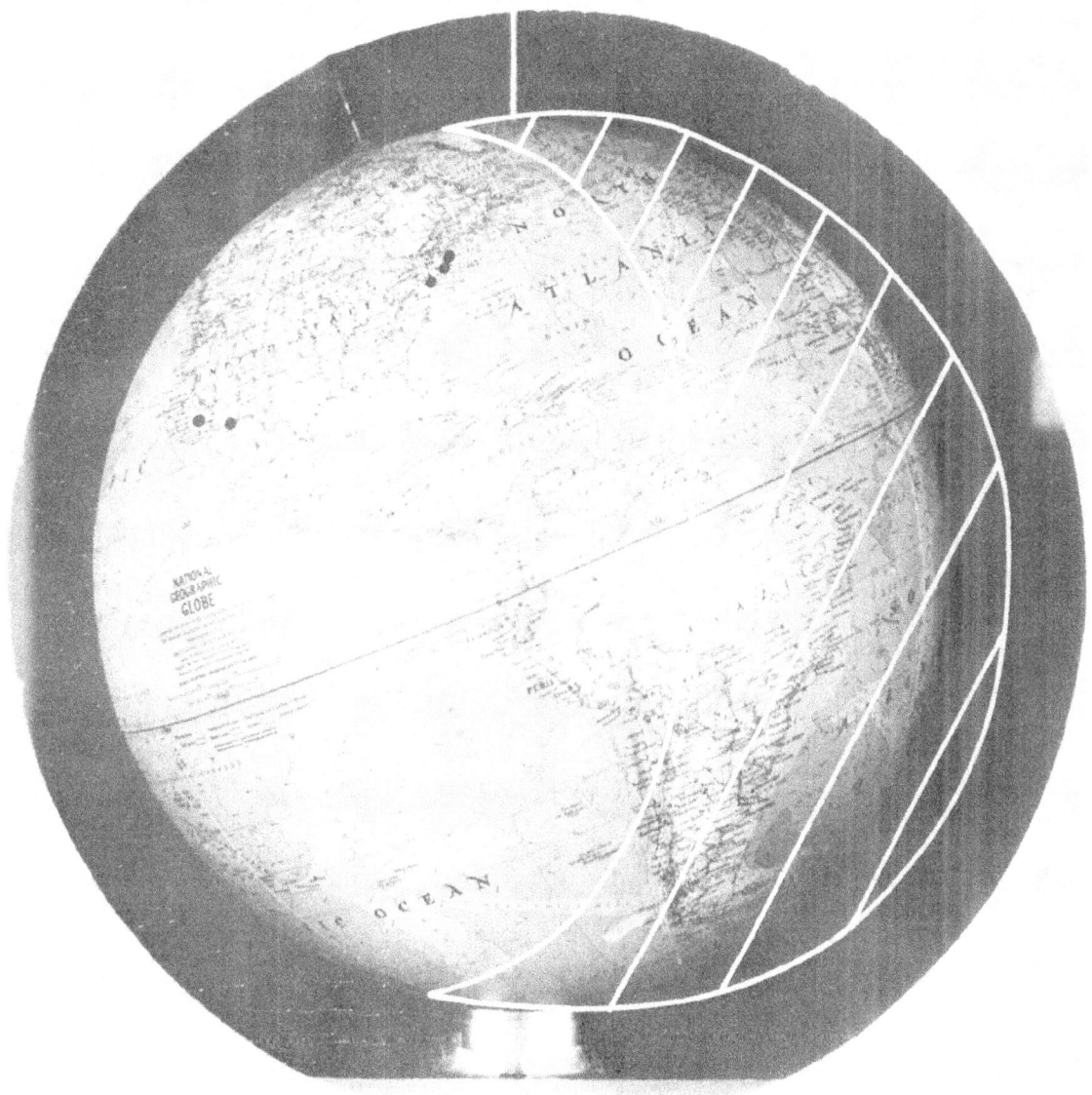

FIGURE 11-1.—This photograph of a globe, with the overexposed crescent indicated by crosshatching, simulates the Earth as seen from Surveyor VII at 09:00 GMT on January 20, 1968. The station locations are indicated by black dots, and permit ready identification of the origin of the two laser beams in figure 11-2 as Table Mountain Observatory near Los Angeles, Calif., and Kitt Peak National Observatory, near Tucson, Ariz. Simulations similar to this photograph were prepared in advance by J. J. Rennilson, JPL, for each period of attempted laser detection.

400 SURVEYOR: PROGRAM RESULTS

FIGURE 11-2.—Laser beams with powers of approximately 1 watt each appear as starlike images comparable in brightness to Sirius (magnitude, −2.5) in this narrow-angle, f/4, 3-second exposure of the Earth. The crescent of the Sun-illuminated Earth is distorted because of overexposure. This was one of the first pictures in which the beams were readily visible (Jan. 20, 1968, about 09:06 GMT).

discussed above also served to verify the detection of the beams. Full confirmation was obtained only with the subsequent, detailed study of correlations in projected enlargements from high-quality photographic negatives reproduced from the video tape recordings by a kinescope film recorder. A positive print of one of the negatives, enhanced using a high-contrast process, is shown in figure 11-2. The spread of the images over several of the video scan lines is caused by aberrations in the optics (electron and visible) of the camera and also in the ground reproducing equipment.

Each of the stations detected was transmitting about 1 watt. Wrightwood was systematically scanning about the position of Surveyor VII and was limited by atmospheric "seeing," while Tucson had deliberately spread the beam. The spots appeared with an approximate star magnitude of -1, as originally calculated. Detection of these stations was accomplished again visually with about the same intensity during the second run on January 20 and 21. The approximate magnitude of the detected beams was determined by comparing pictures of the laser beams with those of Jupiter.

By digitization of the video pictures, it has been possible to increase the sensitivity of detection considerably beyond the visual. It is estimated that, by stretching the digitization in regions near station locations, intensities of laser beams directed to illuminate Surveyor VII can be detected with $\frac{1}{25}$ the intensity displayed by Tucson and Wrightwood. This technique enabled easy detection of the Tucson beam on January 19 (see fig. 11-3). (Wrightwood was not operating on that day.) Figure 11-4 shows a computer printout of the region around the laser spot shown in figure 11-3.

A search for beams from the east-coast station has been made with the equipment at the University of Maryland developed for visual scan of bubble-chamber pictures. No positive results were found. Examination of the stretched digitized printouts has not given positive indication as yet, but the work is continuing with the technique of averaging successive frames for enhancement and looking for correlations at predicted locations. Although local weather conditions and structural obscurations interfered with transmission from east-coast stations (especially in the Boston area), there were periods when contact with Surveyor VII seemed possible.

Conclusions

The primary value of these tests lies in the experience gained in a variety of techniques for tracking and pointing laser beams with

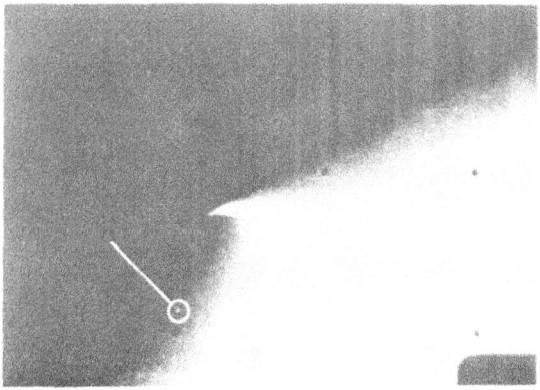

FIGURE 11-3. — A comparison of these two pictures clearly reveals the Tucson laser beam. The picture was digitized by the JPL Image Processing Laboratory, and the contrast then "stretched." (a) Jan. 19, 1968, 08:41:53 GMT. (b) Portion of the digitized, stretched picture; the spot produced by the laser is indicated.

52	51	45	40	45	40	50	44	43	48	40
51	50	50	45	54	39	37	45	40	45	41
47	50	50	46	46	45	39	48	43	47	47
45	40	48	48	43	30	37	42	43	45	44
46	48	48	47	33	4	16	52	52	47	44
46	37	46	49	25	0	0	33	50	46	52
43	45	38	49	43	38	40	51	52	45	51
52	48	41	44	50	53	43	61	42	42	48
49	48	50	49	52	51	55	46	50	52	52
52	62	52	53	45	50	51	50	48	49	55
58	53	59	57	50	50	48	54	48	43	54

FIGURE 11-4.—Computer printout of the region around the laser spot of the digitized, stretched picture shown in figure 11-3. The number at each point represents the optical density, on a scale of 64 gray levels (0=white, 63=black). The laser spot is completely saturated at two points in the picture.

different types of telescopes. A report on this subject by this Surveyor Working Group will be prepared in the future.

The potential value of well-collimated laser beams for space communications is emphasized by noting that the 1-watt laser beams appeared as bright stars, while the uncollimated light from major cities was not detected.

References

11-1. ALLEY, C. O.; BENDER, P. L.; DICKE, R. H.; FALLER, J. E.; FRANKEN, P. A.; PLOTKIN, H. H.; and WILKINSON, D. T.: Optical Radar Using a Corner Reflector on the Moon. J. Geophys. Res., vol. 70, 1965, pp. 2267–2269.

11-2. ALLEY, C. O.; and BENDER, P. L.: Information Obtainable From Laser Range Measurements to a Lunar Corner Reflector, to be published in the Proceedings of the IAU/IUGG Symposium on Continental Drift, Secular Motion of the Pole, and Rotation of the Earth (William Markowitz and B. Guinot, eds.), symposium held at Stresa, Italy, Mar. 21–25, 1967.

11-3. ALLEY, C. O.; BENDER, P. L.; CURRIE, D. G.; DICKE, R. H.; and FALLER, J. E.: Some Implications for Physics and Geophysics of Laser Range Measurements From Earth to a Lunar Retro-Reflector, to be published in the Proceedings of the N.A.T.O. Advanced Study Institute on the Application of Modern Physics to the Earth and Planetary Interiors (S. K. Runcorn, ed.), symposium held at the University of Newcastle upon Tyne, Mar. 29–Apr. 4, 1967.

11-4. MACDONALD, G. J. F.: Implications for Geophysics of the Precise Measurement of the Earth's Rotation. Science, vol. 157, July 21, 1967, pp. 204–205.

11-5. RENNILSON, J. J.; DRAGG, J. L.; MORRIS, E. C.; SHOEMAKER, E. M.; and TURKEVICH, A.: Lunar Surface Topography. Surveyor I Mission Report. Part II: Scientific Data and Results. Tech. Rept. 32-1023, Jet Propulsion Laboratory, Pasadena, Calif., Sept. 10, 1966, pp. 7–44.

ACKNOWLEDGMENTS

We wish to thank BENJAMIN MILWITZKY, Surveyor Program Manager, and STEPHEN DWORNIK, Surveyor Program Scientist, for endorsing the laser pointing test and for providing the support of the National Aeronautics and Space Administration, which made the test possible.

Because of the short time between the initiation and the execution of the laser pointing test, many people and organizations voluntarily participated under adverse conditions and without compensation at each of the six stations. The essential contributions of the following persons are gratefully acknowledged:

Tucson: DR. JAMES BRAULT, staff astronomer of the Kitt Peak National Observatory, and PROF. S. K. POULTNEY of the University of Maryland, using a Spectra-Physics laser loaned by the Aerospace Corp.

Wrightwood: M. S. SHUMATE, JPL, and J. W. YOUNG, Table Mountain Observatory, using a laser constructed and loaned by Hughes Research Laboratories.

Waltham: PROF. J. E. FALLER, Wesleyan University, using a laser from the laboratory of DR. GEORGE DE MARS, Raytheon Research Laboratory. Ten undergraduate students from Wesleyan University, led by D. BURSTEIN, and M. HULETT assisted Professor Faller.

Lexington: DR. ROBERT KINGSTON and DR. HOYT BOSTICK, Lincoln Laboratories, using a laser loaned by Spacerays, Inc.

Greenbelt: DR. H. H. PLOTKIN, H. RICHARD, and W. CARRION of the Optical Systems Branch, GSFC, using an existing laser satellite tracking system incorporating an RCA laser.

Norwalk: H. WISHNIA and DR. MORLEY LIPSETT of the Perkin-Elmer Corp., using a Perkin-Elmer laser and R. PERKIN'S telescope.

The test would not have been possible without the integration mode of the Surveyor vidicon camera. Work that led to the incorporation of the integration mode into the camera was initiated by L. H. ALLEN, JPL, who also performed the vidicon sensitivity measurements at the ruby wavelength and assisted in the overall test operations.

Appreciation is expressed to all members of the Surveyor Project at the Jet Propulsion Laboratory for technical help in the organization and performance of the tests. Special appreciation is extended to the following JPL personnel involved in the television aspects of the mission: J. STRAND, T. H. BIRD, J. J. RENNILSON, D. L. SMYTHE, and C. CHOCOL; and to DR. R. NATHAN and E. T. JOHNSON of the JPL Image Processing Laboratory.

12. Astronomy: Solar Corona Observations

R. H. Norton

Observations of the solar corona were made, using the spacecraft television camera, during postsunset periods of the Surveyor I, V, VI, and VII missions (table 12-1). The method of observation from each spacecraft was essentially the same. After the upper limb of the solar disk had set behind the western horizon, a sequence of timed pictures was taken with gradually increasing exposures that ranged from $f/5.6$, 1.2 seconds, for the innermost K-corona at 2 solar radii to $f/4$, 30 minutes, for the outer F-corona beyond 20 solar radii (refs. 12-1 through 12-4).

Because photometric reduction of all pictures has not been completed, no final conclusions can be made at this time. Although all data will be digitized and reduced, primary emphasis will be placed on the data derived from Surveyor VII, since it represents the best set of solar corona observations obtained. Earth-based (telescope) eclipse observations permit measurements of coronal radiance out to 8 or 10 solar radii; spacecraft observations have extended this measurement to about 15 solar radii. Measurements from observations of the zodiacal light have been made inward to about 50 solar radii. Surveyor VII will provide a determination of the way in which the far solar corona merges into the zodiacal light. The brightness of the coronal/zodiacal light in this region is due to scattering from particulate matter. Measurements of this brightness, polarization, and variation with solar distance will, in turn, permit determinations of the distribution and density of particulate matter in the inner solar system.

TABLE 12-1. *Solar corona observations*

Mission	GMT of sunset		Number of frames	Filters used	Duration of observations, hr	Corona coverage ($R_\odot$) solar radii
	Date	Hr:min				
Surveyor I	June 14, 1966	15:49	46	Clear	0 to 0.5	2 to 4
Surveyor V	Sept. 24, 1967	10:57	37	Green	0 to 3.5	2 to 30
Surveyor VI	Nov. 24, 1967	13:40	44	Polarizing	0.6 to 6.0	2 to 30
Surveyor VII	Jan. 23, 1968	06:06	7	Polarizing	8.5 to 14.5	15 to 40

References

12-1. NORTON, R. H.; GUNN, J. E.; LIVINGSTON, W. C.; NEWKIRK, G. A.; and ZIRIN, H.: Astronomy. Surveyor I Mission Report. Part II: Scientific Data and Results, Tech. Rept. 32-1023, Jet Propulsion Laboratory, Pasadena, Calif., Sept. 10, 1966, pp. 87-91.

12-2. NORTON, R. H.; GUNN, J. E.; LIVINGSTON, W. C.; NEWKIRK, G. A.; and ZIRIN, H.: Astronomy. Surveyor V Mission Report. Part II: Science Results, Tech. Rept. 32-1246, Jet Propulsion Laboratory, Pasadena, Calif., Nov. 1, 1967, pp. 115-118.

12-3. NORTON, R. H.; GUNN, J. E.; LIVINGSTON, W. C.; NEWKIRK, G. A.; and ZIRIN, H.: Astronomy. Surveyor VI Mission Report. Part II: Science Results, Tech. Rept. 32-1262, Jet Propulsion Laboratory, Pasadena, Calif., Jan. 10, 1968, p. 125.

12-4. NORTON, R. H.; GUNN, J. E.; LIVINGSTON, W. C.; NEWKIRK, G. A.; and ZIRIN, H.: Astronomy. Surveyor VII Mission Report. Part II: Science Results, Tech. Rept. 32-1264, Jet Propulsion Laboratory, Pasadena, Calif., Mar. 15, 1968.

Appendix A

Effects of Lunar Particles on Spacecraft Mirror Surfaces

L. D. Jaffe and J. J. Rennilson

During some of the Surveyor missions, spacecraft surfaces apparently were affected by lunar particles thrown against them by the exhaust from the vernier-engine firings close to the lunar surface. These affected spacecraft surfaces included the television camera mirror on Surveyor III, which landed with its vernier engines firing through touchdown (refs. A–1 through A–4); the side of the Surveyor V alpha-scattering-instrument sensor head, affected during the static firing of the vernier engines after landing (ref. A–5); and the Surveyor VI photometric target, affected during the hop made by the spacecraft when the vernier engines were fired (refs. A–6 through A–8). Two auxiliary mirrors on Surveyor VII apparently were affected by lunar material ejected by a footpad or crushable block during landing (refs. A–9 and A–10).

Some lunar material was observed as a coating on the Surveyor VI photometric target and on one of the Surveyor VII auxiliary mirrors. It has not been determined, however, whether the observed effects on the mirror surfaces of Surveyors III and V were caused by fine lunar particles adhering to the mirrors, by sandblasting of the mirrors by such particles, or possibly by deposition of condensed products of the engine exhaust.

Prefiring and postfiring television observations of the gold plating on the Surveyor V sensor head help to determine the cause of these effects. The sides of the sensor head were made of polished aluminum alloy on which a layer of gold, nominally 1 micron thick, was electroplated for temperature control with an underlayer of copper 25 microns thick. Before the vernier-engine firing (fig. A–1(a)), an image of the sensor-head circular plate and of the lunar surface for a few centimeters inboard was clearly visible in the gold plating; after the firing (fig. A–1(b)), no image was visible when the near side of the sensor head was in sunlight. Some image could be seen in a part of the side in shadow (fig. A–1(c)). The side itself appeared matte when sunlit and the lower few centimeters of it were darkened; lunar fragments were observed on the circular plate. Before the firing, the sensor head was about 70 cm from the nearest engine, which was fired for 0.55 ± 0.05 second at a thrust level of $1.2 \times 10^7 \pm 0.2 \times 10^7$ dynes; the nozzle was 39 cm above the lunar surface. The exhaust from the vernier-engine firing caused the sensor head to move horizontally about 10 cm and to rotate 15°.

No pictures of the sensor head were taken through the color filters before the engines were fired; after the firing, however, a set of wide-angle pictures was taken through the red, green, and blue filters of the television camera (ref. A–11). These pictures were taken on September 14, 1967, at 03:55:43 (fig. A–2), 03:56:07, and 03:57:05 GMT. Areas in these pictures were selected for color determination as follows:

Area 1: Lunar surface material illuminated by direct sunlight.

Area 2: Lunar surface material illuminated by direct sunlight and by sunlight reflected from the right face of the sensor head. (This face was shielded from the vernier engines.)

Area 3: Upper part of sensor-head side exposed to engine exhaust.

Area 4: Upper part of sensor-head side exposed to engine exhaust.

FIGURE A-1.—Narrow-angle pictures of the inboard face of the sensor head on Surveyor V.
    (a) Before the static firing of the vernier engines (Sept. 12, 1967, 05:06:27 GMT).
    (b) After the static firing of the vernier engines (Sept. 14, 1967, 07:15:09 GMT).
    (c) After the static firing of the vernier engines (Sept. 21, 1967, 08:33:14 GMT).

FIGURE A-2.—Wide-angle, digitized picture of Surveyor V sensor head and surrounding lunar surface, after engine firing. Squares and numbers indicate the areas selected for color determination. Taken through the green filter (Sept. 14, 1967, 03:55:43 GMT).

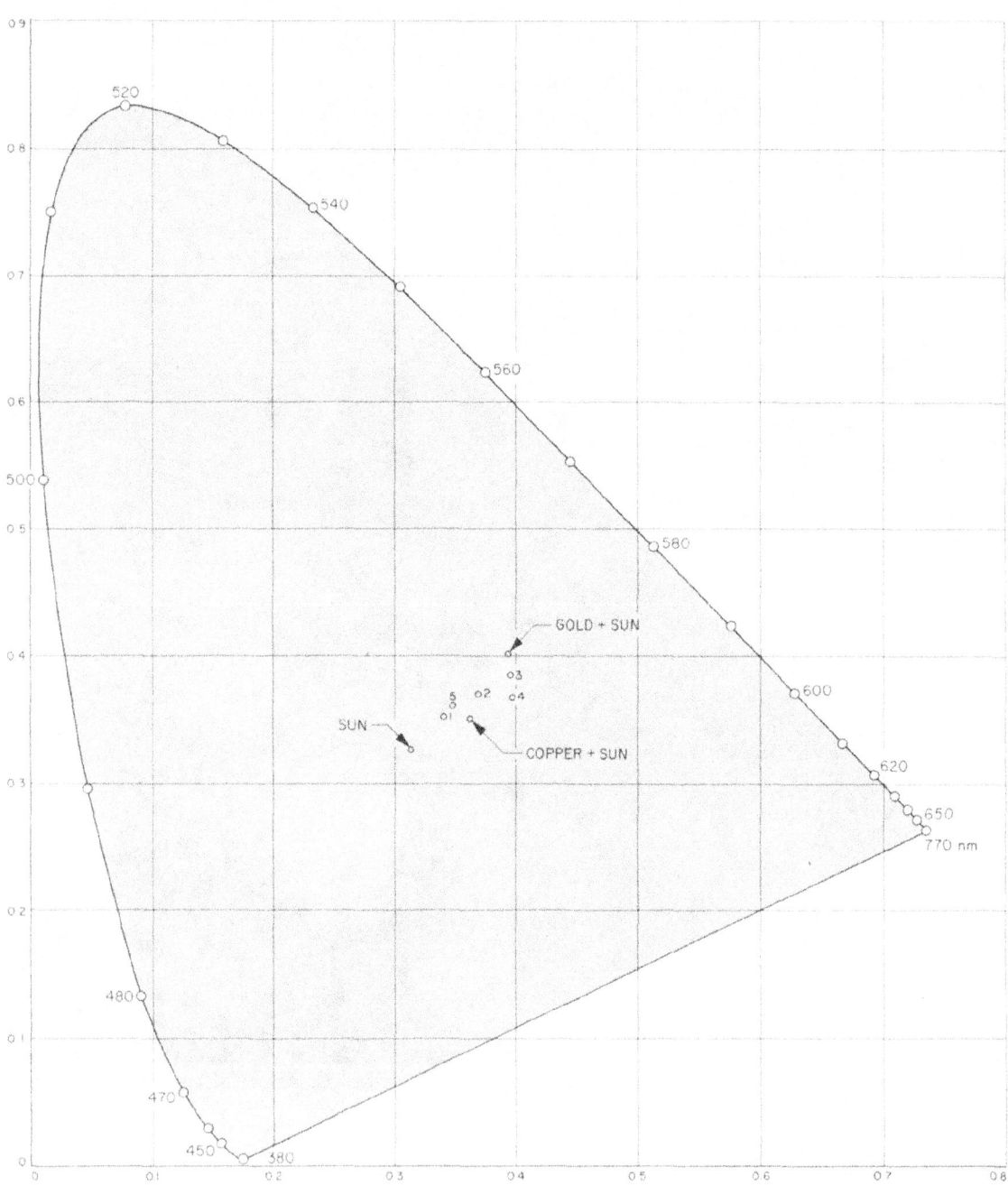

FIGURE A-3.—Chromaticity coordinates of selected points on the Surveyor V sensor head and on the nearby lunar surface, after the engine firing. Coordinates are also shown for extraterrestrial sunlight, for electroplated gold, and for electroplated copper.

Area 5: Lower, darker part of sensor-head side exposed to engine exhaust.

Each area contained 5 by 5 picture elements.

During the mission, the camera video signals from the three frames were recorded directly onto magnetic tape. These analog recordings were then converted into digital units, dividing the black-to-white video level into 64 equal divisions. Corrections for the camera's light-transfer characteristics enabled the digital units to be converted to luminous efficiency values for each frame. The colors of the areas can be specified by chromaticity coordinates in the international color system (ref. A-1). The CIE[1] tristimulus values (color coordinates) are linearly proportional to the luminous efficiency values. During preflight camera calibration, the proportionality factors were determined by exposing the camera to colors of known chromaticities. For added confidence, these factors may be computed from the colors on the photometric targets while in the lunar environment. The results presented here, however, originate from the preflight calibration only.

The differences among the five areas can best be shown by plotting their mean chromaticities on the CIE diagram (fig. A-3). Included in this figure, for comparison, are the calculated chromaticities of typical electroplated gold and electroplated copper illuminated by extraterrestrial sunlight. The chromaticity of sunlight illuminating a neutral gray surface is also included.

The color of area 1, the lunar surface in sunlight, differs somewhat from a neutral gray chromaticity. Nevertheless, it would still be considered gray. Area 2 shows the additive mixing of the surface color with that of reflected sunlight by the gold right side of the sensor head. The surface luminances (areas 1 and 2) indicate that the right side of the sensor head reflects light as well on the lunar surface as it did during preflight tests. The measurements on the upper part of the sensor head (areas 3 and 4) are within the expected color range for electroplated gold. Area 5, on the lower part of the sensor head, resembles closely the color of the surface (area 1).

It seems clear, then, that the darker, lower part of the sensor-head side (area 5) was covered with lunar material, adhering to a nearly vertical surface. On the other hand, the upper part of the sensor-head side (areas 3 and 4) was essentially gold, apparently roughened by impact of fine lunar particles. Lunar particles thrown by the vernier engines onto the spacecraft surface evidently both sandblast and coat the surfaces.

References

A-1. SHOEMAKER, E. M.; BATSON, R. M.; HOLT, H. E.; MORRIS, E. C.; RENNILSON, J. J.; and WHITAKER, E. A.: Television Observations From Surveyor III. Surveyor III Mission Report. Part II: Scientific Results, Tech. Rept. 32-1177, Jet Propulsion Laboratory, Pasadena, Calif., June 1, 1967, pp. 9-67

A-2. CHRISTENSEN, E. M.; BATTERSON, S. A.; BENSON, H. E.; CHOATE, R.; JAFFE, L. D.; JONES, R. H.; KO, H. Y.; SPENCER, R. L.; SPERLING, F. B.; and SUTTON, G. H.: Lunar Surface Mechanical Properties. Surveyor III Mission Report. Part II: Scientific Results, Tech. Rept. 32-1177, Jet Propulsion Laboratory, Pasadena, Calif., June 1, 1967, pp. 111-153.

A-3. GAULT, D.; COLLINS, R.; GOLD, T.; GREEN, J.; KUIPER, G. P.; MASURSKY, H.; O'KEEFE, J.; PHINNEY, R.; and SHOEMAKER, E. M.: Lunar Theory and Processes. Surveyor III Mission Report. Part II: Scientific Results, Tech. Rept. 32-1177, Jet Propulsion Laboratory, Pasadena, Calif., June 1, 1967, pp. 195-213.

A-4. JAFFE, L. D.; BATTERSON, S. A.; BROWN, W. E., JR.; CHRISTENSEN, E. M.; GAULT, D. E.; LUCAS, J. W.; NORTON, R. H.; SCOTT, R. F.; SHOEMAKER, E. M.; SUTTON, G. H.; and TURKEVICH, A. L.: Principal Scientific Results of the Surveyor III Mission. Surveyor III Mission Report. Part II: Scientific Results, Tech. Rept. 32-1177, Jet Propulsion Laboratory, Pasadena, Calif., June 1, 1967, pp. 3-7.

A-5. CHRISTENSEN, E. M.; BATTERSON, S. A.; BENSON, H. E.; CHOATE, R.; HUTTON, R. E.; JAFFE, L. D.; JONES, R. H.; KO, H. Y.; SCHMIDT, F. N.; SCOTT, R. F.; SPENCER, R. L.; and SUTTON, G. H.: Lunar Surface Mechanical Properties. Surveyor V Mission Report. Part II: Science Results, Tech. Rept. 32-1246, Jet Propulsion Laboratory, Pasadena, Calif., Nov. 1, 1967, pp. 43-88.

A-6. MORRIS, E. C.; BATSON, R. M.; HOLT, H. E.; RENNILSON, J. J.; SHOEMAKER, E. M.; and WHITAKER, E. A.: Television Observations

---

[1] Commission Internationale d'Eclairage (International Commission on Illumination).

From Surveyor VI. Surveyor VI Mission Report. Part II: Science Results, Tech. Rept. 32-1262, Jet Propulsion Laboratory, Pasadena, Calif., Jan. 10, 1968, pp. 9–45.

A-7. CHRISTENSEN, E. M.; BATTERSON, S. A.; BENSON, H. E.; CHOATE, R.; HUTTON, R. E.; JAFFE, L. D.; JONES, R. H.; KO, H. Y.; SCHMIDT, F. N.; SCOTT, R. F.; SPENCER, R. L.; SPERLING, F. B.; and SUTTON, G. H.: Lunar Surface Mechanical Properties. Surveyor VI Mission Report. Part II: Science Results, Tech. Rept. 32-1262, Jet Propulsion Laboratory, Pasadena, Calif., Jan. 10, 1968, pp. 47–108.

A-8. JAFFE, L. D.; BATTERSON, S. A.; BROWN, W. E., JR.; CHRISTENSEN, E. M.; DWORNIK, S. E.; GAULT, D. E.; LUCAS, J. W.; NORTON, R. H.; SCOTT, R. F.; SHOEMAKER, E. M.; STEINBACHER, R. H.; SUTTON, G. H.; and TURKEVICH, A. L.: Principal Science Results From Surveyor VI. Surveyor VI Mission Report. Part II: Science Results, Tech. Rept. 32-1262, Jet Propulsion Laboratory, Pasadena, Calif., Jan. 10, 1968, pp. 5–7.

A-9. CHOATE, R.; BATTERSON, S. A.; CHRISTENSEN, E. M.; HUTTON, R. E.; JAFFE, L. D.; JONES, R. H.; KO, H. Y.; SPENCER, R. L.; and SPERLING, F. B.: Lunar Surface Mechanical Properties. Surveyor VII Mission Report. Part II: Science Results, Tech. Rept. 32-1264, Jet Propulsion Laboratory, Pasadena, Calif., Mar. 15, 1968, pp. 77–134.

A-10. JAFFE, L. D.; ALLEY, C. O.; BATTERSON, S. A.; CHRISTENSEN, E. M.; DWORNIK, S. E.; GAULT, D. E.; LUCAS, J. W.; MUHLEMAN, D. O.; NORTON, R. H.; SCOTT, R. F.; SHOEMAKER, E. M.; STEINBACHER, R. H.; SUTTON, G. H.; and TURKEVICH, A. L.: Principal Science Results From Surveyor VII. Surveyor VII Mission Report. Part II: Science Results, Tech. Rept. 32-1264, Jet Propulsion Laboratory, Pasadena, Calif., Mar. 15, 1968, pp. 5–8.

A-11. SMOKLER, M. I.: Television System. Surveyor V Mission Report. Part III: Television Data, Tech. Rept. 32-1246, Jet Propulsion Laboratory, Pasadena, Calif., July 15, 1968, pp. 3–15.

Appendix B

The Lunar Sunset Phenomenon

L. H. Allen

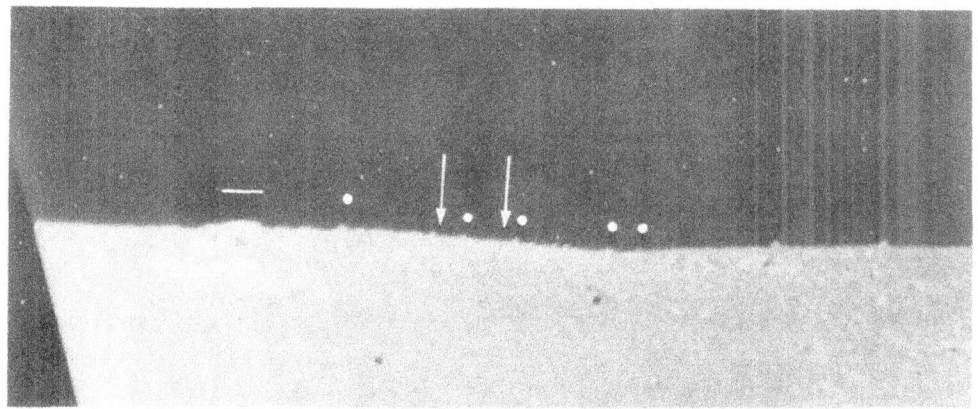

FIGURE B-1.—Surveyor VII daylight picture of the western horizon, including the sunset position (Jan. 10, 1968, 08:16:45).

Lunar Observations

Television pictures, taken by several Surveyors a few minutes after local lunar sunset, showed the lunar western horizon highlighted by the Sun as a thin, bright, jagged, discontinuous line. The Surveyor VII lunar sunset pictures probably are the most interesting in that this line was jagged, its brightest parts shifted with time, and its broken parts changed positions with time. (See figs. B-1 to B-5.)

The brightest parts of this line have sometimes erroneously been called "beads" because they are remotely similar in appearance to "Bailey's beads." The discontinuous portions of the line are gaps and appear to be the result of lunar-surface features that lie beyond (farther west), casting shadows on the local horizon. As a general rule, the gaps appeared to grow larger with time during the Surveyor VI and VII missions, as the Sun moved farther below the horizon.

Because Surveyor VII was at a southern lunar latitude, the gaps appeared to shift slightly to the north as a function of time and by different amounts, which are assumed to be a function of the distance between the individual occulting objects and the point on the horizon being occulted. The brightest parts of the line shifted slowly to the south as a function of time and as a function of Surveyor VII spacecraft's southern latitude on the lunar surface. The lateral shifting effects were noticeable only on Surveyor VII, because it operated much farther from the lunar equator than any of the other spacecraft.

It is interesting to note that on the left (south) side of figures B-1 to B-5, the horizon

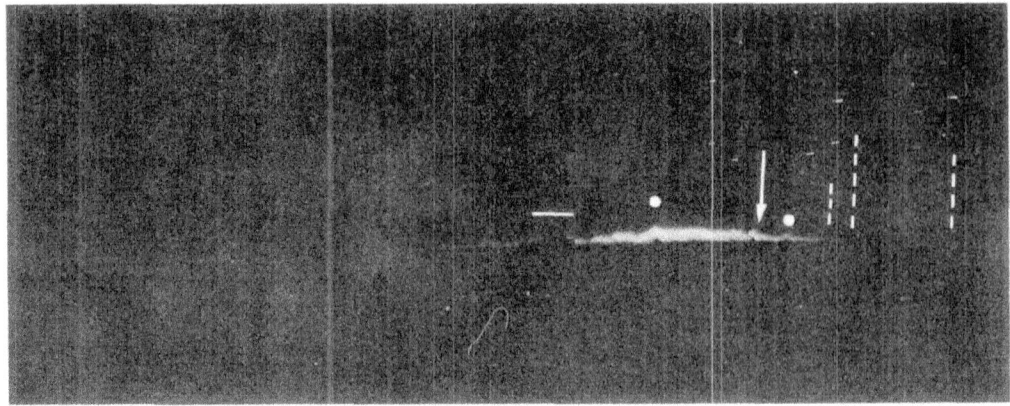

FIGURE B-2.—Discontinuous, bright line in the western horizon. The brightness of the line decreased with time. This narrow-angle Surveyor VII picture was taken about 17 minutes after local sunset (Jan. 23, 1968, 06:20:28 GMT).

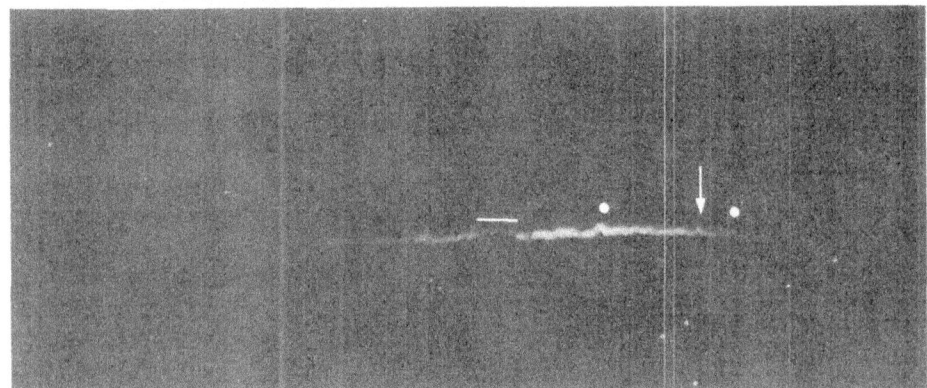

FIGURE B-3.—Narrow-angle Surveyor VII picture of the line taken about 50 minutes after local sunset (Jan. 23, 1968, 06:52:11 GMT).

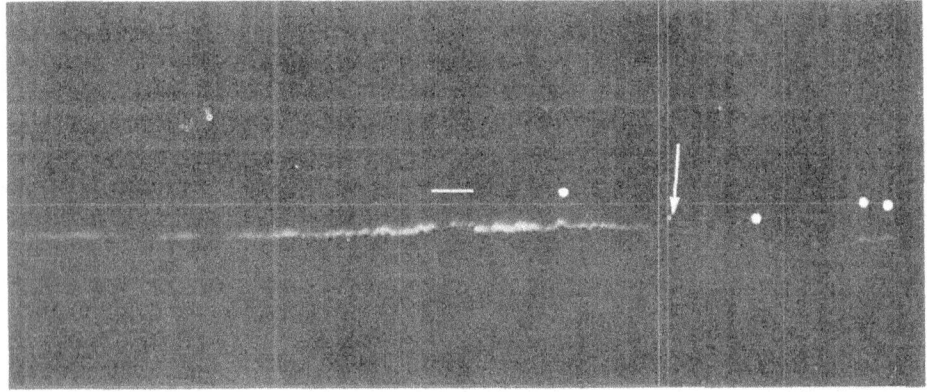

FIGURE B-4.—Narrow-angle Surveyor VII picture of the line taken about 90 minutes after local sunset (Jan. 23, 1968, 07:31:52 GMT).

FIGURE B-5.—View of the sunset position taken over 14 hours after sunset by the Surveyor VII television camera. This wide-angle picture, made in the integration mode of the camera, shows the faint, residual solar corona and the lunar surface illuminated by earthlight. The bright spot in the sky is Mercury, spread by the camera's white spread function and elongated by lunar rotation during the 30-minute exposure time (Jan. 23, 1968, 20:45:55 GMT).

is formed by a close ridge; on the extreme right (north) it is formed by another ridge that is farther away (farther west). In the pictures taken following sunset (and at different exposures), the right-hand ridge is illuminated. However, looking farther south (going to the left), the nearer ridge obstructs the view of the more distant ridge. The more distant ridge occults (shadows) the closer ridge for a short distance and causes a large gap in the bright line.

Figure B-6 shows two other ridges (to the southeast) that reveal a similar condition; these two ridges should provide a better understanding of the relationship of the western ridges. In figure B-6, the closer ridge casts shadows on the other one, resulting in what appears to be a discontinuous line if viewed from its other side. In turn, this closer ridge is itself shadowed by still another ridge (not shown) whose features are shadow profiled on it.

Figure B-7, a Surveyor VI picture, shows not only the bright line but also a faint streak appearing in the lower foreground, which is the result of the line internally reflected by the camera's lens. The resultant image is the same size, inverted, and attenuated. Only the image width (thickness), not the length of elements between the gaps, has been reduced. This effect has also been confirmed by taking pictures at various exposures.

Camera Characteristics

The apparent thickness, or width, of the line in the television pictures is the result of the television camera's white spread function: the brighter the object viewed, the greater the image will be spread. This spreading, as confirmed in tests at the Jet Propulsion Laboratory's advanced imaging laboratory, with a slit target whose width was made considerably smaller than the limiting resolution of the

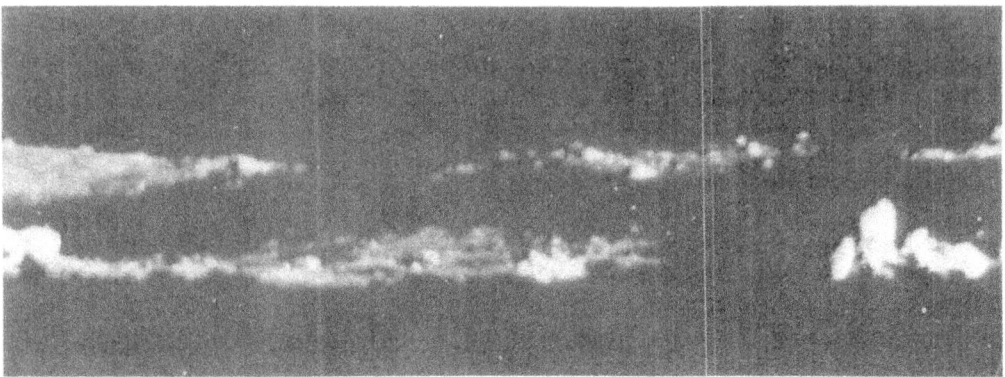

FIGURE B-6.—Two ridges to the southeast of Surveyor VII. These ridges were only partially illuminated as the Sun set. The closer ridge cast its shadow on the other, resulting in the discontinuous line. The closer ridges showed features, in shadow outline, of another ridge (not visible) to the west of it (Jan. 23, 1968, 09:25:57 GMT).

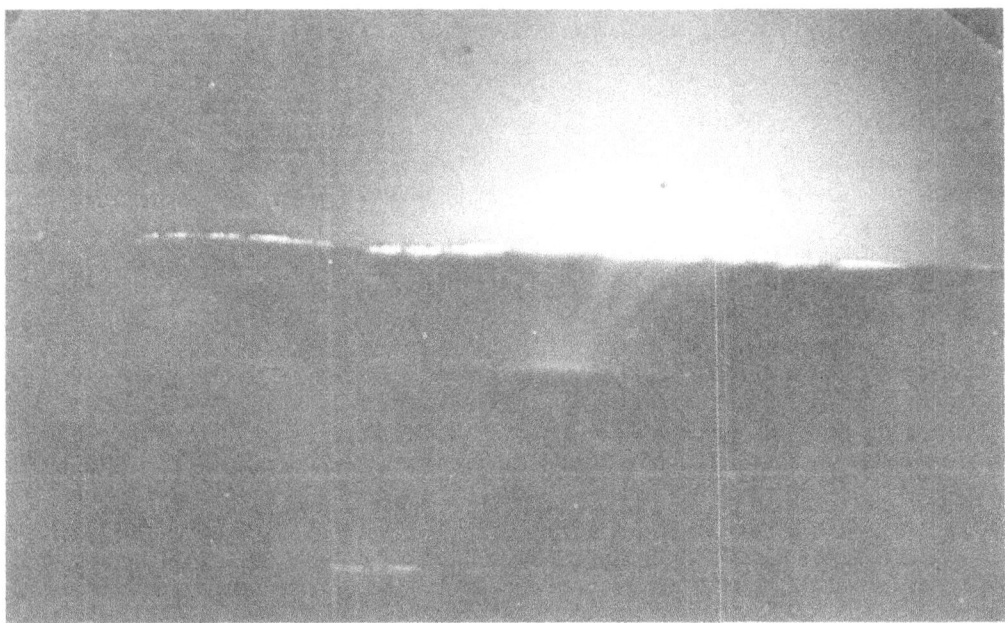

FIGURE B-7.—Narrow-angle picture of the bright line taken by the Surveyor VI television camera. The faint streak that appears in the lower foreground is an internal lens reflection of the strip, inverted and attenuated (Nov. 24, 1967, 14:15:26 GMT).

television camera, resulted in a smooth, almost straight-line linear spreading relationship as a function of scene luminance. This function was observed to hold true even with scene luminance four times greater than that required for video-signal saturation. Thus, this spreading can be used advantageously for relative photometry measurements of the line, because the true width of the line is considered by the author to be less than the limiting resolution of the television camera. This can be appreciated by observing the line's width at, or near, its extremities in figures B-2, B-3, B-4, and B-7. A more accurate analysis of the photometry involved with the line may afford not only an understanding of the limits of particle size but rather a complete size-frequency distribution function.

Factors contributing to the shape distortion in the resultant images of a thin white line target on a black background are vidicon lateral leakage white spread function (halation; as a function of differential electron charge remaining within the vidicon target's photoconductive material) and the introduction of a clamping action (with its resultant overshoot) whenever the video exceeded a certain threshold level in the white direction. The lateral leakage is essentially symmetrical, and the actual position of the original line in the field of view would be at the center of the resultant band if this were the only anomaly.

The overshoot, with such a line target and with the bright sunset line, was the introduction of a zone of "blacker than black" signals trailing the initial transition, and makes the center of the resultant band appear to be closer to the leading edge of the transition than it actually is. In the Surveyor VII pictures, this could give the illusion that the lunar bright line is suspended above the surface when compared to the daylight picture of the same area. (This illusion is even more pronounced if one takes into account that lateral leakage white spreading existing in the daylight picture makes the horizon appear slightly raised.)

The landed orientation of the Surveyor VII spacecraft on the lunar surface is such that the television camera's line-scan direction, in the Surveyor VII pictures of the line, is sweeping from the black sky, down through the line, and ending in the dark-gray portion of the lunar surface. The Surveyor VI orientation at its location on the lunar surface is such that the sunset pictures (fig. B-7) resulted in the line-scan direction from the dark-gray surface, up through the bright line, and ending in the sky.

Laboratory Simulations

In an attempt to appreciate the diffraction-scattering effect, laboratory tests were conducted with sandpaper. Preliminary experiments with various sizes of sandpaper (garnet cabinet paper) have led to the following conclusions. For a given threshold of detectability, or for a given amount of masking such as by earthshine, a maximum angular length of line is observable for a given size of "scattering particles." This angle appears to be rather constant and does not seem to be dependent on the distance from the observer to the line of "scattering particles." The smaller the particle, the larger the angular length of the bright line; however, its brightness level is decreased.

Inasmuch as the sandpaper-scattering particles were not of uniform spherical size, but were graded only by the bulk size of the grains, the irregular shapes of some of the grain edges gave the effect of having a distribution that contained various particle sizes limited only by the size of the larger graded grains.

The forward-scattering processes involved with sandpaper when illuminated by a non-coherent, near point source, white light probably sufficiently duplicates the processes involved with the generation of the lunar sunset line to make it a useful model except for the remote possibility of particle suspension. Preliminary measurements using various sandpaper models did not reveal any polarized light component along the bright line at the large phase angles involved with its generation.

Interpretations of Optical Phenomenon

The bright line must, at least in part, be the result of diffraction of sunlight by the horizon or edge of lunar-surface material that is viewed by the camera and illuminated by the Sun, even though sunset at the television camera's location has occurred up to 2 hours earlier. It seems very likely that diffraction, refraction, and forward scattering may all be involved (ref. B-1); further study is required, however, before any explanation is considered firm.

Because the Sun serves as an illuminating source subtending a solid angle of ½° for the full disk, the effective source size diminishes almost to a point source as the Sun completely sets, and the accompanying reduction in penumbra angle should be considered. (Such penumbra changes are usually not observable during Earth sunsets because of atmospheric conditions.)

Interpretations of Lunar Processes

Regarding possible explanations of the existence of the line, it should be noted that the diffraction of sunlight by sunlit particles at the horizon, or skyline, exists regardless of whether electrostatic suspension of these particles does or does not exist, or regardless of whether the frequency and distribution of ejecta particles in ballistic trajectory, resulting from micrometeorite bombardment impacts, is sufficient to make any significant contribution.

The jagged appearance of the line was not an anomaly, but a departure from the mean or offset of the line, with a high positive correlation with the jagged stones that appear on the horizon in figure B-1. In the author's opinion, it is extremely improbable that individual particles of dust ejected by micrometeorite impacts and suspended either electrostatically or in ballistic trajectory would assume a distribution that would so closely match the jaggedness of individual rocks. Instead, ballistic trajectories, if sufficient impacting did exist, would occupy a band whose height and apparent density variations would probably be described by the familiar mass-particle frequency function and whose position would be above the mean surface (horizon, in this case). Particles held by an electrostatic-suspension mechanism (if it were capable of being sustained or even existing) would be constrained by net charge and mass functions by a mean net charge gradient above the mean surface (horizon). In either case, the jagged effect would not be retained. There was no camera anomaly present that would introduce the jagged effect, and there is certainly nothing that would match the jaggedness of the individual rocks.

Because the electrostatic suspension of particles hypothesis is said to be dependent on sunlight as the source of energy to generate and sustain it, and because particles are suspended for a few seconds only, according to the proposed mechanism, it appears that electrostatic suspension would be highly improbable at sunset when the only sunlight present would be that of grazing incidence.

Because of the general appearance and seemingly varying thickness of the line, as observed by the television camera, the presence of the line, with the solar corona visible in the background, has caused some speculation that a lunar atmosphere exists. In reality, the strip's apparent thickness is the result of the camera's white spread function. (See "Camera Characteristics," p. 415.)

Possible Future Observations

Based on the information obtained from analyzing Surveyor television pictures, that the line was detectable for 2 hours, or 1° of lunar rotation, the author predicts that a continuous ring or halo, simultaneously showing a ring of lunar sunrise and sunset, should be observable from a spacecraft positioned in the lunar umbra and on the centerline of the lunar umbra at distances from about 100 000 to 375 000 km from the Moon. At distances much closer than about 100 000 km, the intensity of this ring will probably be so faint that earthshine will probably mask its detection, since the lunar disk half angle will exceed 1°. At distances much larger than about 375 000 km, the presence of Bailey's beads will probably mask its presence as the solid angle subtended by the lunar disk is reduced to that of the solid angle subtended by the solar disk where the umbra ends. For all except the closer distances, where very small gaps may appear because of occultation by lunar-surface features, the ring should appear to be very thin, but bright and uniform in intensity. Its brightness should increase as a function of distance up to a point and should exceed the brightness of both the solar F- and

K-coronas, even though the solar corona probably will always be detectable during all conditions that permit the ring to be viewed. The ring will probably reveal a small amount of jaggedness as it outlines surface features such as mountains. (There is a possibility that some of the total eclipse pictures made from Earth-based telescopes reveal the presence of this ring. In such cases, it would have been considered part of the solar corona.)

If viewed from positions other than the centerline of the umbra, but still within the umbra at distances greater than about 145 000 km from the Moon, the ring will still be continuous, but will no longer be uniform in brightness. At closer distances, there will be positions within the umbra where the ring will no longer be continuous, but will show the same type of gaps that exist in the Surveyor pictures.

Similarly, it appears logical to assume that any spherical body will appear to have a halo around it if it is viewed from its umbra and at distances such that the body subtends an angle larger than that of the illuminating source, but no larger than 2°.

Reference

B-1. Surveyor VII: A Preliminary Report. NASA SP-173, National Aeronautics and Space Administration, Washington, D.C., 1968, pp. 71, 72, 274, and 276.

Appendix C

Surveyor Science Teams and Cognizant Personnel

Analyses of the scientific data for the Surveyor missions were conducted by the Surveyor Scientific Evaluation Advisory Team, Investigator Teams, and Working Groups. Membership for these groups, at the time of the Surveyor VII mission, was as follows:

Surveyor Scientific Evaluation Advisory Team

L. D. JAFFE, *Chairman*	Jet Propulsion Laboratory
C. O. ALLEY	University of Maryland
S. A. BATTERSON	Langley Research Center
E. M. CHRISTENSEN	Jet Propulsion Laboratory
S. E. DWORNIK	NASA Headquarters
D. E. GAULT	Ames Research Center
J. W. LUCAS	Jet Propulsion Laboratory
D. O. MUHLEMAN	California Institute of Technology
R. H. NORTON	Jet Propulsion Laboratory
R. F. SCOTT	California Institute of Technology
E. M. SHOEMAKER	U.S. Geological Survey
R. H. STEINBACHER	Jet Propulsion Laboratory
G. H. SUTTON	University of Hawaii
A. L. TURKEVICH	University of Chicago

Investigator Teams

Television

E. M. SHOEMAKER, *Principal investigator*	U.S. Geological Survey
R. A. ALTENHOFEN	U.S. Geological Survey
R. M. BATSON	U.S. Geological Survey
H. E. HOLT	U.S. Geological Survey
G. P. KUIPER	University of Arizona
E. C. MORRIS	U.S. Geological Survey
J. J. RENNILSON	Jet Propulsion Laboratory
E. A. WHITAKER	University of Arizona

Alpha Scattering

A. L. TURKEVICH, *Principal investigator*	University of Chicago
E. J. FRANZGROTE	Jet Propulsion Laboratory
J. H. PATTERSON	Argonne National Laboratory

Soil Mechanics Surface Sampler

R. F. SCOTT, *Principal investigator*	California Institute of Technology
R. HAYTHORNWAITE	Pennsylvania State University
R. LISTON	Detroit Arsenal

Working Groups

Lunar Surface Thermal Properties

J. W. Lucas, *Chairman*	Jet Propulsion Laboratory
J. E. Conel	Jet Propulsion Laboratory
D. Greenshield	Manned Spacecraft Center
R. R. Garipay	Hughes Aircraft Co.
W. A. Hagemeyer	Jet Propulsion Laboratory
H. C. Ingrao	Harvard College Observatory
B. P. Jones	Marshall Space Flight Center
J. M. Saari	The Boeing Co.
G. Vitkus	Northrop Corp.

Lunar Surface Electromagnetic Properties

D. O. Muhleman, *Chairman*	California Institute of Technology
W. E. Brown, Jr.	Jet Propulsion Laboratory
L. H. Davids	Hughes Aircraft Co.
J. Negus de Wys	Jet Propulsion Laboratory
G. B. Gibson	Manned Spacecraft Center
W. H. Peake	Ohio State University
V. J. Poehls	Ryan Aeronautical Co.

Lunar Surface Mechanical Properties

E. M. Christensen, *Chairman*	Jet Propulsion Laboratory
S. A. Batterson	Langley Research Center
H. E. Benson	Manned Spacecraft Center
R. Choate	Jet Propulsion Laboratory
R. E. Hutton	TRW Systems
L. D. Jaffe	Jet Propulsion Laboratory
R. H. Jones	Hughes Aircraft Co.
H. Y. Ko	University of Colorado
R. F. Scott	California Institute of Technology
F. Schmidt	Bellcomm, Inc.
R. L. Spencer	Jet Propulsion Laboratory
F. B. Sperling	Jet Propulsion Laboratory
G. H. Sutton	University of Hawaii

Astronomy

R. H. Norton, *Chairman*	Jet Propulsion Laboratory
J. E. Gunn	Jet Propulsion Laboratory
W. C. Livingston	Kitt Peak National Observatory
G. A. Newkirk	High Altitude Observatory
H. Zirin	Mount Wilson and Palomar Observatories

Lunar Theory and Processes

D. E. Gault, *Chairman*	Ames Research Center
J. B. Adams	Jet Propulsion Laboratory
R. J. Collins	University of Minnesota
T. Gold	Cornell University
J. Green	McDonnell-Douglas Corp.
G. P. Kuiper	University of Arizona
H. Masursky	U.S. Geological Survey
J. A. O'Keefe	Goddard Space Flight Center
R. A. Phinney	Princeton University
E. M. Shoemaker	U.S. Geological Survey
H. E. Urey	University of California, San Diego

APPENDIX C

Laser Tests

C. O. Alley, *Chairman*	University of Maryland
L. H. Allen	Jet Propulsion Laboratory
H. Bostick	Lincoln Laboratories
J. Brault	Kitt Peak National Observatory
D. G. Currie	University of Maryland
J. E. Faller	Wesleyan University
H. Plotkin	Goddard Space Flight Center
S. Poultney	University of Maryland
M. Shumate	Jet Propulsion Laboratory

The cognizant personnel of the various science instrument aspects of the Surveyor Program, at the time of the Surveyor VII mission, were

Program and Project Scientists

S. E. Dwornik	Program Scientist
L. D. Jaffe	Project Scientist
R. H. Steinbacher	Associate Project Scientist
E. M. Christensen	Assistant Project Scientist

Cognizant Scientists and Science Staff

F. I. Roberson	Cognizant Scientist, Soil Mechanics Surface-Sampler Experiment
E. J. Franzgrote	Cognizant Scientist, Alpha-Scattering Experiment
R. E. Parker	Alpha-Scattering Experiment
T. H. Bird	Cognizant Scientist, Television Experiment
J. J. Rennilson	Television Experiment
D. L. Smyth	Television Experiment
M. Benes	Television Experiment
J. N. Strand	Television Science Data Handling
E. T. Johnson	Cognizant Engineer, Image Processing Laboratory
S. L. Grotch	Non-Television Science Data Handling
C. H. Goldsmith	Surveyor Experiment Test Laboratory
A. L. Filice	Landing Sites

Instrument Development

D. H. Le Croissette	Manager, Instrument Development
C. E. Chandler	Project Engineer, Instruments
R. J. Holman	Cognizant Engineer, Alpha-Scattering Instrument
M. I. Smokler	Supervisor and Cognizant Engineer, Television Instrument
E. R. Rouze	Cognizant Engineer, Soil Mechanics Surface Sampler

Space Science Analysis and Command

J. N. Lindsley	Director
D. D. Gordon	Assistant Director
R. C. Heyser	Director, Television Performance Analysis and Command
D. L. Smyth	Director, Television Science Analysis and Command
E. J. Franzgrote	Director, Alpha-Scattering Analysis and Command
F. I. Roberson	Director, Surface-Sampler Analysis and Command
C. R. Heinzen	Command Controller

Appendix D
Surveyor Management Organization

National Aeronautics and Space Administration

J. E. Naugle	Associate Administrator for Space Science and Applications
O. W. Nicks	Deputy Associate Administrator for Space Science and Applications
L. R. Scherer	Director, Apollo Lunar Exploration Office
B. Milwitzky	Surveyor Program Manager
W. Jakobowski	Program Engineer
F. A. Zihlman	Program Engineer
S. E. Dwornik	Program Scientist

Jet Propulsion Laboratory

W. H. Pickering	Director
R. J. Parks	Assistant Laboratory Director for Flight Projects
H. H. Haglund	Surveyor Project Manager
K. S. Watkins	Assistant Project Manager
R. G. Forney	Spacecraft Systems Manager
L. D. Jaffe	Project Scientist
R. H. Steinbacher	Assistant Project Scientist
N. A. Renzetti	Tracking and Data System Manager

Hughes Aircraft Co.

J. H. Richardson	Senior Vice President
C. G. Carlson	Manager, Space Systems Division
R. E. Sears	Surveyor Program Manager
R. R. Gunter	Assistant Manager, Test and Operations
J. D. Cloud	Assistant Manager, Engineering and Manufacturing
S. C. Shallon	Chief Scientist

www.ingramcontent.com/pod-product-compliance
Lightning Source LLC
Chambersburg PA
CBHW081715170526
45167CB00009B/3589